Lecture Notes in Computer Science   13479

More information about this series at https://link.springer.com/bookseries/558

Erika Ábrahám · Marco Paolieri (Eds.)

Quantitative Evaluation of Systems

19th International Conference, QEST 2022
Warsaw, Poland, September 12–16, 2022
Proceedings

 Springer

Editors
Erika Ábrahám
RWTH Aachen University
Aachen, Germany

Marco Paolieri
University of Southern California
Los Angeles, CA, USA

ISSN 0302-9743   ISSN 1611-3349   (electronic)
Lecture Notes in Computer Science
ISBN 978-3-031-16335-7   ISBN 978-3-031-16336-4   (eBook)
https://doi.org/10.1007/978-3-031-16336-4

This Springer imprint is published by the registered company Springer Nature Switzerland AG
The registered company address is: Gewerbestrasse 11, 6330 Cham, Switzerland

Preface

This volume contains the papers presented at the 19th International Conference on Quantitative Evaluation of SysTems (QEST 2022), hosted within CONFEST 2022 and held as a hybrid event, both online and in-person, during September 12–16, 2022. The event was co-located with CONCUR, FORMATS, FMICS, and other workshops.

The QEST conference series has a long and rich history, as can be seen at www.qest. org. Most recently, QEST was held in Paris (France, virtually), Vienna (Austria, virtually), Glasgow (UK), Beijing (China), and Berlin (Germany). Further information on QEST 2022 can be found on the conference webpage at www.qest.org/qest2022.

The 36 members of the international Program Committee (PC) helped to provide at least three reviews for each of the 44 submitted contributions. Based on the reviews and PC discussions, 19 high-quality papers (three of them as tool papers) were accepted to be presented during the conference. The overall acceptance rate for the conference was 43%. The contributions were organized into seven thematic sessions, covering the following topics on the verification and evaluation of systems: Program Analysis; Parameter Synthesis; Markovian Agents and Population Models; Dynamical Systems; Tools; Applications; and Automata Theory and Applications.

These contributions appear as papers in the ensuing proceedings. The program chairs plan to edit a special issue of the journal ACM TOMACS, where the authors of selected papers will be invited to contribute significantly extended versions of their manuscripts containing new results.

Best Paper awards were presented according to QEST policies and tradition. The Artifact Evaluation process introduced in 2021 was adopted also for this edition of QEST: a total of 14 submissions participated in the artifact evaluation (artifact submissions were mandatory only for tool papers), eight of which were found to be repeatable. A special badge marks accepted papers with a valid artifact.

A highlight of QEST 2022 was the presence of two invited speakers, Marta Kwiatkowska and Pedro D'Argenio. A short contribution on the topics of the keynote of Marta Kwiatkowska appears in these proceedings.

A few words of acknowledgment are due. First and foremost, we thank the authors for entrusting their best work to QEST 2022. The review process clearly showed that the conference was able to set a high bar for acceptance, which makes us proud. Our thanks go to the QEST steering committee and previous conference chairs as well, for their help and feedback on the organization process. We were particularly pleased with the interest in the repeatability evaluation, and we thank the repeatability evaluation committee and chairs, Arnd Hartmanns and David Safranek, for their work, and all the authors who participated in this process. We would like to give special thanks to the local organizing committee of CONFEST and to the steering committee of the QEST conference series, in particular to its chair Enrico Vicario. Finally, we wish to thank all the PC members and additional reviewers for their hard work in ensuring the quality of the contributions to QEST 2022, along with all the participants for contributing to this memorable event.

August 2022   Erika Ábrahám
                                                                Marco Paolieri

Organization

Local Organization

Piotrek Hofman	University of Warsaw, Poland
Bartek Klin	University of Oxford, UK
Sławek Lasota	University of Warsaw, Poland

Program Committee Chairs

Erika Ábrahám	RWTH Aachen University, Germany
Marco Paolieri	University of Southern California, USA

Artifact Evaluation Chairs

Carlos Budde	University of Trento, Italy
Arnd Hartmanns	University of Twente, The Netherlands

Publicity Chair

David Safranek	Masaryk University, Czech Republic

Steering Committee

Benoît Barbot	Université Paris-Est Créteil, France
Ezio Bartocci	TU Vienna, Austria
Peter Buchholz	TU Dortmund, Germany
Giuliano Casale	Imperial College London, UK
Jane Hillston	University of Edinburgh, UK
Jorge Júlvez	University of Zaragoza, Spain
Joost-Pieter Katoen	RWTH Aachen University, Germany
Gethin Norman	University of Glasgow, UK
Anne Remke	University of Münster, Germany
Enrico Vicario	University of Florence, Italy
Katinka Wolter	FU Berlin, Germany
Lijun Zhang	Chinese Academy of Sciences, China

Program Committee

Alessandro Abate	University of Oxford, UK
Ebru Aydin Gol	Middle East Technical University, Turkey
Ezio Bartocci	TU Vienna, Austria
Borzoo Bonakdarpour	Michigan State University, USA

Luca Bortolussi	University of Trieste, Italy
Peter Buchholz	TU Dortmund, Germany
Laura Carnevali	University of Florence, Italy
Giuliano Casale	Imperial College London, UK
Gianfranco Ciardo	Iowa State University, USA
Liliana Cucu-Grosjean	Inria Paris, France
Pedro D'Argenio	Universidad Nacional de Córdoba, Argentina
Carina da Silva	University of Münster, Germany
Susanna Donatelli	University of Turin, Italy
Maryam Elahi	Mount Royal University, Canada
Jean-Michel Fourneau	University of Versailles Saint-Quentin-en-Yvelines, France
Marco Gribaudo	Polytechnic University of Milan, Italy
András Horváth	University of Turin, Italy
David N. Jansen	Chinese Academy of Sciences, China
Nils Jansen	Radboud University, The Netherlands
Sebastian Junges	Radboud University, The Netherlands
Peter Kemper	College of William & Mary, USA
William Knottenbelt	Imperial College London, UK
Jan Křetínský	Technical University of Munich, Germany
Andrea Marin	Ca' Foscari University of Venice, Italy
Gethin Norman	University of Glasgow, UK
David Parker	University of Birmingham, UK
Elizabeth Polgreen	University of Edinburgh, UK
Anne Remke	University of Muenster, Germany
Sabina Rossi	Ca' Foscari University of Venice, Italy
Gerardo Rubino	Inria Rennes, France
Miklós Telek	Budapest University of Technology and Economics, Hungary
Mirco Tribastone	IMT School for Advanced Studies, Italy
Benny Van Houdt	University of Antwerp, Belgium
Verena Wolf	Saarland University, Germany
Katinka Wolter	FU Berlin, Germany
Paolo Zuliani	Newcastle University, UK

Artifact Evaluation Committee

Lukas Armborst	University of Twente, The Netherlands
Oyendrila Dobe	Michigan State University, USA
James Fox	University of Oxford, UK
Michaela Klauck	University of Saarland, Germany
Mathis Niehage	University of Münster, Germany
Riccardo Pinciroli	Gran Sasso Science Institute, Italy
Luciano Putruele	Universidad Nacional de Río Cuarto, Argentina
Riccardo Reali	University of Florence, Italy
Stefan Schupp	TU Wien, Austria

Jip Spel	RTWH Aachen, Germany
Matej Troják	Masaryk University, Czechia
Alexander Weigl	Karlsruher Institute for Technology, Germany

Additional Reviewers

Michael Backenköhler	Saarland University, Germany
Thom Badings	Radboud University, The Netherlands
Francesca Cairoli	University of Trieste, Italy
Ginevra Carbone	University of Trieste, Italy
Joanna Delicaris	University of Muenster, Germany
Oyendrila Dobe	Michigan State University, USA
Joschka Groß	Saarland University, Germany
Timo P. Gros	Saarland University, Germany
Hosein Hasanbeig	University of Oxford, UK
Benjamin L. Kaminski	Saarland University, Germany
Andrey Kofnov	TU Vienna, Austria
Pasquale Malacaria	Queen Mary University of London, UK
Kristóf Marussy	Budapest University of Technology and Economics, Hungary
Ali Nasir	University of Central Punjab, Pakistan
Andrew Nguyen	University of Oxford, UK
Francesco Pontiggia	TU Vienna, Austria
Diptarkó Roy	University of Oxford, UK
Gaia Saveri	University of Trieste, Italy
Stella Simic	University of Oxford, UK
Sadegh Soudjani	Newcastle University, UK
Marnix Suilen	Radboud University, The Netherlands
Matthias Volk	University of Twente, The Netherlands
Drishti Yadav	TU Vienna, Austria
Eshita Zaman	Iowa State University, USA

Robustness Guarantees for Bayesian Neural Networks (Invited Extended Abstract of a Keynote Speaker)

Marta Kwiatkowska[iD]

Department of Computer Science, University of Oxford, Oxford, UK
marta.kwiatkowska@cs.ox.ac.uk
http://www.cs.ox.ac.uk/people/marta.kwiatkowska/

Abstract. Bayesian neural networks (BNNs), a family of neural networks with a probability distribution placed on their weights, have the advantage of being able to reason about uncertainty in their predictions as well as data. Their deployment in safety-critical applications demands rigorous robustness guarantees. This paper summarises recent progress in developing algorithmic methods to ensure certifiable safety and robustness guarantees for BNNs, with the view to support design automation for systems incorporating BNN components.

Keywords: Bayesian neural networks · Probabilistic safety · Adversarial robustness · Certification

1 Introduction

Neural networks (NNs) are being introduced across many domains, including robotics, autonomous vehicles, security and healthcare, but their deployment in safety-critical scenarios demands rigorous robustness guarantees in the presence of uncertainty, which are lacking for NNs. *Bayesian neural networks* (BNNs) [6] are a family of neural networks that place distributions over their weights, instead of viewing them as fixed values, and can thus account for uncertainty in data and predictions. Starting with a prior distribution and a given likelihood, the application of Bayes' theorem results in posterior probability distribution over the BNN weights conditional on the observed data. This induces posterior predictive distribution on the BNN outputs, with the final BNN prediction selected from this distribution according to Bayesian decision theory. BNNs therefore combine the high capacity of NNs while enabling (Bayesian) probabilistic reasoning, since they can be viewed as stochastic processes.

This invited paper describes recent progress in developing methods to provide robustness guarantees for Bayesian neural networks. These include certifiable adversarial training, statistical evaluation of probabilistic safety, and certified lower

This project was funded by the ERC under the European Union's Horizon 2020 research and innovation programme (FUN2MODEL, grant agreement No. 834115, and ELSA: European Lighthouse on Secure and Safe AI project, grant agreement No. 101070617 under UK guarantee).

bounding of safety probability. The discussed methods draw on probabilistic reacha-bility analysis, sampling, statistical model checking and convex relaxation, and con-stitute part of an effort to develop probabilistic verification and synthesis methodologies for systems incorporating BNN components.

2 Background on Bayesian Neural Networks

A feed-forward neural network (NN) is a function $f^w : \mathbb{R}^m \to \mathbb{R}^n$, parametrised by a vector $w \in \mathbb{R}^{n_w}$ that includes all the weights of the network (for simplicity assume no bias). We work in a supervised learning scenario, where we are given a dataset $\mathcal{D} = \{(x_i, y_i)\}_{i=1}^{n_{\mathcal{D}}}$ of pairs of inputs and ground truth labels, with $x_i \in \mathbb{R}^m$, and where each target output $y \in \mathbb{R}^n$ is either a one-hot class vector for classification or a real-valued vector for regression.

A Bayesian neural network (BNN) [6] is an NN with a distribution placed over the network parameters w, and can thus be viewed as a stochastic process $f^{\mathbf{w}}$ (vector of random variables $\mathbf{w}$ associated to the weights) indexed by the input space. Note that, for a weight vector w sampled from the distribution of $\mathbf{w}$, the BNN induces a (deter-ministic) NN f^w with weights fixed to w. We employ Bayesian learning to infer the weight parameters, starting with a prior distribution $p_{\mathbf{w}}(w)$ over $\mathbf{w}$ and likelihood $p(\mathcal{D}|w) = \prod_{i=1}^{n_{\mathcal{D}}} p(y_i|x_i, w)$, to compute the posterior distribution $p_{\mathbf{w}}(w|\mathcal{D})$ of parame-ters conditioned on data by applying the Bayes formula, i.e., $p_{\mathbf{w}}(w|\mathcal{D}) \propto p(\mathcal{D}|w)p_{\mathbf{w}}(w)$. This induces the distribution over outputs called the posterior predictive distribution defined for an unseen point x^* by $p(y^*|x^*, \mathcal{D}) = \int p(y^*|x^*, w)p_{\mathbf{w}}(w|\mathcal{D})dw$. The final prediction is obtained based on Bayesian decision theory and is the value $\hat{y}$ that minimizes the Bayesian risk of an incorrect prediction according to the posterior predictive distribution and a loss function $\mathcal{L}$, computed as $\hat{y} = argmin_y \int_{\mathbb{R}^n} \mathcal{L}(y, y^*)p(y^*|x^*, \mathcal{D})dy^*$. For classification decisions, we typically work with 0-1 loss and the optimal decision is then the class that maximises the predictive distribution, whereas for regression ℓ_2 loss is used and the optimal decision the expected value of the BNN output over the posterior distribution.

Unfortunately, the computation of the posterior distribution $p_{\mathbf{w}}(w|\mathcal{D})$ over weights cannot be computed analytically and is generally intractable [6]. Instead, approximate inference methods have been developed for BNNs, of which Hamiltonian Monte Carlo (HMC) [6] and Variational Inference (VI) [1] are commonly used. HMC considers Hamiltionian dynamics to speed up the exploration, working with a Markov chain whose invariant distribution is $p_{\mathbf{w}}(w|\mathcal{D})$, and is asymptotically correct [6]. The result of HMC is a set of samples that approximates $p_{\mathbf{w}}(w|\mathcal{D})$. VI proceeds by finding a Gaussian approximating distribution over the weight space $q(w) \sim p_{\mathbf{w}}(w|\mathcal{D})$, where $q(w)$ depends on some hyperparameters that are then iteratively optimized by mini-mizing a divergence measure between $q(w)$ and $p_{\mathbf{w}}(w|\mathcal{D})$, thus trading off approxi-mation accuracy against scalability. Samples can then be efficiently extracted from $q(w)$.

3 Certifiable Adversarial Robustness

Though the ability of Bayesian neural networks to capture uncertainty is appealing for safety-critical applications, they are susceptible to adversarial attacks. In [7], a principled Bayesian approach was proposed for incorporating adversarial robustness in the posterior inference procedure of BNNs. To this end, the robustness requirement is formulated as the worst-case prediction over an adversarial input ball of radius $\varepsilon \geq 0$ induced by a user-defined probability density function p_ε, and the standard cross-entropy likelihood model was extended by marginalising the network output over p_ε called *robust likelihood*. Further, for any $\varepsilon > 0$, certified lower bounds to the robust likelihood can be computed by employing interval bound propagation techniques. This novel adversarial training procedure adapts naturally to the main approximate inference techniques employed for training of BNNs, including HMC and VI. An experimental evaluation in [7] demonstrated that the robust likelihood can double the maximal safe radius for the standard model and results in better calibrated uncertainty when predicting out-of-distribution samples.

4 Probabilistic Safety Evaluation

Safe decision making is important in autonomous scenarios, where it can benefit from uncertainty estimates being propagated through the decision pipeline. In [5], a setting involving an end-to-end BNN autonomous driving controller based on NVIDIA's PilotNet was considered, which can be viewed as a discrete-time stochastic process, and a framework was proposed for evaluating safety of the controller's decisions. Two properties were considered, probabilistic safety, i.e., the probability that the controller will maintain the safety of the car for a given time horizon, and real-time decision confidence, i.e., the probability that the BNN is certain of a given decision. We remark that probabilistic safety represents a probabilistic variant of the notion of safety [3] commonly used to certify deterministic NNs. A statistical model checking framework based on [2] is employed to evaluate robustness of these properties to changes in weather, location and observation noise with a priori confidence interval guarantees (using Chernoff bounds) in a simulated scenario. Here, we exploit the fact that sampling BNN weights results in a deterministic NN, which can be checked using conventional methods for NNs, and the proportion of sampled NNs that are safe yields a probability estimate of BNN safety. [5] also shows how to quantify the uncertainty of the controller's decisions and utilise uncertainty thresholds in order to guarantee the safety of the self-driving car with high probability. Separately, [4] study infinite-time horizon robustness properties for BNNs.

5 Certified Bounds on Safety Probability

Probabilistic safety evaluation based on [2] can only provide guarantees in the form of confidence intervals, which may not be sufficient for highly safety-critical systems. [8] considered certification of (lower bounds on) the safety probability. The method is based on observing that probabilistic safety translates into computing the probability that adversarial perturbations of an input cause small variations in the BNN output. For BNNs, this involves working with posterior probability and showing that the computation of probabilistic safety for BNNs is equivalent to computing the measure, w.r.t. BNN posterior, of the set of weights for which the resulting deterministic NN is safe, i.e., robust to adversarial perturbations. Once the set of such weights is computed, relaxation techniques from non-linear optimisation (interval bound propagation and linear bound propagation) are employed to check whether all the networks instantiated by these weights are safe. This yields lower bounds on the probability for the case of BNNs trained with VI, but the method extends to other approximate Bayesian inference techniques. Experimental evaluation on the VCAS collision-avoidance case study demonstrates the practicality of the method. In follow-on work, [9] consider also synthesis of certified policies for BNNs.

6   Conclusion and Further Work

We have provided an overview of algorithmic techniques developed to ensure certified guarantees of safety and adversarial robustness for BNNs. Certification of BNNs is more involved than for NNs, because of the need to consider weight intervals instead of single values, and presents significant computational challenges that have so far been tackled using a combination of numerical, statistical and symbolic techniques. Despite encouraging progress, much remains to be done, including upper bounding of safety, certified bounds on decision probability, temporal logic specifications, strategy synthesis and explanations for BNNs.

References

1. Blundell, C., Cornebise, J., Kavukcuoglu, K., Wierstra, D.: Weight uncertainty in neural networks. In: ICML (2015)
2. Cardelli, L., Kwiatkowska, M., Laurenti, L., Paoletti, N., Patane, A., Wicker, M.: Statistical guarantees for the robustness of Bayesian neural networks. In: IJCAI (2019)
3. Huang, X., Kwiatkowska, M., Wang, S., Wu, M.: Safety verification of deep neural networks. In: CAV (2017)
4. Lechner, M., Zikelic, D., Chatterjee, K., Henzinger, T.A.: Infinite time horizon safety of Bayesian neural networks. In: NeurIPS (2021)
5. Michelmore, R., Wicker, M., Laurenti, L., Cardelli, L., Gal, Y., Kwiatkowska, M.: Uncertainty quantification with statistical guarantees in end-to-end autonomous driving control. In: ICRA (2020)
6. Neal, R.M.: Bayesian learning for neural networks. Springer Science & Business Media (2012)
7. Wicker, M., Laurenti, L., Patane, A., Chen, Z., Zhang, Z., Kwiatkowska, M.: Bayesian inference with certifiable adversarial robustness. In: AISTATS (2021)
8. Wicker, M., Laurenti, L., Patane, A., Kwiatkowska, M.: Probabilistic safety for Bayesian neural networks. In: UAI (2020)
9. Wicker, M., Laurenti, L., Patane, A., Paoletti, N., Abate, A., Kwiatkowska, M.: Certification of iterative predictions in Bayesian neural networks. In: UAI (2021)

Contents

Dynamical Systems

Tools

Applications

Automata Theory and Applications

Program Analysis

Moment-Based Invariants for Probabilistic Loops with Non-polynomial Assignments

Andrey Kofnov[1]([✉]), Marcel Moosbrugger[2], Miroslav Stankovič[2], Ezio Bartocci[2], and Efstathia Bura[1]

[1] Applied Statistics, Faculty of Mathematics and Geoinformation,
TU Wien, Vienna, Austria
andrey.kofnov@tuwien.ac.at
[2] Faculty of Informatics, TU Wien, Vienna, Austria

Abstract. We present a method to automatically approximate moment-based invariants of probabilistic programs with non-polynomial updates of continuous state variables to accommodate more complex dynamics. Our approach leverages polynomial chaos expansion to approximate non-linear functional updates as sums of orthogonal polynomials. We exploit this result to automatically estimate state-variable moments of all orders in Prob-solvable loops with non-polynomial updates. We showcase the accuracy of our estimation approach in several examples, such as the turning vehicle model and the Taylor rule in monetary policy.

Keywords: Probabilistic programs · Prob-solvable loops · Polynomial Chaos Expansion · Non-linear updates

1 Introduction

Probabilistic programs (PPs) are becoming widely employed in many areas including AI applications, security/privacy protocols or modeling stochastic dynamical systems. The study of the properties of these processes requires knowledge of their distribution; that is, the distribution(s) of the random variable(s) generated by executing the probabilistic program.

The characterization of many distributions can be accomplished via their moments. In [2] the authors introduced a class of probabilistic programs, *Prob-solvable loops*, for which moment-based invariants over the state variables of the programs are automatically computed as a closed-form expression. A Prob-solvable loop consists of an initialization section and a non-nested loop where the variables can be updated by drawing from distributions determined by their moments (e.g., Bernoulli, Normal) and using polynomial arithmetic. However,

Supported by the Vienna Science and Technology Fund (WWTF ICT19-018), the TU Wien Doctoral College (SecInt), the FWF research projects LogiCS W1255-N23 and P 30690-N35, and the ERC Consolidator Grant ARTIST 101002685.

ⓒ Springer Nature Switzerland AG 2022
E. Ábrahám and M. Paolieri (Eds.): QEST 2022, LNCS 13479, pp. 3–25, 2022.
https://doi.org/10.1007/978-3-031-16336-4_1

modeling complex dynamics often requires the use of non-polynomial updates, such as in the *turning vehicle* example in Fig. 1. An open research question is how to leverage the class of *Prob-solvable loops* to estimate moment-based invariants as closed-form expressions for probabilistic loops with updates governed by non-polynomial non-linear functions.

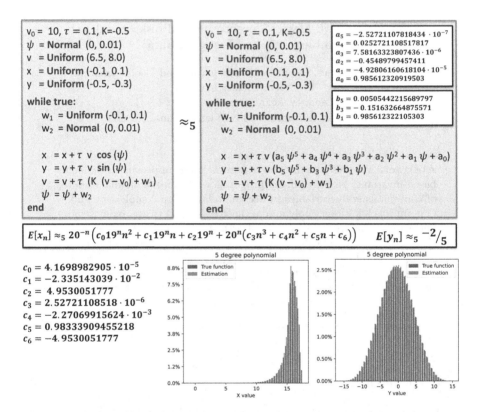

Fig. 1. On the top left a probabilistic loop modeling the behaviour of a turning vehicle [25] using non-polynomial (cos, sin) updates in the loop body. On top right a Prob-Solvable loop obtained by approximating the cos, sin functions using polynomial chaos expansion (up to 5th degree). In the middle the expected position (x, y) computed automatically from the Prob-Solvable loop as a closed-form expression in the number of the loop iterations n. In the bottom center and right the comparison among the true and the estimated distribution for a fixed iteration (we execute the loop for $n = 20$ iterations and $8 \cdot 10^5$ repetitions).

At the heart of our approach is the decomposition of a random function into a linear combination of basis functions that are orthogonal polynomials. We accommodate non-polynomial updates of program variables to allow for more complicated dynamics. By expressing the non-linear functionals of the updates as sums of orthogonal polynomials, we can apply the approach in [2,3] to automatically estimate the moments of all the program random variables. Our approach is within the framework of general polynomial chaos expansion (gPCE) [33]. As such, it converges to the truth with guaranteed recovery of the moments of random variables with complex distributions. We focus on state variables with continuous distributions with updates that are square-integrable functionals and use general polynomial chaos expansion to represent them. In Fig. 1 we illustrate our approach via the turning vehicle example, where we estimate the expected position of a vehicle. In this example, we approximate the original cosine and sine functions with $5th$ degree polynomials and obtain a Prob-solvable loop. This enables the automatic computation of the moments in closed-form at each loop iteration (n) using the approach proposed in [2].

Related Work. [25] proposed polynomial forms to represent distributions of state variables. Polynomial forms are generalizations of affine forms [4] and use the Taylor series expansion to represent functions of random variables generated in a PP. Functions can be only approximated in a small interval around a fixed point, otherwise high order derivatives are required to guarantee sufficient accuracy of the approximation. As a consequence, functions with unbounded support cannot be handled with this approach.

So-called Taylor models have been proposed in [5,19,23] for reachability analysis of (non-probabilistic) non-linear dynamical systems. Taylor models combine polynomials and error intervals to capture the set of reachable states after some fixed time horizon. Application of Taylor series expansions for generalized functions of probabilistic distributions can also be found in [27,30].

[17] introduced trigonometric and mixed-trigonometric-polynomial moments to obtain an exact description of the moments of uncertain states for nonlinear autonomous and robotic systems over the planning horizon. This approach can only handle systems encoded in PPs, where all nonlinear transformations take standard, trigonometric, or mixed-trigonometric polynomial forms.

Polynomial chaos expansion based methods have been extensively used for uncertainty quantification in engineering problems of solid and fluid mechanics (e.g. [9,13,16]), computational fluid dynamics (e.g., [18]), flow through porous media [11,12], thermal problems [15], analysis of turbulent velocity fields [6, 20], differential equations (e.g., [31,33]), and, more recently, geosciences and meteorology (e.g., [7,10,14]).

Outline. Section 2 reviews the notion of *Prob-solvable Loop* and the general theory of *Polynomial Chaos Expansion* (PCE). Section 3 presents our PCE algorithm and the conditions under which it produces accurate approximations to general random functions in probabilistic program loops. Section 4 combines general PCE with Prob-solvable loops to automatically compute moments of all orders of state variables. There, we also characterize the structure a probabilistic program ought to have in order to be compatible with the Prob-solvable loops approach for computation of moments. Section 5 demonstrates the accuracy and feasibility of our approach on different benchmarks as compared with the state-of-the-art. We conclude in Sect. 6.

2   Preliminaries

2.1   Prob-Solvable Loops

[2] defined the class of *Prob-solvable loops* for which moments of all orders of program variables can be computed symbolically: given a Prob-solvable loop and a program variable x, their method computes a closed-form solution for $\mathbb{E}(x_n^k)$ for arbitrary $k \in \mathbb{N}$, where n denotes the nth loop iteration. Prob-solvable loops are restricted to polynomial variable updates.

Definition 1 (Prob-solvable loops). *Let* $m \in \mathbb{N}$ *and* $x_1, \ldots x_m$ *denote real-valued program variables. A Prob-solvable loop with program variables* $x_1, \ldots x_m$ *is a loop of the form*

$$I; \text{\textit{while(true)}}\{U\},$$

where

- *I is a sequence of initial assignments over a subset of $\{x_1, \ldots, x_m\}$. The initial values of x_i can be drawn from a known distribution. They can also be real constants.*
- *U is the loop body and is a sequence of m random updates, each of the form:*

$$x_i = Dist \quad or \quad x_i := a x_i + P_i(x_1, \ldots x_{i-1})$$

where $a \in \mathbb{R}$, $P_i \in \mathbb{R}[x_1, \ldots, x_{i-1}]$ is a polynomial over program variables $x_1, \ldots, x_{i-1}$ and Dist is a distribution independent from program variables with computable moments.

Many real-life systems exhibit non-polynomial dynamics and require more general updates, such as, for example, trigonometric or exponential functions. In this work, we develop a method that allows approximation of non-polynomial assignments in probabilistic loops by polynomial assignments. In doing so, we can use the methods for Prob-solvable loops to compute the moments for a broader class of stochastic systems.

The programming model we use (Definition 1) is a simplified version of the Prob-solvable model as introduced in [2]. Our approach, described in the following sections, is not limited to this simple fragment of the Prob-solvable and can be used for Prob-solvable loops as originally defined as well as other more general probabilistic loops. The only requirement is that the loops satisfy the conditions in Sect. 3.1.

2.2   Polynomial Chaos Expansion

Polynomial chaos expansion recovers a random variable in terms of a linear combination of functionals whose entries are known random variables, sometimes called germs, or, basic variables. Let $(\Omega, \Sigma, \mathbb{P})$ be a probability space, where Ω is the set of elementary events, Σ is a σ-algebra of subsets of Ω, and $\mathbb{P}$ is a probability measure on Σ. Suppose X is a real-valued random variable defined on $(\Omega, \Sigma, \mathbb{P})$, such that

$$\mathbb{E}(X^2) = \int_\Omega X^2(\omega)d\mathbb{P}(\omega) < \infty. \tag{1}$$

The space of all random variables X satisfying (1) is denoted by $L^2(\Omega, \Sigma, \mathbb{P})$. That is, the elements of $L^2(\Omega, \Sigma, \mathbb{P})$ are real-valued random variables defined on $(\Omega, \Sigma, \mathbb{P})$ with finite second moments. If we define the inner product as

$$\mathbb{E}(XY) = (X, Y) = \int_\Omega X(\omega)Y(\omega)d\mathbb{P}(\omega) \tag{2}$$

and norm $||X|| = \sqrt{\mathbb{E}(X^2)} = \sqrt{\int_\Omega X^2(\omega)d\mathbb{P}(\omega)}$, then $L^2(\Omega, \Sigma, \mathbb{P})$ is a Hilbert space; i.e., an infinite dimensional linear space of functions endowed with an inner product and a distance metric. Elements of a Hilbert space can be uniquely specified by their coordinates with respect to an orthonormal basis of functions, in analogy with Cartesian coordinates in the plane. Convergence with respect to $|| \cdot ||$ is called *mean-square convergence*. A particularly important feature of a Hilbert space is that when the limit of a sequence of functions exists, it belongs to the space.

The elements in $L^2(\Omega, \Sigma, \mathbb{P})$ can be classified in two groups: *basic* and *generic* random variables, which we want to decompose using the elements of the first set of basic variables. [8] showed that the basic random variables that can be used in the decomposition of other functions have finite moments of all orders with continuous probability density functions (pdf).

The σ-algebra generated by the basic random variable Z is denoted by $\sigma(Z)$. Suppose we restrict our attention to decompositions of a random variable $X = g(Z)$, where g is a function with $g(Z) \in L^2(\Omega, \sigma(Z), \mathbb{P})$ and the basic random variable Z determines the class of orthogonal polynomials $\{\phi_i(Z), i \in \mathbb{N}\}$,

$$\langle \phi_i(Z), \phi_j(Z) \rangle = \int_\Omega \phi_i(Z(\omega))\phi_j(Z(\omega))d\mathbb{P}(\omega)$$

$$= \int \phi_i(x)\phi_j(x)f_Z(x)dx = \begin{cases} 1 & i = j \\ 0 & i \neq j \end{cases} \tag{3}$$

which is a polynomial chaos basis. If Z is normal with mean zero, the Hilbert space $L^2(\Omega, \sigma(Z), \mathbb{P})$ is called *Gaussian* and the related set of polynomials is represented by the family of Hermite polynomials (see, for example, [33]) defined on the whole real line. Hermite polynomials form a basis of $L^2(\Omega, \sigma(Z), \mathbb{P})$. Therefore, every random variable X with finite second moment can be approximated by the truncated PCE

$$X^{(d)} = \sum_{i=0}^{d} c_i \phi_i(Z), \tag{4}$$

for suitable coefficients c_i that depend on the random variable X. The truncation parameter d is the highest polynomial degree in the expansion. Since the polynomials are orthogonal,

$$c_i = \frac{1}{||\phi_i||^2}\langle X, \phi_i \rangle = \frac{1}{||\phi_i||^2}\langle g, \phi_i \rangle = \frac{1}{||\phi_i||^2}\int_{\mathbb{R}} g(x)\phi_i(x)f_Z(x)dx. \tag{5}$$

The truncated PCE of X in (4) converges in mean square to X [8, Sec. 3.1]. The first two moments of (4) are determined by

$$\mathbb{E}(X^{(d)}) = c_0, \tag{6}$$

$$\mathrm{Var}(X^{(d)}) = \sum_{i=1}^{d} c_i^2 ||\phi_i||^2. \tag{7}$$

Representing a random variable by a series of Hermite polynomials in a countable sequence of independent Gaussian random variables is known as Wiener-Hermite polynomial chaos expansion. In applications of Wiener-Hermite PCEs, the underlying Gaussian Hilbert space is often taken to be the space spanned by a sequence $\{Z_i, i \in \mathbb{N}\}$ of independent standard Gaussian basic random variables, $Z_i \sim \mathcal{N}(0, 1)$. For computational purposes, the countable sequence $\{Z_i, i \in \mathbb{N}\}$ is restricted to a finite number $k \in \mathbb{N}$ of random variables. The Wiener-Hermite polynomial chaos expansion converges for random variables with finite second moment. Specifically, for any random variable $X \in L^2(\Omega, \sigma(\{Z_i, i \in \mathbb{N}\}), \mathbb{P})$, the approximation (4) satisfies

$$X_k^{(d)} \to X \quad \text{as } d, k \to \infty \tag{8}$$

in mean-square convergence. The distribution of X can be quite general; e.g., discrete, singularly continuous, absolutely continuous as well as of mixed type.

3   Polynomial Chaos Expansion Algorithm

3.1   Random Function Representation

In this section, we state the conditions under which the estimated polynomial is an unbiased and consistent estimator and has exponential convergence rate.

Suppose k continuous random variables $Z_1, \ldots, Z_k$ are used to introduce stochasticity in a PP, with corresponding cumulative distribution functions (cdf) F_{Z_i} for $i = 1, \ldots, k$. Also, suppose all k distributions have probability density functions, and let $\mathbf{Z} = (Z_1, \ldots, Z_k)$ with cdf $F_{\mathbf{Z}}$. We assume that the elements of $\mathbf{Z}$ satisfy the following conditions:

(A) $Z_i, i = 1, \ldots, k$, are independent.
(B) We consider functions g such that $g(\mathbf{Z}) \in L^2(\mathcal{Q}, F_{\mathbf{Z}})$, where $\mathcal{Q}$ is the support of the joint distribution of $\mathbf{Z} = (Z_1, \ldots, Z_k)^1$.

---

[1] Ω is dropped from the notation as the sample space is not important in our formulation.

(C) All random variables Z_i have distributions that are uniquely defined by their moments.[2]

Under condition (A), the joint cdf of the components of $\mathbf{Z}$ is $F_{\mathbf{Z}} = \prod_{i=1}^{k} F_{Z_i}$. To ensure the construction of unbiased estimators with optimal exponential convergence rate (see [8,33]) in the context of probabilistic loops, we further introduce the following assumptions:

(D) g is a function of a fixed number of basic variables (arguments) over all loop iterations.

(E) If $\mathbf{Z}(j) = (Z_1(j), \ldots, Z_k(j))$ is the stochastic argument of g at iteration j, then $F_{Z_i(j)}(x) = F_{Z_i(l)}(x)$ for all pairs of iterations (j, l) and x in the support of F_{Z_i}.

If Conditions (D) and (E) are not met, then the polynomial coefficients in the PCE need be computed for each loop iteration individually to ensure optimal convergence rate. It is straightforward to show the following proposition.

Proposition 1. *If functions f and g satisfy conditions (B) and (D), and $\mathbf{Z} = (Z_1, \ldots, Z_{k_1})$, $\mathbf{Y} = (Y_1, \ldots, Y_{k_2})$ satisfy conditions (A), (C) and (E) and are mutually independent, then their sum, $f(\mathbf{Z}) + g(\mathbf{Y})$, and product, $f(\mathbf{Z}) \cdot g(\mathbf{Y})$, also satisfy conditions (B) and (D).*

3.2 PCE Algorithm

Let $Z_1, \ldots, Z_k$ be independent continuous random variables, with respective cdfs F_i, satisfying conditions (A), (B) and (C), and $\mathbf{Z} = (Z_1, \ldots, Z_k)^T$ with cdf $\mathbf{F} = \prod_{i=1}^{k} F_i$ and support $\mathcal{Q}$. The function $g : \mathbb{R}^k \to \mathbb{R}$, with $g \in L^2(\mathcal{Q}, \mathbf{F})$ can be approximated with the truncated orthogonal polynomial expansion, as described in Fig. 2,

$$g(\mathbf{Z}) \approx \hat{g}(\mathbf{Z}) = \sum_{\substack{d_i \in \{0, \ldots, \bar{d}_i\}, \\ i = 1, \ldots, k}} c(d_1, \ldots, d_k) z_1^{d_1} \cdots z_k^{d_k} = \sum_{j=1}^{L} c_j \prod_{i=1}^{k} \bar{p}_i^{d_{ji}}(z_i), \qquad (9)$$

where

- $\bar{p}_i^{d_{ji}}(z_i)$ is a polynomial of degree d_{ji}, and belongs to the set of orthogonal polynomials with respect to F_{Z_i} that are calculated with the Gram-Schmidt orthogonalization procedure[3];
- $\bar{d}_i = \max\limits_{j}(d_{ji})$ is the highest degree of the univariate orthogonal polynomial, for $i = 1, \ldots, k$;

---

[2] Conditions that ascertain this are given in Theorem 3.4 of [8].

[3] Generalized PCE typically entails using orthogonal basis polynomials specific to the distribution of the basic variables, according to the Askey scheme of [32,33]. We opted for the most general procedure that can be used for any basic variable distribution.

- $L = \prod_{i=1}^{k} (1 + \bar{d}_i)$ is the total number of multivariate orthogonal polynomials and equals the truncation constant;
- c_j are the Fourier coefficients.

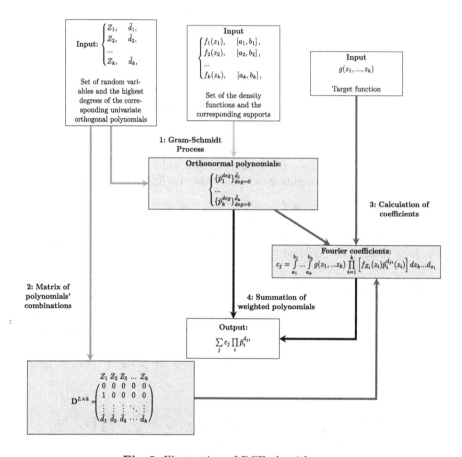

Fig. 2. Illustration of PCE algorithm

The Fourier coefficients are calculated using

$$c_j = \int_{\mathcal{Q}} g(z_1, \ldots, z_k) p_i^{d_{ji}}(z_i) d\mathbf{F} = \int \cdots \int_{\mathcal{Q}} g(z_1, \ldots, z_k) \prod_{i=1}^{k} \bar{p}_i^{d_{ji}}(z_i) dF_{Z_k} \ldots dF_{Z_1},$$

(10)

by Fubini's theorem.

Example 1. Returning to the Turning Vehicle model in Fig. 1, the non-polynomial functions to approximate are $g_1 = cos$ and $g_2 = sin$ from the updates of program variables x, y, respectively. In both cases, we only need to consider a single basic random variable, $Z \sim \mathcal{N}(0, 0.01)$ (ψ in Fig. 1).

For the approximation, we use polynomials of degree up to 5. Eq. (9) has the following form for the two functions,

$$\hat{g}_1(z) = cos(\psi) = a_0 + a_1\psi + ... + a_5\psi^5 \tag{11}$$

and

$$\hat{g}_2(z) = sin(\psi) = b_0 + b_1\psi + ... + b_5\psi^5. \tag{12}$$

We compute the coefficients a_i, b_i in Eqs. (11)–(12) using (10) to obtain the values shown in Fig. 1.

Complexity. Assuming the expansion is carried out up to the same polynomial degree d for each basic variable, $\bar{d}_i = d, \forall i = 1, ..., k$. This implies $d = \sqrt[k]{L} - 1$. The complexity of the scheme is $\mathcal{O}(sd^2k + s^kd^k)$, where $\mathcal{O}(s)$ is the complexity of computing univariate integrals.

The complexity of our approximation scheme is comprised of two parts: (1) the orthogonalization process and (2) the calculation of coefficients. Regarding (1), we orthogonalize and normalize k sets of d basic linearly independent polynomials during the Gram-Schmidt process. For degree $d = 1$, we need to calculate one integral, the inner product with the previous polynomial. Additionally, we need to compute one more integral, the norm of itself (for normalization). For each subsequent degree d', we must calculate d' additional new integrals. The computation of each integral has complexity $\mathcal{O}(s)$. Regarding (2), the computation of the coefficients requires calculating $L = (d + 1)^k$ integrals with k-variate functions as integrands.

We define the approximation error to be

$$se(\hat{g}) = \sqrt{\int_{\mathcal{Q}} (g(z_1, ..., z_k) - \hat{g}(z_1, ..., z_k))^2 \, dF_{Z_1} \ldots dF_{Z_k}} \tag{13}$$

since $\mathbb{E}(\hat{g}(Z_1, ..., Z_k)) = g(Z_1, \cdots, Z_k)$ by construction.

The implementation of this algorithm may become challenging when the random functions have complicated forms and the number of parametric uncertainties is large. In this case, the calculation of the PCE coefficients involves high dimensional integration, which may prove difficult and time prohibitive for real-time applications [26].

4   Prob-Solvable Loops for General Non-polynomial Functions

PCE[4] allows incorporating non-polynomial updates into Prob-solvable loop programs and use the algorithm in [2] and exact tools, such as POLAR [21], for moment (invariant) computation. We identify two classes of programs based on how the distributions of the random variables generated by the programs vary.

---

[4] We provide further details about PCE computation in Appendix 2 and 3.

4.1   Iteration-Stable Distributions of Random Arguments

Let $\mathcal{P}$ be an arbitrary Prob-solvable loop and suppose that a (non-basic) state variable $x \in \mathcal{P}$ has a non-polynomial L^2-type update $g(\mathbf{Z})$, where $\mathbf{Z} = (Z_1, ..., Z_k)^T$ is a vector of (basic) continuous, independent, and identically distributed random variables *across iterations*. That is, if $f_{Z_j(n)}$ is the pdf of the random variable Z_j in iteration n, then $f_{Z_j(n)} \equiv f_{Z_j(n')}$, for all iterations n, n' and $j = 1, \ldots, k$. The basic random variables $Z_1, \ldots, Z_k$ and the update function g satisfy conditions (A)–(E) in Sect. 3.1. For the class of Prob-solvable loops where all variables with non-polynomial updates satisfy these conditions, the computation of the Fourier coefficients in the PCE approximation (9) can be carried out as explained in Sect. 3.2. In this case, the convergence rate is optimal.

4.2   Iteration Non-stable Distribution of Random Arguments

Let $\mathcal{P}$ be an arbitrary Prob-solvable loop and suppose that a state variable $x \in \mathcal{P}$ has a non-polynomial L^2-type update $g(\mathbf{Z})$, where $\mathbf{Z} = (Z_1, ..., Z_k)^T$ is a vector of continuous independent but *not necessarily identically* distributed random variables across iterations. For this class of Prob-solvable loops, conditions (A)–(C) in Sect. 3.1 hold, but (D) and/or (E) may not be fulfilled. In this case, we can ensure optimal exponential convergence by fixing the number of loop iterations. For unbounded loops, we describe an approach converging in mean-square and establish its convergence rate next.

Conditional Estimator Given Number of Iterations. Let N be an a priori fixed finite integer representing the maximum iteration number. The set $\{1, ..., N\}$ is a finite sequence of iterations for the Prob-solvable loop $\mathcal{P}$.

Iterations are executed sequentially for $n = 1, \ldots, N$, which allows the estimation of the final functional that determines the target state variable at each iteration $n \in \{1, ..., N\}$ and its set of supports. Knowing these features, we can carry out N successive expansions. Let $P(n)$ be a PCE of $g(\mathbf{Z})$ for iteration n. We introduce an additional program variable c that counts the loop iterations. The variable c is initialized to 0 and incremented by 1 at the beginning of every loop iteration. The final estimator of $g(\mathbf{Z})$ can be represented as

$$\hat{g}(\mathbf{Z}) = \sum_{n=1}^{N} P(n) \left[\prod_{j=1, j \neq n}^{N} \frac{(c-j)}{n-j} \right]. \tag{14}$$

Replacing non-polynomial functions with (14) results in a program with only polynomial-type updates and *constant* polynomial structure; that is, polynomials with coefficients that remain constant across iterations. Moreover, the estimator is unbiased with optimal exponential convergence on the set of iterations $\{1, ..., N\}$ [33].

Unconditional Estimator. Here the iteration number is unbounded. Without loss of generality, we consider a single basic random variable Z; that is, $k = 1$. The function $g(Z)$ is scalar valued and can be represented as a polynomial of *nested L^2 functions*, which depend on polynomials of the argument variable. Each nested functional argument is expressed as a sum of orthogonal polynomials yielding the final estimator, which is itself a polynomial.

Since PCE converges to the function it approximates in mean-square (see [8]) on the whole interval (argument's support), PCE converges on any sub-interval of the support of the argument in the same sense.

Let us consider a function g with sufficiently large domain, and a random variable Z with known distribution and support. For example, $g(Z) = e^Z$, with $Z \sim N(\mu, \sigma^2)$. The domain of g and the support of Z are the real line. We can expand g into a PCE with respect to the distribution of Z as

$$g(Z) = \sum_{i=0}^{\infty} c_i p_i(Z). \tag{15}$$

The distribution of Z is reflected in the polynomials in (15). Specifically, p_i, for $i = 0, 1, \ldots$, are Hermite polynomials of special type in that they are orthogonal (orthonormal) with respect to $N(\mu, \sigma^2)$. They also form an orthogonal basis of the space of L^2 functions. Consequently, any function in L^2 can be estimated arbitrarily closely by these polynomials. In general, any continuous distribution with finite moments of all orders and sufficiently large support can also be used as a model for basic variables in order to construct a basis for L^2 (see [8]).

Now suppose that the distribution of the underlying variable Z is unknown with pdf $f(Z)$ that is continuous on its support $[a, b]$. Then, there exists another basis of polynomials, $\{q_i\}_{i=0}^{\infty}$, which are orthogonal on the support $[a, b]$ with respect to the pdf $f(Z)$. Then, on the interval $[a, b]$, $g(Z) = \sum_{i=0}^{\infty} k_i q_i(Z)$, and $\mathbb{E}_f[g(Z)] = \mathbb{E}_f\left[\sum_{i=0}^{M} k_i q_i(Z)\right]$, $\forall M \geq 0$.

Since $[a, b] \subset \mathbb{R}$, the expansion $\sum_{i=0}^{\infty} c_i p_i(Z)$ converges in mean-square to $g(Z)$ on $[a, b]$. In the limit, we have $g(Z) = \sum_{i=0}^{\infty} c_i p_i(Z)$ on the interval $[a, b]$. Also, $\mathbb{E}_f(g(Z)) = \mathbb{E}_f(\sum_{i=0}^{\infty} c_i p_i(Z))$ for the true pdf f on $[a, b]$. In general, though, it is not true that $\mathbb{E}_f(g(Z)) = \mathbb{E}_f\left(\sum_{i=0}^{M} c_i p_i(Z)\right)$ for any arbitrary $M \geq 0$ and any pdf $f(Z)$ on $[a, b]$, as the estimator is biased.

To capture this discrepancy, we define the approximation error as

$$e(M) = \mathbb{E}_f\left[g(Z) - \sum_{i=0}^{M} c_i p_i(Z)\right]^2 = \mathbb{E}_f\left[\sum_{i=M+1}^{\infty} c_i p_i(Z)\right]^2. \tag{16}$$

Computation of Error Bound. Assume the true pdf f_Z of Z is supported on $[a, b]$. Also, assume the domain of g is $\mathbb{R}$. The random function $g(Z)$ has PCE on the whole real line based on Hermite polynomials $\{p_i(Z)\}_{i=0}^{\infty}$ that are orthogonal with respect to the standard normal pdf ϕ. The truncated expansion estimate of (15) with respect to a normal basic random variable is

$$\hat{g}(Z) = \sum_{i=0}^{M} c_i p_i(Z). \tag{17}$$

We compute an upper-bound for the approximation error for our scheme in Theorem 1.

Theorem 1. *Suppose Z has density f supported on $[a, b]$, $g : \mathbb{R} \to \mathbb{R}$ is in L^2, and ϕ denotes the standard normal pdf. Under (15) and (17),*

$$\|g(Z) - \hat{g}(Z)\|_f^2 = \int_a^b (g(z) - \hat{g}(z))^2 f_Z(z) dz$$

$$\leq \left(\frac{2}{\min(\phi(a), \phi(b))} + 1 \right) \mathbb{Var}_\phi(g(Z)). \tag{18}$$

The upper bound in (18) depends only on the support of f, the pdf of Z, and the function g. If Z is standard normal ($f = \phi$), then the upper bound in (18) equals $\mathbb{Var}_\phi(g(Z))$. We provide the proof of Theorem 1 in Appendix 1.

Remark 1. The approximation error inequality in [22, Lemma 1],

$$\left\| g(Z) - \sum_{i=0}^{T} c_i p_i(Z) \right\| \leq \frac{\|g(Z)^{(k)}\|}{\prod_{i=0}^{k-1} \sqrt{T - i + 1}}, \tag{19}$$

is a special case of Theorem 1 when $Z \sim \mathcal{N}(0, 1)$ and $f = \phi$, and the polynomials p_i are Hermite. In this case, the left hand side of (19) equals $\sqrt{\sum_{i=n+1}^{\infty} c_i^2}$.

Although Theorem 1 is restricted to distributions with bounded support, the approximation in (17) also converges for distributions with unbounded support.

5   Evaluation

In this section, we evaluate our approach on four benchmarks from the literature. We use our method based on PCE to approximate non-polynomial functions. After PCE, all benchmark programs fall into the class of Prob-solvable loops. We use the static analysis tool POLAR [21] on the resulting Prob-solvable loops to compute the moments of the program variables parameterized by the loop iteration n. All experiments were run on a machine with 32 GB of RAM and a 2.6 GHz Intel i7 (Gen 10) processor.

Taylor Rule Model. Central banks set monetary policy by raising or lowering their target for the federal funds rate. The Taylor rule[5] is an equation intended to describe the interest rate decisions of central banks. The rule relates the target of the federal funds rate to the current state of the economy through the formula

$$i_t = r_t^* + \pi_t + a_\pi(\pi_t - \pi_t^*) + a_y(y_t - \bar{y}_t),$$

---

[5] It was proposed by the American economist John B. Taylor as a technique to stabilize economic activity by setting an interest rate [29].

```
a_p = 0.5, a_y = 0.5, y=1, y_1=1
p=0.01, p_1=0.01, i=0.02, r=0.015

while true:
    dp  = Normal (0, 0.01)
    dy  = Exponential (100)
    p = p_1
    p_1= p + dp
    y_1= 0.01 + 1.02 y
    y  = y_1 – dy
    l_y= log(1 + y)
    l_y1= log(1 + y_1)
    i = r + p + a_p (p-p_1)+ a_y (l_y - l_y1)
end
                                          A
```

```
angles= [10,60,110,160,140,100,60,20,100, 0]

x = TruncNormal (0, 0.0025,-0.5, 0.5)
y = TruncNormal (0, 0.01,-0.5,0.5)

while true:
    for θ in angles
        d  = Uniform (0.98, 1.02)
        t  = (θπ/180) (1+TruncNormal (0, 0.0001,-0.05,0.05))
        x = x + d cos (t)
        y = y + d sin (t)
    end
                                          B
```

```
t = π/6, γ_0 = 2π/90, σ = π/120, c2theta = 0.75
x = Uniform (-0.1, 0.1)

while true:
    w  = TruncNormal (γ_0, σ^2, γ_0 − 0.05 π, γ_0 + 0.05 π)
    β_1 = t/2 + w
    β_2 = t/2 − w
    update_1 = 1 − cos (β_1)
    update_2 = 1 − cos (β_2)
    x = c2theta (x + 20 update_1) - 20 update_2
end
                                          C
```

Fig. 3. Probabilistic loops: (A) Taylor rule [29], (B) 2D Robotic Arm [4] (in the figure we use the inner loop as syntax sugar to keep the program compact), (C) Rimless Wheel Walker [28].

where i_t is the nominal interest rate, r_t^* is the equilibrium real interest rate, $r_t^* = r$, π_t is inflation rate at t, π_t^* is the short-term target inflation rate at t, $y_t = \log(1 + Y_t)$, with Y_t the real GDP, and $\bar{y}_t = \log(1 + \bar{Y}_t)$, with $\bar{Y}_t$ denoting the potential real output.

Highly-developed economies grow exponentially with a sufficiently small rate (e.g., according to the World Bank,[6] the average growth rate of the GDP in the USA in 2001–2020 equals 1,73%). Therefore, we set the growth rate of the potential output to 2%. Moreover, we follow [1] and model inflation as a martingale process; that is, $\mathbb{E}_t [\pi_{t+1}] = \pi_t$. The Taylor rule model is described by the program in Fig. 3, A.

Figure 4 illustrates the performance of our approach as a function of the polynomial degree of our approximation. The approximations to the true first

---

[6] https://data.worldbank.org/indicator/NY.GDP.MKTP.KD.ZG?locations=US.

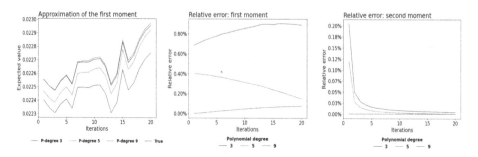

Fig. 4. The approximations and their relative errors for the Taylor rule model.

moment (in red) are plotted in the left panel and the relative errors for the first and second moments are in the middle and the right panels, respectively, over iteration number. The y-axis in both middle and right panels shows relative errors calculated as $rel.err = |est - true|/true$. All plots show that the approximation error is low and that it deteriorates as the polynomial degree increases from 3 to 9, across iterations. The drop is sharper for the second moment.

Turning Vehicle Model. The Turning vehicle model is described by the program in Fig. 1. The model was introduced in [25] and depicts the position of a vehicle, as follows. The state variables are (x, y, v, ψ), where (x, y) is the vehicle's position with velocity v and yaw angle ψ. The vehicle's velocity is stabilized around $v_0 = 10$ m/s. The dynamics are modelled by the equations $x(t + 1) = x(t) + \tau v \cos(\psi(t))$, $y(t + 1) = y(t) + \tau v \sin(\psi(t))$, $v(t + 1) = v(t) + \tau(K(v(t) - v_0) + w_1(t + 1))$, and $\psi(t + 1) = \psi(t) + w_2(t + 1)$. The disturbances w_1 and w_2 have distributions $w_1 \sim U[-0.1, 0.1]$, $w_2 \sim N(0, 0.1)$. Moreover, as in [25], we set $K = -0.5$. Initially, the state variables are distributed as: $x(0) \sim U[-0.1, 0.1]$, $y(0) \sim U[-0.5, -0.3]$, $v(0) \sim U[6.5, 8.0]$, $\psi(0) \sim N(0, 0.01)$. We allow all normally distributed parameters take values over the entire real line, in contrast to [25] who could not accommodate distributions with infinite support and required the normal variables to take values over finite intervals.

This program requires the approximation of trigonometric functions for the computation of the location of the vehicle at time t. We used PCEs of degree 3, 5 and 9, built on a basic standard normal random variable, to estimate the dynamics of x and y, the first and second coordinates of the vehicle's location. For all three PCEs for sin, the Prob-solvable loops tool estimates the first moment of y to be the same, namely $-2/5$. We report the value of the first moment of x in Table 1. The *polynomial form* of [25] can not be applied to approximate any moments of x and y. We see that our PCE based estimate is very close to the "true" first moment of x, with the 9th degree PCE being the closest, as expected.

Rimless Wheel Walker. The *Rimless wheel walker* [25, 28] is a system that describes a human walking. The system models a rotating wheel consisting of n_s spokes, each of length L, connected at a single point. The angle between

Table 1. Evaluation of our approach on 4 benchmarks. Poly form = the interval for the target as reported in [25]; Sim = target approximated through 10^6 samples; Deg. = maximum degrees used for the approximation of the non-linear functions; Result = result of our method per degree; Runtime = execution time of our method in seconds (time of PCE + time of POLAR).

Benchmark	Target	Poly form	Sim.	Deg.	Result	🕐 Runtime
Taylor rule model	$\mathbb{E}\,(i_n)$ $n{=}20$	×	0.02298	3	0.02278	$0.4s{+}0.5s$
				5	0.02295	$0.5s{+}5.0s$
				9	0.02300	$5.9s{+}34.6s$
Turning vehicle model	$\mathbb{E}\,(x_n)$ $n{=}20$	×	15.69792	3	14.44342	$0.6s{+}3.6s$
				5	15.43985	$1.4s{+}9.2s$
				9	15.60595	$15.6s{+}16.1s$
Turning vehicle model (trunc.)	$\mathbb{E}\,(x_n)$ $n{=}20$	for deg. 2 $[-3 \cdot 10^5, 3 \cdot 10^5]$ 🕐 1117s	15.69882	3	14.44342	$0.6s{+}3.6s$
				5	15.43985	$1.4s{+}9.1s$
				9	15.60595	$15.6s{+}19.0s$
Rimless wheel walker	$\mathbb{E}\,(x_n)$ $n{=}2000$	for deg. 2 $[1.791, 1.792]$ 🕐 46.1s	1.79155	1	1.79159	$0.2s{+}0.5s$
				2	1.79159	$0.3s{+}0.5s$
				3	1.79159	$0.6s{+}0.5s$
Robotic arm model	$\mathbb{E}\,(x_n)$ $n{=}100$	for deg. 2 $[268.87, 268.88]$ 🕐 35.8s	268.853	1	268.85236	$1.3s{+}0.3s$
				2	268.85227	$2.5s{+}0.6s$
				3	268.85227	$4.8s{+}0.7s$

consecutive spokes is $\theta = 2\pi/n_s$. We set $L = 1$ and $\theta = \pi/6$. The Rimless wheel walker is modeled by the program in Fig. 3, C. For more details we refer to [25].

Robotic Arm Model. Proposed and studied in [4, 24, 25], this system models the position of a 2D robotic arm. The arm moves through translations and rotations. Moreover, at every step, errors in movement are modeled with probabilistic noise. The Robotic arm model is described by the program in Fig. 3, B.

The *Rimless wheel walker* and the *Robotic arm model* are the only two benchmarks from [25] containing non-polynomial updates. In [25], polynomial forms of degree 2 were used to compute bounding intervals for $\mathbb{E}(x_n)$ (for fixed n) for the *Rimless wheel walker* and the *Robotic arm model*. Their tool does not support the approximation of logarithms (required for the *Taylor rule model*) and distributions with unbounded support (required for the *Turning vehicle model*). To facilitate comparison to polynomial forms, our set of benchmarks is augmented with a version of the *Turning vehicle model* using truncated normal distributions instead of normal distributions with unbounded support (*Turning vehicle model (trunc.)* in Table 1). We note that the technique in [25] supports more general probabilistic loops than Prob-solvable loops. However, as already mentioned in Sect. 2.1, we emphasize that our results in Sects. 2.2–4 are not limited to

Prob-solvable loops and can be applied to approximate non-linear dynamics for more general probabilistic loops.

Table 1 summarizes the evaluation of our approach on these four benchmarks and of the technique based on *polynomial forms* of [25] on the directly comparable *Turning vehicle model (trunc.)*, *Rimless wheel walker* and the *Robotic arm* models. Our results illustrate that our method is able to accurately approximate general non-linear dynamics for challenging programs. Specifically, for the *Rimless wheel walker* model, our first moment estimate is reached upon with a first degree approximation, is close to the truth up to the fourth decimal and falls in the interval estimate of the polynomial forms technique. For the *Robotic arm model*, our results lie outside the interval predicted by the polynomial forms technique, yet are closer to the simulation ("truth") calculated with 10^6 samples. Moreover, our simulation agrees with the estimation provided in [25].

Our experiments also demonstrate that our method provides suitable approximations in a fraction of the time required by the technique based on polynomial forms. While polynomial forms additionally provide an error interval, they need to be computed on an iteration-by-iteration basis. In contrast, our method based on PCE and Prob-solvable loops computes an expression for the target parameterized by the loop iteration $n \in \mathbb{N}$ (cf. Fig. 1). As a result, increasing the target iteration does *not* increase the runtime of our approach.

Both *Robotic arm* and *Rimless wheel walker* models contain no stochastic accumulation: each basic random variable is iteration-stable and can be estimated using the scheme in Sect. 4.1. Therefore, for these two benchmarks, our estimation converges exponentially to the true values. On the other hand, the *Taylor rule model* and the *Turning vehicle model* contain stochasticity accumulation, which leads to the instability of the distributions of basic random variables. We apply the scheme in Sect. 4.2 for these two examples.

6   Conclusion

We present an approach to compute the moments of the distribution of random outputs in probabilistic loops with non-linear, non-polynomial updates. Our method is based on polynomial chaos expansion to approximate non-polynomial general functional assignments. The approximations produced by our technique have optimal exponential convergence when the parameters of the general non-polynomial functions have distributions that are stable across all iterations. We derived an upper bound on the approximation error for the case of unstable parameter distributions. Our methods can accommodate non-linear, non-polynomial updates in classes of probabilistic loops amenable to automated moment computation, such as the class of Prob-solvable loops. Moreover, our techniques can be used for moment approximation for uncertainty quantification in more general probabilistic loops. Our experiments demonstrate the ability of our methods to characterize non-polynomial behavior in stochastic models from various domains via their moments, with high accuracy and in a fraction of the time required by other state-of-the-art tools.

Appendix 1. Proof of Theorem 1

Proof (Theorem 1). Since $f(z) = 0 \ \forall z \notin [a, b]$,

$$\left\| g(z) - \sum_{i=0}^{T} c_i p_i(z) \right\|_f^2 = \int_a^b \left(g(z) - \sum_{i=0}^{T} c_i p_i(z) \right)^2 f(z) dz$$

$$= \int_{-\infty}^{\infty} \left(g(z) - \sum_{i=0}^{T} c_i p_i(z) \right)^2 f(z) dz$$

$$= \int_{-\infty}^{a} \left(g(z) - \sum_{i=0}^{T} c_i p_i(z) \right)^2 f(z) dz + \int_b^{\infty} \left(g(z) - \sum_{i=0}^{T} c_i p_i(z) \right)^2 f(z) dz$$

$$+ \int_a^b \left(g(z) - \sum_{i=0}^{T} c_i p_i(z) \right)^2 f(z) dz$$

$$\leq \int_{-\infty}^{a} \left(g(z) - \sum_{i=0}^{T} c_i p_i(z) \right)^2 \phi(z) dz + \int_b^{\infty} \left(g(z) - \sum_{i=0}^{T} c_i p_i(z) \right)^2 \phi(z) dz$$

$$+ \int_a^b \left(g(z) - \sum_{i=0}^{T} c_i p_i(z) \right)^2 \phi(z) dz + \int_a^b \left(g(z) - \sum_{i=0}^{T} c_i p_i(z) \right)^2 (f(z) - \phi(z)) dz$$

$$= A + B + C + D \tag{20}$$

Since $f(z) - \phi(z) \leq \phi(z) + f(z)$, D satisfies

$$\int_a^b \left(g(z) - \sum_{i=0}^{T} c_i p_i(z) \right)^2 (f(z) - \phi(z)) dz$$

$$\leq \int_a^b \left(g(z) - \sum_{i=0}^{T} c_i p_i(z) \right)^2 dz \int_a^b (\phi(z) + f(z)) dz$$

$$= (1 + \Phi(b) - \Phi(a)) \times \int_a^b \left(g(z) - \sum_{i=0}^{T} c_i p_i(z) \right)^2 dz,$$

with $(1 + \Phi(b) - \Phi(a)) < 2$. Now,

$$1 \leq \frac{\phi(z)}{\min(\phi(a), \phi(b))} \quad \forall z \in [a, b],$$

and hence

$$\int_a^b \left(g(z) - \sum_{i=0}^T c_i p_i(z) \right)^2 dz$$

$$\leq \min\left(\phi(a), \phi(b)\right)^{-1} \int_a^b \left(g(z) - \sum_{i=0}^T c_i p_i(z) \right)^2 \phi(x) dz$$

$$\leq \min\left(\phi(a), \phi(b)\right)^{-1} C. \tag{21}$$

By (21) and (15), (20) satisfies

$$A + B + C + D \leq \left(\frac{2}{\min\left(\phi(a), \phi(b)\right)} + 1 \right) \left\{ \int_{-\infty}^a \left(g(z) - \sum_{i=0}^T c_i p_i(z) \right)^2 \phi(z) dz \right.$$

$$\left. + \int_b^\infty \left(g(z) - \sum_{i=0}^T c_i p_i(z) \right)^2 \phi(z) dz + \int_a^b \left(g(z) - \sum_{i=0}^T c_i p_i(z) \right)^2 \phi(z) dz \right\}$$

$$= \left(\frac{2}{\min\left(\phi(a), \phi(b)\right)} + 1 \right) \int_{-\infty}^\infty \left(g(z) - \sum_{i=0}^T c_i p_i(z) \right)^2 \phi(z) dz$$

$$= \left(\frac{2}{\min\left(\phi(a), \phi(b)\right)} + 1 \right) \sum_{i=T+1}^\infty c_i^2 \leq \left(\frac{2}{\min\left(\phi(a), \phi(b)\right)} + 1 \right) \mathbb{V}\mathrm{ar}_\phi\left[g(Z)\right]$$

since $\mathbb{V}\mathrm{ar}_\phi(g(Z)) = \sum_{i=1}^\infty c_i^2$. In consequence, the error (16) can be upper bounded by (18).

Appendix 2. Computation Algorithm in Detail

We let $\mathbf{D} \in \mathbb{Z}^{L \times k}$ be the matrix with each row $j = 1, \ldots, L$ containing the degrees of Z_i (in column i) of the corresponding polynomial in (9). For example, the first row corresponds to the constant polynomial (1), and the last row to $\bar{p}_1^{\bar{d}_1}(z_1) \ldots \bar{p}_k^{\bar{d}_k}(z_k)$. That is,

$$\mathbf{D} = (d_{ji})_{j=1,\ldots,L,\ i=1,\ldots,k} = \begin{array}{c} \begin{matrix} Z_1\ Z_2\ Z_3\ \ldots\ Z_k \end{matrix} \\ \begin{pmatrix} 0 & 0 & 0 & 0 & 0 \\ 0 & 0 & 0 & 0 & 1 \\ \vdots & \vdots & \vdots & \ddots & \vdots \\ \bar{d}_1 & \bar{d}_2 & \bar{d}_3 & \cdots & \bar{d}_k \end{pmatrix} \end{array}$$

The computer implementation of the algorithm computes c_j in (10) for each j combination (row) of degrees of the corresponding polynomials.

We apply the above described computation to the following example. Suppose that X has a truncated normal distribution with parameters $\mu = 2$, $\sigma = 0.1$, and is supported over $[1, 3]$, and that Y is uniformly distributed over $[1, 2]$. We expand $g(x, y) = \log(x + y)$ along X and Y, as follows. We choose the relevant highest degrees of expansion to be $\bar{d}_X = 2$ and $\bar{d}_Y = 2$. The pdf of Y is $f_Y(y) \equiv 1$, and of X is $f_X(x) = e^{-\frac{(x-2)^2}{0.02}}/0.1\alpha\sqrt{2\pi}$, where the truncation multiplier α equals $\int_1^3 e^{-\frac{(x-2)^2}{0.02}}\,dx/0.1\sqrt{2\pi}$.

The two sets of polynomials, $\{p_i\} = \{1, x, x^2\}$ and $\{q_i\} = \{1, y, y^2\}$, are linearly independent. Applying the Gram-Schmidt procedure to orthogonalize and normalize them, we obtain $\bar{p}_0(x) = 1$, $\bar{p}_1(x) = 10x - 20$, $\bar{p}_2(x) = 70.71067x^2 - 282.84271x + 282.13561$, and $\bar{q}_0(y) = 1$, $\bar{q}_1(y) = 3.4641y - 5.19615$, $\bar{q}_2(y) = 13.41641y^2 - 40.24922y + 29.06888$. In this case, $L = (1 + \bar{d}_X) * (1 + \bar{d}_Y) = 9$, and $\mathbf{D}$ has 9 rows and 2 columns,

$$
\mathbf{D} = (d_{ji}) = \begin{matrix} & X\ Y & \\ \begin{pmatrix} 0 & 0 \\ 0 & 1 \\ 0 & 2 \\ 1 & 0 \\ 1 & 1 \\ 1 & 2 \\ 2 & 0 \\ 2 & 1 \\ 2 & 2 \end{pmatrix} & \begin{matrix} \leftarrow c_1 = & 1.2489233 \\ \leftarrow c_2 = & 0.0828874 \\ \leftarrow c_3 = & -0.0030768 \\ \leftarrow c_4 = & 0.0287925 \\ \leftarrow c_5 = & -0.0023918 \\ \leftarrow c_6 = & 0.0001778 \\ \leftarrow c_7 = & -0.0005907 \\ \leftarrow c_8 = & 0.0000981 \\ \leftarrow c_9 = & -0.0000109 \end{matrix} \end{matrix}
$$

Iterating through the rows of matrix $\mathbf{D}$ and choosing the relevant combination of degrees of polynomials for each variable, we calculate the Fourier coefficients

$$
c_j = \int_1^3 \int_1^2 \log(x + y)\bar{p}_{d_{j1}}(x)\bar{q}_{d_{j2}}(y)f_X(x)f_Y(y)\,dy\,dx.
$$

The final estimator can be derived by summing up the products of each coefficient and the relevant combination of polynomials:

$$
\log(x + y) \approx \hat{g}(x, y) = \sum_{j=1}^{9} c_j\bar{p}_{d_{j1}}(x)\bar{q}_{d_{j2}}(y) \approx
$$

$$
- 0.01038x^2y^2 + 0.05517x^2y - 0.10031x^2
$$

$$
+ 0.06538xy^2 - 0.37513xy + 0.86515x
$$

$$
- 0.13042y^2 + 0.93998y - 0.59927.
$$

The estimation error is

$$
se(\hat{g}) = \sqrt{\int_1^3 \int_1^2 (\log(x + y)) - \hat{g}(x, y))^2 f_X(x)f_Y(y)\,dy\,dx} \approx 0.000151895
$$

Appendix 3. PCEs of Exponential and Trigonometric Functions

Table 2 lists examples of functions of up to three random arguments approximated by PCE's of different degrees and, correspondingly, number of coefficients. We use $TruncNormal\left(\mu, \sigma^2, [a, b]\right)$ to denote the truncated normal distribution with expectation μ and standard deviation σ on the (finite or infinite) interval $[a, b]$, and $TruncGamma\left(\theta, k, [a, b]\right)$ for the truncated gamma distribution on the (finite or infinite) interval $[a, b]$, $a, b > 0$, with shape parameter k and scale parameter θ. The approximation error in (13) is reported in the last column. The results confirm (8) in practice: the error decreases as the degree or, equivalently, the number of components in the approximation of the polynomial increases.

Table 2. Approximations of 5 non-linear functions using PCE.

Function	Random Variables	Degree / #coefficients	Error
$f(x_1, x_2) = \xi e^{-x_1} + (\xi - \frac{\xi^2}{2})e^{x_2 - x_1}$ $\xi = 0.3$	$x_1 \sim Normal(0, 1)$, $x_2 \sim Normal(2, 0.01)$	1 / 4	3.076846
		2 / 9	1.696078
		3 / 16	0.825399
		4 / 25	0.363869
		5 / 36	0.270419
$f(x_1, x_2) = 0.3e^{x_1 - x_2} + 0.6e^{-x_2}$	$x_1 \sim TruncNormal(4, 1, [3, 5])$, $x_2 \sim TruncNormal(2, 0.01, [0, 4])$	1 / 4	0.343870
		2 / 9	0.057076
		3 / 16	0.007112
		4 / 25	0.000709
		5 / 36	0.000059
$f(x_1, x_2) = e^{x_1 x_2}$	$x_1 \sim TruncNormal(4, 1, [3, 5])$ $x_2 \sim TruncGamma(3, 1, [0.5, 1])$	1 / 4	5.745048
		2 / 9	1.035060
		3 / 16	0.142816
		4 / 25	0.016118
		5 / 36	0.001543
$f(x_1, x_2, x_3) = 0.3e^{x_1 - x_2} +$ $0.6e^{x_2 - x_3} + 0.1e^{x_3 - x_1}$	$x_1 \sim TruncNormal(4, 1, [3, 5])$ $x_2 \sim TruncGamma(3, 1, [0.5, 1])$ $x_3 \sim U[4, 8]$	1 / 8	1.637981
		2 / 27	0.303096
		3 / 64	0.066869
$f(x_1) = \psi cos(x_1) + (1 - \psi)sin(x_1)$ $\psi = 0.3$	$x_1 \sim Normal(0, 1)$	1 / 2	0.222627
		2 / 3	0.181681
		3 / 4	0.054450
		4 / 5	0.039815
		5 / 6	0.009115

References

1. Atkeson, A., Ohanian, L.E.: Are Phillips curves useful for forecasting inflation? Q. Rev. **25**(Win), 2–11 (2001). https://ideas.repec.org/a/fip/fedmqr/y2001iwinp2-11nv.25no.1.html

2. Bartocci, E., Kovács, L., Stankovič, M.: Automatic generation of moment-based invariants for prob-solvable loops. In: Chen, Y.-F., Cheng, C.-H., Esparza, J. (eds.) ATVA 2019. LNCS, vol. 11781, pp. 255–276. Springer, Cham (2019). https://doi.org/10.1007/978-3-030-31784-3_15

3. Bartocci, E., Kovács, L., Stankovič, M.: MORA - automatic generation of moment-based invariants. In: TACAS 2020. LNCS, vol. 12078, pp. 492–498. Springer, Cham (2020). https://doi.org/10.1007/978-3-030-45190-5_28

4. Bouissou, O., Goubault, E., Putot, S., Chakarov, A., Sankaranarayanan, S.: Uncertainty propagation using probabilistic affine forms and concentration of measure inequalities. In: Chechik, M., Raskin, J.-F. (eds.) TACAS 2016. LNCS, vol. 9636, pp. 225–243. Springer, Heidelberg (2016). https://doi.org/10.1007/978-3-662-49674-9_13

5. Chen, X., Ábrahám, E., Sankaranarayanan, S.: Taylor model flowpipe construction for non-linear hybrid systems. In: Proceedings of RTSS 2012: the 33rd IEEE Real-Time Systems Symposium, pp. 183–192. IEEE Computer Society (2012). https://doi.org/10.1109/RTSS.2012.70

6. Chorin, A.J.: Gaussian fields and random flow. J. Fluid Mech. **63**(1), 21–32 (1974). https://doi.org/10.1017/S0022112074000991

7. Denamiel, C., Huan, X., Šepić, J., Vilibić, I.: Uncertainty propagation using polynomial chaos expansions for extreme sea level hazard assessment: the case of the eastern adriatic meteotsunamis. J. Phys. Oceanogr. **50**(4), 1005–1021 (2020). https://doi.org/10.1175/JPO-D-19-0147.1

8. Ernst, O.G., Mugler, A., Starkloff, H.J., Ullmann, E.: On the convergence of generalized polynomial chaos expansions. ESAIM: M2AN **46**(2), 317–339 (2012). https://doi.org/10.1051/m2an/2011045

9. Foo, J., Yosibash, Z., Karniadakis, G.E.: Stochastic simulation of riser-sections with uncertain measured pressure loads and/or uncertain material properties. Comput. Methods Appl. Mech. Eng. **196**, 4250–4271 (2007). https://doi.org/10.1016/j.cma.2007.04.005

10. Formaggia, L., et al.: Global sensitivity analysis through polynomial chaos expansion of a basin-scale geochemical compaction model. Comput. Geosci. **17**, 25–42 (2013). https://doi.org/10.1007/s10596-012-9311-5

11. Ghanem, R., Dham, S.: Stochastic finite element analysis for multiphase flow in heterogeneous porous media. Transp. Porous Medias **32**, 239–262 (1998). https://doi.org/10.1023/A:1006514109327

12. Ghanem, R.: Probabilistic characterization of transport in heterogeneous media. Comput. Methods Appl. Mech. Eng. **158**, 199–220 (1998). https://doi.org/10.1016/s0045-7825(97)00250-8

13. Ghanem, R.G., Spanos, P.D.: Stochastic Finite Elements: A Spectral Approach. Springer, New York (1991). https://doi.org/10.1007/978-1-4612-3094-6

14. Giraldi, L., Le Maître, O.P., Mandli, K.T., Dawson, C.N., Hoteit, I., Knio, O.M.: Bayesian inference of earthquake parameters from buoy data using a polynomial chaos-based surrogate. Comput. Geosci. **21**(4), 683–699 (2017). https://doi.org/10.1007/s10596-017-9646-z

15. Hien, T.D., Kleiber, M.: Stochastic finite element modelling in linear transient heat transfer. Comput. Methods Appl. Mech. Eng. **144**(1), 111–124 (1997). https://doi.org/10.1016/S0045-7825(96)01168-1

16. Hou, T.Y., Luo, W., Rozovskii, B., Zhou, H.M.: Wiener chaos expansions and numerical solutions of randomly forced equations of fluid mechanics. J. Comput. Phys. **216**, 687–706 (2006). https://doi.org/10.1016/j.jcp.2006.01.008

17. Jasour, A., Wang, A., Williams, B.C.: Moment-based exact uncertainty propagation through nonlinear stochastic autonomous systems. CoRR abs/2101.12490 (2021). https://arxiv.org/abs/2101.12490

18. Knio, O.M., Maître, O.P.L.: Uncertainty propagation in CFD using polynomial chaos decomposition. Fluid Dyn. Res. **38**(9), 616–640 (2006). https://doi.org/10.1016/j.fluiddyn.2005.12.003

19. Makino, K., Berz, M.: Taylor models and other validated functional inclusion methods. Int. J. Pure Appl. Math. **6**, 239–316 (2003)

20. Meecham, W.C., Jeng, D.T.: Use of the Wiener-Hermite expansion for nearly normal turbulence. J. Fluid Mech. **32**(2), 225–249 (1968). https://doi.org/10.1017/S0022112068000698

21. Moosbrugger, M., Stanković, M., Bartocci, E., Kovács, L.: This is the Moment for Probabilistic Loops. arXiv (2022). https://doi.org/10.48550/arXiv.2204.07185

22. Mühlpfordt, T., Findeisen, R., Hagenmeyer, V., Faulwasser, T.: Comments on truncation errors for polynomial chaos expansions. IEEE Control Syst. Lett. **2**(1), 169–174 (2018). https://doi.org/10.1109/LCSYS.2017.2778138

23. Neher, M., Jackson, K.R., Nedialkov, N.S.: On taylor model based integration of ODEs. SIAM J. Numer. Anal. **45**, 236–262 (2007). https://doi.org/10.1137/050638448

24. Sankaranarayanan, S.: Quantitative analysis of programs with probabilities and concentration of measure inequalities. Found. Probab. Program. 259 (2020). https://doi.org/10.1017/9781108770750.009

25. Sankaranarayanan, S., Chou, Y., Goubault, E., Putot, S.: Reasoning about uncertainties in discrete-time dynamical systems using polynomial forms. In: Larochelle, H., Ranzato, M., Hadsell, R., Balcan, M.F., Lin, H. (eds.) Advances in Neural Information Processing Systems, vol. 33, pp. 17502–17513. Curran Associates, Inc. (2020). https://proceedings.neurips.cc/paper/2020/file/ca886eb9edb61a42256192745c72cd79-Paper.pdf

26. Son, J., Du, Y.: Probabilistic surrogate models for uncertainty analysis: dimension reduction-based polynomial chaos expansion. Int. J. Numer. Meth. Eng. **121**(6), 1198–1217 (2020). https://doi.org/10.1002/nme.6262

27. Stanković, B.: Taylor expansion for generalized functions. J. Math. Anal. Appl. **203**, 31–37 (1996). https://doi.org/10.1006/jmaa.1996.0365

28. Steinhardt, J., Tedrake, R.: Finite-time regional verification of stochastic nonlinear systems. Int. J. Robot. Res. **31**(7), 901–923 (2012). https://doi.org/10.1177/0278364912444146

29. Taylor, J.B.: Discretion versus policy rules in practice. In: Carnegie-Rochester Conference Series on Public Policy, vol. 39, no. 1, pp. 195–214 (1993). https://ideas.repec.org/a/eee/crcspp/v39y1993ip195-214.html

30. Triebel, H.: Taylor expansions of distributions. In: The Structure of Functions. Birkhäuser Basel (2001). https://doi.org/10.1007/978-3-0348-8257-6_8

31. Wan, X., Karniadakis, G.E.: An adaptive multi-element generalized polynomial chaos method for stochastic differential equations. J. Comput. Phys. **209**, 617–642 (2005). https://doi.org/10.1016/j.jcp.2005.03.023

32. Xiu, D.: Numerical Methods for Stochastic Computations: A Spectral Method Approach. Princeton University Press, Princeton (2010). http://www.jstor.org/stable/j.ctv7h0skv
33. Xiu, D., Karniadakis, G.E.: The Wiener-Askey polynomial chaos for stochastic differential equations. SIAM J. Sci. Comput. **24**(2), 619–644 (2002). https://doi.org/10.1137/S1064827501387826

Distribution Estimation for Probabilistic Loops

Ahmad Karimi[1], Marcel Moosbrugger[2], Miroslav Stankovič[2], Laura Kovács[2],
Ezio Bartocci[2], and Efstathia Bura[1(✉)]

[1] Applied Statistics, Faculty of Mathematics and Geoinformation,
TU Wien, Vienna, Austria
{ahmad.karimi,efstathia.bura}@tuwien.ac.at
[2] Faculty of Informatics, TU Wien, Vienna, Austria

Abstract. We present an algorithmic approach to estimate the value
distributions of random variables of probabilistic loops whose statistical
moments are (partially) known. Based on these moments, we apply two
statistical methods, Maximum Entropy and Gram-Charlier series, to estimate the distributions of the loop's random variables. We measure the
accuracy of our distribution estimation by comparing the resulting distributions using exact and estimated moments of the probabilistic loop,
and performing statistical tests. We evaluate our method on several probabilistic loops with polynomial updates over random variables drawing
from common probability distributions, including examples implementing
financial and biological models. For this, we leverage symbolic approaches
to compute exact higher-order moments of loops as well as use sampling-
based techniques to estimate moments from loop executions. Our experimental results provide practical evidence of the accuracy of our method
for estimating distributions of probabilistic loop outputs.

Keywords: Probabilistic Loops · Distribution Estimation ·
Quantitative Evaluation

1 Introduction

Probabilistic programs (PPs) are programs with primitives to draw from probability distributions. As such, PPs do not produce a single output but rather a
probability distribution over outputs. In consequence, PPs provide a powerful
framework to model system behavior involving uncertainties.

Many machine and statistical learning techniques leverage PPs for representing and updating data-driven AI systems [17]. Quantifying and modelling distributions arising from PPs, and thus formally capturing PP behavior, is challenging. In this work, we address this challenge and provide an algorithmic approach

This research was partially supported by the WWTF grant ProbInG ICT19-018,
the ERC consolidator grant ARTIST 101002685, the FWF research projects LogiCS
W1255-N23 and P 30690-N35, and the TU Wien SecInt doctoral program.

© Springer Nature Switzerland AG 2022
E. Ábrahám and M. Paolieri (Eds.): QEST 2022, LNCS 13479, pp. 26–42, 2022.
https://doi.org/10.1007/978-3-031-16336-4_2

to effectively estimate the probability distributions of random variables generated by PPs, focusing on PPs with unbounded loops.

While sampling-based techniques are standard statistical approaches to approximate probability distributions in PPs, see e.g. [19], they cannot be applied to infinite-state PPs with potentially unbounded loops. More recently, static program analysis was combined with statistical techniques to infer higher-order statistical moments of random program variables in PPs with restricted loops and polynomial updates [2,30]. Our work complements these techniques with an algorithmic approach to compute the distributions of random variables in PPs with unbounded loops for which (some) higher-order moments are known (see Algorithm 1). Moreover, we assess the quality of our estimation via formal statistical tests.

Our method can be applied to any PP for which some moments are known. We provide an algorithmic solution to the so-called finite-moment problem [14, 23], as follows: *using a finite number of statistical moments of a PP random variable, compute the probability density function (pdf) of the variable to capture the probability of the random variable taking values within a particular range.* In full generality, the finite-moment problem is ill-posed and there is no single best technique available to solve it [11]. We tackle the finite-moment problem for PPs by focusing on two statistical methods, the Gram-Charlier expansion [18] and Maximum Entropy [5,26], in order to estimate the distribution of PP variables. Our approach is further complemented by statistical goodness-of-fit tests for assessing the accuracy of our estimated pdfs, such as the chi-square [34] and Kolmogorov-Smirnov [29] tests.

Motivating Example. We motivate our work with the Vasicek model in finance [37], which describes the evolution of interest rates. This model is defined by the stochastic differential equation,

$$dr_t = a(b - r_t)\, dt + \sigma\, dW_t \tag{1}$$

where W_t is a Wiener process (standard Brownian motion) [31] modeling the continuous inflow of randomness into the system, σ is the standard deviation representing the amplitude of the randomness inflow, b is the long term mean level around which paths evolve in the long term, a is the speed of reversion specifying the velocity at which such trajectories will regroup around b in time. We encode (1) as a PP in Fig. 1, using program constants a, b, σ, as described above and variables r, w to respectively capture the randomness (r_t) and Wiener process (W_t) of the Vasicek model. The PP of Fig. 1 has polynomial loop updates over random variables r, w drawn from a normal distribution with zero mean and unit variance. This PP satisfies the programming model of [2,30], so that higher-order moments of this PP can be computed using [2,30]. Based on the first- and second-order moments of r in the Vasicek model PP, we estimate the distribution of the random variable r using Maximum Entropy and Gram-Charlier expansion (see Sect. 3.1). Data generated by executing the PP repeatedly, at loop iteration 100, are summarized in the histogram in the right panel of Fig. 1, giving a rough estimate of the distribution of r. The black and red lines are kernel density estimates [35] of the pdf of

$a := 0.5$
$b := 0.2$
$\sigma := 0.2$
$w := 0$
$r := 2$
while true **do**
    $w := \text{Normal}(0,1)$
    $r := (1-a)r + ab + \sigma w$
end while

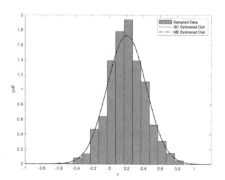

Fig. 1. The Vasicek model (1) describing the evolution of interest rates, modeled as a PP in the left panel. The estimated distributions of r are plotted in the right panel. The (normalized) histogram plots the 1000 sampled x-values generated by running the PP 1000 times at iteration $n = 100$. The kernel density estimates of the Maximum Entropy (black dash-dotted line) and Gram-Charlier (red solid line) pdf estimations follow closely the sampled data based histogram, showcasing that both estimates are effectively estimating the true (sampled) distributions of r (blue histogram). (Color figure online)

r using Maximum Entropy and Gram-Charlier expansion, respectively. Both pdf estimates closely track the histogram, a proxy for the true distribution of r. The close match between our estimated pdfs and the true distribution of r is also supported by the chi-square and Kolmogorov-Smirnov test results (see Sect. 3.2).

Remark 1. To discretize the Vasicek model (1), we set the time step to 1, so that $dt = 1$, $dr_t = r_{t+1} - r_t$, and $dW_t = W_{t+1} - W_t = \mathcal{N}(0,1)$. This leads to $r_{t+1} = (1-a)r_t + ab + \sigma\mathcal{N}(0,1)$, from which the program of Fig. 1 can be constructed.

Related Works. Kernel density estimation is combined with a constrained minimization problem for an (Hausdorff) instance of the finite-moment problem in [1]. This approach is further extended in [14]. This nonparametric estimation method requires constant tuning of the bandwidth parameters, which increase with the number of moments, and is seen to often perform poorly [26]. Along similar lines, the approach of [23] employs spline-based non-parametric density estimation using piecewise polynomial functions to recover target distributions. This method is improved in [11] to an adaptive algorithm dealing with non-equidistant grids. This method, though, assumes a smooth transition at the boundaries of sub-intervals that needs to be checked repeatedly during the reconstruction process [26]. Our work complements these approaches by a tailored algorithmic method to effectively estimate distributions arising from PPs that uses higher-order moments of these PPs in combination with Maximum Entropy and Gram-Charlier expansion. To the best of our knowledge, these statistical methods have not yet been used and evaluated in the setting of probabilistic loops.

Our work is related to emerging efforts in estimating probabilistic distributions/densities in PP, as in [15,16,20,21]. The provided automation, supported by the tools (λ)PSI [15,16], AQUA [21], and DICE [20], yield symbolic frameworks to compute exact (posterior) densities for PPs with bounded loops. Also focusing on bounded loops, sampling-based techniques to approximate the distributions resulting from PPs are exploited in [4,9,10,32]. In contrast, our approach is not restricted to bounded loops but focuses on density estimation for PPs with unbounded loops.

Our Contributions. The main contribution of this paper is the development of an *algorithmic approach to estimate distributions arising from PPs with unbounded loops (Algorithm 1)*, allowing thus to estimate the functional behaviour of probabilistic loops. We employ two formal techniques from mathematical statistics, Maximum Entropy (ME) and Gram-Charlier (GC) expansion, to estimate the distributions of PP variables (Sect. 3.1). We use symbolic approaches to compute exact moments of loop variables as symbolic expressions parameterized by the loop counter. Further, we apply statistical tests to assess the adequacy of the estimated distributions of PP variables (Sect. 3.2). We evaluate our approach on a number of benchmarks and demonstrate the accuracy of our proposed estimation approach (Sect. 4).

Paper Outline. The rest of the paper is organized as follows. Section 2 introduces the necessary prerequisites from probability theory, statistics, and probabilistic programs. In Sect. 3, we introduce the (adjusted) methods we use to estimate the distributions of program variables from their moments. We report on our practical findings in Sect. 4 and conclude the paper in Sect. 5.

2 Preliminaries

This section reviews relevant terminology from statistics and probabilistic programs; for further details we refer to [12,30]. Throughout the paper, $\mathbb{N}$ denotes the set of natural numbers. To facilitate readability, we sometimes write $\exp[t]$ to denote the exponential function e^t, where t is an arbitrary expression/argument.

2.1 PPs and Moments of Random Variables

The result of a PP is not a single output but rather multiple values with different probabilities according to the distribution of the random variables the PP encodes. Different execution paths in a PP are typically selected by draws from commonly used distributions that are fully characterized by their moments, such as the uniform or normal. Yet, since output values of PPs are results of multiple operations, distributions arising from the random variables the PPs encode are most often not common.

The study of distributions is a much explored topic in statistics. Many distributions are fully characterized by their moments and most well known distributions (for example the normal, the Poisson and the uniform) fall in this category.

Definition 1 (Moments). *Let X be a random variable, $c \in \mathbb{R}$ and $k \in \mathbb{N}$. We write $Mom_k[c, X]$ to denote the kth moment about c of X, which is defined as*

$$Mom_k[c, X] = \mathbb{E}((X - c)^k). \tag{2}$$

In this paper we consider *raw moments* about $c = 0$. We note, however, that we can move between moments of X with different centers, c and d, using

$$\mathbb{E}\left((X - d)^k\right) = \sum_{i=0}^{k} \binom{k}{i} \mathbb{E}\left((X - c)^i\right)(c - d)^{k-i}. \tag{3}$$

Moments of Probabilistic Loops. Moments of program variables of probabilistic loops can be approximated by sampling; that is, by executing the loops for an arbitrary but fixed number of iterations (see, e.g. [19,39]). Alternatively, for restricted classes of probabilistic loops with polynomial updates, symbolic methods from algorithmic combinatorics can be used to compute the exact (higher-order) moments of random program variables x by expressing these moments as closed-form expressions over loop iterations and some initial values [2,30]. That is, [2,30] derive the expected value of the kth moment of variable x at loop iteration n, denoted as $\mathbb{E}(x^k(n))$, in closed form, where $x^k(n)$ specifies the value of x^k at loop iteration n, and $k, n \in \mathbb{N}$.

2.2 From Moments to Distributions

Given a finite set of moments of a random variable x, its distribution can be estimated using various statistical approaches. Here, we focus on maximum entropy and Gram-Charlier expansion.

Maximum Entropy (ME). The Maximum Entropy (ME) distribution estimation method is based on the maximization of constraints describing the Shannon information entropy [5,8,26]. Specifically, in order to estimate the unknown distribution f of a PP variable x in our setting, we maximize the Shannon entropy H of f, defined by

$$H[f] = -\int_l^u f(x) \ln(f(x)) dx, \tag{4}$$

subject to its given moments $\mathbb{E}(x^i) = \int x^i f(x) dx$ [3,33]. The ME approximation $f_{\text{ME}}(x)$ of the target probability density function (pdf) takes the form

$$f_{\text{ME}}(x) = \exp\left[-\sum_{j=0}^{N} \xi_j x^j\right], \tag{5}$$

where the Lagrange's multipliers ξ_j, with $j = 0, 1, \ldots, m$, can be obtained from the first $m \in \mathbb{N}$ moments, $\{\mathbb{E}(x), \mathbb{E}(x^2), \ldots, \mathbb{E}(x^m)\}$. To this end, the following system of $m + 1$ nonlinear equations is solved,

$$\int_l^u x^i \exp\left[-\sum_{j=0}^m \xi_j x^j\right] dx = \mathbb{E}(x^i). \tag{6}$$

Gram-Charlier Expansion. The Gram-Charlier series approximates the pdf f of a PP variable x in terms of its cumulants and using a known distribution ψ [6,18]. As an alternative to moments, cumulants of a distribution are defined using the *cumulant-generating function*, which equals the natural logarithm of the characteristic function, $K(t) = \ln\left(\mathbb{E}(e^{itx})\right)$. In what follows, we denote by κ_m the mth cumulant of f, the unknown target distribution to be approximated. The relationship between moments and cumulants can be obtained by extracting coefficients from the expansion. To be precise, we can express the mth cumulant κ_m in terms of the first m moments [7] as

$$\kappa_m = (-1)^{m+1} \det \begin{pmatrix} \mathbb{E}(x) & 1 & 0 & 0 & 0 & \cdots & 0 \\ \mathbb{E}(x^2) & \mathbb{E}(x) & 1 & 0 & 0 & \cdots & 0 \\ \mathbb{E}(x^3) & \mathbb{E}(x^2) & \binom{2}{1}\mathbb{E}(x) & 1 & 0 & \cdots & 0 \\ \mathbb{E}(x^4) & \mathbb{E}(x^3) & \binom{3}{1}\mathbb{E}(x^2) & \binom{3}{2}\mathbb{E}(x) & 1 & \cdots & 0 \\ \vdots & \vdots & \vdots & \vdots & \vdots & \ddots & \vdots \\ \mathbb{E}(x^{m-1}) & \mathbb{E}(x^{m-2}) & \cdots & \cdots & \cdots & \ddots & 1 \\ \mathbb{E}(x^m) & \mathbb{E}(x^{m-1}) & \cdots & \cdots & \cdots & \cdots & \binom{m-1}{m-2}\mathbb{E}(x) \end{pmatrix}, \tag{7}$$

where $\det(\cdot)$ stands for determinant. The first cumulant κ_1 of the random variable x is the mean $\mu = \mathbb{E}(x)$; the second cumulant κ_2 of f is the variance σ^2, and the third cumulant κ_3 is the same as the third central moment[1]. Higher-order cumulants of f, however, are in general not equal to higher moments. Using the cumulants κ_m computed from exact moments of x together with the cumulants of the known distribution ψ, the pdf f of a random variable x can approximated by the Gram-Charlier (type-)A expansion $f_{\mathrm{GC}}(x)$, as in [24], and given by

$$f_{\mathrm{GC}}(x) = \psi(x) \sum_{m=0}^{\infty} \frac{1}{m!\sigma^m} B_m(0, 0, \kappa_3, \ldots, \kappa_m) He_m\left(\frac{x - \mu}{\sigma}\right), \tag{8}$$

where ψ is the normal pdf with mean $\mu = \kappa_1$ and variance $\sigma^2 = \kappa_2$, and B_m and He_m are respectively the Bell and Hermite polynomials [6]. Derivation details of (8) can be found in [38].

---

[1] The ith central moment of x is defined as $\mathbb{E}\left((x - \mathbb{E}(x))^i\right)$.

Algorithm 1. Effective Distribution Estimation

Input: Probabilistic loop $\mathcal{P}$ with program variable(s) x;
     set M of exact moments of x for loop iteration n of $\mathcal{P}$

Output: Estimated distributions f_{ME} and f_{GC} of x, with respective accuracy $\mathcal{A}_{\mathrm{ME}}$ and $\mathcal{A}_{\mathrm{GC}}$

Parameters: Loop iteration $n \in \mathbb{N}$; number of executions $e \in \mathbb{N}$ of $\mathcal{P}$

Initialization:

1: Choose a subset of exact moments $S_{\mathrm{M}} = \{\mathbb{E}(x), \mathbb{E}(x^2), \ldots, \mathbb{E}(x^{|S_{\mathrm{M}}|})\} \subset M$

2: Collect $Sample_{Data}$ by sampling e many $\mathcal{P}$ variable values at the nth loop iteration

Distribution Estimation:

3: Compute f_{ME} using S_{M} and ME as in (5)

4: Compute f_{GC} using S_{M} and GC as in (9)

χ^2 Test:

5: Split $Sample_{Data}$ into bins and compute observed frequencies O_i

6: Calculate expected frequencies $E_{\mathrm{ME},i}$ and $E_{\mathrm{GC},i}$ from f_{ME} and f_{GC}, as in (11)

7: Compute χ^2 test statistics for f_{ME} and f_{GC} as in (12) and compare them to the critical value CV_{χ^2}

K-S Test:

8: Calculate the empirical cdf F_{Sample} of $Sample_{Data}$ and cdfs F_{ME}, F_{GC} using f_{ME} and f_{GC}

9: Compute K-S test statistics D^*_{ME}, D^*_{GC} of f_{ME} and f_{GC} as in (13) and compare to the critical value $CV_{\mathrm{K\text{-}S}}$

Pdf Accuracy Evaluation:

10: If $\chi^2_{\mathrm{ME}} < CV_{\chi^2}$ or $D^*_{\mathrm{ME}} < CV_{\mathrm{K\text{-}S}}$ then $\mathcal{A}_{\mathrm{ME}} \leftarrow NOT\ REJECTED$
                else $\mathcal{A}_{\mathrm{ME}} \leftarrow REJECTED$

11: If $\chi^2_{\mathrm{GC}} < CV_{\chi^2}$ or $D^*_{\mathrm{GC}} < CV_{\mathrm{K\text{-}S}}$ then $\mathcal{A}_{\mathrm{GC}} \leftarrow NOT\ REJECTED$
                else $\mathcal{A}_{\mathrm{GC}} \leftarrow REJECTED$

---

3 Effective Estimation of Distributions for Probabilistic Loops

We present our estimation approach for the pdf of a random variable x in a probabilistic loop $\mathcal{P}$, provided the first M statistical moments of x are known. We use Maximum Entropy (ME) and Gram-Charlier A (GC) expansion (lines 1–4 of Algorithm 1). We assess the accuracy of our pdf estimates by conducting statistical tests over our distribution estimates (lines 5–11 of Algorithm 1). Our approach is summarized in Algorithm 1 and detailed next.

3.1 Distribution Estimation

Given a variable x of a probabilistic loop $\mathcal{P}$, we use a subset $S_M \subset M$ of its exact moments for estimating the distribution of x through ME and GC expansion. For this, we adjust ME and GC expansion to compute the estimated distributions f_{GC} and f_{ME} of f, the pdf of x, respectively. We use the set $S_M = \{\mathbb{E}(x), \mathbb{E}(x^2), \ldots, \mathbb{E}(x^{|S_M|})\}$ of exact moments of x. For ME, we derive the Lagrange multipliers ξ_j, for $j = 0, 1, \ldots, |S_M|$ in the ME approximation (5), where $|S_M|$ denotes the cardinality of S_M. For GC expansion, we truncate the GC expansion (8) based on $|S_M|$ moments, and estimate the pdf of x with

$$f_{GC}(x) = \psi(x) \sum_{m=0}^{|S_M|} \frac{1}{m! \sigma^m} B_m(0, 0, \kappa_3, \ldots, \kappa_m) He_m\left(\frac{x - \mu}{\sigma}\right). \tag{9}$$

The cumulants κ_m of the pdf of x in (9) are computed from the moments in S_M. The $f_{GC}(x)$ estimate of the pdf of a PP variable x can be computed even when the moments of x are parametric; i.e., their closed-form functional representations depend on the loop iteration n and/or other symbolic values. This is especially useful in the analysis of (probabilistic) loops, as it allows us to encompass *all* loop iterations in a single symbolic estimate using the GC expansion.

Example 1. We use the PP in Fig. 1 to illustrate our approach. The set M in Algorithm 1 contains the exact first two moments of r for an arbitrary loop iteration n. These two moments can be expressed as functions of loop iteration n of the PP (see Sect. 4),

$$\begin{aligned}
\mathbb{E}(r(n)) &= 2^{-n}(2^n + 9)/5, \\
\mathbb{E}(r^2(n)) &= 7/75 + (18)2^{-n}/25 + (239)2^{-2n}/75.
\end{aligned} \tag{10}$$

As a result, we set $S_M = M = \{\mathbb{E}(r(n)), \mathbb{E}(r^2(n))\}$ for loop iteration n (line 1 of Algorithm 1). These functions yield exact moments of r for concrete values of the loop iteration n, by only instantiating the above expressions with the respective values of n. For example, at loop iteration $n = 100$, the two exact moments of r are $\mathbb{E}(r(100)) = 0.20$ and $\mathbb{E}(r^2(100)) = 0.093\bar{3}$.

ME Estimation of the pdf of r. Given (10), we solve the system (6) of nonlinear equations to obtain the respective Lagrange multipliers ξ_j, $j = 0, 1, 2$. For solving (6), we assume that the support of the target pdf of r is a subset of $[l, u]$, where l and u are known scalars and apply the Levenberg–Marquardt numerical minimization algorithm [27,28]. Once the optimal multipliers for (6) are computed, the ME estimate is $f_{ME}(r) = \exp\left[-(\xi_0 + \xi_1 r + \xi_2 r^2)\right]$. Specifically, $f_{ME}(r) = \exp[0.171658 + 3.749999\, r + 9.374997\, r^2]$ at loop iteration $n = 100$. As seen from Fig. 1, this estimate is closely approximating the true (sampled) distribution of r.

GC Estimation of the pdf of r. Given the first two exact moments of r in (10), we apply (7) in order to compute the corresponding cumulants. Using the normal distribution with mean $\mu = \kappa_1 = \mathbb{E}(r)$ and variance $\sigma^2 = \kappa_2 = \mathbb{E}(r^2) - \mathbb{E}^2(r)$ in (9), the GC estimate of the pdf of r is $f_{GC}(r) = (1/\sqrt{2\pi\kappa_2})\exp\big[-(r-\kappa_1)^2/2\kappa_2\big]$. By expressing κ_1 and κ_2 in terms of exact moments, $\kappa_1 = \mathbb{E}(r)$ and $\kappa_2 = \mathbb{E}(r^2) - \mathbb{E}^2(r)$, and using (10), the GC estimate is further expressed as a function of n by

$$f_{GC}(r(n)) = \frac{\eta \,\exp\left[-\dfrac{\left(r(n)-\frac{2^n+9}{5\cdot 2^n}\right)^2}{\frac{36\cdot 2^{-n}}{25}-\frac{2\cdot(2^n+9)^2}{2^{2n}\cdot 25}+\frac{478}{2^{2n}\cdot 75}+\frac{14}{75}}\right]}{\beta\left(\dfrac{18\cdot 2^{-n}}{25}-\dfrac{2^{-2n}(2^n+9)^2}{25}+\dfrac{239\cdot 2^{-2n}}{75}+\dfrac{7}{75}\right)^{1/2}},$$

where $\eta = 2251799813685248$ and $\beta = 5644425081792261$. In particular, for loop iteration $n = 100$, the obtained $f_{GC}(r) = 1.7275\exp[-9.3750(r-0.2)^2]$ closely approximates the true (sampled) distribution of r, as evidenced in Fig. 1.

3.2   Assessing Accuracy of Estimated Distributions

To evaluate the accuracy of the estimated pdfs f_{ME} and f_{GC} of the PP variable x, we would, ideally, compare them to the true underlying pdf of x. This comparison, however, is not possible in general, as the true distribution of x arising from an (unbounded) probabilistic loop is unknown and frequently complex. To overcome this obstacle, we execute the probabilistic loop e times to sample the distribution of x and use the collected $Sample_{Data}$ as proxy of the underlying probability distribution of x (line 2 of Algorithm 1). For this, we carry out a Monte-Carlo "experiment" [19], by executing $\mathcal{P}$ a large number of times (e) and collect the value of x at loop iteration n in $Sample_{Data}{}^2$. Based on $Sample_{Data}$, we evaluate the accuracy of our ME and GC expansion estimations using two statistical tests, namely the chi-square (χ^2) [34] and the Kolmogorov-Smirnov [29] goodness-of-fit tests, as described next.

Chi-Square (χ^2) Goodness-of-Fit Test. The chi-square (χ^2) goodness-of-fit test is used to detect statistically significant differences between observed and expected frequencies [36]. The expected frequencies are computed under the assumption the sample comes from a specific distribution. We use the χ^2 goodness-of-fit test to compare f_{ME} and f_{GC} with the "true" (sampled) empirical distribution of x that is based on the sampled data, $Sample_{Data}$ (lines 5–7 of Algorithm 1).

   We partition our $Sample_{Data}$ into $k \in \mathbb{N}$ non-overlapping intervals I_k (also called bins), such that $Sample_{Data} \subset \bigcup_{i=1}^{k} I_i$. Let $O_i = |I_i|$ denote the number of samples, or *observed* data frequency of x, in the ith interval I_i, with $i \in \{1,\dots,k\}$, and let l_i and u_i respectively denote the lower and upper bounds

---

[2] in our experiments, we use $e = 1000$ and $n = 100$, see Sect. 4.

of I_i. We compute the expected frequencies[3] in I_i, denoted as $E_{ME,i}$ and $E_{GC,i}$, from f_{ME} and f_{GC} as

$$
\begin{aligned}
E_{ME,i} &= |Sample_{Data}| * \int_{l_i}^{u_i} f_{ME}(x)dx, \\
E_{GC,i} &= |Sample_{Data}| * \int_{l_i}^{u_i} f_{GC}(x)dx.
\end{aligned}
\tag{11}
$$

where $|Sample_{Data}|$ denotes the number of elements (points) in $Sample_{Data}$.
The resulting chi-square goodness-of-fit test statistics are

$$
\begin{aligned}
\chi_{ME}^2 &= \sum_{i=1}^{k}(O_i - E_{ME,i})^2/E_{ME,i}, \\
\chi_{GC}^2 &= \sum_{i=1}^{k}(O_i - E_{GC,i})^2/E_{GC,i}.
\end{aligned}
\tag{12}
$$

If the values in (12) exceed the chi-square critical value $CV_{\chi^2} = \chi_{1-\alpha,k-1}^2$, where α is the statistical significance level (parameter) and $k-1$ the degrees of freedom, the hypothesis that the ME or GC estimated distributions, respectively, are the same as the "true" distribution of $Sample_{Data}$ is rejected (lines 10–11 of Algorithm 1). Otherwise, there is not enough evidence to support the claim that the distributions differ significantly.

Example 2. We set $k = 15$ and $\alpha = 0.05$, so that $CV_{\chi^2} = \chi_{1-\alpha,k-1}^2 = 23.685$. Figure 2 shows observed frequencies from $Sample_{Data}$, as well as the expected frequencies from the f_{ME} and f_{GC} pdfs of Example 1, where $Sample_{Data}$ are collected while sampling the PP of Fig. 1 for $e = 1000$ times at the $n = 100$th loop iteration. The test statistic values of (12) are $\chi_{ME}^2 = 17.4018$ and $\chi_{GC}^2 = 17.4017$. Since both are smaller than CV_{χ^2}, we conclude that both f_{ME} and f_{GC} are accurate estimates of the pdf of r. This result is also supported by the close agreement of the plotted frequencies in Fig. 2.

Kolmogorov-Smirnov (K-S) Test. The Kolmogorov-Smirnov (K-S) test [29] compares two cumulative distribution functions (cdfs). We compute the cdfs F_{ME} and F_{GC} of the estimated pdfs f_{ME} and f_{GC}, respectively. We also compute the (empirical) cdf F_{Sample} of $Sample_{Data}$, and let

$$
\begin{aligned}
D_{ME}^* &= \max_x(|F_{ME}(x) - F_{Sample}(x)|), \\
D_{GC}^* &= \max_x(|F_{GC}(x) - F_{Sample}(x)|)
\end{aligned}
\tag{13}
$$

denote the K-S test statistics D_{ME}^* and D_{GC}^*, respectively. To assess the distance of F_{ME} and F_{GC} from F_{Sample} with K-S, we compare D_{ME}^* and D_{GC}^* to the K-S test critical value $CV_{K\text{-}S} = \sqrt{-(1/|Sample_{Data}|)\ln(\alpha/2)}$, where α is the statistical significance level. The K-S test rejects the claim that the compared distributions are the same if $D^* < CV_{K-S}$ (lines 10–11 of Algorithm 1).

---

[3] Alternatively, these frequencies can also be obtained from the cumulative distribution function (cdf) of x.

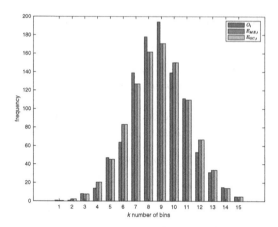

Fig. 2. Observed frequencies (blue) of the $Sample_{Data}$ of Fig. 1, in alignment with the expected frequencies obtained from the f_{ME} (orange) and f_{GC} (yellow) estimated distributions of the PP of Fig. 1. (Color figure online)

Example 3. We set $\alpha = 0.05$, so that $CV_{\text{K-S}} = 0.0608$. We compute the K-S test statistics for the ME and GC estimated distributions of Example 1, to obtain $D^*_{\text{ME}} = 0.03602307$ and $D^*_{\text{GC}} = 0.03602304$. Since both test statistics are smaller than the critical value, we conclude that both f_{ME} and f_{GC} are close to the underlying distribution of r.

4   Experimental Evaluation

In this section we report on our experimental results towards estimating distributions of probabilistic loop variables using Algorithm 1. We describe our benchmark set and present our practical findings using these benchmarks. We also report additional results on evaluating the precision of higher-order moments of loop variables.

Benchmarks. We use four challenging examples of unbounded probabilistic loops from state-of-the-art approaches on quantitative analysis of PPs [2,13,25]; these benchmarks are the first four entries of the first column of Table 1. In addition, we also crafted three new examples, listed in the last three entries of the first column of Table 1, as follows: (i) line 5 of Table 1 specifies a PP loop approximating a uniform distribution; (ii) line 6 of Table 1 refers to the Vasicek model of Fig. 1; and (iii) the last line of Table 1 lists an example encoding a piece-wise deterministic process (PDP) modeling gene circuits based on [22]. In particular, the PDP model can be used to estimate the distribution of protein x and the mRNA levels y in a gene; our PP encoding of this PDP model is given in Fig. 3.

$k_1 := 4,\ k_2 := 40,\ y := 0$
$x := 0,\ a := 0.2,\ b := 4,\ s := 0$
$h := 0.6,\ f := 0.1,\ \rho := 0.5$
while *true* **do**
    if $s = 0$ **then**
        $s = 1\ [\,f\,]\ 0$
    else
        $s = 0\ [\,h\,]\ 1$
    end if
    $k := k_2 * s + k_1 * (1 - s)$
    $y := (1 - \rho)y + k$
    $x := (1 - a)x + by$
end while

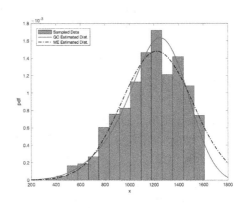

Fig. 3. A PDP model for gene circuits, modeled as a PP in the left, representing a gene controlled by a two-valued, probabilistically updated variable $s \in \{0, 1\}$. We respectively denote with k_0 and k_1 the gene transcription rates in states $s = 0$ and $s = 1$, and let b the translation rate of protein x production in the gene. Further, ρ controls the mRNA level y and a denotes the protein degradation rate. The (normalized) histogram of sampled data of the resulting PP, with execution time/sample size $e = 1000$ and loop iteration $n = 100$ is given on the right, together with the ME and GC expansion estimates of the pdf of x using the first 3 moments of x.

Experimental Setup. All our seven examples in Table 1 implement polynomial loop updates and fall in the class of probabilistic loops supported by [30]. As such, for each example of Table 1, exact higher-order moments of random loop variables can be computed using the algorithmic approach of [30]. Here we use the Polar tool of [30] to derive a *finite* set M of exact higher-order moments for each PP in Table 1, and we set $S_M = M$ in Algorithm 1. Further, we generate our sampled data ($Sample_{Data}$) by executing each PP $e = 1000$ times at loop iteration $n = 100$. For assessing the adequacy of estimated distributions, we set the statistical significance level $\alpha = 0.05$ for both the K-S and chi-square tests. For the latter, we use $k = 15$ bins. The critical test values for the chi-square and K-S tests are respectively $CV_{\chi^2} = 23.685$ and $CV_{\text{K-S}} = 0.0608$. All our numerical computations are conducted in Matlab.

Experimental Results on Distribution Estimation. Table 1 summarizes our experimental results on estimating the distributions of our benchmark programs. The first column of Table 1 names the benchmark. The second column specifies the random variable of the benchmark for which $|S_M|$ moments are derived in order to estimate the distribution of the respective variable. The number $|S_M|$ of moments used is given in column 3 of Table 1. Columns 4–5 respectively give the resulting chi-square test values for the ME and GC expansion estimations of the pdf of the respective PP variable of the benchmark. In addition to the chi-square test values, columns 4–5 also indicate the adequacy of our estimated distribution (as in lines 10–11 of Algorithm 1): we use ✓ to

Table 1. Accuracy of the ME and GC expansion pdf estimates for benchmarks PPs, assessed with the chi-square and K-S statistical tests.

Program	Var	$|S_M|$	χ^2_{ME}	χ^2_{GC}	D^*_{ME}	D^*_{GC}
StutteringP [2]	s	2	18.6432 ✓	16.7943 ✓	0.0181 ✓	0.0213 ✓
Square [2]	y	2	36.9009 ✗	39.1309 ✗	0.0586 ✓	0.0566 ✓
Binomial [13]	x	2	27.4661 ✗	27.4574 ✗	0.0598 ✓	0.0597 ✓
Random Walk 1D [25]	x	2	18.6709 ✓	18.7068 ✓	0.0264 ✓	0.0263 ✓
Uniform(0,1)	u	6	16.3485 ✓	105.276 ✗	0.0214 ✓	0.0658 ✗
Vasicek Model	r	2	17.4018 ✓	17.4017 ✓	0.0360 ✓	0.0360 ✓
PDP Model	x	3	64.4182 ✓	65.9500 ✓	0.0403 ✓	0.0393 ✓

specify that the estimated pdf is an accurate estimate of the true distribution, and write ✗ otherwise. The results reported in columns 6–7 of Table 1 are as in columns 4–5, yet by using the K-S test instead of the chi-square test.

Table 1 indicates that our approach in Algorithm 1, based on GC expansion and ME, for distribution estimation yields accurate estimates for pdfs of continuous random variables, as assessed by the chi-square test and the K-S test; the benchmarks of STUTTERINGP, RANDOM WALK 1D, VASICEK MODEL, and PDP MODEL fall in this category. For estimating the pmf of discrete random variables, as in the SQUARE and BINOMIAL programs, the K-S test infers our method be accurate, but the chi-square test does not. The GC expansion (see Sect. 3.1) expresses a distribution as a series of Gaussian terms. When estimating pdfs of random variables whose distributions are markedly different from the normal, this GC expansion is not adequate. This occurs in the UNIFORM example, where the PP implements a uniform distribution. In contrast, the ME based pdf estimation is accurate, as affirmed by both the chi-square and the K-S test.

Precision Evaluation of Higher-Order Moments of PP Variables. In addition to effectively estimating distributions of a PP variable x, in our experiments we were also interested to compare the higher-order moments of the estimated pdfs of x against the exact moments of x. That is, we were interested to see how the estimated moments we compute from the $|S_M|$ exact moments of x differ from their respective, exact higher-order moments.

For this evaluation setting, we use the loop iteration $n = 100$ as before and apply the following setting: we compute the higher-order moments M'_{ME} and M'_{GC} of the estimated distributions f_{ME} and f_{GC}, and compare them with the set S_M of exact moments of x. In the sequel, we let $\mathbb{E}(x^i)_{ME}$ and $\mathbb{E}(x^i)_{GC}$ denote the ith estimated moment of x computed from f_{ME} and f_{GC}, and write $\mathbb{E}(x^i)$ for the ith exact moment of x from S_M (as in Algorithm 1). For comparing moments, we compute the *absolute estimate error* for the ith moment of x as

Table 2. Precision evaluation of higher-order moments using $|S_M|$ exact moments.

Program	Var	$|S_M|$	Moment	Exact Moment	AE_{Sample} (RE_{Sample})	AE_{ME} (RE_{ME})	AE_{GC} (RE_{GC})
StutteringP	s	2	1	210	$5.69 \times 10^{-2}(0.027\%)$	$8.08 \times 10^{-5}(0.00003\%)$	$4.71 \times 10^{-5}(0.00002\%)$
			2	4.4405×10^4	$2.39 \times 10^1(0.053\%)$	$1.75 \times 10^{-2}(0.00003\%)$	$1.75 \times 10^{-2}(0.00003\%)$
			3	9.4536×10^6	$6.78 \times 10^3(0.0.072\%)$	$9.21 \times 10^1(0.000974\%)$	$9.25 \times 10^1(0.000978\%)$
			4	2.0260×10^9	$1.48 \times 10^6(0.073\%)$	$7.38 \times 10^4(0.00364\%)$	$7.42 \times 10^4(0.00366\%)$
			5	4.3705×10^{11}	$2.17 \times 10^8(0.050\%)$	$3.82 \times 10^7(0.00873\%)$	$3.83 \times 10^7(0.00876\%)$
			6	9.4884×10^{13}	$5.51 \times 10^9(0.0058\%)$	$1.60 \times 10^{10}(0.0168\%)$	$1.60 \times 10^{10}(0.0168\%)$
			7	2.0729×10^{16}	$2.10 \times 10^{13}(0.101\%)$	$5.88 \times 10^{12}(0.0284\%)$	$5.90 \times 10^{12}(0.0285\%)$
			8	4.5570×10^{18}	$1.12 \times 10^{16}(0.245\%)$	$1.99 \times 10^{15}(0.0438\%)$	$2.00 \times 10^{15}(0.0439\%)$
Square	y	2	1	10100	$1.66 \times 10^1(0.16\%)$	$1.11 \times 10^0(0.011\%)$	$7.97 \times 10^{-3}(0.00007\%)$
			2	1.0602×10^8	$2.31 \times 10^5(0.22\%)$	$2.30 \times 10^3(0.00217\%)$	$1.64 \times 10^2(0.00015\%)$
			3	1.1544×10^{12}	$1.27 \times 10^9(0.11\%)$	$1.26 \times 10^7(0.00011\%)$	$2.39 \times 10^9(0.2072\%)$
			4	1.3012×10^{16}	$2.98 \times 10^{13}(0.23\%)$	$1.99 \times 10^{12}(0.0153\%)$	$9.79 \times 10^{13}(0.7581\%)$
			5	1.5157×10^{20}	$1.32 \times 10^{18}(0.87\%)$	$8.83 \times 10^{16}(0.0583\%)$	$2.61 \times 10^{18}(1.7505\%)$
Binomial	x	2	1	50	$2.95 \times 10^{-1}(0.59\%)$	$5.19 \times 10^{-5}(0.000051\%)$	$3.16 \times 10^{-3}(0.0063\%)$
			2	2525	$2.82 \times 10^1(1.12\%)$	$2.74 \times 10^{-3}(0.000108\%)$	$1.87 \times 10^{-1}(0.0074\%)$
			3	128750	$2.04 \times 10^3(1.59\%)$	$1.43 \times 10^{-1}(0.000111\%)$	$1.22 \times 10^1(0.0095\%)$
			4	6.6268×10^6	$1.32 \times 10^5(2.00\%)$	$3.66 \times 10^0(0.000055\%)$	$8.22 \times 10^2(0.0125\%)$
			5	3.4421×10^8	$8.08 \times 10^6(2.35\%)$	$5.52 \times 10^2(0.000160\%)$	$5.53 \times 10^4(0.0161\%)$
			6	1.8038×10^{10}	$4.67 \times 10^8(2.64\%)$	$1.20 \times 10^5(0.000664\%)$	$3.66 \times 10^6(0.0203\%)$
			7	9.5354×10^{11}	$2.74 \times 10^{10}(2.87\%)$	$1.51 \times 10^7(0.001587\%)$	$2.38 \times 10^8(0.0250\%)$
			8	5.0830×10^{13}	$1.54 \times 10^{12}(3.04\%)$	$1.54 \times 10^9(0.0030363\%)$	$1.51 \times 10^{10}(0.0298\%)$
RandomWalk1D	x	2	1	20	$1.57 \times 10^{-1}(0.79\%)$	$1.83 \times 10^{-6}(0.000009\%)$	$4.44 \times 10^{-3}(0.022\%)$
			2	4.2933×10^2	$6.78 \times 10^0(1.58\%)$	$6.76 \times 10^{-5}(0.00001\%)$	$1.90 \times 10^{-1}(0.0044\%)$
			3	9.7516×10^3	$2.45 \times 10^2(2.51\%)$	$8.40 \times 10^0(0.09\%)$	$5.61 \times 10^{-1}(0.0057\%)$
			4	2.3230×10^5	$8.38 \times 10^3(3.61\%)$	$6.48 \times 10^2(0.28\%)$	$3.54 \times 10^2(0.15\%)$
			5	5.7681×10^6	$2.79 \times 10^5(4.48\%)$	$3.37 \times 10^4(0.58\%)$	$2.32 \times 10^4(0.40\%)$
			6	1.4858×10^8	$9.19 \times 10^6(6.19\%)$	$1.48 \times 10^6(0.98\%)$	$1.12 \times 10^6(0.75\%)$
			7	3.9565×10^9	$3.03 \times 10^8(7.66\%)$	$5.94 \times 10^7(1.48\%)$	$4.72 \times 10^7(1.18\%)$
			8	1.0857×10^{11}	$1.01 \times 10^{10}(9.26\%)$	$2.26 \times 10^9(2.04\%)$	$1.85 \times 10^9(1.67\%)$
Uniform(0,1)	u	6	1	0.5	$5.92 \times 10^{-3}(1.18\%)$	$1.47 \times 10^{-9}(3 \times 10^{-7}\%)$	$3.28 \times 10^{-2}(7.03\%)$
			2	0.333333	$4.94 \times 10^{-3}(1.48\%)$	$1.66 \times 10^{-5}(0.00499\%)$	$3.40 \times 10^{-2}(11.36\%)$
			3	0.25	$4.57 \times 10^{-3}(1.83\%)$	$2.50 \times 10^{-5}(0.00998\%)$	$3.46 \times 10^{-2}(16.06\%)$
			4	0.20	$4.57 \times 10^{-3}(2.28\%)$	$3.33 \times 10^{-5}(0.0166\%)$	$3.43 \times 10^{-2}(20.69\%)$
			5	0.166667	$4.66 \times 10^{-3}(2.80\%)$	$4.18 \times 10^{-5}(0.0249\%)$	$3.35 \times 10^{-2}(25.18\%)$
			6	0.142857	$4.74 \times 10^{-3}(3.32\%)$	$4.99 \times 10^{-5}(0.0349\%)$	$3.25 \times 10^{-2}(29.51\%)$
			7	0.125	$4.76 \times 10^{-3}(3.81\%)$	$5.82 \times 10^{-5}(0.00465\%)$	$3.15 \times 10^{-2}(33.65\%)$
			8	0.1111111	$4.73 \times 10^{-3}(4.25\%)$	$6.66 \times 10^{-5}(0.00599\%)$	$3.04 \times 10^{-2}(37.60\%)$
Vasicek Model	r	2	1	0.20	$2.16 \times 10^{-3}(1.08\%)$	$6.63 \times 10^{-11}(3.32 \times 10^{-8}\%)$	$1.40 \times 10^{-8}(7.00 \times 10^{-6}\%)$
			2	0.0933	$5.91 \times 10^{-3}(6.33\%)$	$9.67 \times 10^{-11}(1.04 \times 10^{-7}\%)$	$2.15 \times 10^{-8}(2.31 \times 10^{-5}\%)$
			3	0.0400	$3.13 \times 10^{-3}(7.83\%)$	$2.00 \times 10^{-8}(5.00 \times 10^{-5}\%)$	$3.32 \times 10^{-8}(8.30 \times 10^{-5}\%)$
			4	0.0229	$2.54 \times 10^{-3}(11.09\%)$	$4.02 \times 10^{-8}(1.75 \times 10^{-4}\%)$	$5.11 \times 10^{-8}(2.23 \times 10^{-4}\%)$
			5	0.0131	$1.83 \times 10^{-3}(13.93\%)$	$7.11 \times 10^{-8}(5.42 \times 10^{-4}\%)$	$7.89 \times 10^{-8}(6.01 \times 10^{-4}\%)$
			6	0.0087	$1.61 \times 10^{-3}(18.46\%)$	$1.15 \times 10^{-7}(1.32 \times 10^{-3}\%)$	$1.22 \times 10^{-7}(1.39 \times 10^{-3}\%)$
			7	0.0059	$1.40 \times 10^{-3}(23.49\%)$	$1.83 \times 10^{-7}(3.07 \times 10^{-3}\%)$	$1.88 \times 10^{-7}(3.16 \times 10^{-3}\%)$
			8	0.0044	$1.33 \times 10^{-3}(29.86\%)$	$2.86 \times 10^{-7}(6.43 \times 10^{-3}\%)$	$2.91 \times 10^{-7}(6.53 \times 10^{-3}\%)$
PDP Model	x	3	1	1.1885×10^3	$1.60 \times 10^1(1.35\%)$	$2.84 \times 10^{-1}(0.024\%)$	$5.74 \times 10^0(0.48\%)$
			2	1.4767×10^6	$3.93 \times 10^4(2.66\%)$	$3.95 \times 10^2(0.027\%)$	$1.16 \times 10^4(0.78\%)$
			3	1.8981×10^9	$7.37 \times 10^7(3.88\%)$	$3.19 \times 10^6(0.168\%)$	$2.35 \times 10^7(1.23\%)$
			4	2.5058×10^{12}	$1.28 \times 10^{11}(5.01\%)$	$1.64 \times 10^{10}(0.650\%)$	$4.85 \times 10^{10}(1.90\%)$
			5	3.3804×10^{15}	$2.04 \times 10^{14}(6.03\%)$	$4.97 \times 10^{13}(1.450\%)$	$1.00 \times 10^{14}(2.87\%)$

the difference between the respective estimated (either from M'_{ME} or M'_{GC}) and exact moments (from S_M), i.e.

$$AE_{ME} = |\mathbb{E}(x^i)_{ME} - \mathbb{E}(x^i)|,$$
$$AE_{GC} = |\mathbb{E}(x^i)_{GC} - \mathbb{E}(x^i)|. \qquad (14)$$

In addition, we also compare the exact moments of x against its respective moments obtained from sampling the PP (from $Sample_{Data}$ in Algorithm 1). We write $\mathbb{E}(x^i)_{Sample}$ to denote the ith higher-order moment of x obtained from $Sample_{Data}$. As such, the *absolute sample error* between the sampled and exact moments of x is derived by

$$\text{AE}_{Sample} = |\mathbb{E}(x^i)_{Sample} - \mathbb{E}(x^i)|. \tag{15}$$

The respective *relative errors* RE_{ME} and RE_{GC} of the ME and GC estimation, as well as the relative error RE_{Sample} based on sampled data, are computed as

$$
\begin{aligned}
\text{RE}_{\text{ME}} &= \frac{|\mathbb{E}(x^i)_{\text{ME}} - \mathbb{E}(x^i)|}{\mathbb{E}(x^i)}, \\
\text{RE}_{\text{GC}} &= \frac{|\mathbb{E}(x^i)_{\text{GC}} - \mathbb{E}(x^i)|}{\mathbb{E}(x^i)}, \\
\text{RE}_{Sample} &= \frac{|\mathbb{E}(x^i)_{Sample} - \mathbb{E}(x^i)|}{\mathbb{E}(x^i)}.
\end{aligned}
\tag{16}
$$

Table 2 summarizes our experiments on evaluating the precision of sampled and estimated moments against exact moments, for each PP in Table 1. Columns 1–3 of Table 2 correspond to columns 1–3 of Table 1. Column 4 lists the order of the moment of the random variable in column 2: for each ith moment, we give its exact value (column 5), as well as its absolute (14) and relative estimate errors (16) (in parentheses) using $Sample_{Data}$ (column 6), ME (column 7) and GC expansion (column 8).

Table 2 gives practical evidence of the accuracy of estimating the pdf, and hence moments, using Algorithm 1. The absolute and relative errors listed in Table 2 show that we gain higher precision when computing moments from the estimated pdfs using ME and/or GC expansion when compared to the moments calculated using sampled data. The accuracy of moments calculated from sampled data depends on the quality of the sampling process, which in turn depends on the number of samples (e) and number of loop iterations (n). Our results in Table 2 indicate that computing moments from estimated pdfs provides a more accurate alternative to estimating moments by sampling.

5   Conclusion

We present an algorithmic approach to estimate the distribution of the random variables of a PP using a finite number of its moments. Our estimates are based on Maximum Entropy and Gram-Charlier expansion. The accuracy of our estimation is assessed with the chi-square and Kolmogorov-Smirnov statistical tests. Our evaluation combines static analysis methods for the computation of exact moments of a PP random variable x with the aforementioned statistical techniques to produce estimates of the distribution of x. Extending our approach to support probabilistic inference and quantify the loss of precision in the estimation are future research directions.

References

1. Athanassoulis, G., Gavriliadis, P.: The truncated Hausdorff moment problem solved by using kernel density functions. Probab. Eng. Mech. **17**(3), 273–291 (2002)
2. Bartocci, E., Kovács, L., Stankovič, M.: Automatic generation of moment-based invariants for prob-solvable loops. In: Chen, Y.-F., Cheng, C.-H., Esparza, J. (eds.) ATVA 2019. LNCS, vol. 11781, pp. 255–276. Springer, Cham (2019). https://doi.org/10.1007/978-3-030-31784-3_15
3. Bernardo, J.M., Smith, A.F.: Bayesian Theory. Wiley, Hoboken (2009)
4. Bhat, S., Borgström, J., Gordon, A.D., Russo, C.V.: Deriving probability density functions from probabilistic functional programs. Log. Methods Comput. Sci. **13**(2), 1–32 (2017)
5. Biswas, P., Bhattacharya, A.K.: Function reconstruction as a classical moment problem: a maximum entropy approach. J. Phys. A: Math. Theor. **43**(405003), 1–19 (2010)
6. Brenn, T., Anfinsen, S.N.: A revisit of the Gram-Charlier and Edgeworth series expansions. Preprint, pp. 1–12 (2017). https://hdl.handle.net/10037/11261
7. Broca, D.: Cumulant-moment relations through determinants. Int. J. Math. Educ. Sci. Technol. **35**(6), 917–921 (2004)
8. Buchen, P.W., Kelly, M.: The maximum entropy distribution of an asset inferred from option prices. J. Financ. Quant. Anal. **31**(1), 143–159 (1996)
9. Carette, J., Shan, C.-C.: Simplifying probabilistic programs using computer algebra. In: Gavanelli, M., Reppy, J. (eds.) PADL 2016. LNCS, vol. 9585, pp. 135–152. Springer, Cham (2016). https://doi.org/10.1007/978-3-319-28228-2_9
10. Carpenter, B., et al.: Stan: a probabilistic programming language. J. Stat. Softw. **76**(1), 1–32 (2017)
11. De Souza, L., Janiga, G., John, V., Thévenin, D.: Reconstruction of a distribution from a finite number of moments with an adaptive spline-based algorithm. Chem. Eng. Sci. **65**(9), 2741–2750 (2010)
12. Durrett, R.: Probability: Theory and Examples. Cambridge University Press, Cambridge (2019)
13. Feng, Y., Zhang, L., Jansen, D.N., Zhan, N., Xia, B.: Finding polynomial loop invariants for probabilistic programs. In: D'Souza, D., Narayan Kumar, K. (eds.) ATVA 2017. LNCS, vol. 10482, pp. 400–416. Springer, Cham (2017). https://doi.org/10.1007/978-3-319-68167-2_26
14. Gavriliadis, P., Athanassoulis, G.: Moment information for probability distributions, without solving the moment problem, II: main-mass, tails and shape approximation. J. Comput. Appl. Math. **229**(1), 7–15 (2009)
15. Gehr, T., Misailovic, S., Vechev, M.: PSI: exact symbolic inference for probabilistic programs. In: Chaudhuri, S., Farzan, A. (eds.) CAV 2016. LNCS, vol. 9779, pp. 62–83. Springer, Cham (2016). https://doi.org/10.1007/978-3-319-41528-4_4
16. Gehr, T., Steffen, S., Vechev, M.: λPSI: exact inference for higher-order probabilistic programs. In: Proceedings of the 41st ACM SIGPLAN Conference on Programming Language Design and Implementation (PLDA 2020), pp. 883–897 (2020)
17. Ghahramani, Z.: Probabilistic machine learning and artificial intelligence. Nature **521**(7553), 452–459 (2015)
18. Hald, A.: The early history of the cumulants and the Gram-Charlier series. Int. Stat. Rev. **68**(2), 137–153 (2000)

19. Hastings, W.: Monte Carlo sampling methods using Markov chains and their applications. Biometrika **57**(1), 97–109 (1970)
20. Holtzen, S., Van den Broeck, G., Millstein, T.: Scaling exact inference for discrete probabilistic programs. Proc. ACM Program. Lang. **4**, 1–31 (2020)
21. Huang, Z., Dutta, S., Misailovic, S.: AQUA: automated quantized inference for probabilistic programs. In: Hou, Z., Ganesh, V. (eds.) ATVA 2021. LNCS, vol. 12971, pp. 229–246. Springer, Cham (2021). https://doi.org/10.1007/978-3-030-88885-5_16
22. Innocentini, G.C., Hodgkinson, A., Radulescu, O.: Time dependent stochastic mRNA and protein synthesis in piecewise-deterministic models of gene networks. Front. Phys. **6**, 1–16 (2018)
23. John, V., Angelov, I., Öncül, A., Thévenin, D.: Techniques for the reconstruction of a distribution from a finite number of its moments. Chem. Eng. Sci. **62**(11), 2890–2904 (2007)
24. Kendall, M., Stuart, A.: The Advanced Theory of Statistics. Volume 1: Distribution Theory. Macmillan, New York (1977)
25. Kura, S., Urabe, N., Hasuo, I.: Tail probabilities for randomized program runtimes via martingales for higher moments. In: Vojnar, T., Zhang, L. (eds.) TACAS 2019. LNCS, vol. 11428, pp. 135–153. Springer, Cham (2019). https://doi.org/10.1007/978-3-030-17465-1_8
26. Lebaz, N., Cockx, A., Spérandio, M., Morchain, J.: Reconstruction of a distribution from a finite number of its moments: a comparative study in the case of depolymerization process. Comput. Chem. Eng. **84**, 326–337 (2016)
27. Levenberg, K.: A method for the solution of certain non-linear problems in least squares. Q. Appl. Math. **2**(2), 164–168 (1944)
28. Marquardt, D.W.: An algorithm for least-squares estimation of nonlinear parameters. J. Soc. Ind. Appl. Math. **11**(2), 431–441 (1963)
29. Massey, F.J., Jr.: The Kolmogorov-Smirnov test for goodness of fit. J. Am. Stat. Assoc. **46**(253), 68–78 (1951)
30. Moosbrugger, M., Stankovič, M., Bartocci, E., Kovács, L.: This is the moment for probabilistic loops, pp. 1–25. arXiv preprint arXiv:2204.07185 (2022)
31. Mörters, P., Peres, Y.: Brownian Motion. Cambridge University Press, Cambridge (2010)
32. Narayanan, P., Carette, J., Romano, W., Shan, C., Zinkov, R.: Probabilistic inference by program transformation in Hakaru (system description). In: Kiselyov, O., King, A. (eds.) FLOPS 2016. LNCS, vol. 9613, pp. 62–79. Springer, Cham (2016). https://doi.org/10.1007/978-3-319-29604-3_5
33. O'Hagan, A., Forster, J.J.: Kendall's Advanced Theory of Statistics, Volume 2B: Bayesian Inference. Arnold (2004)
34. Ross, S.M.: Introduction to Probability and Statistics for Engineers and Scientists. Academic Press (2020)
35. Silverman, B.W.: Density Estimation for Statistics and Data Analysis. Chapman & Hall/CRC (1998)
36. Snedecor, G.W., Cochran, G.W.: Statistical Methods. Iowa State University Press (1989)
37. Vasicek, O.: An equilibrium characterization of the term structure. J. Financ. Econ. **5**(2), 177–188 (1977)
38. Wallace, D.L.: Asymptotic approximations to distributions. Ann. Math. Stat. **29**(3), 635–654 (1958)
39. Younes, H.L., Simmons, R.G.: Statistical probabilistic model checking with a focus on time-bounded properties. Inf. Comput. **204**(9), 1368–1409 (2006)

An Automated Quantitative Information Flow Analysis for Concurrent Programs

Khayyam Salehi[1]([⊠]) [iD], Ali A. Noroozi[2] [iD], Sepehr Amir-Mohammadian[3] [iD], and Mohammadsadegh Mohagheghi[4] [iD]

[1] Department of Computer Science, Shahrekord University, Shahrekord, Iran
kh.salehi@sku.ac.ir
[2] Department of Computer Science, University of Tabriz, Tabriz, Iran
noroozi@tabrizu.ac.ir
[3] Department of Computer Science, University of the Pacific, Stockton, CA, USA
samirmohammadian@pacific.edu
[4] Department of Computer Science, Vali-e-Asr University of Rafsanjan,
Rafsanjan, Iran
mohagheghi@vru.ac.ir

Abstract. Quantitative information flow is a rigorous approach for evaluating the security of a system. It is used to quantify the amount of secret information leaked to the public outputs. In this paper, we propose an automated approach for quantitative information flow analysis of concurrent programs. Markovian processes are used to model the behavior of these programs. To this end, we assume that the attacker is capable of observing the internal behavior of the program and propose an equivalence relation, back-bisimulation, to capture the attacker's view of the program behavior. A partition refinement algorithm is developed to construct the back-bisimulation quotient of the program model and then a quantification method is proposed for computing the information leakage using the quotient. Finally, an anonymous protocol, dining cryptographers, is analyzed as a case study to show applicability and scalability of the proposed approach.

Keywords: Information leakage · protocol security · quantitative information flow · confidentiality · Markovian Processes

1 Introduction

Secure information flow is a rigorous technique for evaluating security of a system. A system satisfies confidentiality requirements if it does not *leak* any secret information to its public outputs. However, imposing no leakage policy is too restrictive and in practice the security policy of the system tends to permit minor leakages. For example, a password checking program leaks information about what the password is not when it shows a message indicating that user has entered a wrong password. *Quantitative information flow* has been a well-established attempt to overcome this deficiency. Given a system with *secret*

© Springer Nature Switzerland AG 2022
E. Ábrahám and M. Paolieri (Eds.): QEST 2022, LNCS 13479, pp. 43–63, 2022.
https://doi.org/10.1007/978-3-031-16336-4_3

(high confidentiality) inputs and *public* (low confidentiality) outputs, quantitative information flow addresses the problem of measuring the amount of *information leakage*, i.e., the amount of information that an *attacker* can deduce about the secret inputs by observing the outputs. Quantitative information flow is widely used in analyzing timing attacks [23,24], differential privacy [1], anonymity protocols [13,25,29,30], and cryptographic algorithms [18,19].

Assume a program with a secret input and a public output. Furthermore, assume an *attacker* that executes the program and observes the public output. A common approach for measuring the amount of leaked information is to use the notion of *uncertainty* [31]. Before executing the program, the attacker has an *initial uncertainty* about the secret, which is determined by her prior knowledge of the secret. After executing the program and observing the output, she may infer information about the secret and thus her uncertainty may be reduced. This yields the following intuitive definition of the information leakage [31]: leaked information = initial uncertainty - remaining uncertainty.

In this paper, a practical and automated formal approach is proposed to quantify the information leakage of terminating concurrent programs. The approach considers leakages in intermediate states of the program executions and effect of the scheduling policy.

We assume the program has a secret input h, a public output l, and zero or more neutral variables. Neutral variables specify temporary and/or auxiliary components of the runtime program configuration that do not belong to a certain confidentiality level by nature, e.g., the stack pointer and loop indexes. h is fixed and does not change during program executions. This is the case in any analysis in the context of confidentiality that assumes data integrity to be out of scope, e.g., [2,7]. We also assume that the public and neutral variables have single initial values. Furthermore, a probabilistic attacker is supposed, who has the full knowledge of source code of the concurrent program and is able to choose a scheduler and execute the program under the control of that scheduler. She can observe sequences of values of l during the executions, called *execution traces*. We also assume that the attacker can execute the program an arbitrary number of times and can then *guess* the value of h in a single attempt. This is called *one-try guessing model* [31].

In order to model the concurrent program, *Markov decision processes* (MDPs) are used. MDPs provide a powerful state transition system, capable of modeling probabilistic and nondeterministic behaviors [28]. The scheduler is assumed to be probabilistic, resolving nondeterminism in the MDP and inducing a *Markov chain* (MC). States of an MC contain values of h, l, and possible neutral variables. For computing the leakage, however, MC should capture the attacker's view of the program. The attacker, while executing the program and observing the execution traces, does not know the exact value of h in each step. She can only guess a set of possible values based on the executed program statements and the observed traces. She also cannot distinguish those executions of MC that have the same trace. In this regard, we define an equivalence relation for a given MC, called *back-bisimulation*, to specify these requirements of the threat model. Back-bisimulation induces a *quotient* which models the *attacker's view* of the program. A partition-refinement algorithm is proposed to compute the back-bisimulation quotient.

Each state of the back-bisimulation quotient contains a *secret distribution*, which shows possible values of h in that state, and thus is a determiner of the attacker's uncertainty about h. Each execution trace of the quotient shows a *reduction* of the attacker's uncertainty from the initial state to the final state of the trace. Therefore, secret distribution in the initial state of the quotient determines the attacker's initial uncertainty and secret distributions in the final states determine the remaining uncertainty. In the literature, uncertainty is measured based on the notion of entropy. The entropy of h expresses the difficulty for an attacker to discover its value. Based on the program model and the attacker's observational power, various definitions of entropy have been proposed. As Smith [31] shows, in the context of one-try guessing model, uncertainty about a random variable should be defined in terms of *Renyi's min-entropy*. This yields that the information leakage is computed as the difference of the Renyi's min-entropy of h in the initial state of the quotient and the expected value of the Renyi's min-entropy of h in the final states of the quotient.

We also show a subclass of MCs, called *Markov chains with pseudoback-bisimilar states*, in which back-bisimulation cannot correctly construct the attacker's view of the program behavior. Using back-bisimulation to handle this situation is considered a potential future work. Briefly, our contributions include

- proposing back-bisimulation equivalence, in order to capture the attacker's observation of the program,
- developing an algorithm to compute back-bisimulation quotient of an MC,
- proposing a method to compute the leakage of a concurrent program from the back-bisimulation quotient, and
- analyzing the dining cryptographers problem.

1.1 Paper Outline

The paper proceeds as follows. Section 2 provides a core background on some basics, information theory, Markovian models and probabilistic schedulers. Section 3 presents the proposed approach. It starts with introducing the program and threat models. It then formally defines back-bisimulation and discusses how to compute the program leakage. Finally, it describes how to construct the attacker's view of the program model, the back-bisimulation quotient. Section 4 concludes the paper and proposes future work. Finally, the case study and related work are discussed in Appendix A and Appendix B, respectively.

2 Background

In this section, we provide preliminary concepts and notations required for the proposed approach.

2.1   Basics

A *probability distribution Pr* over a set $\mathcal{X}$ is a function $Pr : \mathcal{X} \to [0,1]$, such that $\sum_{x \in \mathcal{X}} Pr(x) = 1$. We denote the set of all probability distributions over $\mathcal{X}$ by $\mathcal{D}(\mathcal{X})$.

Let S be a set and $\mathcal{R}$ an equivalence relation on S. For $s \in S$, $[s]_{\mathcal{R}}$ denotes the *equivalence class* of s under $\mathcal{R}$, i.e., $[s]_{\mathcal{R}} = \{s' \in S \mid s\mathcal{R}s'\}$. Note that for $s' \in [s]_{\mathcal{R}}$ we have $[s']_{\mathcal{R}} = [s]_{\mathcal{R}}$. The set $[s]_{\mathcal{R}}$ is often referred to as the $\mathcal{R}$-equivalence class of s. The *quotient space* of S under $\mathcal{R}$, denoted by $S/\mathcal{R} = \{[s]_{\mathcal{R}} \mid s \in S\}$, is the set consisting of all $\mathcal{R}$-equivalence classes. A partition for S is a set $\Pi = \{B_1, \ldots, B_k\}$ such that $B_i \neq \varnothing$ (for $0 < i \leq k$), $B_i \cap B_j = \varnothing$ (for $0 < i < j \leq k$) and $S = \cup_{0 < i \leq k} B_i$. $B_i \in \Pi$ is called a *block*. $C \subseteq S$ is a *superblock* of Π if $C = B_{i_1} \cup \cdots \cup B_{i_l}$ for some $B_{i_1}, \ldots, B_{i_l} \in \Pi$. Note that for equivalence relation $\mathcal{R}$ on S, the quotient space $S/\mathcal{R}$ is a *partition* for S.

2.2   Information Theory

Let X denote a random variable with the finite set of values $\mathcal{X}$. *Vulnerability* [31] of X is defined as $Vul(\mathrm{X}) = \max_{x \in \mathcal{X}} Pr(\mathrm{X} = x)$. Vulnerability is defined as the highest probability of correctly guessing the value of the variable in just a single attempt. In order to quantify information leaks, we convert this probability into bits using *Renyi's min-entropy* [31].

Definition 1. *The **Renyi's min-entropy** of a random variable* X *is given by* $\mathcal{H}_{\infty}(\mathrm{X}) = -\log_2 \; Vul(\mathrm{X})$.

2.3   Markovian Models

We use Markov decision processes (MDPs) to model operational semantics of concurrent programs. MDPs are state transition systems that permit both *probabilistic* and *nondeterministic* choices [28]. In any state of an MDP, a nondeterministic choice between probability distributions exists. Once a probability distribution is chosen nondeterministically, the next state is selected in a probabilistic manner. Nondeterminism is used to model concurrency between threads by means of *interleaving*, i.e., all possible choices of the threads are considered. Formally,

Definition 2. *A **Markov decision process (MDP)** is defined as a tuple* $\mathcal{M} = (S, Act, \boldsymbol{P}, \zeta, AP, V)$ *where,*

- S *is a set of states,*
- *Act is a set of actions,*
- $\boldsymbol{P} : S \to (Act \to (S \to [0,1]))$ *is a transition probability function such that for all states* $s \in S$ *and actions* $\alpha \in Act$, $\sum_{s' \in S} \boldsymbol{P}(s)(\alpha)(s') \in \{0,1\}$,
- $\zeta : S \to [0,1]$ *is an initial distribution such that* $\sum_{s \in S} \zeta(s) = 1$.
- *AP is a set of atomic propositions,*
- $V : S \to AP$ *is a labeling function.*

Atomic propositions represent simple known facts about the states. The function V labels each state with atomic propositions. An MDP $\mathcal{M}$ is called *finite* if S, Act, and AP are finite. An action α is *enabled* in state s iff $\sum_{s' \in S} \mathbf{P}(s)(\alpha)(s') = 1$. Let $Act(s)$ denote the set of enabled actions in s. Each state s' for which $\mathbf{P}(s)(\alpha)(s') > 0$ is called an α-*successor* of s. The set of α-successors of s is denoted by $Succ(s, \alpha)$. The set of *successors* of s is defined as $Succ(s) = \bigcup_{\alpha \in Act(s)} Succ(s, \alpha)$. The set of successors of a set of states $\mathcal{S}$ is defined as $Succ(\mathcal{S}) = \bigcup_{s \in \mathcal{S}} Succ(s)$. The set of *predecessors* of s is defined as $Pre(s) = \{s' \in S \mid s \in Succ(s')\}$. The set of labels that are associated with the predecessors of s is defined as $PreLabels(s) = \{V(s') \mid s' \in Pre(s), s' \neq s\}$.

MDP Semantics. The intuitive operational behavior of an MDP $\mathcal{M}$ is as follows. At the beginning, an initial state s_0 is probabilistically chosen such that $\zeta(s_0) > 0$. Assuming that $\mathcal{M}$ is in state s, first a nondeterministic choice between the enabled actions needs to be resolved. Suppose action $\alpha \in Act(s)$ is selected. Then, one of the α-successors of s is selected probabilistically according to the transition function $\mathbf{P}$. That is, with probability $\mathbf{P}(s)(\alpha)(s')$ the next state is s'.

Initial and Final States. The states s with $\zeta(s) > 0$ are considered as the *initial states*. The set of initial states of $\mathcal{M}$ is denoted by $Init(\mathcal{M})$. To ensure $\mathcal{M}$ is non-blocking, we include a self-loop to each state s that has no successor, i.e., $\mathbf{P}(s)(\tau)(s') = 1$. The distinguished action label τ is used to show that the self-loop's action is not of further interest. Then, a state s is called *final* if $Succ(s) = \{s\}$. In the literature, these states are called absorbing [3]. We call them final, because in our program model they show termination of the program. The set of final states of $\mathcal{M}$ is denoted by $final(\mathcal{M})$.

Execution Paths. Alternating sequences of states that may arise by resolving both nondeterministic and probabilistic choices in an arbitrary MDP $\mathcal{M}$ are called *(execution) paths*. More precisely, a finite path fragment $\hat{\sigma}$ of $\mathcal{M}$ is a finite state sequence $s_0 s_1 \ldots s_n$ such that $s_i \in Succ(s_{i-1})$ for all $0 < i \leq n$. A path σ is an infinite state sequence $s_0 s_1 \ldots s_{n-1} s_n^\omega$ such that $s_0 \in Init(\mathcal{M})$, $s_i \in Succ(s_{i-1})$ for all $0 < i \leq n$, ω denotes infinite iteration, and $s_n \in final(\mathcal{M})$, i.e., s_n^ω denotes the infinite iteration over s_n. The final state of σ, i.e. s_n, is given by $final(\sigma)$. The set of execution paths of $\mathcal{M}$ is denoted by $Paths(\mathcal{M})$. The set of finite path fragments starting in s and ending in s' is denoted by $PathFrags(s, s')$.

Traces and Trace Fragments. A *trace* of an execution path is the sequence of atomic propositions of the states of the path. Formally, the trace of a finite path fragment $\hat{\sigma} = s_0 s_1 \ldots s_n$ is defined as $\hat{T} = trace(\hat{\sigma}) = V(s_0) V(s_1) \ldots V(s_n)$. For a path $\sigma = s_0 s_1 \ldots$, $trace_{\ll i}(\sigma)$ is defined as the prefix of $trace(\sigma)$ up to index i, i.e., $trace_{\ll i}(\sigma) = V(s_0) V(s_1) \ldots V(s_i)$. Let $Paths(T)$ be the set of paths that have the trace T, i.e., $Paths(T) = \{\sigma \in Paths(\mathcal{M}) \mid trace(\sigma) = T\}$. We define $final(Paths(T))$ to denote the set of final states that result from the trace T, i.e., $final(Paths(T)) = \{final(\sigma) \mid \sigma \in Paths(T)\}$.

MDPs are suitable for modeling concurrent programs, but since they contain nondeterministic choices, they are too abstract to implement. We need to resolve these nondeterministic choices into probabilistic ones. The result is a Markov chain, which does not contain action and nondeterminism.

Definition 3. *A (discrete-time) Markov chain (MC) is a tuple* $\mathcal{M} = (S, \boldsymbol{P}, \zeta, AP, V)$ *where,*

- S *is a set of states,*
- $\boldsymbol{P} : S \times S \to [0, 1]$ *is a transition probability function such that for all states* $s \in S$, $\sum_{s' \in S} \boldsymbol{P}(s, s') = 1$,
- $\zeta : S \to [0, 1]$ *is an initial distribution such that* $\sum_{s \in S} \zeta(s) = 1$,
- AP *is a set of atomic propositions,*
- $V : S \to AP$ *is a labeling function.*

The function $\boldsymbol{P}$ determines for each state s the probability $\boldsymbol{P}(s, s')$ of a single transition from s to s'. Note that for all states $s \in S$, $\sum_{s' \in S} \boldsymbol{P}(s, s') = 1$.

Reachability Probabilities. We define the probability of *reaching* a state s from an initial state in an MC $\mathcal{M}$ as $Pr(s) = \sum_{\substack{\hat{\sigma} \in PathFrags(s_0, s) \\ s_0 \in Init(\mathcal{M})}} Pr(\hat{\sigma})$, where

$$
Pr(\hat{\sigma} = s_0 s_1 \ldots s_n) = \begin{cases} \zeta(s_0) & \text{if } n = 0, \\ \zeta(s_0). \prod_{0 \le i < n} \boldsymbol{P}(s_i, s_{i+1}) & \text{otherwise.} \end{cases}
$$

Trace Probabilities. The occurrence probability of a trace T is defined as

$$
Pr(T) = \sum_{\sigma \in Paths(T)} Pr(\sigma), \text{ where } Pr(\sigma = s_0 s_1 \ldots s_n^\omega) = Pr(\hat{\sigma} = s_0 s_1 \ldots s_n).
$$

DAG Structure of Program Models. We assumed that the programs always terminate and states indicate the current values of the variables and the program counter. This implies that Markovian models of every terminating program takes the form of a *directed acyclic graph (DAG)*, modulo self-loops in final states. Therefore, reachability probabilities coincide with long-run probabilities [3]. Initial states of the program are represented as roots of the DAG, and final states as leaves. Each state of a Markovian model is located at a level equal to the least distance of that state from an initial state. Level of state s is denoted by $level(s)$.

2.4 Probabilistic Schedulers

A probabilistic scheduler implements the scheduling policy of a concurrent program. It determines the order and probability of execution of threads. When a probabilistic scheduler is applied to a concurrent program, nondeterministic choices are replaced by probabilistic ones. As we modeled concurrency between threads using nondeterminism in MDP, the scheduler is used to resolve the possible nondeterminism in MDP. For demonstration purposes, it suffices to consider a simple but important subclass of schedulers called *memoryless probabilistic schedulers*. Given a state s, a memoryless probabilistic scheduler returns a probability for each action $\alpha \in Act(s)$. This random choice is independent of what has happened in the history, i.e., which path led to the current state. This is why it is called memoryless[1]. Formally,

Definition 4. *Let* $\mathcal{M} = (S, Act, \boldsymbol{P}, \zeta, AP, V)$ *be an MDP. A **memoryless probabilistic scheduler** for* $\mathcal{M}$ *is a function* $\delta : S \to \mathcal{D}(Act)$, *such that* $\delta(s) \in \mathcal{D}(Act(s))$ *for all* $s \in S$.

As all nondeterministic choices in an MDP $\mathcal{M}$ are resolved by a scheduler δ, a Markov chain $\mathcal{M}_\delta$ is induced. Formally,

Definition 5. *Let* $\mathcal{M} = (S, Act, \boldsymbol{P}, \zeta, AP, V)$ *be an MDP and* $\delta : S \to \mathcal{D}(Act)$ *be a memoryless probabilistic scheduler on* $\mathcal{M}$. ***The MC of*** $\mathcal{M}$ ***induced by*** δ *is given by* $\mathcal{M}_\delta = (S, \boldsymbol{P}_\delta, \zeta, AP, V)$ *where* $\boldsymbol{P}_\delta(s, s') = \sum\limits_{\alpha \in Act(s)} \delta(s)(\alpha).\boldsymbol{P}(s)(\alpha)(s')$

3 The Proposed Approach

Suppose a concurrent program P, running under control of a scheduling policy δ. The proposed approach proceeds in three steps: (1) defining an MDP representing P and applying δ to the MDP to resolve the nondeterminism in the MDP (Sect. 3.1), (2) constructing a back-bisimulation quotient (Sect. 3.2), and (3) computing the leakage (Sect. 3.3). Finally, an algorithm for computing the back-bisimulation quotient is presented (Sect. 3.4).

3.1 The Program and Threat Models

It is assumed P has a secret input variable h and a public output variable l and h has a fixed ue during the program executions. If the program has several secret variables, they can be encoded (e.g. concatenated) into one secret variable. The same is done for public and neutral variables. Possible values of l and h are denoted by Val_l and Val_h.

The attacker has a prior knowledge of the secret, which is modeled as a prior probability distribution over the possible values of h, i.e. $Pr(h)$. Here, the attacker is assumed to be probabilistic, i.e., she knows size of the secret, in addition to some accurate constraints about the values of h. For instance, the attacker could know that h is 2 bits long, its value is not 1, the probability that its value is 2 is 0.6, and the probability that its value is 3 is thrice the probability that it is 0. The prior distribution encoding these constraints is $Pr(h) = \{0 \mapsto 0.1, 2 \mapsto 0.6, 3 \mapsto 0.3\}$[2]. A special case of the probabilistic attacker is *ignorant* [6], who has no prior information about the value of h except its size. Thus, the ignorant attacker's initial knowledge is a uniform prior distribution on h.

---

[1] A rather general notion of schedulers is to let them use the full history of execution to make decisions. Here, this general definition is not needed and only makes the program model unnecessarily complex.

[2] Only elements with a positive probability are shown.

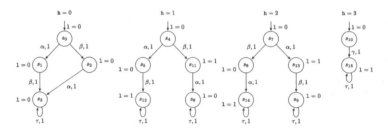

Fig. 1. $\mathcal{M}^{\mathrm{P1}}$: MDP of the program P1, with α denoting 1:=h/2, β denoting 1:=h mod 2, γ denoting 1:=1, and τ denoting termination of the program

Define an MDP Representing P. Operational semantics of the concurrent program P is represented by an MDP $\mathcal{M}^{\mathrm{P}} = (S, Act, \mathbf{P}, \zeta, Val_l, V)$. Each state $s \in S$ is a tuple $\langle \bar{l}, \bar{h}, \bar{n}, pc \rangle$, where $\bar{l}$, $\bar{h}$, and $\bar{n}$ are values of the public, secret, and neutral variables, respectively, and pc is the program counter. Actions Act are program statements of P. The function $\mathbf{P}$ defines probabilities of transitions between states. Atomic propositions are Val_l and the function V labels each state with value of 1 in that state. In fact, a state label is what an attacker observes in a state and traces of $\mathcal{M}^{\mathrm{P}}$ are the sequences of public values that are valid during the execution.

The initial distribution ζ is determined by the prior knowledge of the attacker $Pr(h)$, i.e., $\zeta(s) = Pr(h = \bar{h})$ for all $s \in Init(\mathcal{M}^{\mathrm{P}})$, where $s = \langle ., \bar{h}, ., . \rangle$.

Example 1 (Program P1). Consider the following program, where h is a 2-bit random variable and || denotes the concurrency of the executions:

```
l:=0; if h=3 then l:=1   else (l:=h/2 || l:=h mod 2) (P1)
```

The attacker's prior knowledge is the size of h, yielding a uniform distribution on h, i.e., $Pr(h) = \{0 \mapsto \frac{1}{4}, 1 \mapsto \frac{1}{4}, 2 \mapsto \frac{1}{4}, 3 \mapsto \frac{1}{4}\}$. The MDP $\mathcal{M}^{\mathrm{P1}}$ of the program is shown in Fig. 1. The initial distribution ζ is determined by $Pr(h)$, i.e., $\zeta = \{s_0 \mapsto \frac{1}{4}, s_4 \mapsto \frac{1}{4}, s_7 \mapsto \frac{1}{4}, s_{10} \mapsto \frac{1}{4}\}$. Each state is labeled by the value of 1 in that state. Each transition is labeled by an action (a program statement) and a probability. For instance, the transition from s_0 to s_1 has the action $\alpha : 1:=h/2$ and the probability $\mathbf{P}(s_0)(\alpha)(s_1) = 1$; Or the transition from s_0 to s_2 has the label $\beta : 1:=h \bmod 2$ and the probability $\mathbf{P}(s_0)(\beta)(s_2) = 1$.

Resolve the Nondeterminism in the MDP. The scheduling policy is represented by a memoryless probabilistic scheduler δ. As the MDP $\mathcal{M}^{\mathrm{P}}$ is run under the control of the scheduler δ, all nondeterministic transitions are resolved and an MC $\mathcal{M}^{\mathrm{P}}_{\delta} = (S, \mathbf{P}_{\delta}, \zeta, Val_l, V)$ is produced.

Example 2 (MC of P1). We choose the scheduler to be uniform. The uniform scheduler, denoted by the function *uni*, picks each thread with the same probability. This yields the definition of the scheduler as follows:

$$uni(s_0) = uni(s_4) = uni(s_7) = \{\alpha \mapsto \frac{1}{2}, \beta \mapsto \frac{1}{2}\}, \qquad uni(s_{10}) = \{\gamma \mapsto 1\},$$

$$uni(s_1) = uni(s_5) = uni(s_{13}) = \{\beta \mapsto 1\}, \quad uni(s_2) = uni(s_8) = uni(s_{11}) = \{\alpha \mapsto 1\},$$

$$uni(s_3) = uni(s_6) = uni(s_9) = uni(s_{12}) = uni(s_{14}) = uni(s_{15}) = \{\tau \mapsto 1\}.$$

The MC $\mathcal{M}_{uni}^{\text{P1}}$ of the program **P1** running under control of the uniform scheduler is depicted in Fig. 2. In this Figure, transitions are labeled by the transition probability.

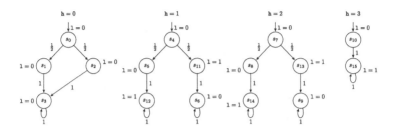

Fig. 2. $\mathcal{M}_{uni}^{\text{P1}}$: MC of the program **P1** with the uniform scheduler

3.2 The Attacker's View of the Program: Back-Bisimulation Quotient

In order to measure the amount of information the attacker can deduce about **h**, we need to construct the attacker's view of the program. First, the attacker can distinguish a final state from a non-final one by observing termination of the program. Second, she cannot discriminate between those paths that have the same trace. For instance, in $\mathcal{M}_{uni}^{\text{P1}}$ (Fig. 2) the attacker only observes the traces $\{\langle 0, 0, 0^\omega \rangle, \langle 0, 0, 1^\omega \rangle, \langle 0, 1, 0^\omega \rangle, \langle 0, 1^\omega \rangle\}$, whereas there are seven different execution paths. The implication is that she cannot distinguish those final states that have the same public values and result from the same traces. Third, she does not know secret values in the final states, but may guess the value of **h** based on a probability distribution that she can compute according to the possible values of **h** in each final state. These three requirements are captured by an equivalence relation, called *back-bisimulation*, denoted by $\sim_b$.

Definition 6. *Let $\mathcal{M}_\delta^P$ be an MC. A **back-bisimulation** for $\mathcal{M}_\delta^P$ is a binary relation $\mathcal{R}$ on S such that for all $s_1 \mathcal{R} s_2$, the following three conditions hold: (1) $V(s_1) = V(s_2)$, (2) if $s_1' \in Pre(s_1)$, then there exists $s_2' \in Pre(s_2)$ with $s_1' \mathcal{R} s_2'$, (3) if $s_2' \in Pre(s_2)$, then there exists $s_1' \in Pre(s_1)$ with $s_1' \mathcal{R} s_2'$.*
States s_1 and s_2 are back-bisimilar, *denoted by $s_1 \sim_b s_2$, if there exists a back-bisimulation $\mathcal{R}$ for $\mathcal{M}_\delta^P$ with $s_1 \mathcal{R} s_2$.*

Condition (1) requires that the states s_1 and s_2 have the same public values. According to condition (2), every incoming transition of s_1 must be matched by an incoming transition of s_2; the reverse is assured by condition (3).

Theorem 1. *Back-bisimulation is an equivalence relation.*[3]

As $\sim_b$ is an equivalence relation, it induces a set of equivalence classes on the state space of an MC. Given MC $\mathcal{M}_\delta^P$, a quotient space $\mathcal{M}_\delta^P/\sim_b$ captures the attacker's view of the program P. The MC $\mathcal{M}_\delta^P/\sim_b$ aggregates same-trace paths of $\mathcal{M}_\delta^P$ into one path.

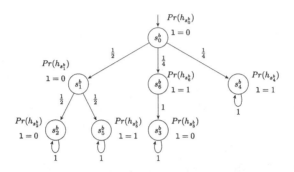

Fig. 3. $\mathcal{M}_{uni}^{P1}/\sim_b$: back-bisimulation quotient of $\mathcal{M}_{uni}^{P1}$

Definition 7. *For MC $\mathcal{M}_\delta^P = (S, \boldsymbol{P}_\delta, \zeta, Val_l, V)$ and back-bisimulation $\sim_b$, the* **back-bisimulation quotient** $\mathcal{M}_\delta^P/\sim_b$ *is defined by* $\mathcal{M}_\delta^P/\sim_b$ *where*
$$\mathcal{M}_\delta^P/\sim_b = (S/\sim_b, \boldsymbol{P}_\delta', s_{init}^b, Val_l, V, Pr(h))$$

- *$S/\sim_b$ is the quotient space of S under $\sim_b$,*
- *$\boldsymbol{P}_\delta' : (S/\sim_b) \times (S/\sim_b) \to [0,1]$ is a probability transition function between equivalence classes of $S/\sim_b$ such that $\forall\ s^b, t^b \in S/\sim_b$. $\boldsymbol{P}_\delta'(s^b, t^b) = \sum_{s \in s^b,\ t \in t^b} Pr(s) * \boldsymbol{P}_\delta(s,t)$, where $Pr(s)$ is the probability of reaching to s in MC $\mathcal{M}_\delta^P$,*
- *$s_{init}^b = Init(\mathcal{M}_\delta^P)$,*
- *$V([s]_{\sim_b}) = V(s)$,*
- *$Pr(h)$ is a mapping from each quotient state s^b to $Pr(h_{s^b})$, where $Pr(h_{s^b})$ is the probability distribution of h in the state s^b and is computed, for all $\overline{h} \in Val_h$, as*

$$Pr(h_{s^b} = \overline{h}) = \frac{\sum_{s_i \in s^b,\ s_i = \langle ., \overline{h}, ., . \rangle} Pr(s_i)}{Pr(s^b)},$$

where $Pr(s^b)$ is the reachability probability of s^b in $\mathcal{M}_\delta^P/\sim_b$.

The public variable has a single initial value. Thus, all of the initial states of $\mathcal{M}_\delta^P$ have the same public value and form a single equivalence class s_{init}^b. Each state s^b is labeled with a probability distribution $Pr(h_{s^b})$ which shows the probabilities of possible values of h in that state.

---

[3] The proofs of the theorems have been omitted due to meet the page limit.

Example 3 (Back-bisimulation quotient of P1). The back-bisimulation quotient $\mathcal{M}_{uni}^{P1}/\sim_b$ is depicted in Fig. 3. Each state of $\mathcal{M}_{uni}^{P1}/\sim_b$ is an equivalence class, containing related states of $\mathcal{M}_{uni}^{P1}$:

$$s_0^b = \{s_0, s_4, s_7, s_{10}\}, \quad s_1^b = \{s_1, s_2, s_5, s_8\}, \quad s_2^b = \{s_3\}, \quad s_3^b = \{s_6, s_9\},$$

$$s_4^b = \{s_{15}\}, \quad s_5^b = \{s_{12}, s_{14}\}, \quad s_6^b = \{s_{11}, s_{13}\}.$$

States are labeled with the value of 1, together with the distribution of h:

$$Pr(h_{s_0^b}) = \{0 \mapsto \frac{1}{4}, 1 \mapsto \frac{1}{4}, 2 \mapsto \frac{1}{4}, 3 \mapsto \frac{1}{4}\}, \qquad Pr(h_{s_3^b}) = \{1 \mapsto \frac{1}{2}, 2 \mapsto \frac{1}{2}\},$$

$$Pr(h_{s_1^b}) = \{0 \mapsto \frac{1}{2}, 1 \mapsto \frac{1}{4}, 2 \mapsto \frac{1}{4}\}, \quad Pr(h_{s_5^b}) = \{1 \mapsto \frac{1}{2}, 2 \mapsto \frac{1}{2}\},$$

$$Pr(h_{s_2^b}) = \{0 \mapsto 1\}, \quad Pr(h_{s_4^b}) = \{3 \mapsto 1\}, \quad Pr(h_{s_6^b}) = \{1 \mapsto \frac{1}{2}, 2 \mapsto \frac{1}{2}\}.$$

The back-bisimulation quotient can be automatically constructed from the MC. This will be discussed in Sect. 3.4. After constructing the quotient, the next step is to compute the program leakage from the back-bisimulation quotient.

3.3   Measuring the Leakage Using Back-Bisimulation Quotient

Let $\mathcal{M}_\delta^P/\sim_b = (S/\sim_b, \mathbf{P}_\delta', s_{init}^b, Val_l, V, Pr(h))$ be the attacker's view of the program P running with the scheduler δ. In each state s^b of $\mathcal{M}_\delta^P/\sim_b$, the secret distribution $Pr(h_{s^b})$ determines the attacker's uncertainty about h. Depending on the program statements that are chosen and executed by the scheduler, and the public values observed by the attacker, the distribution of h changes from state to state along each trace of $\mathcal{M}_\delta^P/\sim_b$. In fact, $\mathcal{M}_\delta^P/\sim_b$ *transforms* a priori distribution of h in the initial state s_{init}^b to posterior distributions in the final states $final(\mathcal{M}_\delta^P/\sim_b)$.

The attacker's uncertainty about h in a state s^b with the secret distribution $Pr(h_{s^b})$ is measured by $\mathcal{H}_\infty(\mathbf{h}_{s^b})$. Thus, the initial uncertainty is measured by $\mathcal{H}_\infty(\mathbf{h}_{s_{init}^b})$.

Since there might be more than one final state with different reachability probabilities and the MC can seen as a discrete probability distribution over all of its final states, the remaining uncertainty is defined as the expectation of uncertainties in all final states: $\sum_{s_f^b \in final(\mathcal{M}_\delta^P/\sim_b)} Pr(s_f^b)\mathcal{H}_\infty(\mathbf{h}_{s_f^b})$, where $Pr(s_f^b)$ is the probability of reaching s_f^b from the initial state s_{init}^b. It now follows that the leakage of the concurrent program P running under control of the scheduler δ is computed as $\mathcal{L}(P_\delta) = \mathcal{H}_\infty(\mathbf{h}_{s_{init}^b}) - \sum_{s_f^b \in final(\mathcal{M}_\delta^P/\sim_b)} Pr(s_f^b).\mathcal{H}_\infty(\mathbf{h}_{s_f^b})$.

Notice that for measuring the leakage of P, we computed min-entropy of initial and final states, and did not consider min-entropy of intermediate states. This is not in contrast with our assumption of taking into account the intermediate values of 1 along the execution paths. This is because in $\mathcal{M}_\delta^P/\sim_b$ distributions of h in the final states result from values of 1 and distributions of h in the intermediate states. Thus, when computing the remaining uncertainty from the final distributions, the intermediate values of 1 are automatically taken into account. The final distributions of h also result from the program statements which are chosen by the scheduler. Therefore, the effect of the scheduler choices is considered, as well.

Moreover, in the literature, the remaining uncertainty is usually measured by the conditional entropy $\mathcal{H}_\infty(h|l)$, but we measure it by the non-conditional entropy $\mathcal{H}_\infty(h)$. These entropies are identical in our program model, because in $\mathcal{M}_\delta^P / \sim_b$ the entropy $\mathcal{H}_\infty(h)$ is computed from final states that result from traces observed by the attacker. This is exactly the same as the conditional entropy $\mathcal{H}_\infty(h|l)$.

Example 4 (Back-bisimulation quotient of P1 is a distribution transformer.). In the initial state s_0^b of $\mathcal{M}_{uni}^{P1} / \sim_b$, the distribution of h is $Pr(h_{s_0^b}) = \{0 \mapsto \frac{1}{4}, 1 \mapsto \frac{1}{4}, 2 \mapsto \frac{1}{4}, 3 \mapsto \frac{1}{4}\}$. This means that before executing the program, the attacker only knows that the value of h belongs to the set $\{0, 1, 2, 3\}$ and if she guesses the value of h, then the likelihood of her being successful is $\frac{1}{4}$. Therefore, $Pr(h_{s_0^b})$ determines the attacker's initial uncertainty about h. Now consider the final state s_3^b, in which the distribution of h is $Pr(h_{s_3^b}) = \{1 \mapsto \frac{1}{2}, 2 \mapsto \frac{1}{2}\}$. In this state, the attacker knows that the value of h belongs to $\{1, 2\}$, and thus her uncertainty about h is reduced. This means that after executing the program and observing the trace $\langle 0, 1, 0^\omega \rangle$, if the attacker guesses the value of h, then the likelihood of her being successful is $\frac{1}{2}$. These considerations imply that the back-bisimulation quotient is a distribution transformer.

Example 5 (Information leakage of P1). The initial uncertainty is quantified as the Renyi's min-entropy of h in the initial state s_0^b, i.e., $\mathcal{H}_\infty(h_{s_0^b}) = -\log_2 \frac{1}{4} = 2$ *(bits)*.

The remaining uncertainty is quantified as the Renyi's min-entropy of h in the final states. There are four final states with different reachability probabilities: $s_2^b, s_5^b, s_3^b, s_4^b$. Consequently, the remaining uncertainty is quantified as the *expectation* of the Renyi's min-entropy of h in these states:

$$\sum_{s^b \in \{s_2^b, s_5^b, s_3^b, s_4^b\}} Pr(s^b).\mathcal{H}_\infty(h_{s^b}) = -\frac{1}{4} * \log_2 1 - \frac{1}{4} * \log_2 \frac{1}{2} - \frac{1}{4} * \log_2 \frac{1}{2}$$

$$-\frac{1}{4} * \log_2 1 = 0.5 \ (bits),$$

where $Pr(s^b)$ denotes the probability of reaching s^b from the initial state s_0^b. Finally, the leakage of the program P1 running with the uniform scheduler is computed as $\mathcal{L}(P1_{uni}) = 2 - 0.5 = 1.5$ *(bits)*.

The following section formally defines the back-bisimulation equivalence and explains how to compute the back-bisimulation quotient.

3.4   Computing Back-Bisimulation Quotient

In this section, we discuss how to compute the back-bisimulation quotient. Before that, we first explain a subclass of MCs, called *Markov chains with pseudoback-bisimilar states*, in which back-bisimulation cannot correctly construct the attacker's view of the program behavior.

Pseudoback-Bisimilar States. In order to compute the states of a back-bisimulation quotient, we need to aggregate back-bisimilar states into one equivalence class. For that, we define *Back-bisimulation signature*, which is defined as a kind of fingerprint for states of a back-bisimulation equivalence class.

Definition 8. *The **back-bisimulation signature** of a state s is defined as*

$$sig_{\sim_b}(s) = \{ \; \big(V(s), [s']_{\sim_b}\big) \mid \exists s' \in Pre(s) \; \}.$$

It asserts that two states that have the same public value and their predecessors belong to the same equivalence class, have the same signature.

Definition 9. *Let $\mathcal{M}_\delta^P$ be an MC. Two states $s_1, s_2 \in S$ are **pseudoback-bisimilar** iff (1) $V(s_1) = V(s_2)$, (2) $level(s_1) = level(s_2)$, (3) $sig_{\sim_b}(s_1) \neq sig_{\sim_b}(s_2)$, (4) $PreLabels(s_1) \cap PreLabels(s_2) \neq \varnothing$. An MC that contains some pseudoback-bisimilar states is denoted by $MC_\mathfrak{p}$ and an MC with no pseudoback-bisimilar state is denoted by $MC_\mathfrak{n}$.*

Stated in words, two states are pseudoback-bisimilar if they have the same label, are at the same level (distance from an initial state), and have different signatures, but intersecting pre-labels. In an $MC_\mathfrak{n}$, states at the same level and with the same label, either have no intersecting pre-labels or have the same pre-labels.

Example 6 (An example $MC_\mathfrak{p}$). Consider the following program:

```
l:=0;
if h=1 then l:=1; l:=2; l:=3; l:=4; l:=5
else (l:=1 || l:=2); l:=3; (l:=4 || l:=5)          (P2)
```

where $Val_h \in \{0,1\}$ and $Pr(h) = \{0 \mapsto \frac{1}{2}, 1 \mapsto \frac{1}{2}\}$. A uniform scheduler is selected for both parallel operators. The MC $\mathcal{M}_{uni}^{P2}$ is shown in Fig. 4a.

In $\mathcal{M}_{uni}^{P2}$, states s_8 and s_9 are pseudoback-bisimilar. They both have the label 3, are at the level 3, and have different signatures:

$$sig_{\sim_b}(s_8) = \Big\{(3, \{s_5, s_7\}), (3, \{s_3\})\Big\}, sig_{\sim_b}(s_9) = \Big\{(3, \{s_5, s_7\})\Big\},$$

but intersecting pre-labels: $preLabels(s_8) = \{2, 1\}, \qquad preLabels(s_9) = \{2\}$.

Back-bisimulation captures the attacker's view for all programs that do not contain pseudoback-bisimilar states, i.e., those final states that have the same public values and result from the same trace are indistinguishable and fall into the same $\sim_b$-equivalence class. Formally,

Theorem 2. *Let $\mathcal{M}_\delta^P$ be an $MC_\mathfrak{n}$. For all paths $\sigma_1, \sigma_2 \in Paths(\mathcal{M}_\delta^P)$ with $\sigma_1 = s_{0,1}s_{1,1}\ldots s_{n-1,1}(s_{n,1})^\omega$, $\sigma_2 = s_{0,2}s_{1,2}\ldots s_{n-1,2}(s_{n,2})^\omega$, and $n \geq 0$ it holds that $s_{n,1} \sim_b s_{n,2}$ iff $trace(\sigma_1) = trace(\sigma_2)$.*

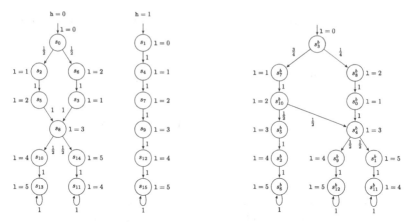

(a) MC of the program P2 with the uniform scheduler

(b) back-bisimulation quotient of $\mathcal{M}^{P2}_{uni}$

Fig. 4. $\mathcal{M}^{P2}_{uni}$ and $\mathcal{M}^{P2}_{uni}/\sim_b$

Theorem 2 argues that same-trace final states are back-bisimilar. A similar argument can be made for non-final states.

Theorem 3. *Let $\mathcal{M}^P_\delta$ be an MC_n. For all paths $\sigma_1, \sigma_2 \in Paths(\mathcal{M}^P_\delta)$ with $\sigma_1 = s_{0,1}s_{1,1}\ldots s_{n-1,1}(s_{n,1})^\omega$, $\sigma_2 = s_{0,2}s_{1,2}\ldots s_{m-1,2}(s_{m,2})^\omega$, $n, m > 0$, and $0 \le i < min(n, m)$ it holds that $s_{i,1} \sim_b s_{i,2}$ iff $trace_{\ll i}(\sigma_1) = trace_{\ll i}(\sigma_2)$.*

Therefore, all paths of $\mathcal{M}^P_\delta$ with the same trace form a single path in $\mathcal{M}^P_\delta/\sim_b$. Stated formally, let $\sigma' \in Paths(\mathcal{M}^P_\delta/\sim_b)$, $s^b_f = final(\sigma') \in final(\mathcal{M}^P_\delta/\sim_b)$, and $T = trace(\sigma')$. The path σ' is the aggregation of all paths $Paths(T) \subseteq Paths(\mathcal{M}^P_\delta)$. All final states of $\mathcal{M}^P_\delta$ that result from the trace T fall into the same $\sim_b$-equivalence class $s^b_f = \{s_f \mid s_f \in final(Paths(T))\}$. For s^b_f, the secret distribution $Pr(h_{s^b_f})$ contains probabilities of possibles values of h that the attacker might be able to guess by observing T.

Pseudoback-bisimilar states do not fall into the same $\sim_b$-equivalence class and thus $\sim_b$ is not able to aggregate all paths with the same trace. For instance, in $\mathcal{M}^{P2}_{uni}/\sim_b$ (Fig. 4b) there are two paths $s^b_3 s^b_7 s^b_{10} s^b_5 s^b_2 (s^b_6)^\omega$ and $s^b_3 s^b_7 s^b_{10} s^b_4 s^b_9 (s^b_{12})^\omega$ with the same trace $\langle 0, 1, 2, 3, 4, 5^\omega \rangle$. The attacker, after observing the trace, cannot discriminate the value of h to be 0 or 1. But, in the attacker's view of the MC constructed by back-bisimulation (Fig. 4b) the value of h is distinguished in the final states of the two paths. Furthermore, the probability of some traces in the back-bisimulation quotient might be different from their probability in the concrete model. For example, the probability of the trace $\langle 0, 1, 2, 3, 4, 5^\omega \rangle$ in $\mathcal{M}^{P2}_{uni}$ is $\frac{5}{8}$, while it is $\frac{9}{16}$ in $\mathcal{M}^{P2}_{uni}/\sim_b$. The implication is that back-bisimulation cannot correctly construct the attacker's view of an MC_p. For MC_ps, we use the trace-exploration-based method, introduced in [25], which computes the program leakage directly from the MC_p $\mathcal{M}^P_\delta$.

Algorithm for Computing the Back-Bisimulation Quotient Space. In this section, an algorithm is proposed for obtaining the back-bisimulation

Algorithm 1. A first iterative quotienting algorithm

Input: finite MC_n $\mathcal{M}_\delta^P$ with state space S
Output: back-bisimulation quotient space $S/\sim_b$

/* *Determine the initial partition* Π_0 */
1: $s_{init}^b := Init(\mathcal{M}_\delta^P)$;
2: $\mathcal{R} := \{(s_1, s_2) \mid V(s_1) = V(s_2)\}$;
3: $\Pi_0 = \{s_{init}^b\} \cup \left((S \setminus Init(\mathcal{M}_\delta^P)) / \mathcal{R}\right)$;

4: $\Pi := \Pi_0$;
5: $\Pi_{old} := \{S\}$; // Π_{old} *contains the previous partition*
/* *loop until no refinement possible* */
6: **while** $\Pi \;!= \Pi_{old}$ **do**
7:    $\Pi_{old} := \Pi$;
/* *search through the blocks of* Π_{old} *to find a splitter candidate for* Π */
8:    **for all** $\mathcal{C} \in \Pi_{old}$ **do**
9:        $\Pi := Refine_b(\Pi, \mathcal{C})$;
10: **return** Π;

quotient space for a finite MC_n. This algorithm is similar to Kanellakis and Smolka's algorithm for computing the bisimulation quotient space [20]. It relies on a *partition refinement* technique, where the state space is partitioned into *blocks*. It starts from an initial partition Π_0 and computes successive refinements of Π_0 until a stable partition is reached. A partition is stable if no further refinements are possible. The obtained partition is $S/\sim_b$, the largest back-bisimulation over the input finite MC_n. The essential steps are outlined in Algorithm 1. The algorithm consists of two main parts: (a) computing the initial partition, and (b) successively refining the partitions.

Computing the Initial Partition. Since back-bisimilar states have the same public value, it is sensible to use this in determining the initial partition Π_0.

All initial states have the same public value and have no predecessors. Consequently, they form a single block $s_{init}^b = Init(\mathcal{M}_\delta^P)$. This block will remain unchanged during the refinements.

For the remaining states $S \backslash s_{init}^b$, each group of states with the same public value forms a block. Same-label blocks can be obtained by the equivalence relation $\mathcal{R} = \{(s_1, s_2) \mid V(s_1) = V(s_2)\}$, which induces the quotient spaces $(S \setminus Init(\mathcal{M}_\delta^P)) / \mathcal{R}$. Thus, the initial partition is obtained as $\Pi_0 = \{s_{init}^b\} \cup \left((S \setminus Init(\mathcal{M}_\delta^P)) / \mathcal{R}\right)$.

Partition Refinement. Since all partitions are a refinement of the initial partition Π_0, each block in these partitions contains states with the same public value. However, blocks of Π_0, except s_{init}^b, do not consider the one-step predecessors of states. This is taken care of in the successive refinement steps, by the refinement operator.

Definition 10. *Let Π be a partition for S and $\mathcal{C}$ be a superblock of Π. Then,*

$$Refine_b(\Pi, \mathcal{C}) = \bigcup_{\mathcal{B} \in \Pi} Refine_b(\mathcal{B}, \mathcal{C}),$$

where $Refine_b(\mathcal{B}, \mathcal{C}) = \{\mathcal{B} \cap Succ(\mathcal{C}), \mathcal{B} \backslash Succ(\mathcal{C})\} \backslash \{\varnothing\}$. Here, $\mathcal{C}$ is called a split-ter for Π, refining blocks of Π to subblocks.

Using $\mathcal{C}$, $Refine_b(\mathcal{B}, \mathcal{C})$ decomposes the block $\mathcal{B}$ into two subblocks, provided that the subblocks are nonempty.

A key step of computing the back-bisimulation quotient space is to determine a splitter $\mathcal{C}$ for a given partition Π. Algorithm 1 uses the blocks of the previous partition Π_{old} as splitter candidates for Π.

Theorem 4. *Algorithm 1 always terminates and correctly computes the back-bisimulation quotient space $S/\sim_b$.*

The following theorem discusses the time complexity of Algorithm 1.

Theorem 5. *The time complexity of Algorithm 1 is $O(|S|.|E|)$, where E denotes the set of transitions of $\mathcal{M}_\delta^P$.*

4   Conclusions and Future Work

In this paper, a quantification approach is proposed for concurrent programs. Back-bisimulation equivalence is defined to model the attacker's view of the program behavior. Then a partition refinement algorithm is developed to compute the back-bisimulation quotient of the program. The back-bisimulation quotient is automatically constructed and contains secret distributions, which are used to compute the information leakage of the program.

The back-bisimulation quotient contains all execution traces which the attacker can observe during executing the program. Thus, it can be used to compute maximal and minimal leakages that might occur during the program executions. Furthermore, the quotient is an abstract model of the program and the quantification analysis is done on a minimized model, most likely saving time and space.

Back-bisimulation equivalence creates a lot of exciting opportunities for future works. It can be used to verify any trace-equivalence-based property, such as *observational determinism* [21,26,33], a widely-studied confidentiality property of secure information flow for concurrent programs. It can also be defined on multi-terminal binary decision diagrams (MTBDDs), in order to improve the scalability of the quantification approach to a great extent. We aim to lift the program-termination restriction and extend the proposed approach to non-terminating concurrent programs. We also aim to study *bounded leakage problem* [25] and *channel capacity* [29] on the back-bisimulation quotient. Probably, using some reductions, such as on-the-fly techniques, can improve the scalability of the problem. Furthermore, handling programs with pseudoback-bisimilar states using back-bisimulation is a possible future work. Another avenue to consider the current work is to perform time analysis of the proposed approach, e.g. on dining cryptographers protocol.

A   Case Study

In this section, we analyze a case study to show applicability and feasibility of the approach.

The Dining Cryptographers Protocol. We consider the *dining cryptographers problem* [11] to show how an attacker can deduce secret information through execution observations. The dining cryptographers problem was first proposed by David Chaum in 1988 as an example of anonymity and identity hiding [11]. In this problem, N cryptographers are sitting at a round table to have dinner at their favorite restaurant. The waiter informs them that the meal has been arranged to be paid by one of the cryptographers or their master. The cryptographers respect each other's right to stay anonymous, but would like to know whether the master is paying or not. So, they decide to take part in the following two-stage protocol:

- Stage 1: Each cryptographer tosses a coin and only informs the cryptographer on the right of the outcome.
- Stage 2: Each cryptographer publicly announces whether the two coins that she can see are the same ('agree') or different ('disagree'). However, if she actually paid for the dinner, then she lies, i.e., she announces 'disagree' when the coins are the same, and 'agree' when they are different.

Let the variable *parity* be exclusive-or (XOR) between all the announcements. If N is odd, then an odd number of 'agree's (parity = 1) implies that none of the cryptographers paid (the master paid), while an even number (parity = 0) implies that one of the cryptographers paid. The latter is reverse for an even N.

The payer can be either

i. one of the cryptographers, i.e., $Val_{payer} = \{c_1, \ldots, c_N\}$, or
ii. the master (m, for short) or one of the cryptographers, i.e., $Val_{payer} = \{m, c_1, \ldots, c_N\}$.

Assume an attacker who tries to find out the payer's identity. The attacker is external, i.e., none of the cryptographers. This attacker can observe the parity and also the announcements of the cryptographers. All observable variables are concatenated to form a single public variable. The program model is an MC_n and we employ the proposed algorithms to compute the leakage.

The experimental results for the cases in which the coin probability is 0.5 are shown in Table 1. In this table, N denotes the number of cryptographers. $\mathcal{M}_{uni}^{DC_N}$ and $\mathcal{M}_{uni}^{DC_N} / \sim_b$ denote the MC of the program run with a uniform scheduler and the back-bisimulation quotient, respectively. Symbols $\#st$ and $\#tr$ denote the number of states and transitions, respectively.

Similar results for the coin probability of 0 or 1 are shown in Table 2. As shown in Tables 1 and 2, back-bisimulation results in impressive reductions of

Table 1. Evaluation results for the dining cryptographers protocol with the coin probability 0.5

Val_{payer}	N	$\mathcal{M}^{DC_N}_{uni}$		$\mathcal{M}^{DC_N}_{uni}/\sim_b$		leakage
		#st	#tr	#st	#tr	(bits)
	3	380	776	26	45	**0.811 (40%)**
$\{m, c_1, \ldots, c_N\}$	4	2165	5720	64	144	**0.721 (31%)**
	5	11850	38772	152	420	**0.65 (25%)**
	6	63063	246820	352	1152	**0.59 (21%)**
	3	285	582	22	36	**0**
$\{c_1, \ldots, c_N\}$	4	1732	4576	56	121	**0**
	5	9875	32310	136	365	**0**
	6	54054	211560	320	1125	**0**

the state space. For example, when the coin probability is 0.5 (Table 1) reductions vary between 92% and 99.5%.

Consider the last three cases of Table 1, where the coin probability is 0.5 and the payer is one of the cryptographers ($Val_{payer} = \{c_1, \ldots, c_N\}$). In these cases, the program leakage is 0. This shows that the attacker cannot identify the payer. This is why the dining cryptographers protocol is said to be secure in the context of anonymity.

The analysis results in Table 2 show that when the probability of the coin is 0 or 1, no matter whoever the payer is, the leakage is $\log_2 |Val_{payer}|$, proving that the secret gets completely leaked and thus the attacker learns the identity of the payer.

B   Related Work

The notion of back-simulation is similar to the notion of backward strong bisimulation considered by De Nicola and Vaandrager [15]. They use a different notion than our definition, as they only allow to move back from a state along the path representing the history that brought one into that state. Högberg et al. [17] defined and considered backward bisimulation minimization on tree automata, Sproston and Donatelli [32] considered a probabilistic version of backward bisimulation and studied the logical properties it preserves, and Cardelli et al. [9] who considered backward bisimulation in the stochastic setting of chemical reaction networks. None of these works use backward bisimulation in quantitative information flow.

Chen and Malacaria [12] model multi-threaded programs as state transition systems. They use Bellman's optimality principle to determine the leakage bounds, i.e., minimal and maximal leakage occurred during possible program executions.

Table 2. Evaluation results for the dining cryptographers protocol with the coin probability 0 or 1

Val_{payer}	N	$\mathcal{M}_{uni}^{DC_N}$		$\mathcal{M}_{uni}^{DC_N}/\sim_b$		leakage (bits)
		#st	#tr	#st	#tr	
$\{m, c_1, \ldots, c_N\}$	3	72	124	21	37	**2 (100%)**
	4	235	525	47	107	**2.32 (100%)**
	5	738	2046	103	286	**2.585 (100%)**
	6	2254	7483	223	729	**2.807 (100%)**
$\{c_1, \ldots, c_N\}$	3	54	93	20	34	**1.585 (100%)**
	4	188	420	46	103	**2 (100%)**
	5	615	1705	102	281	**2.32 (100%)**
	6	1932	6414	222	723	**2.585 (100%)**

Phan et al. [27] propose to use symbolic execution, a verification technique which bounds runtime behavior of the program, thus mitigating state-space explosion problem. In state-space explosion problem, the amount of state-space of the program model gets too huge to store in the memory, thus making the analysis difficult. Phan et al. run symbolic execution to extract all symbolic paths of the program. Then, paths with a direct information flow are labeled. Finally, they use a model counting technique to count the number of inputs that follow direct-labeled paths, to compute *channel capacity*, which is an upper bound of the leakage over *all* possible distributions of the secret input.

Biondi et al. [8] use interval Markov chains to compute the channel capacity of deterministic processes. They reduce the channel capacity computation to entropy maximization, a well-known problem in Bayesian statistics.

Chothia et al. [14] have developed LeakWatch to approximate leakage of Java programs. LeakWatch is based on probabilistic point-to-point information leakage, in which the leakage between any given two points in the program from secret to public variables is computed.

Chadha et al. [10] employ symbolic algorithms to quantify the precise leakage from public to secret variables. They use Binary Decision Diagrams (BDDs) to model the relation between the inputs and outputs of the program. To do so, Moped [16], a symbolic model checker, is exploited to construct BDDs. Chadha et al. have implemented their method into a tool called Moped-QLeak.

Klebanov [22] uses symbolic execution in combination with deductive verification [5] and self-composition [4] to measure residual Shannon entropy and min-entropy of the secret input. Exploitation of deductive verification makes the analysis immune to the state-space explosion problem, but also makes it semi-automatic, as user-supplied invariants are needed for the analysis to proceed.

References

1. Alvim, M.S., Andrés, M.E., Chatzikokolakis, K., Palamidessi, C.: Quantitative information flow and applications to differential privacy. In: Aldini, A., Gorrieri, R. (eds.) FOSAD 2011. LNCS, vol. 6858, pp. 211–230. Springer, Heidelberg (2011). https://doi.org/10.1007/978-3-642-23082-0_8
2. Amir-Mohammadian, S.: A semantic framework for direct information flows in hybrid-dynamic systems. In: Proceedings of the 7th ACM Cyber-Physical System Security Workshop (CPSS 2021), pp. 5–15. Association for Computing Machinery, June 2021
3. Baier, C., Katoen, J.P.: Principles of Model Checking. MIT Press, Cambridge (2008)
4. Barthe, G., D'Argenio, P.R., Rezk, T.: Secure information flow by self-composition. In: Proceedings of the 17th IEEE Workshop on Computer Security Foundations, CSFW 2004, pp. 100–114. IEEE Computer Society (2004)
5. Beckert, B., Hähnle, R., Schmitt, P.H.: Verification of Object-Oriented Software. The KeY Approach. Springer, Heidelberg (2007). https://doi.org/10.1007/978-3-540-69061-0
6. Biondi, F.: Markovian processes for quantitative information leakage. Ph.D. thesis, IT University of Copenhagen (2014)
7. Biondi, F., Legay, A., Malacaria, P., Wasowski, A.: Quantifying information leakage of randomized protocols. Theor. Comput. Sci. **597**, 62–87 (2015)
8. Biondi, F., Legay, A., Nielsen, B.F., Wasowski, A.: Maximizing entropy over Markov processes. J. Log. Algebraic Methods Program. **83**(5), 384–399 (2014)
9. Cardelli, L., Tribastone, M., Tschaikowski, M., Vandin, A.: Forward and backward bisimulations for chemical reaction networks. arXiv preprint arXiv:1507.00163 (2015)
10. Chadha, R., Mathur, U., Schwoon, S.: Computing information flow using symbolic model-checking. In: Raman, V., Suresh, S.P. (eds.) 34th International Conference on Foundation of Software Technology and Theoretical Computer Science (FSTTCS 2014). Leibniz International Proceedings in Informatics (LIPIcs), vol. 29, pp. 505–516. Schloss Dagstuhl-Leibniz-Zentrum fuer Informatik, Dagstuhl, Germany (2014)
11. Chaum, D.: The dining cryptographers problem: Unconditional sender and recipient untraceability. J. Cryptol. **1**(1), 65–75 (1988). https://doi.org/10.1007/BF00206326
12. Chen, H., Malacaria, P.: The optimum leakage principle for analyzing multi-threaded programs. In: Kurosawa, K. (ed.) ICITS 2009. LNCS, vol. 5973, pp. 177–193. Springer, Heidelberg (2010). https://doi.org/10.1007/978-3-642-14496-7_15
13. Chen, H., Malacaria, P.: Quantifying maximal loss of anonymity in protocols. In: Proceedings of the 4th International Symposium on Information, Computer, and Communications Security, pp. 206–217. ACM (2009)
14. Chothia, T., Kawamoto, Y., Novakovic, C.: LeakWatch: estimating information leakage from Java programs. In: Kutyłowski, M., Vaidya, J. (eds.) ESORICS 2014. LNCS, vol. 8713, pp. 219–236. Springer, Cham (2014). https://doi.org/10.1007/978-3-319-11212-1_13
15. De Nicola, R., Vaandrager, F.: Three logics for branching bisimulation. J. ACM (JACM) **42**(2), 458–487 (1995)

16. Esparza, J., Kiefer, S., Schwoon, S.: Abstraction refinement with Craig interpolation and symbolic pushdown systems. J. Satisfiability Boolean Model. Comput. **5**, 27–56 (2008)
17. Högberg, J., Maletti, A., May, J.: Backward and forward bisimulation minimization of tree automata. Theor. Comput. Sci. **410**(37), 3539–3552 (2009)
18. Jurado, M., Palamidessi, C., Smith, G.: A formal information-theoretic leakage analysis of order-revealing encryption. In: Proceedings of the 34th IEEE Workshop on Computer Security Foundations, CSFW 2021. IEEE Computer Society (2021)
19. Jurado, M., Smith, G.: Quantifying information leakage of deterministic encryption. In: Proceedings of the 2019 ACM SIGSAC Conference on Cloud Computing Security Workshop, pp. 129–139 (2019)
20. Kanellakis, P.C., Smolka, S.A.: CCS expressions, finite state processes, and three problems of equivalence. Inf. Comput. **86**(1), 43–68 (1990)
21. Karimpour, J., Isazadeh, A., Noroozi, A.A.: Verifying observational determinism. In: Federrath, H., Gollmann, D. (eds.) SEC 2015. IAICT, vol. 455, pp. 82–93. Springer, Cham (2015). https://doi.org/10.1007/978-3-319-18467-8_6
22. Klebanov, V.: Precise quantitative information flow analysis - a symbolic approach. Theor. Comput. Sci. **538**, 124–139 (2014)
23. Köpf, B., Basin, D.: An information-theoretic model for adaptive side-channel attacks. In: Proceedings of the 14th ACM Conference on Computer and Communications Security, pp. 286–296. ACM (2007)
24. Köpf, B., Smith, G.: Vulnerability bounds and leakage resilience of blinded cryptography under timing attacks. In: Proceedings of 23rd IEEE Computer Security Foundations Symposium (CSF), pp. 44–56. IEEE (2010)
25. Noroozi, A.A., Karimpour, J., Isazadeh, A.: Information leakage of multi-threaded programs. Comput. Electr. Eng. **78**, 400–419 (2019)
26. Noroozi, A.A., Salehi, K., Karimpour, J., Isazadeh, A.: Secure information flow analysis using the PRISM model checker. In: Garg, D., Kumar, N.V.N., Shyamasundar, R.K. (eds.) ICISS 2019. LNCS, vol. 11952, pp. 154–172. Springer, Cham (2019). https://doi.org/10.1007/978-3-030-36945-3_9
27. Phan, Q.S., Malacaria, P., Păsăreanu, C.S., d'Amorim, M.: Quantifying information leaks using reliability analysis. In: Proceedings of the 2014 International SPIN Symposium on Model Checking of Software, pp. 105–108. ACM (2014)
28. Puterman, M.L.: Markov Decision Processes: Discrete Stochastic Dynamic Programming. Wiley, Hoboken (1994)
29. Salehi, K., Karimpour, J., Izadkhah, H., Isazadeh, A.: Channel capacity of concurrent probabilistic programs. Entropy **21**(9), 885 (2019)
30. Salehi, K., Noroozi, A.A., Amir-Mohammadian, S.: Quantifying information leakage of probabilistic programs using the PRISM model checker. In: Emerging Security Information, Systems and Technologies, pp. 47–52. IARIA (2021)
31. Smith, G.: On the foundations of quantitative information flow. In: de Alfaro, L. (ed.) FoSSaCS 2009. LNCS, vol. 5504, pp. 288–302. Springer, Heidelberg (2009). https://doi.org/10.1007/978-3-642-00596-1_21
32. Sproston, J., Donatelli, S.: Backward bisimulation in Markov chain model checking. IEEE Trans. Softw. Eng. **32**(8), 531–546 (2006)
33. Zdancewic, S., Myers, A.C.: Observational determinism for concurrent program security. In: Proceedings of the 16th IEEE Computer Security Foundations Workshop, CSFW 2003, pp. 29–43. IEEE Computer Society (2003)

Parameter Synthesis

Rate Lifting for Stochastic Process Algebra – Exploiting Structural Properties –

Markus Siegle and Amin Soltanieh[✉]

Department of Computer Science, Bundeswehr University Munich,
Neubiberg, Germany
{markus.siegle,amin.soltanieh}@unibw.de

Abstract. This paper presents an algorithm for determining the unknown rates in the sequential processes of a Stochastic Process Algebra (SPA) model, provided that the rates in the combined flat model are given. Such a rate lifting is useful for model reverse engineering and model repair. Technically, the algorithm works by solving systems of nonlinear equations and – if necessary – adjusting the model's synchronisation structure without changing its transition system. This approach exploits some structural properties of SPA systems, which are formulated here for the first time and could be very beneficial also in other contexts.

Keywords: Stochastic Process Algebra · Structural Properties · Markov Chain · Model Repair · Rate Lifting

1 Introduction

Stochastic Process Algebra (SPA) is a family of formalisms widely used in the area of quantitative modelling and evaluation. Typical members of this family are PEPA [6], TIPP [5], EMPA [2], CASPA [10], but also the reactive modules language of tools such as PRISM [11] and STORM [4]. Originally devised for classical performance and dependability modelling, SPA models are now frequently used in probabilistic model checking projects.

This paper presents a solution to the following problem: Given a compositional SPA specification where the transition rates of its components are unknown, but given all transition rates of the associated low-level, flat transition system, find the unknown transition rates for the components of the high-level SPA model. An alternative formulation of the same problem is for a compositional SPA specification with known original transition rates in its components, but given rate modification factors for (a subset of) the transition rates in its flat low-level model. Here the task is to find new transition rates for the components of the high-level SPA model, such that the resulting rates in the flat model will be modified as desired. The first formulation is from the perspective of systems reverse engineering (to be more specific, one could call it rate reverse engineering), whereas the second one pertains to model checking and

© Springer Nature Switzerland AG 2022
E. Ábrahám and M. Paolieri (Eds.): QEST 2022, LNCS 13479, pp. 67–84, 2022.
https://doi.org/10.1007/978-3-031-16336-4_4

model repair [1,3,13]. We will refer to both variants of the problem as "rate lifting problem".

An algorithm that solves the rate lifting problem for SPA models with $n = 2$ components was presented in [15], the equation system involved being studied in [16]. However, developing a rate lifting algorithm for a general number $n \geq 3$ of processes turns out to be a much bigger challenge, since – firstly – SPA models with n components may have a much more complex synchronisation structure than for $n = 2$, and it is the synchronisation structure which plays an essential role during the execution of the algorithm. Secondly, components of SPA models may contain selfloops (meant to synchronise with other components), and – related to this – the transition system underlying a compositional SPA model is actually a flattened multi-transition system [5,6]. These two facts have to be considered during the necessary deconstruction of a flat transition, and they strongly contribute to the complication of the problem. So, in this paper we develop a rate lifting algorithm for an SPA system consisting of n components, where n is arbitrary. The algorithm will assign (new) values to the components' transition rates and – under certain circumstances – it will change the synchronisation structure of the SPA model. The latter means that the algorithm may add actions to certain synchronisation sets and in consequence it will insert additional selfloops at some specific component states, but it will do this in such a way that the set of reachable states and the set of transitions of the overall model are not changed. Only the transition rates of the overall model are set/changed as desired. Technically, the algorithm works by setting up and solving systems of nonlinear (actually multilinear) equations.

It is quite easy to see that an arbitrary assignment of rates to the transitions of the low-level transition system may not always be realisable by suitable rates in the components, i.e. not every instance of the rate lifting problem has a solution. Therefore, naturally, the algorithm presented in this paper will not always succeed. However, it is guaranteed that the algorithm will find a solution, if such a solution exists (see Sect. 6).

We build our algorithm based on certain structural properties of SPA systems, which can be exploited in the course of the algorithm. As an example, for a given transition in one of the SPA components, it is necessary to identify the partners which may or must synchronise with it. To the best of our knowledge, these fundamental properties have not previously been addressed in the literature, which is suprising, since they could be very valuable also in other contexts. For example, in compositional system verification, distinguishing between different types of neighbourhoods of processes or determining the participating set of a transition (see Sect. 2) is the key to establishing dependence/independence relations between processes.

2   Structural Properties of SPA

We consider a simple but fairly general class of Markovian Stochastic Process Algebra models constructed by the following grammar:

Definition 1 (SPA language). *For a finite set of actions Act, let $a \in Act$ and $A \subseteq Act$. Let $\lambda \in \mathbb{R}^{>0}$ be a transition rate. An SPA system Sys is a process of type Comp, constructed according to the following grammar:*

$$Comp := (Comp \,||_A\, Comp) \mid Seq$$
$$Seq := 0 \mid (a, \lambda); Seq \mid Seq + Seq \mid V$$

Seq stands for sequential processes, and *Comp* for composed processes. V stands for a process variable for a sequential process, which can be used to define cyclic behaviour (including selfloops). One could add a recursion operator, the special invisible action τ, hiding and other features, but this is not essential for our purpose. The semantics is standard, i.e. the SPA specification is mapped to the underlying flat transition system (an action-labelled CTMC), see e.g. [5, 6]. We assume multiway synchronisation[1], i.e. the synchronisation of two a-transitions yields another a-transition (whose rate is a function of the two partner transitions, or – even more generally – of the two partner processes), which can then participate in further a-synchronisations, etc.

An SPA system corresponds to a process tree whose internal nodes are labelled by the parallel composition operator, each one parametrized by a set of synchronising actions ($||_A$, with $A \subseteq Act$), and whose leaves are sequential processes of type *Seq*. For a specific action $a \in Act$, we write $||_a$ as an abbreviation to express that a belongs to the set of synchronising actions, and $||_{\neg a}$ that it does not.

Definition 2. *Let Sys be a given SPA system.*

(a) *The set of all sequential processes within Sys is denoted as seqproc(Sys) (i.e. the set of all leaves of the process tree of Sys).*
(b) *The set of all (sequential or composed) processes within Sys is denoted as proc(Sys).*

The set $proc(Sys)$ equals the set of all nodes of the process tree of Sys. Obviously, $seqproc(Sys) \subseteq proc(Sys)$.

Let us denote all actions occurring in the syntactical specification of a sequential process $P \in seqproc(Sys)$ as $Act(P)$. We can extend this definition to an arbitrary process $X \in proc(Sys)$ by writing $Act(X) = \bigcup Act(P_i)$, where the union is over those sequential processes P_i that are in the subtree of X. For a sequential process P, the fact that $a \in Act(P)$ means that P (considered in isolation) can actually at some point in its dynamic behaviour perform action a. However, for a process $X \in proc(Sys) \setminus seqproc(Sys)$, the fact that $a \in Act(X)$ does not necessarily mean that X can actually perform action a. As an example, think of $X = P \,||_a\, Q$, where $a \in Act(P)$ but $a \notin Act(Q)$. As another example, think of the same X where $a \in Act(P)$ and $a \in Act(Q)$ but no combined state is reachable in which both P and Q can perform action a. Therefore, we define $Act_{perf}(X) \subseteq Act(X)$ to be those actions that X (considered in isolation) can actually perform. While $Act(X)$ is a purely syntactical concept, $Act_{perf}(X)$ is a behavioural concept.

---

[1] Unlike, e.g., the process algebra CCS which has two-way synchronisation [12].

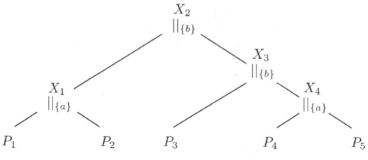

Fig. 1. $(P_1 \,||_{\{a\}}\, P_2) \,||_{\{b\}}\, (P_3 \,||_{\{b\}}\, (P_4 \,||_{\{a\}}\, P_5))$

Given two proceses $X, Y \in proc(Sys)$, we say that X and Y are disjoint if and only if they do not share any part of the process tree of Sys. Inside the disjoint processes X and/or Y, different actions (from $Act(X)$ and $Act(Y)$) may take place, among them the specific action a, say. Synchronisation on action a between X and Y is possible if and only if the root of the smallest subtree containing both X and Y is of type $||_a$. Maximal $||_a$-rooted subtrees are called a-scopes, as formalized in the following definition.

Definition 3. *Let $a \in Act$. An a-scope within an SPA system Sys is a subtree rooted at a node of type $||_a$, provided that on all nodes on the path from that node to the root of Sys there is no further synchronisation on action a (i.e. all nodes on that path, including the root, are of type $||_{\neg a}$).*

Furthermore, as a special case, if $P \in seqproc(Sys)$ and there is no a-synchronisation on the path from P to the root of the process tree of Sys, we say that P by itself is an a-scope.

For example, in the system shown in Fig. 1, subtrees rooted at X_1 and X_4 are a-scopes, and sequential process P_3 is also an a-scope. The only b-scope of this system is at the root of the system, i.e. X_2.

Note that, according to this definition, a-scopes are always maximal, i.e. an a-scope can never be a proper subset of another a-scope. Clearly, if the root node of Sys requires synchronisation on action a, then the whole Sys is a single a-scope. Synchronisation via action a is impossible between two distinct a-scopes. But even within a single a-scope, not all processes can/need to synchronise on action a. The following definition answers the question (from the perspective of a sequential process P) which processes cannot/may/must synchronise with an a-transition in process P.

Definition 4. *For $a \in Act$, consider the a-transitions within process $P \in seqproc(Sys)$. Let $X \in proc(Sys)$ be such that P and X are disjoint (i.e. that P is not part of X). Let r be the root of the smallest subtree that contains both P and X.*

(a) $X \in N_{cannot}(Sys, P, a)$ iff r is of type $||_{\neg a}$.

(b) $X \in N_{may}(Sys, P, a)$ iff r is of type $\|_a$ but on the path[2] from r to X there exists a node of type $\|_{\neg a}$.

(c) $X \in N_{must}(Sys, P, a)$ iff r is of type $\|_a$ and on the path from r to X all nodes are of type $\|_a$.

Remark 1. Note that for a process X it is possible that $X \in N_{may}(Sys, P, a)$ or $X \in N_{must}(Sys, P, a)$ even if $a \notin Act_{perf}(X)$ (or even $a \notin Act(X)$), which of course means that X will never be able to synchronise on action a. Related to this observation, note further that a process $X \in N_{may}(Sys, P, a)$ could actually be forced to synchronise with P on action a (i.e. it could be that X must synchronise with P on a, even though $X \notin N_{must}(Sys, P, a)$). For example, if $Sys = P \|_a (Q \|_{\neg a} R)$ then $Q \in N_{may}(Sys, P, a)$ and $R \in N_{may}(Sys, P, a)$, but if $a \notin Act(Q)$ then P always needs R as a synchronisation partner on action a.

Lemma 1. *(a) The neighbourhood $N_{cannot}(Sys, P, a)$ is disjoint from N_{may} (Sys, P, a) and $N_{must}(Sys, P, a)$.*
(b) Every $X \in N_{may}(Sys, P, a)$ is a subtree of some $Y \in N_{must}(Sys, P, a)$.

Proof. Part (a) follows directly from the definition. Part(b): For given X, one such node Y is the node directly below r on the path from r (as defined in Definition 4) to X.

2.1 Moving Set and Participating Set

Given a system Sys constructed from n sequential processes $P_1, \ldots, P_n$, its global state is a vector $(s_1, \ldots, s_n)$ where s_i is the state of P_i. We follow the convention that the ordering of processes is given by the in-order (LNR) traversal of the process tree of Sys. A transition t in the flat transition system of Sys is given by

$$t = ((s_1, \ldots, s_n) \xrightarrow{a, \lambda_s} (s'_1, \ldots, s'_n))$$

where for at least one $k \in \{1, \ldots, n\}$ we have $s_k \neq s'_k$ and where the transition rate $rate(t) = \lambda_s$ is a function of the rates of the transitions of the participating processes. For such a transition t we introduce the following notation:

$$action(t) = a \qquad rate(t) = \lambda_s$$
$$source(t) = (s_1, \ldots, s_n) \qquad target(t) = (s'_1, \ldots, s'_n)$$
$$source_i(t) = s_i \qquad target_i(t) = s'_i$$

But which are actually the participating processes in the above transition t? For an a-transition t as above, we define the moving set $MS(t)$ as the set of those sequential processes whose state changes, i.e. $MS(t) = \{P_k \mid s_k \neq s'_k\}$. The complement of the moving set is called the stable set $SS(t)$, i.e. $SS(t) = \{P_1, \ldots, P_n\} \setminus MS(t)$.

---

[2] "Path" here means all nodes strictly between r and the root of X.

Since processes may contain selfloops and since synchronisation on self-loops is possible (and often used as a valuable feature to control the context of a transition), the participating set $PS(t)$ of transition t can also include processes which participate in t in an invisible way by performing a selfloop. Therefore $PS(t)$ can be larger than $MS(t)$, i.e. in general we have $MS(t) \subseteq PS(t)$. Processes in $SS(t)$ which *must* synchronise on a with one of the elements of $MS(t)$ must have an a-selfloop at their current state and must belong to $PS(t)$. Furthermore, processes in $SS(t)$ which *may* synchronise on a with one of the elements of $MS(t)$ and have an a-selfloop at their current state may also belong to $PS(t)$, provided that they are not in the N_{cannot}-neighbourhood of one of the processes of $MS(t)$. Altogether we get:

$$PS(t) = MS(t) \cup$$
$$\left\{ P_i \in SS(t) \mid \right.$$
$$\left(\exists P_j \in MS(t) : \right.$$
$$(P_i \in N_{must}(Sys, P_j, a)$$
$$\vee \left(P_i \in N_{may}(Sys, P_j, a) \wedge (\text{selfloop } s_i \xrightarrow{a, \lambda_i} s_i \text{ exists and is enabled in } source(t)) \right) \right)$$
$$\left. \wedge \left(\not\exists P_j \in MS(t) : P_i \in N_{cannot}(Sys, P_j, a) \right) \right\}$$

The condition "selfloop ... is enabled in $source(t)$" means that the selfloop in P_i can actually take place in the source state of transition t, i.e. it is not blocked by any lacking synchronisation partner(s). Note that for the case $P_i \in N_{must}(Sys, P_j, a)$ there obviously exists a selfloop in process P_i, but this existence is implicit, so we do not have to write it down.

Using an example we show why the definition of $PS(t)$ needs to be so complicated, in particular why being in the *may* neighbourhood of a moving component and having a selfloop is not enough to become a participating component. For the system shown in Fig. 2, we wish to find $PS(t)$ where $t = ((1, 1, 3, 1, 2) \xrightarrow{a} (2, 1, 3, 1, 2))$. P_1 is the only moving component, and assume that there are a-selfloops in state 1 in P_2 and also in state 3 in P_3, but that there are no a-selfloops in state 1 of P_4 and in state 2 of P_5. $P_2 \in N_{may}(Sys, P_1, a)$ is in $PS(t)$, since its selfloop can take place without hindrance, whereas $P_3 \in N_{may}(Sys, P_1, a)$ is not included in $PS(t)$, since its selfloop, although it exists, is not enabled in the source state of transition t (it would need P_4 or P_5 as a synchronisation partner).

2.2   Involved Set

In addition to the Participating Set $PS(t)$ of a transition t, we also need to define the Involved Set $IS(t)$ which can be larger than $PS(t)$, since it also contains those processes which may synchronise on action a with one of the processes in $PS(t)$ (in another transition t'), and so on, inductively. Formally:

Definition 5. *For a transition t with $action(t) = a$ we define the Involved Set*
$$IS(t) = PS(t) \cup \left\{ P_k \in seqproc(Sys) \mid \exists P_j \in IS(t) : \left(P_k \in N_{may}(Sys, P_j, a) \right) \right\}.$$

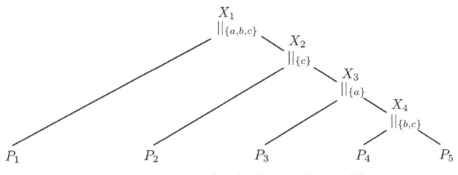

Fig. 2. $P_1 \parallel_{\{a,b,c\}} (P_2 \parallel_{\{c\}} (P_3 \parallel_{\{a\}} (P_4 \parallel_{\{b,c\}} P_5)))$

So $IS(t)$ represents the convex hull of the N_{may}-neighbourhood of one of the participating processes. That's why after the existential quantor in the definition we have to write $P_j \in IS(t)$ instead of only $P_j \in PS(t)$.

In some cases $IS(t) = PS(t)$, but it can be easily shown by example that $IS(t)$ may be a strict superset of $PS(t)$. Consider the system

$$Sys = (P_1 \parallel_{\neg a} P_2) \parallel_a (P_3 \parallel_a P_4)$$

and the transition $t = ((s_1, s_2, s_3, s_4) \xrightarrow{a, \lambda_s} (s_1', s_2, s_3', s_4))$. The moving set is $MS(t) = \{P_1, P_3\}$, and there is obviously a selfloop in P_4 of the form $s_4 \xrightarrow{a, \lambda_4} s_4$, so the participating set is $PS(t) = \{P_1, P_3, P_4\}$. However, P_2 is also (indirectly) involved since it is possible that in some other transition P_3 (and P_4) will synchronise on action a with P_2. More concretely: The transitions $s_3 \xrightarrow{a, \lambda_3} s_3'$ (in P_3) and $s_4 \xrightarrow{a, \lambda_4} s_4$ (in P_4) may synchronise with $s_2' \xrightarrow{a, \lambda_2} s_2''$ (in P_2) (for some states s_2' and s_2'' of P_2). Therefore we get $IS(t) = \{P_1, P_2, P_3, P_4\}$. This will be important for our rate lifting algorithm (Sect. 4), since if we didn't take the involvement of P_2 into account, we might change some rates in P_3 and/or P_4 which would have side effects on other transitions. This means that we have to set up a system of equations involving all four processes.

The following lemma establishes the connection between a transition's Involved Set $IS(t)$ (a behavioural concept) and the a-scope from Definition 3, which latter is a structural concept.

Lemma 2. *For a transition t of the SPA system Sys, with $action(t) = a$, let r be the root of the smallest tree containing all processes of $IS(t)$.*

(i) Then r is a node of type $\parallel_a$.
(ii) There is no other node of type $\parallel_a$ "above" r (i.e. on the path from r to the root of Sys).
(iii) The Involved Set $IS(t)$ is exactly the set of all sequential processes in the subtree rooted at r.
(iv) The Involved Set $IS(t)$ is exactly the set of sequential processes in the a-scope rooted at r. So, in a sense, the Involved Set and the a-scope are equal.

Proof. (i) Assume that r was of type $||_{\neg a}$. Then no process $P_l \in IS(t)$ in the
left subtree of r could synchronise (on action a) with any process $P_r \in IS(t)$
in the right subtree of r, which contradicts the fact that the set $IS(t)$ contains
processes in both subtrees of r.

(ii) Furthermore, assume that there is another node r_2 of type $||_a$ on the path
from r to the root of Sys. Then any a transition in one of the processes of
$IS(t)$ would have to synchronise with some process in the other subtree of
r_2, which means that the subtree rooted at r does actually not contain all
processes of $IS(t)$, which is a contradiction.

(iii) Assume that there is a sequential process P_{not} in the left subtree of the tree
rooted at r such that $P_{not} \notin IS(t)$. We know that there exists a sequential
process P_r in the right subtree of the tree rooted at r such that $P_r \in IS(t)$.
Then, since according to (i) r is of type $||_a$, either $P_{not} \in N_{may}(Sys, P_r, a)$
or $P_{not} \in N_{must}(Sys, P_r, a)$. But from this it follows that P_{not} would have
to be in $IS(t)$, which is a contradiction. A symmetric argument holds if
we assume that there is a sequential process P_{not} in the right subtree of
the tree rooted at r. Furthermore, any sequential process P_{not} not in the
subtree rooted at r cannot be in $IS(t)$ because according to (ii) there is no
a-synchronisation above r.

(iv) This is an immediate consequence of (i)–(iii).

Lemma 3. *For two transitions t_1 and t_2 with $action(t_1) = action(t_2)$, if
$IS(t_1) \cap IS(t_2) \neq \emptyset$ then $IS(t_1) = IS(t_2)$.*

Proof. This follows directly from the closure property of the IS definition.

3 "Parallel" Transitions and Relevant Selfloop Combinations

3.1 Multi-transition System

It is well known that the semantic model underlying an SPA specification is
actually a *multi*-transition system [5,6]. This is usually flattened to an ordinary
transition system by adding up the rates of "parallel" transitions, i.e. transitions
which have the same source state, the same target state and the same action
label. Thus a transition within Sys may be the aggregation of more than one
transition. As an example, consider the system $Sys = P \,||_a (Q \,||_{\neg a} R)$ and the
transition $t = ((s_1, s_2, s_3) \xrightarrow{a, \lambda_s} (s_1', s_2, s_3))$. The moving set is $MS(t) = \{P\}$,
but the participating set must be larger. Assume that Q has a selfloop $s_2 \xrightarrow{a, \lambda_2} s_2$
and that R has a selfloop $s_3 \xrightarrow{a, \lambda_3} s_3$. Since Q and R do not synchronise on a,
only one of those two selfloops synchronises with $s_1 \xrightarrow{a, \lambda_1} s_1'$ at a time, but
both selfloops may synchronise with the a-transition in P. This yields the two
"parallel" transitions

$$((s_1, s_2, s_3) \xrightarrow{a, \lambda_{12}} (s_1', s_2, s_3)) \quad \text{and} \quad ((s_1, s_2, s_3) \xrightarrow{a, \lambda_{13}} (s_1', s_2, s_3))$$

(where λ_{12} is a function of λ_1 and λ_2, and likewise for λ_{13}) which are aggregated to the single transition $((s_1, s_2, s_3) \xrightarrow{a, \lambda_{12}+\lambda_{13}} (s_1', s_2, s_3))$, so $\lambda_s = \lambda_{12} + \lambda_{13}$. As an anticipation of Eq. 1 in Sect. 4, let us mention that in this situation our rate lifting algorithm would create the equation $x_{s_1 s_1'}^{(P)} x_{s_2 s_2}^{(Q)} + x_{s_1 s_1'}^{(P)} x_{s_3 s_3}^{(R)} = \lambda_s \cdot f$.

3.2   Calculating Relevant Selfloop Combinations

In the simple (and most common) case that none of the sequential processes in the SPA specification of Sys has any selfloops (and also no "parallel" transitions), we know that any transition of the flat transition system has only one single semantic derivation. In consequence, for the considered flat transition t it then holds that $PS(t) = MS(t)$. However, as discussed above, in the general case the flat transition system underlying a compositional SPA specification is actually a multi-transition system which gets flattened to an ordinary transition system by amalgamating "parallel" transitions. In order to cover this general case, in the lifting algorithm (see Sect. 4) we have to do the opposite: Instead of amalgamation, we need to deconstruct a flat transition into its constituents. I.e., given a flat transition (which is possibly amalgamated from parallel transitions), we need to find out the contributing transitions, in order to be able to construct the correct equation in the lifting algorithm (Eq. 1 in Sect. 4).

Consider the flat transition $t := ((s_1, \ldots, s_n) \xrightarrow{c, \gamma \cdot f} (s_1', \ldots, s_n'))$. We can determine its (non-empty) moving set $MS(t)$ and its participating set $PS(t)$, where we know that $MS(t) \subseteq PS(t)$. We are particularly interested in the processes from the set $(PS(t) \cap SS(t)) \setminus \bigcup_{P \in MS(t)} N_{must}(Sys, P, c)$, since these are exactly the processes that may (but not must) contribute to transition t. Certain combinations of these processes (which have selfloops, otherwise they wouldn't be in $PS(t)$) contribute to transition t. We call these combinations "relevant selfloop combinations (rslc)". Note that there are also selfloops in $\bigcup_{P \in MS(t)} N_{must}(Sys, P, c)$, but they are not part of rslc.

It remains to calculate rslc for transition t. For this purpuse, we define a function $rslc(t)$ which returns a set of sets of sequential processes, each such set describing a relevant selfloop combination. In the process tree of Sys, let r be the root node of the smallest subtree containing $PS(t)$. We know that r is either an inner node of type $||_c$ or a leaf (if r were an inner node of type $||_{\neg c}$, the participating set $PS(t)$ couldn't span both subtrees of r). Calling the recursive function in Fig. 3 by the top-level call $RSLC(t, r)$ delivers all the relevant selfloop combinations[3]. RSLC is called from the main rate lifting algorithm in Parts C and D (see Sect. 4).

Example: Consider the SPA specification $Sys = ((P \ ||_{\neg c} Q) \ ||_c (R \ ||_{\neg c} S)) \ ||_c (T \ ||_{\neg c} U)$ whose process tree is shown in Fig. 4, and the transition $t := ((s_P, s_Q, s_R, s_S, s_T, s_U) \xrightarrow{c, \gamma \cdot f} (s_P', s_Q, s_R, s_S, s_T, s_U))$. Obviously, the moving set is $MS(t) = \{P\}$, and if we assume that there are c-selfloops in

---

[3] $rslc(t)$, called by the algorithm, has one argument (a transition), but the recursive function $RSLC(t, n)$ has two arguments (a transition and a node of the process tree).

1: **Algorithm** RSLC (t, n)
2: // $t \in T$ is a transition
3: // n is a node of the process tree of Sys
4: // The algorithm returns a set of sets of sequential processes
5: // (each representing a relevant selfloop combination contributing to t)
6: **if** type$(n) = $ leaf **then**
7:     **if** $P_n \in (PS(t) \cap SS(t)) \setminus \bigcup_{P \in MS(t)} N_{must}(Sys, P, c)$ **then**
8:         // P_n denotes the process represented by leaf-node n
9:         // $P_n \in PS(t)$ ensures that P_n has a selfloop at its current state
10:         return $\{\{P_n\}\}$
11:         // a set containing a singleton set is returned
12:     **else**
13:         return $\{\emptyset\}$
14:         // the set containing the empty set is returned
15:     **end if**
16: **else if** type$(n) = \|_c$ **then**
17:     return $\{C_1 \cup C_2 \mid C_1 \in RSLC(t, \text{lchild}(n)) \wedge C_2 \in RSLC(t, \text{rchild}(n))\}$
18:     // all combinations of left and right subtree
19: **else**
20:     // it holds that type$(n) = \|_{\neg c}$
21:     return $\{C \mid C \in RSLC(t, \text{lchild}(n)) \vee C \in RSLC(t, \text{rchild}(n))\}$
22:     // the (disjoint) union of left and right subtree
23: **end if**

Fig. 3. Function for computing the relevant selfloop combinations

states s_R, s_S, s_T and s_U (in all of them!), the participating set is $PS(t) = \{P, R, S, T, U\}$. So transition t can be realised as any combination of a self-loop in R or S with a selfloop in T or U, thus the algorithm will find the set of relevant selfloop combinations $\{\{R, T\}, \{R, U\}, \{S, T\}, \{S, U\}\}$.

Anticipating once again Eq. 1 from Sect. 4, this set of relevant selfloop combinations would lead to the desired equation

$$x^{(P)}_{s_P s'_P} x^{(R)}_{s_R s_R} x^{(T)}_{s_T s_T} + x^{(P)}_{s_P s'_P} x^{(R)}_{s_R s_R} x^{(U)}_{s_U s_U} + x^{(P)}_{s_P s'_P} x^{(S)}_{s_S s_S} x^{(T)}_{s_T s_T} + x^{(P)}_{s_P s'_P} x^{(S)}_{s_S s_S} x^{(U)}_{s_U s_U} = \gamma \cdot f$$

Alternatively, if we assumed that the participating set was smaller, say $PS(t) = \{P, R, S, T\}$ (i.e. if there were no c-selfloop at s_U), then the algorithm would find a smaller set of relevant selfloop combinations, namely $\{\{R, T\}, \{S, T\}\}$, leading to the simpler equation $x^{(P)}_{s_P s'_P} x^{(R)}_{s_R s_R} x^{(T)}_{s_T s_T} + x^{(P)}_{s_P s'_P} x^{(S)}_{s_S s_S} x^{(T)}_{s_T s_T} = \gamma \cdot f$.

4   Lifting Algorithm

Our new lifting algorithm processes the transitions whose rates are to be modified in a one by one fashion. It is, however, not strictly one by one, since in many situations a whole set of "related" transitions is taken into account together with the currently processed transition. The algorithm consists of four parts named A, B, C and D as shown in Fig. 5. In part A, for a transition whose involved set

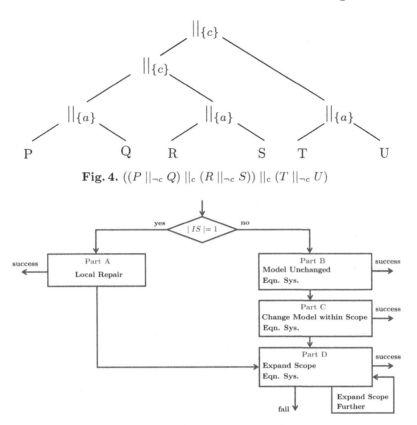

Fig. 4. $((P \, ||_{\neg c} \, Q) \, ||_c \, (R \, ||_{\neg c} \, S)) \, ||_c \, (T \, ||_{\neg c} \, U)$

Fig. 5. Overview of the algorithm

consists of only one single sequential process, the algorithm first tries to change its rate by local repair, which means changing the rate locally in exactly this sequential process. Local repair will fail, however, if two flat transitions which both originate from the same local transition have different modification factors. In part B, which is the starting point for transitions whose involved set contains at least two processes, the algorithm creates a system of nonlinear equations and tries to solve it. This system of equations covers all transitions with the same action label and the same involved set, i.e. all these transitions are dealt with simultaneously in one system of equations. The basic idea behind the system of equations is to consider all involved local rates as variables whose values are to be determined. Part C, entered upon failure of Part B, is the first part where the system specification is modified by augmenting some synchronisation sets and inserting selfloops, all within the current c-scope. These modifications are done in such a way that the global transition system is not changed. Again, like in part B, the algorithm creates a set of nonlinear equations (but now the system of equations is larger since the model has been modified) and tries to solve it. If the previous steps have failed, Part D tries to expand the scope, by modifying the system in a larger scope than the current involved set. This means that the involved set is artificially augmented by adding action c to the synchronisation set at a higher node. Again, a similar but even larger system of equations of the same type is constructed. However, even this system of equations may not have a solution, in which case the desired rate lifting has turned out to be impossible.

4.1 Spurious Transitions

As we have seen, in certain situations the rate lifting algorithm needs to change the synchronisation structure of the given system, i.e. it will change an inner node of type $||_{\neg c}$ to a node of type $||_c$. Clearly, this needs to be done with great care, since such a step will – in general – change the behaviour of the system. Therefore the algorithm, before adding action c to a synchronisation set, has to ensure that no spurious transitions will be generated. Spurious transitions (sp. tr.) are extra, superfluous transitions not present in the original system, and therefore incorrect. Furthermore, after action c has been added to a synchronisation set, the algorithm also has to ensure that all transitions in the original system are still possible (it could easily be that a previously existing transition now lacks a synchronisation partner in the newly synchronised system). For that purpose, the algorithm inserts selfloops into the sequential components, wherever necessary. There are actually two types of spurious transitions:

(A) Superfluous transitions which appear when two previously c-non-synchronised components become synchronised over action c.
(B) Superfluous transitions which appear when a new c-selfloop is inserted into a sequential process which is in the c-must- or c-may-neighbourhood of another process.

Overall, the algorithm guarantees that even though the synchronisation structure of the system may be altered and artificial selfloops are inserted, the set of reachable states and the set of transitions remain the same.

4.2 Detailed Description of the Algorithm

The full pseudocode of the new rate lifting algorithm can be found in the technical report [14]. The arguments of the algorithm are the SPA system Sys and its flat transition system T (a set of transitions), the set of transitions whose rate is to be modified $T_{mod} \subseteq T$ as well as a function $factor$ that returns, for each transition in T_{mod}, its modification factor[4]. For transitions not in T_{mod}, the modification factor is supposed to be 1.

   In each iteration of the outer while-loop, the algorithm picks one of the remaining transitions from T_{mod}, called $\hat{t}$ with action label called c, and processes it (possibly together with other transitions that have the same action label).

---

[4] Thus, this presentation of the algorithm addresses model repair rather than rate reverse engineering (cf. Sect. 1).

(Part A) Local repair: If the involved set of the currently processed transition $\hat{t}$ consists of only a single process, the algorithm tries to adjust the rate of exactly one transition in that process. But this will only work if all transitions where this process makes the same move have the same, common modification factor.

(Part B) System of equations for T_c: If the involved set of the currently processed transition $\hat{t}$ consists of two or more processes, all transitions with the same involved set and the same action as $\hat{t}$ are processed together. This set of transitions is denoted T_c. For every transition $t \in T_c$, the algorithm determines its participating set $PS(t)$, calculates the relevant selfloop combinations (rslc) and from this information creates a nonlinear equation

$$\sum_{C \in rslc(t)} \prod_{P \in MS(t)} x^{(P)}_{s_P s'_P} \prod_{\substack{Q \in PS(t) \setminus MS(t) \\ \wedge \, \exists P \in MS(t): \, Q \in N_{must}(Sys,P,c)}} x^{(Q)}_{s_Q s_Q} \prod_{R \in C} x^{(R)}_{s_R s_R} = \gamma \cdot f \quad (1)$$

where the x's are the unknown rates of the participating processes (some of which are rates of selfloops, if such exist in the system). The superscript of variable $x^{(\cdot)}_{\cdot\cdot}$ identifies the sequential process, and the subscript denotes the source/target pair of states. The equation reflects the fact that the rates of all synchronising processes are multiplied[5], and that the total rate is obtained as the sum over all possible relevant selfloop combinations. Afterwards, this system of equations is solved, and if a solution exists, all c-transitions in the current c-scope have been successfully dealt with. We would like to point out that, if for some transition t the participating set $PS(t)$ is equal to its moving set $MS(t)$, then the resulting equation has a much simpler form

$$x^{(P_1)}_{s_1 s'_1} \cdot x^{(P_2)}_{s_2 s'_2} \cdot \, \cdots \, \cdot x^{(P_k)}_{s_k s'_k} = \gamma \cdot f$$

(assuming that $|MS(t)| = k$), since in this case, there are no selfloops involved, and therefore also no combinations of selfloops to be considered.

(Part C) Expanding the Context of c-Transitions by Synchronising with More Processes and Inserting Artificial Selfloops Within the Current c-Scope: If the system of equations constructed in part (B) for the set T_c had no solution, it is the strategy of the algorithm to involve more processes (for the moment only from the current c-scope), since this opens up more opportunity for controlling the context of these c-transitions, and thereby controlling their rates. In this part of the algorithm, $action(\hat{t}) = c$ is added to the synchronisation set at each node of type $||_{\neg c}$ in the current c-scope, except where this would lead to spurious transitions (of type A or type B). These tasks of the algorithm are outsourced to a function TRYSYNC. Checking for spurious transitions of type

---

[5] Multiplication of rates is a de facto standard for Markovian SPAs, as implemented, for example, by the tools PRISM and STORM. If the rate resulting from the synchronisation of two or more processes were defined other than the product of the participating rates, the equation would have to be changed accordingly, but apart from this change, the lifting algorithm would still work in the same way.

A is done by checking all source states of transitions in the current T_c, making sure that there are no concurrently enabled c-transitions in newly synchronised subprocesses. After adding action c to some synchronisation sets, we also have to make sure that all transitions originally in T_c can still occur, i.e. that they have not been disabled by the new synchronisations. This is also done in function TRYSYNC, by inserting the necessary selfloops in those processes which are now newly synchronising on action c, provided that those new selfloops do not lead to the existence of spurious transitions (of type B). The steps just described guarantee that the modified system Sys' has exactly the same set of transitions as the original system Sys (qualitatively), but it remains to find the correct rates of all c-transitions in the involved processes. For this purpose, a similar (but larger) system of equations as in part (B) is set up and solved.

(Part D) Expanding the Involved Set by Moving the Current Root Upwards: It is possible that the systems of equations constructed in part (B) and thereafter in part (C) both have no solution. In this case, the algorithm seeks to expand the current c-scope by moving its root up by one level (unless the root of the overall system has already been reached). Again, it needs to be ensured that no spurious transitions would be created from this step, which is again done with the help of function TRYSYNC.

5   Experimental Result: Cyclic Server Polling System

This section considers – as a case study – the Cyclic Server Polling System from the PRISM CTMC benchmarks, originally described in [7] as a GSPNs. It is a system where a single server polls N stations and provides service for them in cyclic order. The SPA representation of this system is:

$$Sys = Server \,||_{\Sigma_s} (Station_1 \,||\, Station_2 \,||\ldots||\, Station_N)$$

where $\Sigma_s = \{loop_{ia}, loop_{ib}, serve_i \mid i = 1\ldots N\}$. Assume that for each $loop_{1a}$-transition t in the combined flat model, a modification factor $f(t) \neq 1$ is given. Using our new lifting algorithm, we lift this model repair information to the components. The modification factors f are chosen in such a way that Part B of the algorithm will not find a solution. In part C of the algorithm, it turns out that action $loop_{1a}$ can be added to all $||_{\neg loop_{1a}}$-nodes of the process tree, since it does not cause spurious transitions. Consequently, $loop_{1a}$-selfloops are added to all the states of $Station_2, \ldots, Station_N$, leading to the modified SPA model

$$Sys' = Server||_{\Sigma_s}(Station_1||_{loop_{1a}}Station_2'||_{loop_{1a}} \cdots ||_{loop_{1a}}Station_N')$$

where the stations with added selfloops are shown by $Station_i'$. With the chosen modification factors, a solution can be found in Part C of the algorithm. This example is a scalable model where the state space increases with the number of stations N. Note that the model contains symmetries, but the considered rate lifting problem is not symmetric, since only $Station_1$ and the $Server$ participate in the $loop_{1a}$-transitions.

N	6	7	8	9	10	11
Total number of states	576	1344	3072	6912	15360	33792
Total number of transitions	2208	5824	14848	36864	89600	214016
Number of $loop_{1a}$-transitions	32	64	128	256	512	1024

Table 1. Model statistics of the combined model for different numbers of stations

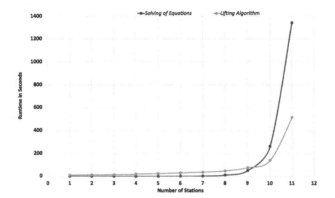

Fig. 6. Runtime comparison for different number of stations

Table 1 shows the model statistics for different numbers of stations. The last row of the table (number of $loop_{1a}$-transitions) equals the number of equations, each of the 2^{N-1} equations containing the product of $N+1$ unknown variables. The whole system of equations has $(N-1)*2+2$ variables stemming from $(N-1)*2$ newly added $loop_{1a}$-selfloops plus two original $loop_{1a}$-transitions (in the *Server* and *Station$_1$*). Figure 6 shows the required times to run our rate lifting algorithm (implemented as a proof-of-concept prototype in Matlab [8]) and to solve the system of equations (done by Wolfram Mathematica [9]) for different values of N^6. For large N, the time for equation solving by far dominates the runtime of our algorithm (by a factor of 2.61 for $N = 11$). As shown in the figure, the runtimes grow exponentially, which is not surprising since the number of equations increases exponentially.

6 Correctness and Existence Considerations

Correctness: It must be guaranteed that a solution found by the algorithm is correct, which means that the modified system (with the calculated rates, possibly modified synchronisation sets and added selfloops) possesses the same transition system, just with the transition rates modified as desired. Once a solution has been found by the algorithm, it is easy to check its correctness by simply constructing the flat transition system for the modified system and comparing

---

^{6} Executed on a standard laptop with Intel Core i7-8650U CPU@ 1.90 GHz–2.11 GHz.

it to the desired transition system. However, one can also constructively argue for the correctness of the algorithm: In Part A, if the condition for local repair is fulfilled, only the transition rate of a single transition in one of the sequential processes is changed, resulting in the change of a well-defined set of global flat transitions, all having the same modification factor, which is the intended result. Parts B, C and D each work by setting up and solving a nonlinear system of equations relating to the original (Part B) resp. carefully modified SPA system (Parts C and D). Each of these systems of equations precisely reflects the synchronisation of sequential SPA processes within a certain scope, taking into account all transitions with action label c that take place in that scope, and making sure that the resulting rates of those transitions are all as desired (thereby considering all relevant selfloop combinations). All procedures in Parts B, C and D affect only a certain scope of the overall SPA system, so it is enough to ensure correctness for such a local context. If in Parts C and D the model is adjusted (by augmenting synchronisation sets and inserting artificial selfloops), care is taken that this will not affect the structure of the low-level transition system. Thus, since each individual step of the algorithm is correct, we can conclude by induction that the total effect of multiple steps is also correct.

Existence: If the lifting algorithm doesn't find a solution, is it really guaranteed that none exists? We briefly give the basic line of argument: If we start with part A of the algorithm and if local repair fails, this happens because the same local transition (involving only a single sequential process) should be executed in different contexts with different modification factors (i.e. different rates), which is not possible in the unmodified system. To solve this problem, some "controlling" context needs to be added, so we synchronise the process with its neighbouring processes (where selfloops are added at specific states) in a subtree of a certain height, which leads to a set of equations in part D of the algorithm. We keep expanding the context until either a solution has been found or the root of the system has been reached, which means that the algorithm uses its full potential. Alternatively, if we start with part B (because the involved set of the currently processed transition is already larger than one), we first search for a solution in the "local" context, i.e. in the current involved set, which is a subtree of the system. First we try to leave the model unchanged, which also leads to a system of equations. If it turns out that this system of equations has no solution, we need to include more degrees of freedom into the equations. This is first done within the current scope (by synchronising with as many processes as possible, albeit all from within this same scope) in part C. If this also fails, even more degrees of freedom can be added by expanding the current scope, leading us again to part D. In total, the algorithm uses all possible degrees of freedom, since at every step it involves all processes, except those whose involvement would cause damage (in the sense that spurious transitions would occur).

7   Conclusion

In this paper, we have studied some novel structural concepts of Markovian SPA, which enabled us to formulate an algorithm for the lifting of rate information from the flat low-level transition system of a general SPA model to its components. The algorithm works for SPA specifications with an arbitrary structure and any number of components. We have also presented a small case study that illustrates the practical use of the algorithm and remarked on the correctness and optimality of the algorithm. As future work, we are planning to develop improved implementation strategies for the algorithm. Another important point for future work is to characterise a priori the set of problem instances for which a solution to the rate lifting problem exists.

References

1. Bartocci, E., Grosu, R., Katsaros, P., Ramakrishnan, C.R., Smolka, S.A.: Model repair for probabilistic systems. In: Abdulla, P.A., Leino, K.R.M. (eds.) TACAS 2011. LNCS, vol. 6605, pp. 326–340. Springer, Heidelberg (2011). https://doi.org/10.1007/978-3-642-19835-9_30
2. Bernardo, M.: Theory and application of extended Markovian process algebra. Ph.D. thesis, University of Bologna (1999)
3. Chen, T., Hahn, E.M., Han, T., Kwiatkowska, M., Qu, H., Zhang, L.: Model repair for Markov decision processes. In: 2013 International Symposium on Theoretical Aspects of Software Engineering, pp. 85–92 (2013)
4. Dehnert, C., Junges, S., Katoen, J.-P., Volk, M.: A storm is coming: a modern probabilistic model checker. In: Majumdar, R., Kuncak, V. (eds.) Computer Aided Verification. CAV 2017. LNCS, vol .10427. Springer, Cham (2017). https://doi.org/10.1007/978-3-319-63390-9_31
5. N. Götz. Stochastische Prozessalgebren: integration von funktionalem Entwurf und Leistungsbewertung Verteilter Systeme. Ph.D. thesis, University of Erlangen-Nuremberg (1994)
6. Hillston, J.: A Compositional Approach to Performance Modelling. Cambridge University Press, New York (1996)
7. Ibe, O., Trivedi, K.: Stochastic Petri net models of polling systems. IEEE J. Select. Areas Commun. **8**(9), 1649–1657 (1990)
8. The MathWorks, Inc., Matlab, Version 9.10 (2021)
9. Wolfram Research, Inc., Mathematica, Version 12.2.0.0. Champaign, IL (2021)
10. Kuntz, M., Siegle, M., Werner, E.: Symbolic performance and dependability evaluation with the tool CASPA. In: Núñez, M., Maamar, Z., Pelayo, F.L., Pousttchi, K., Rubio, F. (eds.) FORTE 2004. LNCS, vol. 3236, pp. 293–307. Springer, Heidelberg (2004). https://doi.org/10.1007/978-3-540-30233-9_22
11. Kwiatkowska, M., Norman, G., Parker, D.: PRISM 4.0: verification of probabilistic real-time systems. In: Gopalakrishnan, G., Qadeer, S. (eds.) CAV 2011. LNCS, vol. 6806, pp. 585–591. Springer, Heidelberg (2011). https://doi.org/10.1007/978-3-642-22110-1_47
12. Milner, R. (ed.): A Calculus of Communicating Systems. LNCS, vol. 92. Springer, Heidelberg (1980). https://doi.org/10.1007/3-540-10235-3

13. Pathak, S., Ábrahám, E., Jansen, N., Tacchella, A., Katoen, J.-P.: A greedy app-roach for the efficient repair of stochastic models. In: Havelund, K., Holzmann, G., Joshi, R. (eds.) NFM 2015. LNCS, vol. 9058, pp. 295–309. Springer, Cham (2015). https://doi.org/10.1007/978-3-319-17524-9_21
14. Siegle, M., Soltanieh, A.: Rate lifting for stochastic process algebra: exploiting structural properties. https://arxiv.org/abs/2206.14505 (2022)
15. Soltanieh, A., Siegle, M.: It sometimes works: a lifting algorithm for repair of stochastic process algebra models. In: Hermanns, H. (ed.) MMB 2020. LNCS, vol. 12040, pp. 190–207. Springer, Cham (2020). https://doi.org/10.1007/978-3-030-43024-5_12
16. Soltanieh, A., Siegle, M.: Solving systems of bilinear equations for transition rate reconstruction. In: Hojjat, H., Massink, M. (eds.) FSEN 2021. LNCS, vol. 12818, pp. 157–172. Springer, Cham (2021). https://doi.org/10.1007/978-3-030-89247-0_11

End-to-End Statistical Model Checking
for Parametric ODE Models

David Julien[1,2], Guillaume Cantin[1], and Benoît Delahaye[1(✉)]

[1] Nantes Université, École Centrale Nantes, CNRS,
LS2N, UMR 6004, 44000 Nantes, France
`benoit.delahaye@univ-nantes.fr`
[2] École Normale Supérieure de Lyon, 69007 Lyon, France

Abstract. We propose a simulation-based technique for the verification of structural parameters in Ordinary Differential Equations. This technique is an adaptation of Statistical Model Checking, often used to verify the validity of biological models, to the setting of Ordinary Differential Equations systems. The aim of our technique is to search the parameter space for the parameter values that induce solutions that best fit experimental data under variability, with any metrics of choice. To do so, we discretize the parameter space and use statistical model checking to grade each individual parameter value w.r.t experimental data. Contrary to other existing methods, we provide statistical guarantees regarding our results that take into account the unavoidable approximation errors introduced through the numerical resolution of the ODE system performed while simulating. In order to show the potential of our technique, we present its application to two case studies taken from the literature, one relative to the growth of a jellyfish population, and another concerning a prey-predator model.

Keywords: Statistical Model Checking · ODE models · structural parameters

1 Introduction

All scientific branches share the common concept of modeling. When a scientist studies a real-life system, the first step he or she goes through is to build a model that gathers all the existing knowledge of the target system. This model is then used as a proxy of the system it represents in order to analyze it, perform simulation or predictions. In several fields, such as Biology, Chemistry, Physics or Engineering, models do not represent a single system but are instead an abstraction for a *family* of systems that share common traits but might exhibit some internal variability. This internal variability can either be left out by considering that the model represents the "average" individual in the family, or taken into account inside of the model through the use of non-determinism, probabilities or parametricity.

© Springer Nature Switzerland AG 2022
E. Ábrahám and M. Paolieri (Eds.): QEST 2022, LNCS 13479, pp. 85–106, 2022.
https://doi.org/10.1007/978-3-031-16336-4_5

When considering parametric models, scientists have to go through a phase of parameterization, which consists in confronting the model with experimental observations of the (family of) system(s) it represents in order to find the parameter values that best fit this (family of) system(s). In most cases, parameterization techniques are *deterministic* [21,24]. They lead to deterministic parameter values that best fit the experimental data, i.e. producing the best fit for the "average" individual. In this paper, we instead focus on a technique that allows to select parameter values that best fit *under variability*, i.e. that produce the best *probabilistic* fit for the *whole family*.

Parameterization, or *parameter synthesis* has been the topic of many works in the context of probabilistic systems [6,9–11,13]. Symbolic techniques such as parametric model checking [1,5] are often difficult to use in practice because they require automata-based models while real-life models are often expressed either with computer programs or with differential equation models. Statistical Model Checking (SMC) [14], on the other hand, is a simulation-based technique that allows to estimate, with formal guarantees, the probability that a given (probabilistic) model satisfies a given property. Because it is simulation-based, it can be applied to any stochastic model for which simulations can be performed. SMC has been successfully applied to perform parameterization of real-life models expressed using several formalisms such as parametric Markov chains [2], parametric Python programs [20], or even parametric Ordinary Differential Equation systems (ODEs) [15]. Unfortunately, the formal guarantees obtained through SMC are linked to the simulation space (i.e. the produced traces) and not to the original model itself. When the model consists in sets of ODEs, as in [15], numerical resolution methods are used in order to solve the ODEs and perform simulations, which means that the formal guarantees obtained through SMC cannot apply to the original ODE model.

In this paper, our main contribution is to bridge the gap between the original ODE model and the results of the parameterization procedure by *combining the statistical guarantees of SMC with the global approximation error of standard numerical resolution methods*. As in [15], we consider ODE models with structural parameters. We assume that these models represent families of real-life systems that need to match some experimental data through simulation. We build on the logic proposed in [15] to express our properties of interest and also consider expected reward properties that might be of interest in practice. We use SMC to grade parameter values by estimating the expectation of a given reward function for these values while taking internal variability into account. Contrarily to what is done in [15], the accuracy of this estimation is guaranteed w.r.t. the original ODE model.

To illustrate our results, we perform the parameterization of two state-of-the-art models taken from the literature using our technique. In this context, and because modelers are often interested by this information in practice, we propose a global evaluation of the parameter space that allows us to get a complete picture of the adequacy of the parameter values w.r.t. the given experimental data. This choice is done by interest only, since our results are generic and could be applied to any search technique, such as the local ones performed in [15].

Intuition. To give an intuition of our contribution, we provide an informal summary of the method we present in this paper. Recall that, given a dataset relative to an experiment and a parametric ODE system, the objective is to find a solution to a parametric ODE system (i.e. parameter values) that satisfies a property φ w.r.t. the dataset, which is, given a distance $\delta > 0$, "the solution stays in a tunnel of radius δ around the experimental data"; we also want to acquire statistical guarantees on said result. The main issue is that we can only simulate our model by solving the ODE system using numerical resolution methods. Hence, we cannot directly verify whether exact solutions (z) of the system satisfy φ and instead have to rely on approximate solutions (y). We therefore proceed as follows: we start by discretizing the set of parameter values into a grid; we then evaluate each point of this grid using the procedure detailed below; finally, we use the resulting scores to select the "best" parameter values w.r.t φ. The score of a given parameter value λ is computed as follows, and illustrated in Figs. 1 and 2 in the context of the case study presented in Sect. 4.1.

1. We set the parameter value to λ. By a careful study of the ODE system, we give a bound on the distance ε between exact (z) and approximate (y) solutions. We emphasize that this bound depends on (1) the chosen resolution technique and (2) the chosen integration step. We show that this distance is uniformly stable w.r.t. internal variability around λ, but also that it can be uniformly bounded on the global set of solutions (i.e. independently of λ).
2. We propose two new properties φ_1 and φ_2 that will be verified on the approximate solutions y, and depend on the above distance. This amounts to changing the size of the tunnel around the experimental dataset. We compute (estimations of) the respective probabilities p_1 and p_2 and prove that the probability p that z satisfies φ lies between p_1 and p_2.
3. We provide statistical guarantees of our estimation, i.e. a confidence interval for our estimation of p, and use this estimation as the score for parameter value λ.

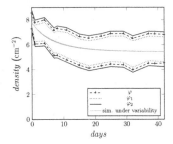
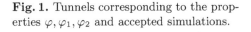

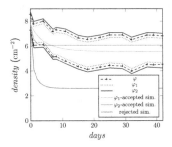

Fig. 1. Tunnels corresponding to the properties $\varphi, \varphi_1, \varphi_2$ and accepted simulations.

Fig. 2. φ-accepted, φ_2-accepted and rejected solutions.

It is worth noting that the underlying theory is generic: the integration method as well as the statistical estimation method can be chosen arbitrarily as long as they provide the usual guarantees. In this paper, we use Runge-Kutta and Monte-Carlo for the sake of example.

Outline. In Sect. 2, we introduce required preliminaries and notations for the rest of the paper. In Sect. 3, we state the main result of the paper, i.e. we compute the approximation error for ODE solutions, show that this error is uniformly stable, and provide the statistical guarantees for the estimation of the probabilities. In Sect. 4, we illustrate our approach on two case studies taken from the literature. Finally, we conclude in Sect. 5 and give perspectives for future work.

2   Background and Notations

In this section, we present the basic notations and definitions that will be used throughout the paper. More precisely, we recall the definition of an ODE, and present the logic used in the paper. Finally, we extend this logic by introducing reward functions.

2.1   ODE Preliminaries

First, we consider an evolution problem described by an Ordinary Differential Equation (ODE) of the form

$$\frac{\mathrm{d}z}{\mathrm{d}t}(t) = f\big(z(t), \boldsymbol{\lambda}\big), \quad t > 0. \tag{1}$$

In Eq. (1), the unknown function z is defined in $\mathbb{R}^+$ with values in $\mathbb{R}^n$; $\boldsymbol{\lambda} \in \mathbb{R}^m$ is a vector of parameters; f is a function defined on $\mathbb{R}^n \times \mathbb{R}^m$ with values in $\mathbb{R}^n$, whose regularity will be detailed below; n, m are positive integers. In the following, we write $z_i(t)$, $1 \leq i \leq n$, for the projection of $z(t)$ on its ith component. As mentioned in our introduction, Eq. (1) can model various real-world problems arising in life sciences. Our goal is to study some properties of the trajectories determined by Eq. (1), by developing an innovative model-checking framework suitable for the continuous dynamics of ODEs.

Here and for the rest of the paper, we fix an initial condition $z_0 \in \mathbb{R}^n$. Standard results of the theory of differential equations (see for instance [18]) ensure that, for any value of the parameter $\boldsymbol{\lambda} \in \mathbb{R}^m$, the Cauchy problem determined by Eq. (1) and the initial value $z(0) = z_0$ admits a unique solution, provided f is C^1 on $\mathbb{R}^n \times \mathbb{R}^m$; we denote by $z^{\boldsymbol{\lambda}}(t)$ the corresponding trajectory, which we assume to be defined on $[0, T]$ with $T > 0$. If the context is sufficiently clear, we may write $z(t)$ for short. As before, we write $z_i^{\boldsymbol{\lambda}}(t)$ (resp. $z_i(t)$), $1 \leq i \leq n$, for the projection on its ith component. We assume that the component λ_j of the parameter vector $\boldsymbol{\lambda} \in \mathbb{R}^m$ ($1 \leq j \leq m$) satisfies $\lambda_j \in [L_j^{\lambda}, U_j^{\lambda}]$, with real coefficients $L_j^{\lambda} < U_j^{\lambda}$ and we consider the compact sets $\mathbf{W}$ and INIT defined by

$$\mathbf{W} = \prod_{j=0}^{m} [L_j^{\lambda}, U_j^{\lambda}] \qquad (2) \qquad \qquad \mathsf{INIT} = \{z_0\} \times \mathbf{W}. \qquad (3)$$

For $\boldsymbol{\lambda} \in W$, we consider the Euclidean norm defined by $\|\boldsymbol{\lambda}\| = \left(\sum_{j=1}^{m} |\lambda_j|^2\right)^{1/2}$.

We assume that the trajectories of Eq. (1) starting from z_0 admit a rectangular invariant region, uniform w.r.t. the parameter $\boldsymbol{\lambda}$, that is $z_i(t, \boldsymbol{\lambda}) \in [L_i, U_i]$ for $t \in [0, T]$, with real coefficients $L_i < U_i$, for all $\boldsymbol{\lambda} \in \mathbf{W}$ and for $1 \leq i \leq n$. The global invariant region for z is written $\mathbf{V} = \prod_{i=0}^{n} [L_i, U_i]$.

Finally, we write TRAJ for the set of all potential trajectories of the solutions to our ODE system. Formally, $\mathsf{TRAJ} = \{z^{\boldsymbol{\lambda}}(t) \mid \boldsymbol{\lambda} \in \mathbf{W}\}$.

It is well-known that Eq. (1) determines time continuous trajectories, which moreover depend continuously on a variation of the initial condition z_0 and of the parameter $\boldsymbol{\lambda} \in \mathbf{W}$ (see for instance [18]). In Sect. 3, we will be interested in the variation of those trajectories under a variation of the parameter $\boldsymbol{\lambda} \in \mathbf{W}$. We now move to the description of the properties for our models.

2.2   Bounded Linear Time Logic

As explained in the introduction, our aim is to find the parameter values that allow our model to best fit some given experimental data. In the following, we therefore assume that we are given a finite set of experimental observations that correspond to a finite set of time points and a tolerance value $\delta > 0^1$. We write

$$\mathcal{T} = \{0 = t_0, t_1, \ldots, t_N = T\} \tag{4}$$

for a set of time points and assume that we have a finite set of observations $\mathcal{O}_i^t = \{o_{i,1}^t, \ldots, o_{i,k}^t\}$ for each of them and for each coordinate i. We assume, in practice, that $\mathcal{T}$ indeed includes all the time points where experimental observations are available. Remark nonetheless that $\mathcal{T}$ is not necessarily limited to this set, as we could have $\mathcal{O}^t = \emptyset$ for a number of $t \in \mathcal{T}$. In practice, since $\mathcal{T} = \{t_0, \ldots t_N\}$ is finite, we abuse notations and substitute it, when convenient, with the integer set $\mathcal{T} = \{0, \ldots N\}$.

We start by recalling the logic defined in [15], which allows to express our properties of interest, i.e. that a given solution agrees with the experimental observations available at given time points. This logic is a slightly modified version of Bounded LTL, where atomic propositions are of the form (i, l, u) with $L_i \leq l \leq u \leq U_i$, where L_i and U_i are the boundaries of the set $\mathbf{V}$ defined above. The intuition is that, for $q \in \mathcal{T}$, z satisfies the atomic proposition (i, l, u) at time point q if and only if $l \leq z_i(q) \leq u$. Since there is a finite number of time points and a finite number of observations, we only consider the finite number of atomic propositions where $1 \leq i \leq n$ and $l, u \in \cup_{q \in \mathcal{T}} (\mathcal{O}_i^q \cup \mathcal{O}_i^q - \delta \cup \mathcal{O}_i^q + \delta)$. We also allow $l, u = +\infty, -\infty$ to account for timepoints q where $\mathcal{O}^q = \emptyset$.

The rest of the logic is defined as usual:

- every atomic proposition and the constants $true, false$ are BLTL formulas,
- the negation and conjunction of BLTL formulas are BLTL formulas,
- if Ψ and Ψ' are BLTL formulas, then $\Psi \mathcal{U}^q \Psi'$ and $\Psi \mathcal{U}^{\leq q} \Psi'$ are BLTL formulas for any positive integer $q \in \mathcal{T}$,
- if Ψ is a BLTL formula, then $\mathcal{X}\Psi$ is a BLTL formula.

---

[1] Note that the method does not depend on the value of δ. We assume its value is provided by the user.

The interpretation of $\Psi\mathcal{U}^{\leq q}\Psi'$ is standard, i.e. Ψ' must happen before q time points have elapsed, while the interpretation of $\Psi\mathcal{U}^q\Psi'$ is that Ψ must hold for exactly q time points before Ψ' holds. The interpretation of $\mathcal{X}\Psi$ is standard as well, i.e. $\mathcal{X}\Psi \Leftrightarrow true\mathcal{U}^1\Psi$. We invite the interested reader to consult [15] for the formal semantics of this logic.

Given a BLTL formula Ψ, we define $\mathsf{models}(\Psi) = \{z \in \mathsf{TRAJ} \mid z, 0 \models \Psi\}$.

Recall that the properties we want our models to verify are the following: the traces of the model need to agree with the given experimental data. One way to rephrase this property is as follows: at all time points where experimental data is available, the trace of our model needs to be between the lower and upper values taken from the experimental data with a given tolerance $\delta > 0$. This is easily expressed in BLTL as the property

$$\Psi^* = \bigwedge_{1 \leq i \leq m} \psi_i^0 \wedge \mathcal{X}\left(\psi_i^1 \wedge \mathcal{X}(\psi_i^2 \wedge \cdots \wedge \mathcal{X}\psi_i^N)\ldots\right) \tag{5}$$

where $\psi_i^q = (i, \min(\mathcal{O}_i^q) - \delta, \max(\mathcal{O}_i^q) + \delta)$. Since our aim is to consider variability on the ODE models of interest, we may use statements of the form $\mathbb{P}_{\geq p}(\Psi^*)$, whose interpretation is expressed as follows: "the probability that a trajectory in TRAJ is in $\mathsf{models}(\Psi^*)$ is greater than p". In this regard, we need to define a probability measure $\mathbb{P}$ over TRAJ.

We start by noticing that each parameter value λ completely determines the trajectory $z^\lambda \in \mathsf{TRAJ}$, since the initial condition $z_0 \in \mathbb{R}^n$ has been fixed. As a consequence, TRAJ can be completely identified with INIT (see Eq. (3)). Formally, we define $\mathsf{Models}(\Psi) \subseteq \mathsf{INIT}$ as the set

$$\{(z_0, \lambda) \in \mathsf{INIT} \mid z^\lambda(t) \in \mathsf{models}(\Psi)\} \tag{6}$$

and consider the Σ-algebra $\mathscr{B}$ generated by the m-dimensional open intervals of INIT. As expected, it is shown in [15] that $\mathscr{B}$ is an adequate support to prove the measurability of $\mathsf{Models}(\Psi)$ for any BLTL formula Ψ.

In the following, we will consider a number of probability distributions $\mathbb{P}^\lambda$ on $\mathscr{B}$ (one for each parameter value λ), and use these probability distributions to evaluate whether our ODE model meets a specification of the form $\mathbb{P}_{\geq p}^\lambda(\Psi)$. This will amount to checking whether $\mathbb{P}^\lambda(\mathsf{Models}(\Psi)) \geq p$. We will refer to the formulas such as $\mathbb{P}_{\geq p}^\lambda(\Psi)$ as PBLTL formulas.

In our context, each parameter value λ in $\mathbf{W}$ will give rise to a probability distribution $\mathbb{P}^\lambda$ taking into account internal variability. This probability distribution will be used for evaluating the model against the property Ψ^*, which will yield a score $\mathsf{grade}(\lambda)$ that represents the adequacy of parameter value λ w.r.t. the given experimental data while taking into account internal variability.

However, it might happen that many of the values λ in INIT have a maximal score $\mathsf{grade}(\lambda) = 1$, i.e. satisfy the PBLTL property $\mathbb{P}_{\geq 1}^\lambda(\Psi^*)$. This could be the case for example if all the traces generated using $\mathbb{P}^\lambda$ satisfy the property Ψ^*. In this case, we will need to consider more complex properties to filter those values and rank them. To this purpose, we introduce the notion of reward function.

2.3 Reward Function

The purpose of statistical model checking in general, and Monte-Carlo in particular, which will be presented in detail in Sect. 3.2, is to estimate with formal guarantees the expected value of a given function on a measurable set. In the context of model checking, this procedure is used to estimate the probability that a given model satisfies a property. To do this, each sample of the system is checked against the property and a Boolean reward is computed accordingly (i.e. 1 if the property is satisfied and 0 otherwise). Statistical model checking then amounts to estimating the expected value of this particular reward function on the measurable set of traces of the model at hand.

In our case, this boils down to defining a reward function $r_{\Psi^*} : \mathsf{TRAJ} \to \{0, 1\}$ that evaluates to 1 if the trajectory satisfies Ψ^* and 0 otherwise. Statistical model checking will then compute an estimation of the expected value of r_{Ψ^*} on the set of traces TRAJ under the probability distribution $\mathbb{P}^\lambda$, which in the end will be an estimation of the measure of $\mathsf{Models}(\Psi^*)$ for the parameter value λ. Remark that this construction would work for any other BLTL property Ψ.

In order to grade the parameter values in a more discriminating way, we allow the use of non-Boolean reward functions. This will allow expressing more powerful properties than those that can be defined using the BLTL logic. For instance, one can use those reward functions in order to measure the number of time points for which the current trace does not agree with the given experimental data, or to measure the cumulative distance between the trace and the experimental data at all time points.

In the following, we will therefore consider a given reward function $r : \mathsf{TRAJ} \to \mathbb{R}$ and use statistical model checking to estimate its expected value on the trajectories of our model under a given probability distribution $\mathbb{P}$. When convenient, we will identify a given BLTL property Ψ with its associated reward function obtained through the above construction r_Ψ.

3 Global Statistical Guarantees

In this section, we state our main result, which provides statistical guarantees on the verification of specific properties. Namely, given a property Ψ (resp. the corresponding reward function r_Ψ) on the trajectories of Eq. (1), we will establish confidence intervals regarding the estimation of the *probability* of satisfaction of that property (resp. the *expected value* of r_Ψ), which shall be computed using *approximate solutions* to Eq. (1), as well as a bound on the errors w.r.t. the *exact probability* corresponding to the *exact solutions* to Eq. (1).

We start by recalling how ODE numerical resolution methods work, and we propose a definition for the approximation errors introduced in the process. Next, we introduce a method for estimating the probability p that exact solutions of our ODE system satisfy a given property Ψ using an estimator $\hat{p}$ that takes the approximation error into account. Finally, we explain how those results along with their statistical guarantees can be extended to the estimation of the expected values of given reward functions.

3.1   Approximation Method for the Numerical Integration of the ODE

We recall that an approximation method, which determines the approximate solution y^λ to the ODE induced by parameter λ, can be written

$$y^\lambda(0) = z_0, \quad y^\lambda(\tau_{j+1}) = y^\lambda(\tau_j) + h\Phi(\tau_j, y^\lambda(\tau_j), \lambda, h), \quad 0 \le j < J, \quad (7)$$

where Φ is a continuous function defined in $[0, T] \times \mathbb{R}^n \times \mathbf{W} \times \mathbb{R}$ with values in $\mathbb{R}^n$, τ_j are the discrete points of definition of y^λ, and $h \in \mathbb{R}$. Intuitively, those methods compute each point thanks to the previous one. In this paper, we use the well-known Runge-Kutta 4 method, which is a standard method for ODE resolution.

For the sake of simplicity, we focus in the following on the theoretical study of 1-dimensional systems ($n = 1$), but our method can be adapted to larger systems ($n \ge 2$) as shown in our second case study presented in Sect. 4.2, mostly by adapting the definition of distance introduced below.

As explained in Sect. 2, we consider a set γ of observation data samples, recorded at $(N + 1)$ time points forming a set $\mathcal{T}$ (see Eq. (4)) with values in $\mathbb{R}^n$ ($N > 0$). We start by defining a notion of distance between functions that will, in the end, allow us to compare the solutions of our ODE model with the given experimental data. Given any two functions $y, \tilde{y}$ in the set $F_{\mathcal{T}} = \{g : I \to \mathbb{R} \mid \mathcal{T} \subseteq I\}$, where I denotes an interval included in $\mathbb{R}$, we consider the distance d defined by

$$d(y, \tilde{y}) = \max_{t \in \mathcal{T}} |y(t) - \tilde{y}(t)|. \quad (8)$$

Note that d is rigorously only a *pseudo*-distance, since two functions y and $\tilde{y}$ defined on $[0, T]$, that are distinct on $[0, T]$, might coincide on the finite set $\mathcal{T}$, thus could satisfy $d(y, \tilde{y}) = 0$. Nevertheless, since our purpose is to measure the distance to the dataset γ, we do not need to distinguish such two functions. Moreover, one may use any (pseudo-)distance of their choice, since all norms are equivalent in the finite-dimensional space $\mathbb{R}$ ($\mathbb{R}^n$ in the general setting). In the rest of the paper, we will abuse notations and use d to compare a given function $y \in F_{\mathcal{T}}$ to γ, even though γ is only defined on $\mathcal{T}$ and not on a continuous subinterval of $\mathbb{R}$.

In most ODE resolution methods, the approximation error depends on an integration step. We therefore introduce a discretization $\mathcal{D}_h$ of the time interval $[0, T]$, which we assume, for simplicity, to admit a constant step $h > 0$:

$$\mathcal{D}_h = \{0 = \tau_0, \tau_1, \tau_2, \ldots, \tau_J = T\}, \quad (9)$$

with $J > 0$ and $\tau_{j+1} - \tau_j = h$ for all $0 \le j < J$.

For each parameter value $\lambda \in \mathbf{W}$, the chosen approximation method will compute an approximate solution to the ODE, which we denote y^λ. Recall that the initial condition $z_0 \in \mathbb{R}$ ($\mathbb{R}^n$ in the general setting) has been fixed and that for any $\lambda \in \mathbf{W}$, the exact solution to Eq. (1) such that $z(0) = z_0$ is written z^λ.

For the sake of measuring the approximation error between y^λ and z^λ, we use a finer notion of distance than the one proposed above. Indeed, standard resolution methods provide guarantees that depend on the integration step in the sense that choosing a finer integration step enhances the quality of the approximation. Our aim here is to be able to take advantage of this fact, which could not be captured if we used the distance d from Eq. (8).

Definition 1 (Global approximation error). *Let $h > 0$ be the integration step of the chosen resolution method. The* global approximation error $\epsilon_h(\boldsymbol{\lambda})$ *between the approximate solution $y^{\boldsymbol{\lambda}}$ and the exact solution $z^{\boldsymbol{\lambda}}$ is defined as follows:*

$$\epsilon_h(\boldsymbol{\lambda}) = \max_{\tau \in \mathcal{D}_h} \left| z^{\boldsymbol{\lambda}}(\tau) - y^{\boldsymbol{\lambda}}(\tau) \right|. \tag{10}$$

In the rest of the paper, we make two important assumptions on the approximation method. First, we assume that the set $\mathcal{T}$ of time points given by Eq. (4), at which the observation data γ are recorded, satisfies $\mathcal{T} \subset \mathcal{D}_h$. This assumption is quite natural as there are a finite number of experimental data, therefore a sufficiently small h can always be chosen accordingly. Our second assumption is that the approximation method is *convergent*, which guarantees that for all $\boldsymbol{\lambda} \in \mathbf{W}$, the global approximation error $\epsilon_h(\boldsymbol{\lambda})$ converges to 0 when h gets smaller. This latter assumption is directly satisfied for usual approximation methods (such as, e.g., Runge-Kutta; see for instance [3]).

3.2 Monte-Carlo Method

We now move to our main result, i.e. providing an estimation of the probability that the original ODE system, with a given parameter value $\boldsymbol{\lambda}^*$, agrees with the experimental data with statistical guarantees. For the sake of simplicity, we focus in this section on standard BLTL properties as introduced in Sect. 2.2. We then show in Sect. 3.3 how these results can be extended to reward functions.

Let $\boldsymbol{\lambda}^* \in \mathbf{W}$ be a parameter value. In order to take the internal variability of our system into account, we will consider that $\boldsymbol{\lambda}^*$ can slightly vary. In order to do this, we set a constant $\rho > 0$ and define the open ball

$$B(\boldsymbol{\lambda}^*, \rho) = \{\boldsymbol{\lambda} \in \mathbb{R}^m \mid \|\boldsymbol{\lambda} - \boldsymbol{\lambda}^*\| < \rho\}, \tag{11}$$

where $\|\cdot\|$ is the Euclidean norm defined in Sect. 2.1.

We start by recalling the Monte Carlo procedure for estimation. This procedure aims at taking advantage of the Central Limit Theorem and the Law of Large Numbers. In order to estimate the probability that our system (where $\boldsymbol{\lambda}$ can vary inside of $B(\boldsymbol{\lambda}^*, \rho)$) satisfies the given BLTL property Ψ^* (see Eq. (5)), we will generate a set of n samples of values for $\boldsymbol{\lambda}$ inside of $B(\boldsymbol{\lambda}^*, \rho)$, and use these values to provide n solutions to the ODE system. Each solution will be evaluated, yielding a score of 1 if it satisfies Ψ^* and 0 otherwise. Informally, the Central Limit Theorem (Theorem 1) states that the mean value of the samples $\hat{p}$ is a good estimator for the probability p that our system (i.e. the ODE system, where the parameter value is set to $\boldsymbol{\lambda}^*$, with internal variability) satisfies Ψ^*. Moreover, it also provides a confidence interval that solely depends on the number of samples—provided this number is large enough—and the variance of the initial distribution.

Theorem 1 (Central Limit Theorem [19]). *Let $X_1, X_2, \ldots$ be a sequence of independent and identically distributed random variables of mean μ and variance σ^2. Then, the distribution of $\frac{\sum_{i=1}^n X_i - n\mu}{\sigma\sqrt{n}}$ tends to the standard normal distribution as $n \to \infty$. That is, for any $a \in \mathbb{R}$,*

$$\lim_{n \to \infty} \mathbb{P}\left(\frac{\sum_{i=1}^n X_i - n\mu}{\sigma\sqrt{n}} \leq a \right) = \frac{1}{\sqrt{2\pi}} \int_{-\infty}^a e^{-x^2/2} dx.$$

Because we cannot evaluate the exact solutions of the ODE system but instead have to rely on approximate solutions, we will define two auxiliary properties φ_1^ε and φ_2^ε (not expressed in BLTL) that take into account the global approximation error defined above, use the Monte Carlo procedure to estimate two probabilities $\hat{p}_1^\varepsilon$ and $\hat{p}_2^\varepsilon$ using those properties and the approximate solutions, and finally propose an estimation of $\hat{p}$ that relies on $\hat{p}_1^\varepsilon$ and $\hat{p}_2^\varepsilon$. We will finally use $\hat{p}$ in order to rate the chosen (central) parameter value $\boldsymbol{\lambda}^*$.

Let $\mathcal{T}$ be a set of time points as described earlier. Let γ be the set of experimental data and $\delta > 0$ be a precision (tolerance) w.r.t. γ. Let $\boldsymbol{\lambda}^* \in \mathbf{W}$ be a parameter value, let $\rho > 0$ be a variability setting. Consider the ball $\mathcal{B}_{\boldsymbol{\lambda}^*} = B(\boldsymbol{\lambda}^*, \rho)$ and let $\mathbb{P}^{\boldsymbol{\lambda}^*}$ be the uniform distribution on this ball.

Given a function $g \in F_{\mathcal{T}}$, we write $\varphi(g) := d(g, \gamma) \leq \delta$ the property that means "the distance between g and γ is less than δ". Note that this property can easily be written in BLTL (see Eq. (5) above). For convenience, if y^λ is an approximate solution to Eq. (1) induced by the parameter $\boldsymbol{\lambda} \in \mathcal{B}_{\boldsymbol{\lambda}^*}$, we will identify $\varphi(\boldsymbol{\lambda})$ to $\varphi(y^\lambda)$.

Given $\varepsilon > 0$, we introduce the properties:

$$\varphi(z^\lambda) := d(z^\lambda, \gamma) \leq \delta,$$
$$\varphi_1^\varepsilon(y^\lambda) := d(y^\lambda, \gamma) + \varepsilon \leq \delta, \quad \varphi_2^\varepsilon(y^\lambda) := d(y^\lambda, \gamma) - \varepsilon \leq \delta.$$

The translation of φ in BLTL is the property of interest Ψ^* defined in Eq. (5). Our aim is to provide an estimation $\hat{p}$ for $\mathbb{P}^{\boldsymbol{\lambda}^*}(\Psi^*)$. For convenience, we write $\mathbb{P}$ for $\mathbb{P}^{\boldsymbol{\lambda}^*}$ in the rest of this section.

In order to do that, we show in Lemma 1 that for a small enough integration step h, we have $\epsilon_h(\boldsymbol{\lambda}) \leq \varepsilon$ for all $\boldsymbol{\lambda} \in \mathcal{B}_{\boldsymbol{\lambda}^*}$, and therefore

$$\varphi_1^\varepsilon(y^\lambda) \Rightarrow \varphi(z^\lambda) \Rightarrow \varphi_2^\varepsilon(y^\lambda). \tag{12}$$

Lemma 1. *Let $(h_i)_{i \in \mathbb{N}} \in \mathbb{R}^+$ be a sequence of integration steps, such that $\lim_{i \to \infty} h_i = 0$. Then for all $\varepsilon > 0$, there exists $i^* > 0$ such that*

$$\epsilon_{h_i}(\boldsymbol{\lambda}) < \varepsilon, \quad \forall i \geq i^*, \forall \boldsymbol{\lambda} \in \mathcal{B}_{\boldsymbol{\lambda}^*}. \tag{13}$$

In other words, the global error $\epsilon_h(\boldsymbol{\lambda})$ can be uniformly bounded in the closure $\overline{\mathcal{B}_{\boldsymbol{\lambda}^*}}$ of the open ball $\mathcal{B}_{\boldsymbol{\lambda}^*}$. The proof of this lemma is given in Appendix A, along with a method to compute h_{i^*}.

Now, we define the probabilities

$$p = \mathbb{P}\big(\varphi(z_{\lambda^*})\big), \quad p_1^\varepsilon = \mathbb{P}\big(\varphi_1^\varepsilon(y_{\lambda^*})\big), \quad p_2^\varepsilon = \mathbb{P}\big(\varphi_2^\varepsilon(y_{\lambda^*})\big). \tag{14}$$

Note that p, p_1, p_2 implicitly depend on δ. However, we omit this dependence in order to lighten our notations. Next, it is straightforward that

$$p_1^\varepsilon \le p \le p_2^\varepsilon, \quad \forall \varepsilon > 0. \tag{15}$$

Estimators $\hat{p}_1^\varepsilon$, $\hat{p}_2^\varepsilon$ of the probabilities p_1^ε and p_2^ε respectively can be determined using the Monte-Carlo procedure, involving a precision α and a risk θ. Our main result, given in Theorem 2 below, establishes a statistical guarantee on the probability p of interest with respect to these estimators $\hat{p}_1^\varepsilon$, $\hat{p}_2^\varepsilon$.

Theorem 2 (Main theorem)
Let $\boldsymbol{\lambda}^* \in \mathbf{W}$, $\rho > 0$, $\delta > 0$, $\varepsilon > 0$. For any risk $\xi \in (0,1)$, we define $\theta = 1 - \sqrt{1 - \xi}$. Then, for any precision $\alpha > 0$, the probabilities p_1^ε and p_2^ε defined in Eq. (14) satisfy

$$\mathbb{P}\big(p_1^\varepsilon \in [\hat{p}_1^\varepsilon - \alpha, \hat{p}_1^\varepsilon + \alpha]\big) \ge 1 - \theta, \quad \mathbb{P}\big(p_2^\varepsilon \in [\hat{p}_2^\varepsilon - \alpha, \hat{p}_2^\varepsilon + \alpha]\big) \ge 1 - \theta, \tag{16}$$

where the estimators $\hat{p}_1^\varepsilon$ and $\hat{p}_2^\varepsilon$ can each be determined after performing a number $N' = \frac{\log(2/\theta)}{2\alpha^2}$ (and hence a total number $N = 2 \times \frac{\log(2/\theta)}{2\alpha^2}$) of simulations of Eq. (1) induced by parameter values $\boldsymbol{\lambda}$ sampled in $\mathcal{B}_{\boldsymbol{\lambda}^*}$.

Furthermore, there exist $\varepsilon_0 > 0$ and $h_0 > 0$ sufficiently small such that, for any integration step $h \le h_0$ and any $\varepsilon < \varepsilon_0$, the following statements hold:

– *the probability p defined in Eq. (14) satisfies the estimation*

$$\mathbb{P}\big(p \in [\hat{p}_1^\varepsilon - \alpha, \hat{p}_2^\varepsilon + \alpha]\big) \ge 1 - \xi, \tag{17}$$

– *the distance between $\hat{p}_1^\varepsilon$ and $\hat{p}_2^\varepsilon$ satisfies:*

$$\mathbb{P}\left(|\hat{p}_1^\varepsilon - \hat{p}_2^\varepsilon| \le 3\alpha\right) \ge 1 - \xi. \tag{18}$$

We emphasize that estimations (17) and (18) imply a confidence interval of width 5α for p and require a number of samples $N = 2 \times \frac{\log(2/\theta)}{2\alpha^2}$. If the analysis was performed directly on the exact solutions of the ODE, we would have a confidence interval of width 2α and only require $\frac{\log(2/\xi)}{2\alpha^2}$ samples.

The proof of Theorem 2, given in Appendix B, is divided in three main steps. First, using the Central Limit Theorem and the Law of Large Numbers, we determine estimators $\hat{p}_1^\varepsilon$ and $\hat{p}_2^\varepsilon$ of p_1^ε and p_2^ε, respectively. Then, Eq. (14) and the independence of simulations lead to the confidence interval of p. Finally, Lemma 1 guarantees that proper values of h and ε can be found, in order to control the distance between $\hat{p}_1^\varepsilon$ and $\hat{p}_2^\varepsilon$. It is worth noting that, for some resolution methods (such as Runge-Kutta 4 for example), a value for h can be explicitly determined to guarantee Lemma 1 for a given ε and therefore Eq. (17). However, the convergence speed of $|\hat{p}_1^\varepsilon - \hat{p}_2^\varepsilon|$ is not known in general, therefore we can only guarantee the existence of a sufficiently small value for ε to ensure Eq. (18) but not compute it.

3.3 Model Checking Extension Through Reward Functions

As explained in Sect. 2.3, our method can be extended to non-Boolean reward functions. Indeed, these functions may provide not only qualitative results—"does the property hold?"—but also *quantitative* ones—"*how well* does the property hold?". In our case, this allows to distinguish the good parameters that induce a suitable solution from the best ones that induce the solutions closest to the data.

To use such a real-valued reward function r, some conditions are required. First, it must be assumed that two other reward functions r_1 and r_2 can be found, such that the following estimation holds for any $\lambda \in \mathcal{B}_{\lambda^*}$:

$$r_1(\lambda) \leq r(\lambda) \leq r_2(\lambda). \tag{19}$$

Second, the law of the unconscious statistician must be applicable to these lower and upper reward functions, i.e. the computation of the expected value[2] must be applicable, so that estimators $\hat{r}_1$ and $\hat{r}_2$ of r_1 and r_2 respectively, can be computed.

Moreover, and most importantly, the reward function must be *compatible* with the global error defined in Eq. (10). Indeed, since we compute score based on approximated solutions, said computations must take this approximation into account to provide any significance to the resulting score. It is worth noting that these conditions are satisfied by all the reward functions we have considered in this work, such as the total accumulated/maximal/average distance to γ or the number of time points where γ is not respected.

Similarly to Eq. (18), the distance between $\hat{r}_1$ and $\hat{r}_2$ must be controlled. Depending on the order of the approximation method used to compute approximate solutions to the ODEs, this may be easy to ensure. For instance, in our case the integration method Runge-Kutta 4 ensures that the approximation error—and thus, the global error as defined in Eq. (10)—is of order 5: all derivatives of the integration functions converge at most linearly w.r.t. h^5, where h is the integration step.

4 Case Studies

In this section, we apply our method to two case studies [17,22] taken from the literature to show its potential. After presenting the studies and their results, we will display our results and discuss them. We implemented our technique in C++ to validate the approach. The experiments were realized on a 2.1 GHz Intel Xeon Silver 4216 processor, running g++ version 7.5.0 on Ubuntu 18.04. The code is available at https://gitlab.com/davidjulien/smc_for_ode.git, and the experiments can be reproduced using the right branches, i.e. *compute_aurelia* to run the experiment from Sect. 4.1 and *compute_prey* to run the experiment from Sect. 4.2. We used the Runge-Kutta 4 method to compute approximate solutions, a SMC precision $\alpha = 0.05$ and a risk $\xi = 0.05$.

---

[2] See Eq. (27) in Appendix B.

First, we briefly recall the experiment. After discretizing the value space $\mathbf{W}$ defined in Eq. (2) for the parameter $\boldsymbol{\lambda}$, we will grade every value in order to select the best ones w.r.t. the experimental data γ. In order to take the internal variability of the model into account, each chosen parameter value $\boldsymbol{\lambda}^*$ is associated with the open ball $\mathcal{B}_{\boldsymbol{\lambda}^*}$ as defined in Eq. (11). Once the SMC parameters α and ξ, as well as a small enough value for ε are chosen, we can compute an integration step h, as well as a required number N of samples such that Theorem 2 holds. Then, we sample N values $\boldsymbol{\lambda} \in \mathcal{B}_{\boldsymbol{\lambda}^*}$, compute the approximated solutions to the induced ODEs, and compare them with the experimental data γ. For each $\boldsymbol{\lambda}^* \in \mathbf{W}$, we thus estimate the probabilities $\hat{p}_1^\varepsilon$ and $\hat{p}_2^\varepsilon$ defined in the previous section, and use them to define $\mathsf{grade}(\boldsymbol{\lambda}^*) = \frac{\hat{p}_1^\varepsilon + \hat{p}_2^\varepsilon}{2}$. In order to better discriminate the best parameter values, we also estimate the expected value of the reward function $r : \boldsymbol{\lambda} \mapsto d(z_{\boldsymbol{\lambda}}, \gamma)$ that measures the distance between the ODE simulations and the experimental data.

4.1 Case Study 1: A Study on *Aurelia Aurita* Population Growth [17]

In 2014, Melica et al. [17] published a paper studying the growth of *Aurelia Aurita*, a species of jellyfish that is very common in Adriatic Sea. In this paper, they compared experimental data, resulting from the culture of *Aurelia Aurita* polyps, to simulation models based on the following ODE:

$$x'(t) = ax(t)(1 - x(t)/b) \tag{20}$$

where t is time, x is the population density, a is the maximum rate of population growth, and b is the positive equilibrium. The authors show that the dynamics of a *Aurelia Aurita* polyps population can, indeed, be modeled by the density-dependent, or Verhulst [23], ODE presented above and compute the values for a and b that ensure the best fitting w.r.t. the experimental data. These values are recalled in Table 1.

Table 1. Estimation of parameters of the logistic curve fitting the laboratory experimental data [17].

	HD	LD
b	$5.35 \pm 0.11 (^{***}p < 0.001)$	$1.81 \pm 0.08(^{***}p < 0.001)$
$x(0)$	$7.59 \pm 0.21(^{***}p < 0.001)$	$0.081 \pm 0.017(^{***}p < 0.001)$
a	$0.130 \pm 0.033(^{**}p = 0.002)$	$0.137 \pm 0.012(^{***}p < 0.001)$
χ^2	0.775	0.056

Remark 1. HD and LD represent the studies for *High* and *Low Density*, respectively, which were both ran by the original authors. Here, we focused on the *High Density* case.

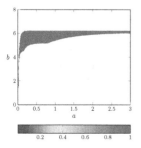 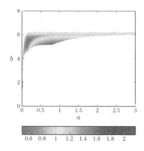 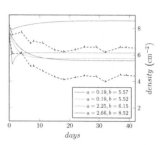

Fig. 3. Heatmap of the score of the parameters. (Color figure online)

Fig. 4. Heatmap of the distance to γ. (Color figure online)

Fig. 5. Solution to the ODE from Eq. (20). (Color figure online)

In order to illustrate our method, we applied it to the same case study, using Eq. (20) as the ODE system. We evaluated parameter values in the ranges $a \in [0,3], b \in [0,9]$, and discretized this space with a parameter step of 0.01. We set the internal variability of the parameters $\rho = 0.005$ and performed $N = 874$ simulations for each parameter value on the discretized space, therefore ensuring a statistical precision of $\alpha = 0.05$ and risk of $\xi = 0.05$.

In Fig. 3, we represent the score of the best parameter values, where the white zones are zones where $\mathsf{grade}(\lambda^*) = 0$. One can see that there is a small gradient in the area where the score is positive, but this is not enough to discriminate between the parameter values in this zone. In order to refine the result, we present in Fig. 4 the estimation of the expected value of the reward function $r : \lambda \mapsto d(z_\lambda, \gamma)$. Figure 4 shows a tighter area of values that induce solutions that are very close to the data (down to 0.50 polyps on average), plotted in red, which contains the parameter value estimated by [17]: it comforts us in saying that our method provides tangible results. The best parameter found using our method is the pair $(a, b) = (0.19, 5.57)$. It induces the red curve in Fig. 5.

4.2 Case Study 2: A Prey-Predator Model for Lynx and Hares [22]

In 2010, Restrepo and Sánchez [22] published a paper describing a genetic algorithm, which aimed at estimating the best parameters for prey-predator models. The first model, which we will study in the following, is a basic prey-predator interaction model defined by the following ODE system:

$$P' = aP - bPD, \quad D' = -cD + dPD \tag{21}$$

where t is time, P and D are the two time-dependent variables representing the quantity of individuals in each group: $P(t)$ for prey and $D(t)$ for predators; a, b, c, d are positive constants, a and c indicating the birth rate of prey and death rate of predators respectively, and b and d representing the rates of predation and reproduction of predators. Note that even if the model is a standard way to

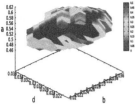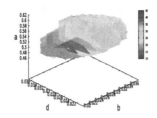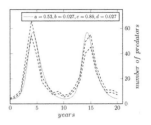

Fig. 6. Heatmap of the score (for $c = 0.89$).

Fig. 7. Heatmap of the distance to γ (for $c = 0.89$).

Fig. 8. Best solution to Eq. (21).

describe, on first approximation, such an interaction between two populations, its simplicity might make it imprecise – which is why other, more complex models are studied in [22]. The best values for parameters (a, b, c, d) w.r.t. experimental data are given in [22]: $(a, b, c, d) = (0.55, 0.027, 0.83, 0.026)$.

Again, we applied our method to this case study, using Eq. (21) as the ODE system. We evaluated parameter values in the ranges $a \in [0.48, 0.68]$, $b \in [0.015; 0.04]$, $c \in [0.78, 0.9]$, $d \in [0.01, 0.05]$, and discretized this space with a parameter step of 0.001. We set the internal variability of the parameters $\rho = 0.0005$ and performed $N = 874$ simulations for each parameter value on the discretized space, therefore ensuring a statistical precision of $\alpha = 0.05$ and risk of $\xi = 0.05$. Note that the parameter ranges have been tightened according to the paper results, since cyclic models can be very sensitive to parameter values.

We encountered two issues with this study. First, the adequacy of the model with the data was improvable, thus computing a fitting solution was challenging. We had to loosen the property we verify: instead of Ψ^*, which enforces the solution to always stay in the tunnel, we verified a property $\Psi^\dagger$, which allows the solutions to step out of the tunnel a total of 5 times (for a total of 22 time points) before rejecting them. This explains why the best solution displayed in Fig. 8 does not perfectly fit inside of the tunnel. Remark that $\Psi^\dagger$ can easily be expressed in BLTL, and is therefore compatible with our theory. Second, since system Eq. (21) involves 4 parameters, displaying the results with heatmaps is more difficult than in case study 1 (Sect. 4.1). Nonetheless, locking a parameter value (here, $c = 0.89$) allows the plot of a 3-dimensional heatmap.

Figure 6 shows a local subset of solutions fitting the data with a certain quality. Notice that the score only goes up to 0.5; because of the internal variability we impose and the sensitivity of the model, very few simulations stayed in the inner tunnel (corresponding to φ_1^ε in Sect. 3), yielding $\hat{p}_1^\varepsilon = 0$ in most cases. That said, Fig. 7 displays the distance for the same subset. We notice that some solutions are, at most, at a 10 individuals distance from the data. The subset contains our best candidate $(a, b, c, d) = (0.52, 0.027, 0.89, 0.027)$, whose corresponding curve is displayed in Fig. 8. We see that the general shape of the curve is satisfying but does not perfectly fit inside of the tunnel. This may be explained by the fact that cyclic ODE systems like prey-predator models can be very sensitive to the non-linear terms, i.e. bPD and dPD. Again, our goal here was to prove the concept rather than describe a phenomenon with the upmost precision: while this is satisfying as far as we are concerned, a more thorough study of the parameters, along with a better quality of the data (with e.g. several observations for each time point t, allowing for more robust data and observable data tunnels) would help getting results closer to the actual experiments.

5   Conclusion

In this paper, we have proposed a statistical method for synthesizing the best parameter values w.r.t. given experimental data for an ODE system with internal variability, while providing formal statistical guarantees that for the first time (to the best of our knowledge) take into account the approximation error introduced through the numerical resolution of the ODEs. To do that, we discretize the parameter space and define balls around the resulting (finite) set of parameter values to take internal variability into account. We then use the Monte-Carlo technique to estimate the probability that exact solutions of the ODE system are close to the experimental data for each resulting parameter ball, and use the result of this estimation to select the best (central) parameter values. Our main contribution is Theorem 2 which guarantees the precision of our estimation despite the fact that it is performed using numerical resolution techniques that do not give us access to exact solutions of the ODE system. In contrast with other existing works on parameter estimation for ODE systems, like [15], where this problem is left aside, we show that the number of simulations required for a given precision and risk of the statistical estimation is (more than) twice the one needed when working with exact solutions. We also show that an upper bound on the integration step of the chosen integration technique exists (and can be computed for standard integration techniques) in order to make sure that a given statistical precision and risk are respected.

One of the limitations of our work is that, in order to prove our results and perform parameter synthesis in practice, we rely on a setting ε that represents the maximal admissible distance between exact and approximate solutions to the ODE system. Although it is possible, for most integration techniques[3], to compute an integration step that will guarantee that a given value for ε is respected, our results only show the existence of a suitable value for ε for any statistical setting, but do not provide any method to compute this value in practice. This is due to our lack of guarantees on the convergence speed of the distance between the two estimators $\hat{p}_1^\varepsilon$ and $\hat{p}_2^\varepsilon$ that appear in Eq. (17) and Eq. (18). What we do in practice is that we set small values for ε, perform experiments and then estimate the value of $|\hat{p}_1^\varepsilon - \hat{p}_2^\varepsilon|$. If the resulting value is too large, then we start over the experiment with a smaller value for ε.

Although the only BLTL property that we verify in this paper is the property Ψ^* defined in Eq. (5), we believe that our reasoning can be easily extended to

---

[3] See Remark 2 in Appendix A, and [16].

other BLTL property following the definition given in Sect. 2.2. This is, in our opinion, a straightforward extension that we will address in the near future.

As said in the introduction, our results are generic and could therefore be combined with any exploration strategy for the parameter space. The global exploration we perform in this paper is obviously costly but yields global information that is precious when analysing a complex system. In the future, we plan on combining a coarse global exploration to identify interesting zones in the parameter space with more efficient and detailed search algorithms (such as the one from [15]) limited to those zones.

Appendices

In these appendices, we provide the complete proofs of Lemma 1 and Theorem 2.

A   Proof of Lemma 1

First, we recall the definition of *stability* of an approximation method.

Definition 2 (Method stability). *We say that the approximation method determined by Eq. (7) is* stable *if there exists a constant $\mathcal{K} > 0$, called* stability constant, *such that, for any two sequences $(y_k)_{0 \leq k \leq J}$ and $(\widetilde{y}_k)_{0 \leq k \leq J}$ defined as $y_{k+1} = y_k + h\,\Phi(\tau_k, y_k, \boldsymbol{\lambda}, h)$ and $\widetilde{y}_{k+1} = \widetilde{y}_k + h\,\Phi(\tau_k, \widetilde{y}_k, \boldsymbol{\lambda}, h) + \eta_k$ respectively, $(0 \leq k < J)$, with $\boldsymbol{\lambda} \in \mathbf{W}$ and $\eta_k \in \mathbb{R}$, we have*

$$\max_{0 \leq k \leq J} |y_k - \widetilde{y}_k| \leq \mathcal{K}\big(|y_0 - \widetilde{y}_0| + \sum_{0 \leq k \leq J} |\eta_k|\big). \tag{22}$$

It is well-known that if Φ is κ-Lipschitz w.r.t. y, *i.e.* $\forall t \in [0, T]$, $\forall y, \widetilde{y} \in \mathbb{R}$, $\forall \boldsymbol{\lambda} \in \mathbf{W}$ and $\forall h \in \mathbb{R}$, $|\Phi(t, y, \boldsymbol{\lambda}, h) - \Phi(t, \widetilde{y}, \boldsymbol{\lambda}, h)| \leq \kappa |y - y_2|$, then stability is ensured (see for instance [3] or [4]).

Now, we fix $\boldsymbol{\lambda}^* \in \mathbf{W}$ and $\boldsymbol{\lambda}_1, \boldsymbol{\lambda}_2 \in \mathcal{B}_{\boldsymbol{\lambda}^*}$, and we consider the approximate solutions $y^{\boldsymbol{\lambda}_1}, y^{\boldsymbol{\lambda}_2}$ to Eq. (1) relative to $\boldsymbol{\lambda}_1$ and $\boldsymbol{\lambda}_2$ and starting from z_0.

$$\begin{cases} y_0^{\boldsymbol{\lambda}_1} = z_0, \\ y_{k+1}^{\boldsymbol{\lambda}_1} = y_k^{\boldsymbol{\lambda}_1} + h\,\Phi(t_k, y_k^{\boldsymbol{\lambda}_1}, \boldsymbol{\lambda}_1, h), \end{cases} \qquad \begin{cases} y_0^{\boldsymbol{\lambda}_2} = z_0, \\ y_{k+1}^{\boldsymbol{\lambda}_2} = \widetilde{y}_k + h\,\Phi(t_k, y_k^{\boldsymbol{\lambda}_2}, \boldsymbol{\lambda}_2, h). \end{cases}$$

We recall that the exact solutions to Eq. (1) relative to $\boldsymbol{\lambda}_1$ and $\boldsymbol{\lambda}_2$ and starting from z_0 are denoted $z^{\boldsymbol{\lambda}_1}$ and $z^{\boldsymbol{\lambda}_2}$ respectively. For $i \in \{1, 2\}$ and $0 \leq k \leq J$, we introduce the consistency error on $y^{\boldsymbol{\lambda}_i}$ at step k:

$$\epsilon_{h,k}(\boldsymbol{\lambda}_i) = |z^{\boldsymbol{\lambda}_i}(\tau_k) - y^{\boldsymbol{\lambda}_i}(\tau_k)|. \tag{23}$$

The consistency errors satisfy $\epsilon_h(\boldsymbol{\lambda}_i) = \max_{0 \leq k \leq J} \epsilon_{h,k}(\boldsymbol{\lambda}_i)$, for $i \in \{1, 2\}$, where $\epsilon_h(\boldsymbol{\lambda}_i)$ is the global approximation error (defined by Eq. (10)). The proof of Lemma 1 can be derived from the following theorem.

Theorem 3 (Stability with respect to consistency error). *Assume that the function Φ defined in Eq. (7) is κ_1-Lipschitz w.r.t. $\boldsymbol{\lambda}$ and κ_2-Lipschitz continuous w.r.t. y. Then the approximation method is stable w.r.t. the consistency error, i.e. there exists $\mathcal{K} > 0$ such that*

$$\forall \boldsymbol{\lambda}_1, \boldsymbol{\lambda}_2 \in \mathcal{B}_{\boldsymbol{\lambda}^*}, \max_{0 \le k \le J} |\epsilon_{h,k}(\boldsymbol{\lambda}_1) - \epsilon_{h,k}(\boldsymbol{\lambda}_2)| \le \mathcal{K} \|\boldsymbol{\lambda}_1 - \boldsymbol{\lambda}_2\|, \tag{24}$$

where $\| \cdot \|$ is the Euclidean norm defined in Sect. 2.1.

Proof (of Theorem 3). By assumption, Φ is κ_1-Lipschitz continuous w.r.t. $\boldsymbol{\lambda}$:

$$\forall t, y, h \in \mathbb{R}, \forall \boldsymbol{\lambda}_1, \boldsymbol{\lambda}_2 \in \mathcal{B}_{\boldsymbol{\lambda}^*}, |\Phi(t, y, \boldsymbol{\lambda}_1, h) - \Phi(t, y, \boldsymbol{\lambda}_2, h)| \le \kappa_1 |\boldsymbol{\lambda}_1 - \boldsymbol{\lambda}_2|.$$

It follows that

$$\begin{aligned}
|y_{k+1}^{\boldsymbol{\lambda}_1} - y_{k+1}^{\boldsymbol{\lambda}_2}| &\le |y_k^{\boldsymbol{\lambda}_1} - y_k^{\boldsymbol{\lambda}_2}| + h|\Phi(t_k, y_k^{\boldsymbol{\lambda}_1}, \boldsymbol{\lambda}_1, h) - \Phi(t_k, y_k^{\boldsymbol{\lambda}_2}, \boldsymbol{\lambda}_2, h)| \\
&\le |y_k^{\boldsymbol{\lambda}_1} - y_k^{\boldsymbol{\lambda}_2}| + h|\Phi(t_k, y_k^{\boldsymbol{\lambda}_1}, \boldsymbol{\lambda}_1, h) - \Phi(t_k, y_k^{\boldsymbol{\lambda}_1}, \boldsymbol{\lambda}_2, h)| \\
&\quad + h|\Phi(t_k, y_k^{\boldsymbol{\lambda}_1}, \boldsymbol{\lambda}_2, h) - \Phi(t_k, y_k^{\boldsymbol{\lambda}_2}, \boldsymbol{\lambda}_2, h)| \\
&\le (1 + h\kappa_2)|y_k^{\boldsymbol{\lambda}_1} - y_k^{\boldsymbol{\lambda}_2}| + h\kappa_1 \|\boldsymbol{\lambda}_1 - \boldsymbol{\lambda}_2\|,
\end{aligned}$$

for $0 \le k \le J$. We write $|y_k^{\boldsymbol{\lambda}_1} - y_k^{\boldsymbol{\lambda}_2}| = \Delta_{y,k}$ and $\|\boldsymbol{\lambda}_1 - \boldsymbol{\lambda}_2\| = \Delta_{\boldsymbol{\lambda}}$, and we get

$$\Delta_{y,k+1} \le (1 + h\kappa_2)\Delta_{y,k} + h\kappa_1 \Delta_{\boldsymbol{\lambda}}. \tag{25}$$

Applying the discrete Gronwall lemma (see for instance [7], VIII.2.3), we deduce

$$\max_{0 \le k \le J} \Delta_{y,k} \le e^{\kappa_2 T} \Big(\Delta_{y,0} + \sum_{0 \le j \le k-1} h\kappa_1 \Delta_{\boldsymbol{\lambda}} \Big)$$

which leads to

$$\max_{0 \le k \le J} |y_k^{\boldsymbol{\lambda}_1} - y_k^{\boldsymbol{\lambda}_2}| \le e^{\kappa_2 T} T \kappa_1 \|\boldsymbol{\lambda}_1 - \boldsymbol{\lambda}_2\|,$$

since $y_0^{\boldsymbol{\lambda}_1} = y_0^{\boldsymbol{\lambda}_2} = z_0$ and $hJ = T$.

Furthermore, it is proved in [4] that if Φ is Lipschitz continuous w.r.t. $\boldsymbol{\lambda}$, then the exact solution $z_{\boldsymbol{\lambda}}$ is also Lipschitz continuous w.r.t. $\boldsymbol{\lambda}$ that is, there exists $\kappa_3 > 0$ such that

$$\forall \boldsymbol{\lambda}_1, \boldsymbol{\lambda}_2 \in \mathcal{B}_{\boldsymbol{\lambda}^*}, \forall t \in [0, T], |z^{\boldsymbol{\lambda}_1}(t) - z^{\boldsymbol{\lambda}_2}(t)| \le \kappa_3 \|\boldsymbol{\lambda}_1 - \boldsymbol{\lambda}_2\|. \tag{26}$$

Finally, we have

$$\begin{aligned}
|\epsilon_{h,k}(\boldsymbol{\lambda}_1) - \epsilon_{h,k}(\boldsymbol{\lambda}_2)| &\le |z^{\boldsymbol{\lambda}_1}(\tau_k) - y^{\boldsymbol{\lambda}_1}(\tau_k) - z^{\boldsymbol{\lambda}_2}(\tau_k) - y^{\boldsymbol{\lambda}_2}(\tau_k)| \\
&\le |z^{\boldsymbol{\lambda}_1}(\tau_k) - z^{\boldsymbol{\lambda}_2}(\tau_k)| + |y^{\boldsymbol{\lambda}_1}(\tau_k) - y^{\boldsymbol{\lambda}_2}(\tau_k)| \\
&\le \mathcal{K} \|\boldsymbol{\lambda}_1 - \boldsymbol{\lambda}_2\|,
\end{aligned}$$

with $\mathcal{K} = \kappa_3 + T\kappa_1 e^{\kappa_2 T}$, which completes the proof of Theorem 3. $\qquad\square$

It remains to show that Theorem 3 implies Lemma 1.

Proof (of Lemma 1). Let $(h_i)_{i\geq 0}$ be a sequence of discretization steps such that $\lim_{i\to\infty} h_i = 0$. Since the approximation method given by (7) is assumed to be convergent, each function $\epsilon_{h_i}(\cdot)$ defined in Eq. (23) is pointwise convergent to 0.

Furthermore, we recall that Φ is Lipschitz continuous w.r.t. $\boldsymbol{\lambda} \in \mathbf{W}$. Hence, Theorem 3 implies that the functions $\big(\epsilon_{h_i}(\cdot)\big)_{i\geq 0}$ defined in Eq. (23) are also Lipschitz continuous, with uniform Lipschitz constant $\mathcal{K}$:

$$|\epsilon_{h_i}(\boldsymbol{\lambda}_1) - \epsilon_{h_i}(\boldsymbol{\lambda}_2)| \leq \mathcal{K}\,\|\boldsymbol{\lambda}_1 - \boldsymbol{\lambda}_2\|, \quad \forall \boldsymbol{\lambda}_1, \boldsymbol{\lambda}_2 \in \mathcal{B}_{\boldsymbol{\lambda}^*}, \quad \forall i \in \mathbb{N}.$$

Consequently, the functions $\big(\epsilon_{h_i}(\cdot)\big)_{i\geq 0}$ are uniformly equicontinuous. Hence, Arzelà-Ascoli Theorem [8] implies that the sequence $\big(\epsilon_{h_i}(\cdot)\big)_{i\geq 0}$ converges uniformly to 0 on $\mathcal{B}_{\boldsymbol{\lambda}^*}$, thus $\forall \varepsilon > 0,\ \exists i^* \in \mathbb{N},\ \forall i \geq i^*,\ \forall \boldsymbol{\lambda} \in \mathcal{B}_{\boldsymbol{\lambda}^*},\ \epsilon_{h_i}(\boldsymbol{\lambda}) < \varepsilon$, and Lemma 1 is proved. $\qquad\square$

Remark 2 (Computation of a sufficiently small integration step). We emphasize that Lemma 1 can be supplemented by an explicit choice of a sufficiently small integration step h, provided the integration method comes with appropriate estimates of their global error. Notably, the accuracy of the Runge-Kutta 4 method, which we use for the numerical treatment of our case studies, has been thoroughly studied (see [16] for instance), and it is known that its inherent error can be bounded in terms of the successive derivatives of the function f involved in Eq. (1), up to order 4.

B   Proof of Theorem 2

First Step. We begin the proof of Theorem 2 by showing how to compute an estimator $\hat{p}_1^\varepsilon$ of the probability p_1^ε defined in (14).

Let $(\boldsymbol{\lambda}_i)_{\mathbb{N}}$ be a sequence of values in the ball $\mathcal{B}_{\boldsymbol{\lambda}^*}$. We write B_i the random variable corresponding to the test "$\varphi_1^\varepsilon(\boldsymbol{\lambda}_i)$ holds": all the B_i are i.i.d. variables and follow a Bernoulli's law of parameter p_1^ε. We write b_i the evaluation of B_i. We introduce the transfer function $g_1 : \mathcal{B}_{\boldsymbol{\lambda}^*} \to \{0,1\}$ corresponding to the test regarding $\varphi_1^\varepsilon(\boldsymbol{\lambda}_i)$, defined by $g_1(\boldsymbol{\lambda}_i) = 1$ if $\varphi_1^\varepsilon(\boldsymbol{\lambda}_i)$ holds, 0 otherwise. Next, we consider

$$G = \mathbb{E}(g_1(X)) = \int_{\mathcal{B}_{\boldsymbol{\lambda}^*}} g_1(x) f_X(x)\mathrm{d}x, \qquad (27)$$

where f_X is defined by a uniform distribution, that is, $f_X(x) = \frac{1}{|\mathcal{B}_{\boldsymbol{\lambda}^*}|}$, $x \in \mathcal{B}_{\boldsymbol{\lambda}^*}$. We produce a sample $(x_1, x_2, \ldots, x_N)$ of the variable X in $\mathcal{B}_{\boldsymbol{\lambda}^*}$, and use it to compute the Monte-Carlo estimator G. By virtue of the Law of Large Numbers, the sample mean satisfies: $\overline{g}_N = \frac{1}{N}\sum_{i=1}^N g_1 x_i$. The Central Limit Theorem states that the variable $Z = \frac{\overline{g}_N - G}{\sigma_{\overline{g}_N}}$ approximately follows a Standard Normal Distribution $\mathcal{N}(0,1)$; hence, for a risk θ, we can bound the error $|\alpha_N|$ of swapping G with $\overline{g}_N$ by building confidence intervals:

$$\mathbb{P}\left(|\alpha_N| \leq \chi_{1-\frac{\theta}{2}}\frac{\sigma_{g_1}}{\sqrt{N}}\right) = 1 - \theta, \qquad (28)$$

where $\chi_{1-\frac{\theta}{2}}$ is the quantile of the Standard Normal Distribution $\mathcal{N}(0,1)$ and σ_{g_1} is the variance of g_1.

Since we are interested in finding p_1^ε with a certain confidence, we can perform this process after setting the desired target error α and risk θ, knowing how many simulations must be ran using Hoeffding's inequality [12]:

$$\theta = \mathbb{P}(\overline{g}_N \notin [p_1^\varepsilon - \alpha, p_1^\varepsilon + \alpha]) \leq 2\exp(-2\alpha^2 N),$$

or equivalently $N \geq \frac{\log(2/\theta)}{2\alpha^2}$. Here, it is worth emphasizing that N can be chosen independently of ε.

Further, the variance of $\overline{g}_N$ can be expressed with the variance of $g_1(X)$:

$$\sigma_{g_1}^2 = \mathbb{E}\left([g_1(X) - \mathbb{E}(g_1(X))]^2\right) = \int_{\mathcal{B}_{\lambda^*}} (g_1(x))^2 f_X(x)\mathrm{d}x - G^2.$$

We consider i.i.d. samples, hence $\sigma_{g_1}^2$ can be estimated with the variance $S_{g_1}^2$:

$$\sigma_{g_1}^2 \simeq S_{g_1}^2 = \frac{1}{N}\sum_{i=1}^{N}(g_1(\boldsymbol{\lambda}_i)^2 - \overline{g}_N^2).$$

It follows that σ_{g_1} can be estimated with its empirical counterpart $\hat{\sigma}_{g_1} = \sqrt{S_{g_1}^2}$, which shows that the error displays a $1/\sqrt{N}$ convergence.

Finally, after estimating σ_{g_1}, we can find $\hat{p}_1^\varepsilon$ using the variance of Bernoulli's law $\hat{\sigma}_{g_1}^2 = \hat{p}_1^\varepsilon \times (1 - \hat{p}_1^\varepsilon)$. We conclude that the probability that $\varphi_1^\varepsilon(\boldsymbol{\lambda})$ holds is estimated by $\hat{p}_1^\varepsilon = \frac{1}{2}\left(1 \pm \sqrt{1 - 4\hat{\sigma}_{g_1}^2}\right)$, with an error α and a risk θ, provided we perform $N \geq \frac{\log(2/\theta)}{2\alpha^2}$ simulations. It follows that

$$\mathbb{P}(p_1^\varepsilon \in [\hat{p}_1^\varepsilon - \alpha, \hat{p}_1^\varepsilon + \alpha]) \geq 1 - \theta. \tag{29}$$

Similarly, we determine an estimator $\hat{p}_2^\varepsilon$ of p_2^ε by running $N \geq \frac{\log(2/\theta)}{2\alpha^2}$ additional simulations, and obtain a confidence interval satisfying

$$\mathbb{P}(p_2^\varepsilon \in [\hat{p}_2^\varepsilon - \alpha, \hat{p}_2^\varepsilon + \alpha]) \geq 1 - \theta. \tag{30}$$

Second Step. Now, let us show how a confidence interval for the probability p can be derived from the confidence intervals given in (29), (30), involving the estimators $\hat{p}_1^\varepsilon$ and $\hat{p}_2^\varepsilon$ respectively. The independence of the samples used to determine the estimators $\hat{p}_1^\varepsilon$, $\hat{p}_2^\varepsilon$ guarantees that

$$\mathbb{P}(p \in [\hat{p}_1^\varepsilon - \alpha, \hat{p}_2^\varepsilon + \alpha]) = \mathbb{P}(\{p \geq \hat{p}_1^\varepsilon - \alpha\}) \times \mathbb{P}(\{p \leq \hat{p}_2^\varepsilon + \alpha\}).$$

By virtue of (29), we have $\mathbb{P}(p_1^\varepsilon \geq \hat{p}_1^\varepsilon - \alpha) \geq 1 - \theta$. Next, the estimate (15) implies $\mathbb{P}(p \geq \hat{p}_1^\varepsilon - \alpha) \geq \mathbb{P}(p_1^\varepsilon \geq \hat{p}_1^\varepsilon - \alpha) \geq 1 - \theta$. Similarly, we have $\mathbb{P}(p \leq \hat{p}_2^\varepsilon + \alpha) \geq 1 - \theta$, and finally $\mathbb{P}(p \in [\hat{p}_1^\varepsilon - \alpha, \hat{p}_2^\varepsilon + \alpha]) \geq (1 - \theta)^2 = 1 - \xi$, since $\theta = 1 - \sqrt{1 - \xi}$.

Third Step. Finally, let us prove how Lemma 1 guarantees that proper values of h and ε can be found, in order to control the distance between $\hat{p}_1$ and $\hat{p}_2$.

Indeed, the continuity of the probability measure $\mathbb{P}$ ensures that there exists $\varepsilon_0 > 0$ such that $|p_1^\varepsilon - p_2^\varepsilon| \leq \alpha$, for $\varepsilon < \varepsilon_0$. Next, we write

$$|\hat{p}_1^\varepsilon - \hat{p}_2^\varepsilon| \leq |\hat{p}_1^\varepsilon - p_1^\varepsilon| + |\hat{p}_2^\varepsilon - p_2^\varepsilon| + |p_1^\varepsilon - p_2^\varepsilon|,$$

hence we have, for $\varepsilon < \varepsilon_0$:

$$\mathbb{P}\left(|\hat{p}_1^\varepsilon - \hat{p}_2^\varepsilon| \leq 3\alpha\right) \geq \mathbb{P}(|\hat{p}_1^\varepsilon - p_1^\varepsilon| \leq \alpha) \times \mathbb{P}(|\hat{p}_2^\varepsilon - p_2^\varepsilon| \leq \alpha) \times \mathbb{P}(|p_1^\varepsilon - p_2^\varepsilon| \leq \alpha)$$
$$\geq (1 - \theta)^2 \times 1 = 1 - \xi.$$

In parallel, Lemma 1 guarantees that for h sufficiently small, the global stability error can be uniformly bounded on $\mathcal{B}_{\lambda^*}$ by ε_0. The proof is complete. $\square$

References

1. Baier, C., de Alfaro, L., Forejt, V., Kwiatkowska, M.: Model checking probabilistic systems. In: Clarke, E., Henzinger, T., Veith, H., Bloem, R. (eds.) Handbook of Model Checking, pp. 963–999. Springer, Cham (2018). https://doi.org/10.1007/978-3-319-10575-8_28

2. Bao, R., Attiogbe, C., Delahaye, B., Fournier, P., Lime, D.: Parametric statistical model checking of UAV flight plan. In: Pérez, J.A., Yoshida, N. (eds.) FORTE 2019. LNCS, vol. 11535, pp. 57–74. Springer, Cham (2019). https://doi.org/10.1007/978-3-030-21759-4_4

3. Butcher, J.C.: Numerical Methods for Ordinary Differential Equations, 3rd edn. Wiley, Hoboken (2016). https://doi.org/10.1002/9781119121534

4. Crouzeix, M., Mignot, A.L.: Analyse numérique des équations différentielles, vol. 1. Masson (1984)

5. Daws, C.: Symbolic and parametric model checking of discrete-time Markov chains. In: Liu, Z., Araki, K. (eds.) ICTAC 2004. LNCS, vol. 3407, pp. 280–294. Springer, Heidelberg (2005). https://doi.org/10.1007/978-3-540-31862-0_21

6. Dehnert, C., et al.: PROPhESY: a PRObabilistic ParamEter SYnthesis tool. In: Kroening, D., Păsăreanu, C.S. (eds.) CAV 2015. LNCS, vol. 9206, pp. 214–231. Springer, Cham (2015). https://doi.org/10.1007/978-3-319-21690-4_13

7. Demailly, J.: Analyse numérique et équations différentielles. Grenoble Sciences, EDP Sciences (2012)

8. Dunford, N., Schwartz, J.: Linear Operators. 1. General Theory. A Wiley Interscience Publication. Interscience Publishers (1967)

9. Gainer, P., Hahn, E.M., Schewe, S.: Accelerated model checking of parametric Markov chains. In: Lahiri, S.K., Wang, C. (eds.) ATVA 2018. LNCS, vol. 11138, pp. 300–316. Springer, Cham (2018). https://doi.org/10.1007/978-3-030-01090-4_18

10. Gyori, B.M., Liu, B., Paul, S., Ramanathan, R., Thiagarajan, P.S.: Approximate probabilistic verification of hybrid systems. In: Abate, A., Šafránek, D. (eds.) HSB 2015. LNCS, vol. 9271, pp. 96–116. Springer, Cham (2015). https://doi.org/10.1007/978-3-319-26916-0_6

11. Han, T., Katoen, J.P., Mereacre, A.: Approximate parameter synthesis for probabilistic time-bounded reachability. In: 2008 Real-Time Systems Symposium, pp. 173–182. IEEE (2008). https://doi.org/10.1109/RTSS.2008.19

12. Hoeffding, W.: Probability inequalities for sums of bounded random variables. J. Am. Stat. Assoc. **58**(301), 13–30 (1963). https://doi.org/10.1080/01621459.1963.10500830

13. Katoen, J.P.: The probabilistic model checking landscape. In: Proceedings of the 31st Annual ACM/IEEE Symposium on Logic in Computer Science, pp. 31–45 (2016). https://doi.org/10.1145/2933575.2934574

14. Legay, A., Delahaye, B., Bensalem, S.: Statistical model checking: an overview. In: Barringer, H., et al. (eds.) RV 2010. LNCS, vol. 6418, pp. 122–135. Springer, Heidelberg (2010). https://doi.org/10.1007/978-3-642-16612-9_11

15. Liu, B., Gyori, B.M., Thiagarajan, P.S.: Statistical model checking-based analysis of biological networks. In: Liò, P., Zuliani, P. (eds.) Automated Reasoning for Systems Biology and Medicine. CB, vol. 30, pp. 63–92. Springer, Cham (2019). https://doi.org/10.1007/978-3-030-17297-8_3. https://doi.org/10.48550/arXiv.1812.01091

16. Lotkin, M.: On the accuracy of Runge-Kutta's method. Math. Tables Other Aids Comput. **5**(35), 128–133 (1951). https://doi.org/10.1090/S0025-5718-1951-0043566-3

17. Melica, V., Invernizzi, S., Caristi, G.: Logistic density-dependent growth of an Aurelia aurita polyps population. Ecol. Model. **291**, 1–5 (2014). https://doi.org/10.1016/j.ecolmodel.2014.07.009

18. Perko, L.: Differential Equations and Dynamical Systems. Texts in Applied Mathematics, Springer, New York (2013). https://doi.org/10.1007/978-1-4613-0003-8

19. Petrov, V.V.: Sums of Independent Random Variables. De Gruyter (2022). https://doi.org/10.1515/9783112573006

20. Ramondenc, S., Eveillard, D., Guidi, L., Lombard, F., Delahaye, B.: Probabilistic modeling to estimate jellyfish ecophysiological properties and size distributions. Sci. Rep. **10**(1), 1–13 (2020). https://doi.org/10.1038/s41598-020-62357-5

21. Ramsay, J.O., Hooker, G., Campbell, D., Cao, J.: Parameter estimation for differential equations: a generalized smoothing approach. J. Roy. Stat. Soc. Ser. B (Stat. Methodol.) **69**(5), 741–796 (2007). https://doi.org/10.1111/j.1467-9868.2007.00610.x

22. Restrepo, J.G., Sánchez, C.M.V.: Parameter estimation of a predator-prey model using a genetic algorithm. In: 2010 IEEE ANDESCON, pp. 1–4, September 2010. https://doi.org/10.1109/ANDESCON.2010.5633365

23. Vandermeer, J.H., Goldberg, D.E.: Population ecology. In: Population Ecology. Princeton University Press (2013)

24. Varah, J.M.: A spline least squares method for numerical parameter estimation in differential equations. SIAM J. Sci. Stat. Comput. **3**(1), 28–46 (1982). https://doi.org/10.1137/0903003

POMDP Controllers with Optimal Budget

Jip Spel[(✉)], Svenja Stein, and Joost-Pieter Katoen

RWTH Aachen University, Aachen, Germany
`jip.spel@cs.rwth-aachen.de`

Abstract. Parametric Markov chains (pMCs) have transitions labeled with functions over a fixed set of parameters. They are useful if the exact transition probabilities are uncertain, e.g., when checking a model for robustness. This paper presents a simple way to check whether the expected total reward until reaching a given target state is monotonic in (some of) the parameters. We exploit this monotonicity together with parameter lifting to find an ε-close bound on the optimal expected total reward. Our results are also useful to automatically synthesise controllers with a fixed memory structure for partially observable Markov decision processes (POMDPs), a popular model in AI planning. We experimentally show that our approach can successfully find ε-optimal controllers for optimal budget in such POMDPs.

1 Introduction

POMDPs. Partial-observable Markov Decision Processes (POMDPs, for short) are models that extend probabilistic and non-deterministic behaviour with partial observability [30,36]. Rather than knowing precisely in which state the process is, one only knows the state's observation. As multiple states may have the same observation, the current state cannot be uniquely identified. POMDPs are key models in AI and planning [36], e.g., for robots that only have a partial perception of their environment. POMDP controllers (aka: schedulers or policies) resolve the non-determinism based on the observation history so far, rather than on the state history as is the case in MDPs. This limited information makes optimal decision making harder. Indeed finding a controller to maximise reachability probabilities is undecidable for POMDPs whereas this is a polynomial-time problem for MDPs.

Parametric MCs. A parametric Markov chain (pMC) has state-transition functions that are functions over a fixed set of parameters, e.g., p and $1-p$ for unknown $0 < p < 1$. To relate POMDPs to pMCs, actions in observation-equivalent states are considered as parameters [27, Ch. 7]; e.g., in case of three enabled actions a_1, a_2 and a_3, we map them onto the fresh parameters p_1, p_2

Supported by DFG RTG 2236 "UnRAVeL" and DFG 1462/4-1 "PASYWI".

E. Ábrahám and M. Paolieri (Eds.): QEST 2022, LNCS 13479, pp. 107–130, 2022.
https://doi.org/10.1007/978-3-031-16336-4_6

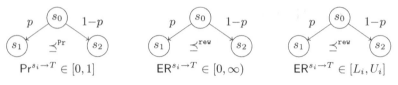

Fig. 1. Different situations for ordering states

and $1-p_1-p_2$. If action a_3 leads to some state with probability $1/4$, then the corresponding function becomes $(1-p_1-p_2)/4$. Finding a POMDP controller with optimal budget is polynomially equivalent to finding an optimal parameter valuation for the expected total reward in the corresponding pMC. The state-of-the-art technique to find an ε-optimal parameter valuation is to use parameter lifting [35]. Put in a nutshell, this approach yields upper- and lower bounds by analysing an abstraction—obtained by relaxing parameter dependencies—of the pMC. As this method does not scale, we take inspiration from [41] and *boost parameter lifting with monotonicity checking* for expected total rewards.

Problem Setting and Approach. To check for monotonicity (see Sect. 3), we impose an order $\preceq^{\mathrm{Pr}}$ on states such that [39]: $s \preceq^{\mathrm{Pr}} s'$ iff $\mathrm{Pr}^{s \to T} \leq \mathrm{Pr}^{s' \to T}$, where $\mathrm{Pr}^{s \to T}$ denotes the probability to reach some state in set T from state s. Note that in our parametric setting $\mathrm{Pr}^{s \to T}$ is a function over the parameters. Consider Fig. 1(a) and assume $s_1 \preceq^{\mathrm{Pr}} s_2$. As p and $(1-p)$ represent probabilities and

$$\mathrm{Pr}^{s_0 \to T} = p \cdot \mathrm{Pr}^{s_1 \to T} + (1-p) \cdot \mathrm{Pr}^{s_2 \to T},$$

it immediately follows $s_1 \preceq^{\mathrm{Pr}} s_0 \preceq^{\mathrm{Pr}} s_2$. Unfortunately, when considering the expected total reward for reaching T from s, denoted $\mathrm{ER}^{s \to T}$, ordering s_0 is not so straightforward. Let $\preceq^{\mathrm{rew}}$ be such that $s \preceq^{\mathrm{rew}} s'$ iff $\mathrm{ER}^{s \to T} \leq \mathrm{ER}^{s' \to T}$. Consider Fig. 1(b) and assume $s_1 \preceq^{\mathrm{rew}} s_2$. As p and $(1-p)$ denote probabilities and

$$\mathrm{ER}^{s_0 \to T} = r(s_0) + p \cdot \mathrm{ER}^{s_1 \to T} + (1-p) \cdot \mathrm{ER}^{s_2 \to T},$$

it immediately follows $s_1 \preceq^{\mathrm{rew}} s_0$. However, ordering s_0 w.r.t. s_2 is in general not possible if $r(s_0) > 0$. Note that for $r(s_0) = 0$ we do obtain $s_1 \preceq^{\mathrm{rew}} s_0 \preceq^{\mathrm{rew}} s_2$.

In order to remedy this deficiency—our aim is to order as many states as possible—we exploit parameter lifting [35], a technique that, with very low overhead, provides bounds for each state s such that $L(s) \leq \mathrm{ER}^{s \to T} \leq U(s)$. This helps to order s_0 w.r.t. s_2; e.g., if $U(s_0) \leq L(s_2)$, $s_0 \preceq^{\mathrm{rew}} s_2$, and similarly if $U(s_2) \leq L(s_0)$, $s_2 \preceq^{\mathrm{rew}} s_0$. More involved ordering possibilities exist that exploit bounds from parameter lifting; this is further detailed in the paper.

Main Contribution. The main contribution of this paper is a set of algorithms to obtain monotonicity and the reward order, and their usage to tackle the ε-optimal problem for expected total reward properties. We realised the monotonicity checking for expected total reward properties on top of the Storm [24] model checker.

Experiments show that monotonicity checking enables the ε-optimal synthesis of POMDP controllers for expected total rewards for benchmarks with up to hundreds of observations whereas plain (vanilla) parameter lifting fails.

Organisation of the Paper. Section 2 provides the technical background and formalises the problem. Section 3 defines monotonicity and explains how to obtain monotonicity with help of a reward order. Section 4 describes how we combine parameter lifting and monotonicity to find an ε-optimal parameter valuation, and presents the used heuristics. Section 5 reports on the experimental results. Finally, Sect. 6 discusses related work while Sect. 7 concludes.

2   Preliminaries

A *probability distribution* over a countable set X is a function $\mu \colon X \to [0, 1] \subseteq \mathbb{R}$ with $\sum_{x \in X} \mu(x) = 1$. Let $Distr(X)$ denote the set of all probability distributions on X. For the set of n real-valued *parameters* (or *variables*) $V = \{p_1, \ldots, p_n\}$, let $\mathbb{Q}[V]$ denote the set of multivariate polynomials with rational coefficients over V. For a polynomial f and variable x, we write $x \in f$ if the variable occurs in the polynomial f. An *instantiation* for a finite set V of real-valued variables is a function $u \colon V \to \mathbb{R}$. We often denote u as a vector $\vec{u} \in \mathbb{R}^n$ with $u_i := u(x_i)$ for $x_i \in V$. A polynomial f can be interpreted as a function $f \colon \mathbb{R}^n \to \mathbb{R}$, where $f(\vec{u})$ is obtained by substitution, i.e., $f[\vec{x} \leftarrow \vec{u}]$, where each occurrence of x_i in f is replaced by u_i.

Definition 1 (pMC). *A* parametric Markov Chain (pMC) *is a tuple* $\mathcal{M} = (S, s_I, T, V, \mathcal{P})$ *with a finite set S of* states, *an* initial state $s_I \in S$, *a finite set* $T \subseteq S$ *of* target states, *a finite set V of real-valued variables* (parameters) *and a transition function* $\mathcal{P} \colon S \times S \to \mathbb{Q}[V]$.

A pMC $\mathcal{M}$ is *well-defined* if the transition function yields *well-defined* probability distributions, i.e., $\mathcal{P}(s, \cdot) \in Distr(S)$ for each $s \in S$. A well-defined pMC $\mathcal{M}$ is a Markov chain (MC) if V is empty. Applying an *instantiation* $\vec{u}$ to a pMC $\mathcal{M}$ yields $\mathcal{M}[\vec{u}]$ by replacing each $f \in \mathbb{Q}[V]$ in $\mathcal{M}$ by $f(\vec{u})$. An instantiation $\vec{u}$ is *well-defined* for $\mathcal{M}$ if $\mathcal{M}[\vec{u}]$ is an MC. A well-defined instantiation $\vec{u}$ is *graph-preserving* for $\mathcal{M}$ if the topology is preserved, i.e., $\mathcal{P}(s, s') \neq 0$ implies $\mathcal{P}(s, s')(\vec{u}) \neq 0$ for all states $s, s' \in S$. A set of instantiations is called a *region*. A region R is well-defined (graph-preserving) if $\vec{u}$ is well-defined (graph-preserving) for all $\vec{u} \in R$. In this paper, we only consider graph-preserving regions. Finally, let $\mathsf{occur}(s)$ be the set of variables $\{x \in V \mid \exists s' \in S. \ x \in \mathcal{P}(s, s')\}$. A state s is called *parametric*, if $\mathsf{occur}(s) \neq \emptyset$; we write $\mathsf{occur}(s) = x$ if $\{x\} = \mathsf{occur}(s)$.

Example 1. Figure 2(a) depicts pMC $\mathcal{M}$ with a single parameter p. Region $R = [0.1, 0.9]$ is graph-preserving, $R' = [0, 0.9]$ is not graph-preserving but well-defined, and $R'' = [-0.1, 0.9]$ is not well-defined.

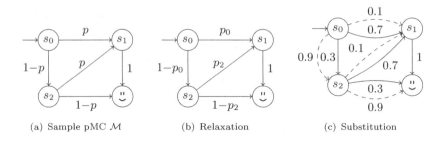

(a) Sample pMC $\mathcal{M}$ (b) Relaxation   (c) Substitution

Fig. 2. Obtaining bounds for a pMC

2.1   Rewards for pMCs

We consider expected total reward properties for pMCs. We let $Paths(s)$ (see e.g., [3, Ch. 10]) denote the set of all *infinite paths* starting in s. A *finite path* is a finite, non-empty prefix of an infinite path. Let *state reward function* $r\colon S \to \mathbb{R}_{\geq 0}$ associate a non-negative reward to each state in a pMC. The cumulative reward for the finite path $\hat{\pi} = s_0 s_1 \ldots s_n$ is defined by:

$$r(\hat{\pi}) = r(s_0) + r(s_1) + \ldots + r(s_{n-1}).$$

Rewards are thus earned when leaving a state. For infinite path $\pi = s_0 s_1 s_2 \cdots$, the reward to eventually reach T in $\mathcal{M}$ is the cumulated reward until reaching T for the first time. Formally:

$$r(\pi, \Diamond T) = \begin{cases} r(s_0 s_1 \ldots s_n) & \text{if } s_i \notin T \text{ for } 0 \leq i < n \text{ and } s_n \in T \\ \infty & \text{if } \pi \not\models \Diamond T. \end{cases}$$

Let $\mathrm{Pr}(\hat{\pi}) = \sum_{i=0}^{n-1} \mathcal{P}(s_i, s_{i+1})$ be the probability of a finite path $\hat{\pi} = s_0 s_1 \ldots s_n$. This can be lifted to the probability of an infinite path $(\mathrm{Pr}(\pi))$ via a cylinder set construction [3, Ch. 10].

Definition 2 (Expected total reward). *The* expected total reward *until reaching T from $s \in S$ for an MC $\mathcal{M}$ is defined as follows:*

$$\mathrm{ER}^s_{\mathcal{M}}(\Diamond T) = \sum_{\pi \in Paths(s)} \mathrm{Pr}(\pi) \cdot r(\pi, \Diamond T).$$

As state rewards are non-negative, expected total rewards are in $\mathbb{R}_{\geq 0}$ or equal ∞.

The expected total reward for an MC is defined by $\mathrm{ER}_{\mathcal{M}}(\Diamond T) = \mathrm{ER}^{s_I}_{\mathcal{M}}(\Diamond T)$. The expected total reward for a pMC $\mathcal{M}$ is defined analogously, however it represents a *function* $\mathrm{ER}^{s \to T}_{\mathcal{M}}\colon V \to \mathbb{R}$, given by $\mathrm{ER}^{s \to T}_{\mathcal{M}}(\vec{u}) = \mathrm{ER}^s_{\mathcal{M}[\vec{u}]}(\Diamond T)$. For conciseness we typically omit the subscript $\mathcal{M}$. On a graph-preserving region, the function $\mathrm{ER}^{s \to T}$ is always continuously differentiable [35] and admits a closed form as a rational function over V [14,21].

Remark 1. We restrict ourselves to pMCs where all infinite paths eventually reach T. Otherwise, $\mathsf{ER}^{s_I \to T} = \infty$.

Example 2. Reconsider the pMC $\mathcal{M}$ from Fig. 2(a) with target state ☺ and state reward function $r(s_i) = i$. Note that the reward for ☺ can be ignored as ☺ is the target state. The expected total reward function $\mathsf{ER}^{s_0 \to ☺}$ is given by $1 \cdot p + 2 \cdot (1-p)^2 + 3 \cdot (1-p) \cdot p$.

2.2 Problem Statement

This paper is concerned with the following questions for a given pMC $\mathcal{M}$ with target states T and region R:

> *Optimal synthesis.* Find the instantiation $\vec{u}^*$ such that
>
> $$\vec{u}^* = \arg\max_{\vec{u} \in R} \mathsf{ER}_{\mathcal{M}[\vec{u}]}(\lozenge T)$$
>
> *ε-Optimal synthesis.* Given tolerance $\varepsilon \geq 0$, find an instantiation $\vec{u}^*$ such that
>
> $$\max_{\vec{u} \in R} \mathsf{ER}_{\mathcal{M}[\vec{u}]}(\lozenge T) \cdot (1-\varepsilon) \leq \mathsf{ER}_{\mathcal{M}[\vec{u}^*]}(\lozenge T) \leq \max_{\vec{u} \in R} \mathsf{ER}_{\mathcal{M}[\vec{u}]}(\lozenge T) .$$

The optimal synthesis problem is co-ETR-hard [44][1], the same applies to ε-optimal synthesis, as computing $\max_{\vec{u} \in R} \mathsf{ER}_{\mathcal{M}[\vec{u}]}(\lozenge T)$ is ETR-complete.

2.3 Parameter Lifting

The state-of-the-art approach for tackling the ε-optimal synthesis problem is by using parameter lifting [35]. Parameter lifting computes lower and upper bounds for the expected total reward at the states in the pMC $\mathcal{M}$. For state s and region R, parameter lifting computes bounds $L_R(s)$ and $U_R(s)$ satisfying

$$L_R(s) \leq \mathsf{ER}^s_{\mathcal{M}[\vec{u}]}(\lozenge T) \leq U_R(s) \quad \text{for all } \vec{u} \in R .$$

It does so by first removing all parameter dependencies in the pMC $\mathcal{M}$. This is accomplished by replacing all parameters by fresh ones, yielding the so-called relaxation, the pMC $\mathsf{relax}(\mathcal{M})$ (see Fig. 2(b)). The relaxed pMC is in fact an abstraction of the pMC $\mathcal{M}$ as parameter dependencies have been dropped. After this, the relaxed pMC is transformed into a *parameter-less* MDP $\mathcal{M}'$ by substituting all transitions by non-deterministic choices for the probabilistic transitions (see Fig. 2(c)). To find the maximal/minimal expected total reward in the MDP $\mathcal{M}'$, standard MDP verification techniques are employed.

---

[1] ETR = Existential Theory of the Reals. ETR-complete decision problems are as hard as finding the roots of a multivariate polynomial. ETR lies inbetween NP and PSPACE.

Example 3. Reconsider Example 2 and assume region $R = [0.1, 0.7]$. Removing all parameter dependencies yields the relaxed pMC relax($\mathcal{M}$) of Fig. 2(b). As $p \in [0.1, 0.7]$ we have $p_0, p_2 \in [0.1, 0.7]$.

Figure 2(c) shows the MDP $\mathcal{M}'$ in which all parametric transitions in the pMC relax($\mathcal{M}$) are replaced by a non-deterministic choice for the value of the parameter, i.e., we have a transition for both the lower bound (dashed arrow) and the upper bound (solid arrow) of the parameter. We now apply standard MDP model checking to maximize the expected total reward in $\mathcal{M}'$. At state s_0, the dashed arrow, representing the lower bound for parameter p_0, will be chosen. At state s_2 however, the solid arrow, representing the upper bound for parameter p_2, is chosen. This yields $\mathsf{ER}^{s_0 \to \smile} \leq 0.1{\cdot}1 + 0.9{\cdot}(2 + 0.7{\cdot}1) = 2.53 = U_R(s_0)$. Similarly, when minimizing, we obtain $L_R(s_0) = 0.7{\cdot}1 + 0.3{\cdot}(2 + 0.1{\cdot}1) = 1.33 \leq \mathsf{ER}^{s_0 \to \smile}$.

The obtained bounds on region R can now be used to find an ε-optimal solution in R as follows. Recall that we need to find an instantiation $\vec{u}^*$ with:

$$\max_{\vec{u} \in R} \mathsf{ER}_{\mathcal{M}[\vec{u}]}(\Diamond T){\cdot}(1 - \varepsilon) \leq \mathsf{ER}_{\mathcal{M}[\vec{u}^*]}(\Diamond T).$$

By construction, it holds for the initial state s_I: $\max_{\vec{u} \in R} \mathsf{ER}_{\mathcal{M}[\vec{u}]}(\Diamond T) \leq U_R(s_I)$. If now, for some $\vec{u}^* \in R$ we have that $\mathsf{ER}_{\mathcal{M}[\vec{u}^*]}(\Diamond T) \geq U_R(s_I){\cdot}(1 - \varepsilon)$, then $\vec{u}^*$ is a solution to the ε-optimal synthesis problem.

This suggests the following procedure to find a $\vec{u}$ in region R: 1) Initially, we add region R with (trivial) bound ∞ to the (initially empty) queue of regions, then we 2) pick a region R from the queue, 3) pick a point $\vec{u} \in R$, 4) update the current maximum seen (CurMax) and store the instantiation at $\vec{u}^*$, 5) use parameter lifting to compute $U_R(s_I)$. If CurMax $\geq U_{R'}(s_I){\cdot}(1 - \varepsilon)$ for all regions R' in the queue, then we stop, otherwise we split R, e.g., at its centre point, into regions $R_1, \ldots, R_n$, add the regions to the queue with bound U_R and continue at 2). Now $\vec{u}^*$ is an ε-optimal maximum. Similarly, one could use the lower bound L to obtain an ε-optimal minimum.

This procedure is relatively simple, but does however not scale to practical cases. As indicated in Table 2 on page xx (columns vanilla), *none* of the POMDP benchmarks could be handled. Not even the benchmarks with three or nine parameters, and less than hundred states. The key idea of this paper—inspired by our earlier work [41] on reachability probabilities—is to exploit monotonicity, i.e., identifying parameters for which the function $\mathsf{ER}^{s \to T}$ is monotonic. To that end, we first define monotonicity for expected total reward properties (Sect. 3.1). Secondly, in Sect. 3.2 we tackle the main challenge of our approach: building a reward order which contains as many states as possible, see also Sect. 1.

3   Monotonicity

In this section, we define monotonicity for the expected total reward function in a pMC. Furthermore, we show how to find monotonic parameters with the help of a reward order on the state space of a pMC.

3.1 Defining Monotonicity

We distinguish between *local* and *global* monotonicity. Whereas local monotonicity only takes into consideration a given state s and the monotonicity of the transition functions for the direct successors of s, global monotonicity focusses on monotonicity on the entire pMC. We adapt the definitions of [39] for expected total rewards.

Definition 3 (Global monotonicity). *A continuously differentiable function* f *on region* R *is* monotonic increasing *in parameter* p, *denoted* $f \uparrow_p^R$, *if* $\frac{\partial}{\partial p} f(\vec{u}) \geq 0$ *for all* $\vec{u} \in R$.[2] *The pMC* $\mathcal{M} = (S, s_I, T, V, \mathcal{P})$ *is* monotonic increasing *in parameter* $p \in V$ *on graph-preserving region* R, *written* $\mathcal{M} \uparrow_p^R$, *if* $\mathsf{ER}^{s_I \to T} \uparrow_p^R$.

Monotonic *decreasing*, written $\mathcal{M} \downarrow_x^R$, is defined analogously.
    Let $\mathsf{succ}(s) = \{s' \in S \mid \mathcal{P}(s, s') \neq 0\}$ be the set of direct successors of s. Given the recursive equation $\mathsf{ER}^{s \to T} = r(s) + \sum_{s' \in \mathsf{succ}(s)} \mathcal{P}(s, s') \cdot \mathsf{ER}^{s' \to T}$, we have

$$\mathcal{M} \uparrow_p^R \quad \text{iff} \quad \frac{\partial}{\partial p} \left(\sum_{s' \in \mathsf{succ}(s_I)} \mathcal{P}(s_I, s') \cdot \mathsf{ER}^{s' \to T} \right)(\vec{u}) \geq 0 \,,$$

for all $\vec{u} \in R$. Note that the state reward $r(s_I)$ vanished, as it is constant.
    Checking for global monotonicity is co-ETR hard [39]. Therefore, we focus on local monotonicity of a parameter at a given state s.

Definition 4 (Locally monotonic increasing). $\mathsf{ER}^{s \to T}$ *is locally monotonic increasing in parameter* p *(at* s*) on region* R, *denoted* $\mathsf{ER}^{s \to T} \uparrow_p^{\ell, R}$, *if:*

$$\left(\sum_{s' \in \mathsf{succ}(s)} \left(\frac{\partial}{\partial p} \mathcal{P}(s, s') \right) \cdot \mathsf{ER}^{s' \to T} \right)(\vec{u}) \geq 0 \quad \text{for all } \vec{u} \in R.$$

For conciseness, we typically omit superscript R for $\uparrow^R$ and $\uparrow^{\ell, R}$.

Example 4. Reconsider pMC $\mathcal{M}$ of Fig. 2(a) with reward function $r(s_i) = i$ and target state $\ddot{\smile}$. Observe that $\mathsf{ER}^{s_1 \to \ddot{\smile}} < \mathsf{ER}^{s_2 \to \ddot{\smile}}$ and $\mathsf{ER}^{\ddot{\smile} \to \ddot{\smile}} < \mathsf{ER}^{s_1 \to \ddot{\smile}}$. From Definition 4 we obtain: $\mathsf{ER}^{s_0 \to \ddot{\smile}} \downarrow_p^{\ell}$ and $\mathsf{ER}^{s_2 \to \ddot{\smile}} \uparrow_p^{\ell}$.

If a parameter p is locally monotonic increasing (decreasing) at all states, the reward function $\mathsf{ER}^{s_i \to T}$ is globally monotonic increasing (decreasing) in p:

Theorem 1.
$$\left(\forall s \in S. \; \mathsf{ER}^{s \to T} \uparrow_p^{\ell} \right) \quad \text{implies} \quad \mathsf{ER}^{s_I \to T} \uparrow_p.$$

A proof sketch can be found in Appendix C.1.

---

[2] To be precise, on the interior of the closed set R.

Example 5. Reconsider pMC relax($\mathcal{M}$) of Fig. 2(b) with reward function $r(s_i) = i$ and target state $\smile$. Following Example 4 we obtain $\mathsf{ER}^{s_0 \to T} \downarrow^\ell_{p_0}$. For all $s \neq s_0$,

we obtain $\sum\limits_{s' \in \mathsf{succ}(s)} \left(\dfrac{\partial}{\partial p_0} \mathcal{P}(s, s') \right) \cdot \mathsf{ER}^{s \to T} = 0$, as p_0 doesn't occur at those states.

Therefore, $\mathsf{ER}^{s \to T} \downarrow^\ell_{p_0}$ for all $s \in S$. Thus for relax($\mathcal{M}$): $\mathsf{ER}^{s_I \to \smile} \downarrow_{p_0}$. Similarly, we obtain $\mathsf{ER}^{s_I \to \smile} \uparrow_{p_2}$.

Remark 2. Here, the state rewards are constants. Monotonicity in the presence of parametric rewards can be accommodated by extending the notion of local monotonicity by including $\dfrac{\partial}{\partial p} r(s)$ in the sum.

Local Monotonicity and Parameter Lifting. Local monotonicity simplifies parameter lifting. This can be seen as follows. Assume for given pMC $\mathcal{M}$, we want to obtain an upper bound with help of parameter lifting. Recall that in parameter lifting, the parametric transition for a state s is replaced by a non-deterministic choice between the lower and upper bound of the parameters occurring at the transition. If s is locally monotonic increasing in p, we know that picking the upper bound for p in the region R will maximize the expected total reward for s. Thus the non-deterministic choice between l_p and u_p at state s in the MDP can be replaced by a deterministic one for u_p.

3.2   Reward Order

To determine local monotonicity at state s, we check for *sufficient conditions* for monotonicity at s. These conditions are based on constructing a pre-order on the states of the pMC, the so-called *reward order*. This is again an adaptation of [39] for expected total rewards.

Definition 5 (Reward order). *An ordering relation $\preceq^{rew}_{R,T} \subseteq S \times S$ is a reward order w.r.t. target set $T \subseteq S$ and region R if for all $s, t \in S$:*

$$s \preceq^{rew}_{R,T} t \quad \text{implies} \quad \forall \vec{u} \in R.\ \mathsf{ER}^{s \to T}(\vec{u}) \leq \mathsf{ER}^{t \to T}(\vec{u}).$$

The order $\preceq^{rew}_{R,T}$ is called exhaustive *if the reverse implication holds too.*

For conciseness, we omit R and T if they are clear from the context.

If the direct successor states of s are ordered, and the transition functions are monotonic, we can obtain local monotonicity:

Lemma 1. *Let $\mathsf{succ}(s) = \{s_1, \ldots, s_n\}$, $f_i = \mathcal{P}(s, s_i)$, and $\forall j > i.\ s_j \preceq^{rew} s_i$. Then:*

$$\exists i \in [1, \ldots, n]. \left(\forall j \leq i.\ f_j \uparrow_p \text{ and } \forall j > i.\ f_j \downarrow_p \right) \text{ implies } \mathsf{ER}^{s \to T}_\mathcal{M} \uparrow^\ell_p.$$

As the rewards are constant, the proof follows directly from [40, Lemma 1].

Example 6. Reconsider pMC $\mathcal{M}$ of Fig. 2(a) with reward function $r(s_i) = i$ and target state $\smile$. From Definition 5, it follows $\smile \preceq^{\text{rew}} s_1$. Furthermore, $\mathcal{P}(s_2, s_1)\!\uparrow_p$ and $\mathcal{P}(s_2, \smile)\!\downarrow_p$. From this we obtain $\text{ER}^{s_2 \to T}\!\uparrow_p^\ell$.

As it suffices for local monotonicity to only have the direct successors of states ordered, we focus on a so-called "sufficient" reward order:

Definition 6 (Sufficient reward order). *Reward order $\preceq^{\text{rew}}$ is sufficient for $s \in S$ if for all $s_1, s_2 \in \text{succ}(s)$: $(s_1 \preceq^{\text{rew}} s_2 \vee s_2 \preceq^{\text{rew}} s_1)$ holds. The reward order is sufficient for pMC $\mathcal{M}$ if it is sufficient for all parametric states in $\mathcal{M}$.*

Phrased differently, the reward order $\preceq^{\text{rew}}$ is sufficient for $p \in V$ if $(\text{succ}(s), \preceq^{\text{rew}})$ is a total order for all $s \in S$ with $p \in \text{occur}(s)$. Definition 6 and Lemma 1 yield:

Corollary 1. *Let pMC $\mathcal{M}$ be such that for each $s \in S$, $|\text{succ}(s)| \leq 2$ and $\mathcal{P}(s, s')$ is monotonic for each $s' \in \text{succ}(s)$. Then: $\preceq^{\text{rew}}$ is sufficient for s implies $\text{ER}^{s \to T}$ is locally monotonic increasing/decreasing on any region R in all parameters.*

3.3   Constructing Reward Orders

Our aim now is to construct a sufficient reward order. The construction is inspired by an observation, graph rules, assumptions, and rules for bounds on states. To be able to construct the reward order, let us introduce some notation.

Let $\prec^{\text{rew}}_{R,T}$ denote the irreflexive variant of $\preceq^{\text{rew}}_{R,T}$ and $s \equiv^{\text{rew}}_{R,T} s'$ denote $\forall \vec{u} \in R$. $\text{ER}^{s \to T}(\vec{u}) = \text{ER}^{s' \to T}(\vec{u})$. Let $[s]^{rew}_{[R,T]}$ denote the set of all $s' \in S$ such that $s \equiv^{\text{rew}}_{R,T} s'$. We omit R and T for conciseness. For $X \subseteq S$, $\text{ub}(X) = \{s \in S \mid X \preceq^{\text{rew}} s\}$ and $\text{lb}(X) = \{s \in S \mid s \preceq^{\text{rew}} X\}$ denote the upper and lower bounds of X w.r.t. reward order $\preceq^{\text{rew}}$. Furthermore, let $\min(X) = \{x \in X \mid \nexists x' \in X.x' \preceq^{\text{rew}} x\}$, and $\max(X) = \{x \in X \mid \nexists x' \in X.x \preceq^{\text{rew}} x'\}$.

Observation. By definition, $\text{ER}^{t \to T} = 0$ for all $t \in T$. Furthermore, for all $s \in S \setminus T$, we have $r(s) \geq 0$, thus $\text{ER}^{s \to T} \geq 0$. From this we obtain $t \preceq^{\text{rew}} s$ for all $s \in S \setminus T$. Note that, as all states have T as lower bound, $\text{lb}(X) \neq \emptyset$ for each $X \subseteq S$.

Graph Rules. We use the graph structure of the pMC to obtain the reward order.

Lemma 2. *For $s \in S$:*

1. *if $r(s) > 0$, then $\text{lb}(\text{succ}(s)) \prec^{rew} s$,*
2. *if $r(s) = 0$ and $\text{ub}(\text{succ}(s)) \neq \emptyset$, then either $\text{succ}(s) \subseteq [s]^{rew}$ or $\text{lb}(\text{succ}(s)) \prec^{rew} s \prec^{rew} \text{ub}(\text{succ}(s))$.*

To deal with cycles we introduce the following lemma:

Lemma 3. *For any state s with $\text{succ}(s) = \{s_1, s_2\}$ the following holds:*

1. *if $s_1 \equiv^{rew} s$ and $r(s) = 0$, then $s_2 \equiv^{rew} s$,*
2. *if $s_1 \equiv^{rew} s$ and $r(s) > 0$, then $s_2 \prec^{rew} s$,*

3. *if* $s \prec^{rew} s_1$, *then* $s_2 \prec^{rew} s$.
4. *if* $s_1 \prec^{rew} s$ *and* $r(s) = 0$, *then* $s \prec^{rew} s_2$.

The idea is that if a state s has only two successors and the relation between state s and one of its successors is known, we can also infer in many cases the relation of s to the other successor.

Remark 3. Lemma 2 case 2 is equivalent to [39, Lemma 3]. Lemma 3 for $r(s) = 0$ is equivalent to [39, Lemma 4]. As explained in the introduction of the paper, the main challenge lies at the cases where $r(s) > 0$.

Appendix A contains the algorithm (Algorithm 1) to obtain a reward order based on Lemmas 2 and 3.

Lemma 4. *Algorithm 1 returns a set with one reward order.*

Assumptions. The above approach does not necessarily yield a sufficient reward order. It might be that we need assumptions to obtain a sufficient reward order.

Definition 7 (Order with assumptions). *Let* $\preceq^{rew}$ *be a reward order, and* $\mathcal{A} = (\mathcal{A}_{\prec^{rew}}, \mathcal{A}_{\equiv^{rew}})$ *with assumptions* $\mathcal{A}_{\prec^{rew}}, \mathcal{A}_{\equiv^{rew}} \subseteq S \times S$. *Then* $(\preceq^{rew,\mathcal{A}}, \mathcal{A})$ *is an order with assumptions where* $\preceq^{rew,\mathcal{A}} = \left(\preceq^{rew} \cup \mathcal{A}_{\prec^{rew}} \cup \mathcal{A}_{\equiv^{rew}}\right)^*$.

Lemma 5. *If assumptions* $\mathcal{A} = (\mathcal{A}_{\prec^{rew}}, \mathcal{A}_{\equiv^{rew}})$ *satisfy:*

$$(s,t) \in \mathcal{A}_{\prec^{rew}} \quad implies \; \forall \vec{u} \in R. \; \mathsf{ER}^{s \to T}(\vec{u}) < \mathsf{ER}^{t \to T}(\vec{u}), \; and$$
$$(s,t) \in \mathcal{A}_{\equiv^{rew}} \quad implies \; \forall \vec{u} \in R. \; \mathsf{ER}^{s \to T}(\vec{u}) = \mathsf{ER}^{t \to T}(\vec{u}),$$

then $\preceq^{rew,\mathcal{A}}$ *is a reward order, and we call* $\mathcal{A}$ *(globally) valid.*

The following results refer to Algorithms 1 and 2 which are found in Appendix A. Algorithm 2 extends Algorithm 1 such that we obtain sufficient reward orders. It is based on Lemma 5.

Theorem 2. *For every order with assumptions* $(\preceq^{rew,\mathcal{A}}, \mathcal{A})$ *computed by Algorithms 1 and 2 in Appendix A: If* $\preceq^{rew,\mathcal{A}}$ *is a reward order, then it is sufficient.*

Theorem 3. *If all orders computed by Algorithms 1 and 2 in Appendix A are witnesses for a parameter to be monotonic increasing (decreasing), then the parameter is indeed monotonic increasing (decreasing).*

Using Bounds. We assume for state s and region R to have bounds $L_R(s)$ and $U_R(s)$ at our disposal satisfying

$$L_R(s) \leq \mathsf{ER}^s_{\mathcal{M}[\vec{u}]}(\lozenge T) \leq U_R(s) \quad \text{for all } \vec{u} \in R .$$

Such bounds can be trivially assumed to be 0 and ∞ respectively, but the idea is to obtain tighter bounds by exploiting parameter lifting. A simple observation on these bounds yields a cheap rule (provided these bounds can be easily obtained) to order states.

Lemma 6. *For $s_1, s_2 \in S$ and region R: $U_R(s_2) \leq L_R(s_1)$ implies $s_2 \preceq^{rew} s_1$.*

We can also use the bounds and $r(s)$ to order state s w.r.t. its successors.

Lemma 7. *For any state s with $\mathsf{succ}(s) = \{s_1, s_2\}$, $f = \mathcal{P}(s, s_1)$, region R and $s_1 \preceq^{rew} s_2$: if for all $\vec{u} \in R$*

1. *$r(s) \geq f(\vec{u}) \cdot (U_R(s_2) - L_R(s_1))$ then $s_2 \preceq^{rew} s$, and*
2. *$r(s) \leq f(\vec{u}) \cdot (L_R(s_2) - U_R(s_1))$ then $s \preceq^{rew} s_2$.*

The proof is found in Appendix C.2.

Example 7. Reconsider the situation in Fig. 1(c), let $r(s_0) = 5$ and $p \in [0.1, 0.7]$. Assume $L(s_1) = 3$, $U(s_1) = 4$, $L(s_2) = 5$, and $U(s_2) = 10$. From this we have $L(s_0) = 8.6$, $U(s_0) = 14.4$. To order s_0 w.r.t. s_2, we cannot apply Lemma 6 as $U(s_2) > L(s_0)$ and $U(s_0) > L(s_2)$. However, by Lemma 7.1 we obtain $s_2 \preceq^{rew} s_0$.

We extend Lemma 7 to several successor states as follows:

Lemma 8. *Let $\mathsf{succ}(s) = \{s_1, \ldots, s_n\}$, $j \in \{1, \ldots n\}$, and $f_j = \mathcal{P}(s, s_j)$. If for all $1 \leq i < j$. $s_i \preceq^{rew} s_j$ then:*

$$r(s) \geq (1-f_j) \cdot (U_R(s_j) - L_R(s_1)) \text{ implies } s_j \preceq^{rew} s$$

The proof is similar to the proof of Lemma 7. Note that the dual of Lemma 8, i.e.,

$$r(s) \leq (1-f_j) \cdot (L_R(s_j) - U_R(s_n)) \text{ implies } s_j \preceq^{\texttt{rew}} s$$

does not help building the order. As $s_j \preceq^{\texttt{rew}} s_n$, we obtain $(L_R(s_j) - U_R(s_n)) \leq 0$ yielding a reward of at most 0.

We extend Algorithm 1 in Appendix A by Algorithm 3 which is based on Lemmas 6 to 8.

3.4   Reachability Probabilities

To conclude this section, we show that the presented approach is a conservative extension of our earlier work [39] on reachability probabilities. The probability to reach some state $\smile$ from state s in pMC $\mathcal{M}$, denoted $\mathsf{Pr}_{\mathcal{M}}^{s \to \smile}$, can be obtained using expected total rewards. In fact, we can create a pMC $\mathcal{M}'$, a mild variant of $\mathcal{M}$, such that $\mathsf{Pr}_{\mathcal{M}}^{s_I \to \smile} = \mathsf{ER}_{\mathcal{M}'}^{s_I \to \smile'}$. This goes as follows. We obtain $\mathcal{M}'$ by adding a sink state $\smile'$ to S (yielding S') with for any $s \in S$, $\mathcal{P}(s, \smile') = 1$ if $\mathcal{P}(s, s) = 1$ and $\mathcal{P}(s, \smile') = 0$ otherwise. Furthermore, all states $\mathcal{M}'$ are equipped with reward zero, except that $r(\smile) = 1$. The quantity $\mathsf{ER}^{s_0 \to \smile'}$ in $\mathcal{M}'$ now equals the reachability probability of eventually reaching $\smile$ in $\mathcal{M}$. The following lemma asserts that the reward order applied to $\mathcal{M}'$ coincides with the reachability order in [39] on $\mathcal{M}$.

Lemma 9. *Let pMC $\mathcal{M} = (S, s_I, T, V, \mathcal{P})$ and region R, and $\mathcal{M}'$ be the equivalent pMC for expected total rewards. Then:*

$$\text{for any } s, t \in S. \quad \underbrace{s \preceq^{rew}_{\mathcal{M}'} t}_{\text{this paper}} \quad \text{iff} \quad \underbrace{s \preceq^{Pr}_{\mathcal{M}} t}_{\text{defined in [39]}} \quad .$$

4   Divide and Conquer

At the end of Secti. 2.3, we argued that the vanilla parameter lifting approach does not scale for ε-optimal synthesis. In Sect. 3.2, we discussed how to build a reward order and how this is used to determine monotonicity. We now bring these things together. This divide-and-conquer approach is highly inspired by [41] that considered reachability properties rather than reward properties. We observe that the approach is similar, up to building the reward order. Furthermore, we improved the heuristics of [41].

We first explain how the divide-and-conquer approach works for parameter lifting with monotonicity. Then, we discuss the heuristics.

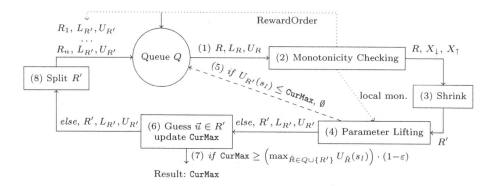

Fig. 3. The symbiosis of monotonicity checking and parameter lifting.

Approach. Figure 3 shows how the extremal value for region R_ι, pMC $\mathcal{M}$, reward property φ and precision ε can be computed using *both parameter lifting and monotonicity* [41]. As explained in Sect. 2.3, the main idea is to analyze regions and split them if the result is inconclusive. The approach uses a queue of regions that need to be checked and the extremal value CurMax found so far. We iteratively check regions and improve both bounds until a satisfactory solution to the ε-optimal synthesis problem is found.

Initially, the queue only contains R_ι. The bounds for this region are set at 0 and ∞. For a selected R from the queue we first check for monotonicity (2). If there is monotonicity, we shrink the region to R' by fixing the monotonic parameters to their bounds (3). We then compute the upper and lower bound ($U_{R'}$ and $L_{R'}$) with parameter lifting. If $U_{R'}$ at the initial state is below CurMax, we can safely discard R' (5). Otherwise, we try to improve CurMax by guessing $\vec{u} \in R'$ and computing $\mathsf{ER}^{s \to T}(\vec{u})$ (6). If $\mathsf{ER}^{s \to T}(\vec{u})$ exceeds CurMax, we update CurMax. We now check if we can terminate. Let the maximum so far be bounded by $\max_{\hat{R} \in Q \cup \{R'\}} U_{\hat{R}}(s_I)$. We multiply this by $(1-\varepsilon)$. If the result is below CurMax, we are done, and return CurMax together with its associated $\vec{u}$ (7). Otherwise, we continue and split R' into smaller regions (8). This is the same procedure as in [41]. However, the monotonicity checking is now tailored to rewards.

Heuristics - Initializing `CurMax`. As in [41], we sample each parameter independently to find out which parameters are definitely not monotonic. We skip parameters already known to be monotonic. We restrict ourselves to ten samples. Non-monotonic parameters are set to the middle point of their interval (as described by the region). For the other parameters, we use the results of the sampling, to determine whether they are possibly monotonic. When maximizing the expected total reward, these parameters are set at the upper (lower) bound if possibly monotonic increasing (decreasing). Furthermore, we store the possibly monotonic parameters for the next two heuristics.

Heuristics - Updating `CurMax`. To update `CurMax`, we consider two points. As first point, we take (like the original approach) the middle point of the region. For the second point, we take the same approach as for initializing `CurMax`. So we set the non-monotonic parameters to their middle point, and the possibly monotonic parameters to their upper/lower bound. Note that we use the initial sampling results, as re-sampling results in overhead.

Heuristics - Creating Copies of the Reward Order. As long as the reward order is sufficient for less than 25% of the states, we continue extending the reward order based on the bounds obtained from parameter lifting.

5   Experimental Validation

We built a prototypical implementation [42] of the techniques advocated in this paper on top of the Storm model checker [24]. The main goal is to compare the plain (vanilla) implementation using parameter lifting (as described in Sect. 2.3) to the integrated implementation using both parameter lifting and monotonicity (as described in Sect. 4).

Setup. We compare the integrated approach to the vanilla approach. Both versions are implemented in Storm and use the same underlying data structures and version of Storm. All experiments were executed on a single core Intel Xeon Platinum 8160 CPU. We did neither use any parallel processing nor randomization. We used a time out of 3600 s and a memory limit of 32 GB. We exclude model-building times from all experiments and emphasize that they coincide for the vanilla and integrated approach. Finally, we set the region for all parameters to [0.1, 0.4], as this increases the number of parameters that are possibly monotonic (Table 1, column $|V|\uparrow^R_?$). This allows us to argue about the influence of the heuristics for `CurMax`.

Benchmarks. We consider all POMDP benchmarks with expected total reward properties from [6]. The `network` benchmark is a unitization example [46], "ps" refers to successfully delivered packets and "dp" refers to dropped packets. The `maze` and `4 × 4 grid` benchmarks are slight extensions of the benchmarks presented in [32]. The `samplerocks` benchmarks [5, 38] model a rover science exploration. The rover has a limited amount of fuel and can only increase this by sampling rocks. The problem however is that the rover doesn't know in advance whether a rock will increase the fuel limit or not. The goal is to reach a given position.

We amended the properties from [6] such that a target state is almost surely reached; this ensures that the expected total reward is finite. The POMDPs are translated into pMCs using the approach in [29]. We excluded typical pMC examples [22], as they only have 2 or 4 parameters. Table 1 provides some statistics of the POMDP benchmarks. The first two columns indicate the POMDPs and the considered instance. The next three columns show respectively the number of states, transitions, and parameters in the pMC after preprocessing. For all benchmarks we assume a fixed memory structure, the last number in the instance column represents how many memory cells are used.

Table 1. POMDP benchmark characteristics and monotonicity checking results

benchmark	instance	# states	# trans	$|V|$	sufficient	$|V|\uparrow^R$	$|V|\uparrow^R?$
network_dp	(2,1,1)	98	229	9	53.1%	0	9
	(2,3,1)	552	2910	84	49.3%	18	65
	(4,1,1)	201	807	36	33.3%	12	24
	(4,3,1)	1230	7708	216	30.2%	23	191
network_ps	(2,1,1)	26	66	14	100.0%	14	0
	(2,5,1)	1085	3731	143	33.5%	20	121
	(4,1,1)	157	607	41	43.3%	9	32
	(4,5,1)	2208	12450	302	26.8%	16	282
maze	(1)	41	82	15	80.5%	7	8
	(2)	134	278	58	26.1%	7	24
	(4)	358	737	131	13.7%	10	118
	(8)	918	1868	296	7.6%	15	269
	(16)	2253	4549	841	4.7%	20	776
	(32)	4812	9678	2049	3.0%	28	1812
	(64)	10682	21429	4373	2.6%	53	3511
samplerocks	(1,1)	282	565	37	41.8%	1	36
	(2,1)	1129	2252	189	21.5%	2	185
4x4grid	(1)	47	106	3	6.4%	0	3
	(2)	139	302	11	4.3%	0	7
	(4)	393	823	26	3.8%	0	25

Results. The last three columns of Table 1 show the results of monotonicity checking. The column "sufficient" lists the percentage of states for which the initial order, i.e., before splitting regions, is sufficient. $|V|\uparrow^R$ lists the number of parameters for which monotonicity was found by our approach, while $|V|\uparrow^R?$ lists the number of parameters that might be monotonic based on sampling. Note that all remaining parameters are certainly not monotonic.

Table 2 shows the results for ε-optimal synthesis for $\varepsilon = 0.1$ and $\varepsilon = 0.05$. For each ε, we consider the time **t** required and the number (**# i**) of iterations that the integrated loop and the baseline require. For the integrated approach, we also provide the number (**# i_b**) of extra parameter lifting invocations needed to assist the monotonicity checker. We also denote **# i_b** without heuristics. Note that we only consider new heuristics, so initializing `CurMax` is enabled for both integrated approaches. Appendix B lists the ε-optimal values.

Discussion. Table 1 shows that for one benchmark (network_ps, (2,1,1)), global monotonicity in all parameters is established. Furthermore, it shows that many parameters are possibly monotonic. Finally, Table 1 shows that the % of states for which the order is sufficient does not directly relate to the number of definitely monotonic parameters. The explanation is twofold. 1) An order can be sufficient for parameter p, however, if for one state p is locally monotonic increasing and

Table 2. Overview of the experimental results comparing vanilla parameter lifting to the integrated approach

name	instance	dir	ε: 0.1 Integrated Heuristic			No heuristic	Vanilla		ε: 0.05 Integrated Heuristic			Vanilla	
			# i	# i_b	t	# i_b	# i	t	# i	# i_b	t	# i	t
network dp	(2,1,1)	min	0	2	<1	2	MO		47	2	<1	MO	
	(2,3,1)				MO		MO				MO	MO	
	(4,1,1)		0	2	<1	2	MO		405	2	<1	MO	
	(4,3,1)				MO		MO				MO	MO	
	(2,1,1)	Max	9	2	<1	6	MO		165	2	<1	MO	
	(2,3,1)				MO		MO				MO	MO	
	(4,1,1)		0	2	<1	2	MO		7	2	<1	MO	
	(4,3,1)				MO		MO				MO	MO	
network ps	(2,1,1)	min	0	2	<1	2	MO		0	2	<1	MO	
	(2,5,1)		43325	2	227	21664	MO				MO	MO	
	(4,1,1)		0	2	<1	2	MO		0	2	<1	MO	
	(4,5,1)				MO		TO				MO	TO	
	(2,1,1)	Max	0	2	<1	2	MO		0	2	<1	MO	
	(2,5,1)				MO		MO				MO	MO	
	(4,1,1)		0	2	<1	2	MO		0	2	<1	MO	
	(4,5,1)				MO		TO				MO	TO	
Maze	(1)	min	9	2	<1	6	MO		13	2	<1	MO	
	(2)		83	2	<1	43	MO		1955	2	3	MO	
	(4)				MO		MO				MO	MO	
	(1)	Max	0	2	<1	2	MO		0	2	<1	MO	
	(2)		0	2	<1	2	MO		0	2	<1	MO	
	(4)		0	2	1	2	MO		0	2	1	MO	
	(8)		0	2	4	2	MO		0	2	4	MO	
	(16)		0	2	28	2	MO		0	2	28	MO	
	(32)		0	2	162	2	MO		0	2	156	MO	
	(64)		0	2	855	2	MO		0	2	805	MO	
sample-rocks	(1,1)	min			MO		MO				MO	MO	
	(1,1)	max	13	2	<1	8	MO		923	2	2	MO	
	(2,1)				MO		MO				MO	MO	
4 × 4 grid	(1)	min	3	3	<1	3	MO		5	4	<1	MO	
	(2)		7	5	<1	5	MO		15	9	<1	MO	
	(4)		377313	188658	858	188658	MO				MO	MO	
	(1)	Max	9	6	<1	6	MO		11	7	<1	MO	
	(2)		59	31	<1	31	MO		571	291	<1	MO	
	(4)				MO		MO				MO	MO	

for another state locally monotonic decreasing, then global monotonicity cannot be obtained. 2) If a parameter occurs at only a few states, the chances that the order is sufficient for that parameter are higher. Also, the chances that the parameter is only locally monotonic increasing (or only decreasing) are higher. Consider e.g. `network_dp`, we observe that a parameter occurs on average at 10.9 and 6.6 states for instances (2,1,1) and (2,3,1) respectively. The order for instance (2,1,1) has a higher sufficiency compared to instance (2,3,1). However, for instance (2,3,1) global monotonicity for 18 parameters is found, whereas for instance (2,1,1) no global monotonicity is found.

Table 2 shows that for vanilla parameter lifting, all benchmarks were out of reach. Due to the number of parameters, many region splits are necessary, causing out of memory errors. As these are the smallest instances of the original POMDP benchmarks, picking benchmarks with less parameters is not possible. For the integrated approach, more results are obtained. First of all, we observe that picking a good initial `CurMax` helps a lot, as for many benchmarks splitting the region is not needed. This can be observed from the fact that $\# \; \mathbf{i} = 0$. Secondly, we observe that also when splitting the region is necessary, e.g., if we want more precise results ($\varepsilon = 0.05$), our approach still works for most benchmarks. Also the heuristics help minimizing the number of extra parameter lifting invocations needed to assist the monotonicity checker. Only when minimizing for `network_ps` instance (2,5,1) and for 4×4 `grid` instance (4) we get memory outs at $\varepsilon = 0.05$. For `network_ps`, at instance (2,1,1) we even have global monotonicity in all parameters. Therefore, the result is *provably optimal*. For $\varepsilon = 0.001$, the results remain mostly the same.

6   Related Work

Partially Observable MDPs. Several techniques exist to find optimal values in POMDPs, including, e.g., the use of approximate value iteration [23], optimisation and search techniques [1,8], dynamic programming [4], Monte Carlo simulation [37], game-based abstraction [45], machine learning [9,10,15], and belief-MDPs [6,7,33]. If the memory structure and size is provided, the main approaches are to either use satisfiability checking and SMT solving [11,43] or transform the POMDP to a pMC (as in this paper) and then use parameter synthesis [29].

Finding ε-optimal Instantiations of Parametric MCs. The first approaches to solve problems for parametric MCs [14,31] focus on computing a closed form for the solution function which maps parameter values to expected total rewards [2,14,16,17,21,26,28]. Chen *et al.* [12] analyze (non-controllable) perturbations in MCs from a robustness perspective. Fast sampling of the parameter space and evaluating the corresponding pMCs is also a preprocessing step to other methods [20,28]. Storm offers optimized routines, and for large numbers of samples, just-in-time compilation is a feasible alternative [18].

Exploiting Monotonicity. Hutschenreiter *et al.* [25] showed that the complexity of model checking (a monotone fragment of) PCTL on monotonic pMC is lower than on general pMCs. Pathak *et al.* [34] provided an efficient greedy approach to repair monotonic pMCs. Spel *et al.* [39] present a graph-based heuristic to determine whether a reachability property for a pMC is monotonic, i.e., whether the gradient w.r.t. some parameter is non-negative on the complete parameter space. Also for other Markov models, research on using monotonicity has been done. Gouberman *et al.* [19] used monotonicity for hitting probabilities in perturbed continuous-time MCs. Furthermore, similar as in this paper, [13] recently defines a partial order between states. They consider MDPs however, and use this order to reduce the amount of non-determinism.

7   Conclusion and Future Work

This paper has presented monotonicity for expected total reward properties. We have shown that with the help of monotonicity, we can tackle the optimal synthesis problem for POMDPs: synthesize a controller that is ε-optimal, thus using ε-optimal budget. We exploit the connection between synthesizing ε-optimal finite-state POMDP controllers and finding an ε-optimal parameter valuation in parametric Markov chains (pMC). The key concept is a deep interplay between parameter lifting, the favourable technique so far for this problem, and monotonicity checking. Experiments showed encouraging results: where with vanilla parameter lifting, no POMDP benchmark was solved, a tight integration with monotonicity checking allows us to check models with hundreds of parameters (thus observations). Future work consists of extending monotonicity to parametric MDPs for both reachability and reward properties.

A   Algorithm

In this section, we describe the algorithmic approach to obtain a sufficient reward order, which is used for monotonicity checking.

Our algorithm takes as input an pMC and bounds for all states. It returns a set of annotated reward orders $(\mathcal{A}, \preceq^{\mathtt{rew},\mathcal{A}})$, where $\mathcal{A}$ is a set of assumptions of the form $s \preceq^{\mathtt{rew}} s'$. The algorithm iteratively computes a set of reward orders. At this stage, both the bounds and the assumptions are not relevant and not used.

Initially, we start with the trivial order $\overset{..}{\smile} \preceq^{\mathtt{rew}} s$ for all $s \in S$. The queue is initialised (l.1) with an empty set of assumptions, the trivial order and all non-target states in S. At each iteration, we pick an order from the queue. If all states are processed (l.5), we are done building the reward order. Otherwise, we pick a state s to process (l.7), and try to order this state based on the reasoning from Lemmas 2 and 3 in Algorithm 1. We pick this state in reversed topological order, as this increases the likelihood that all successor states are contained in the order. Once we processed s, we put the (possibl) extended order in the queue together with the set of assumptions and the states we still need to process

---

Algorithm 1. Construction of a Reward Order

Input: pMC $\mathcal{M} = (S, s_I, T, V, \mathcal{P})$ and bounds $L_R(s)$ and $U_R(s)$ for all $s \in S$
Output: Result = a set of annotated reward orders $\preceq^{\mathrm{rew}, \mathcal{A}}$
1: Result $\leftarrow \emptyset$, Queue $\leftarrow (\mathcal{A} : \emptyset, \preceq^{\mathrm{rew}} : \{(\mathring{\smile}, s)\}, S' : S \setminus \{\mathring{\smile}\})$
2: **while** Queue not empty **do**
3:     $\mathcal{A}, \preceq^{\mathrm{rew}, \mathcal{A}}, S' \leftarrow$ Queue.pop()
4:     **if** $S' = \emptyset$ **then**
5:         Result $\leftarrow$ Result $\cup \{(\mathcal{A}, \preceq^{\mathrm{rew}, \mathcal{A}})\}$.
6:     **else**
7:         select $s \in S'$ with s topologically last w.r.t. $\mathring{\smile}$
8:         **if** $r(s) = 0$ **then**
9:             **if** ub(succ(s)) $\neq \emptyset$ and $\exists s' \in$ succ(s) s.t. succ(s) $\subseteq [s']$ **then**
10:                 extend $\preceq^{\mathrm{rew}, \mathcal{A}}$ with: $s \equiv^{\mathrm{rew}}$ succ(s)   *Lemma 2.2.*
11:             **else if** ub(succ(s)) $\neq \emptyset$ **then**
12:                 extend $\preceq^{\mathrm{rew}, \mathcal{A}}$ with all:
                    $s \prec^{\mathrm{rew}, \mathcal{A}}$ min ub(succ(s)) and max lb(succ(s)) $\prec^{\mathrm{rew}, \mathcal{A}} s$ *Lemma 2.2.*
13:             **else if** succ(s) $= \{s_1, s_2\}$ and $s \equiv^{\mathrm{rew}, \mathcal{A}} s_1$ **then**
14:                 extend $\preceq^{\mathrm{rew}, \mathcal{A}}$ with: $s_2 \equiv^{\mathrm{rew}, \mathcal{A}} s$ *Lemma 3.1.*
15:             **else if** succ(s) $= \{s_1, s_2\}$ and $s \prec^{\mathrm{rew}, \mathcal{A}} s_1$ **then**
16:                 extend $\preceq^{\mathrm{rew}, \mathcal{A}}$ with: $s_2 \prec^{\mathrm{rew}, \mathcal{A}} s$ *Lemma 3.3.*
17:             **else if** succ(s) $= \{s_1, s_2\}$ and $s_1 \prec^{\mathrm{rew}, \mathcal{A}} s$ **then**
18:                 extend $\preceq^{\mathrm{rew}, \mathcal{A}}$ with: $s \prec^{\mathrm{rew}, \mathcal{A}} s_2$ *Lemma 3.4.*
19:         **else**
20:             extend $\preceq^{\mathrm{rew}, \mathcal{A}}$ with: max lb(succ(s)) $\prec^{\mathrm{rew}, \mathcal{A}} s$ *Lemma 2.1.*
21:             **if** succ(s) $= \{s_1, s_2\}$ and $s \equiv^{\mathrm{rew}, \mathcal{A}} s_1$ **then**
22:                 extend $\preceq^{\mathrm{rew}, \mathcal{A}}$ with: $s_2 \prec^{\mathrm{rew}, \mathcal{A}} s$ *Lemma 3.2.*
23:             **else if** succ(s) $= \{s_1, s_2\}$ and $s \prec^{\mathrm{rew}, \mathcal{A}} s_1$ **then**
24:                 extend $\preceq^{\mathrm{rew}, \mathcal{A}}$ with: $s_2 \prec^{\mathrm{rew}, \mathcal{A}} s$ *Lemma 3.3.*
25:         Queue.push($\mathcal{A}, \preceq^{\mathrm{rew}, \mathcal{A}}, S' \setminus \{s\}$)
26: **return** Result

(1.25). Note that, if the state reward is 0 (1.9–18), the algorithm is equivalent to Algorithm 1 extended for treating cycles as found in [39, Sect. 4.1 and 4.3]. As assumptions are not used, Algorithm 1 in fact computes a single reward order; it runs linear in the number of transitions. Note that this order is not necessarily sufficient for the pMC.

Assumptions. To obtain a sufficient reward order, we consider assumptions, as described in Sect. 3.3. We exploit the annotations (called *assumptions*) that were ignored so far. Recall from Definition 6 that a reward order is not sufficient at a parametric state s, if its successors s_1 and s_2 are not totally ordered. To remain sound, we consider *all* possible orderings of s_1 and s_2. We add for each possible ordering a copy of the reward order to the queue, and continue as if the ordering of s_1 and s_2 is known. We extend Algorithm 1 with assumptions by adding the code of Algorithm 2 after lines 18 and 24.

Algorithm 2. Assumption extension (put after l.18 and l.24 in Algorithm 1).

1: **if** $\preceq^{\mathrm{rew},\mathcal{A}}$ is not a total order for succ(s) **then**
2:     pick $s_1, s_2 \in \mathrm{succ}(s)$ s.t. neither $s_1 \preceq^{\mathrm{rew},\mathcal{A}} s_2$ nor $s_2 \preceq^{\mathrm{rew},\mathcal{A}} s_1$
3:     Queue.push$((\mathcal{A}_{\prec^{\mathrm{rew}}} \cup \{(s_1, s_2)\}, \mathcal{A}_{\equiv^{\mathrm{rew}}}), \preceq^{\mathrm{rew},\mathcal{A}}$ extended with $s_1 \prec^{\mathrm{rew}} s_2, S')$
4:     Queue.push$((\mathcal{A}_{\prec^{\mathrm{rew}}} \cup \{(s_2, s_1)\}, \mathcal{A}_{\equiv^{\mathrm{rew}}}), \preceq^{\mathrm{rew},\mathcal{A}}$ extended with $s_2 \prec^{\mathrm{rew}} s_1, S')$
5:     Queue.push$((\mathcal{A}_{\prec^{\mathrm{rew}}}, \mathcal{A}_{\equiv^{\mathrm{rew}}} \cup \{(s_1, s_2)\}), \preceq^{\mathrm{rew},\mathcal{A}}$ extended with $s_1 \equiv^{\mathrm{rew}} s_2, S')$
6:     **continue**

Using Bounds. As creating assumptions might lead to an exponential explosion in the number of orders in the queue, we also consider the situation in which we have bounds $L_R(s)$ and $U_R(s)$ at our disposal satisfying

$$L_R(s) \leq \mathrm{ER}^s_{\mathcal{M}[\vec{u}]}(\Diamond T) \leq U_R(s) \quad \text{for all } \vec{u} \in R .$$

As these bounds are a relatively cheap way to order states, we extend Algorithm 1 by adding Algorithm 3 directly after line 7. This algorithm uses Lemma 6 to order the successor states of s to obtain a sufficient order for s (l.1–5). Furthermore, it uses Lemmas 7 and 8, to order a state relative to its successor states (l. 10–13).

B   Results

Table 3 shows the ε-optimal values for the integrated approach. It shows the results for $\varepsilon = 0.1$ and $\varepsilon = 0.05$. For all entries the result for $\varepsilon = 0.1$ and $\varepsilon = 0.05$ with heuristics is equal. This confirms that our initial guess for CurMax is a good one, even though we do need splitting of the region. The splitting is necessary as the bound found with parameter lifting is not yet tight enough to confirm that CurMax is indeed an ε-optimum. When maximizing the 4×4 grid (2) we find with heuristics a value of 10685 and without heuristics 10679. This can be explained by the new way of updating CurMax.

Algorithm 3. Bounds successor extension (put after l.7 Algorithm 1).

1: **for all** $(s_1, s_2) \in \mathrm{succ}(s) \times \mathrm{succ}(s)$ s.t. neither $s_1 \preceq^{\mathrm{rew},\mathcal{A}} s_2$ nor $s_2 \preceq^{\mathrm{rew},\mathcal{A}} s_1$ **do**
2:     **if** $U_R(s_2) \leq L_R(s_1)$ **then**
3:         extend $\preceq^{\mathrm{rew},\mathcal{A}}$ with: $s_2 \preceq^{\mathrm{rew},\mathcal{A}} s_1$
4:     **else if** $U_R(s_1) \leq L_R(s_2)$ **then**
5:         extend $\preceq^{\mathrm{rew},\mathcal{A}}$ with: $s_1 \preceq^{\mathrm{rew},\mathcal{A}} s_2$
6: **if** succ(s) $= \{s_1, s_2\}$ and $r(s) \geq \mathcal{P}(s, s_1)(\vec{u}) \cdot (U_R(s_2) - L_R(s_1)) \,\forall \vec{u} \in R$ **then**
7:     extend $\preceq^{\mathrm{rew},\mathcal{A}}$ with: $s_2 \preceq^{\mathrm{rew},\mathcal{A}} s$
8: **else if** succ(s) $= \{s_1, s_2\}$ and $r(s) \leq \mathcal{P}(s, s_1)(\vec{u}) \cdot (U_R(s_2) - L_R(s_1)) \,\forall \vec{u} \in R$ **then**
9:     extend $\preceq^{\mathrm{rew},\mathcal{A}}$ with: $s \preceq^{\mathrm{rew},\mathcal{A}} s_2$
10: **if** succ(s) $= \{s_1, \ldots, s_n\}$ **then**
11:     **for all** $j \in \{1, \ldots, n\}$ **do**
12:         **if** $\forall 1 \leq i < j.s_i \preceq^{\mathrm{rew},\mathcal{A}} s_j$ and $r(s) \geq \mathcal{P}(s, s_j)(\vec{u}) \cdot (U_R(s_j) - L_R(s_1))$ **then**
13:         extend $\preceq^{\mathrm{rew},\mathcal{A}}$ with: $s_j \preceq^{\mathrm{rew},\mathcal{A}} s$

C   Proofs

C.1   Proof Sketch of Theorem 1

Theorem 1.
$$\left(\forall s \in S.\ \mathsf{ER}^{s \to T}\!\uparrow_p^\ell\right)\ implies\ \mathsf{ER}^{s_I \to T}\!\uparrow_p.$$

Proof. We sketch the proof of Theorem 1, which follows the lines of [39, Thm. 2]. First of all, we lift the notion of local monotonicity (Definition 4) to local monotonicity for n steps. Secondly, we claim that local monotonicity implies local monotonicity for n steps and from this global monotonicity follows.

We define the *length of a finite path* $|\hat{\pi}| = n$ for $\hat{\pi} = s_0 s_1 \ldots s_n$. Let $Paths^n(s)$ be the set of all paths with length n starting from state $s \in S$.

Definition 8 (Locally monotonic increasing for n steps). $\mathsf{ER}_{\mathcal{M}}^{s \to T}$ is locally monotonic increasing *for n steps in parameter p (at s) on region R,* denoted $\mathsf{ER}_{\mathcal{M}}^{s \to T}\!\uparrow_p^{\ell,n,R}$, *iff for all* $\vec{u} \in R$:

$$\left(\sum_{\hat{\pi} \in Paths^n(s)} \left(\frac{\partial}{\partial p}\mathrm{Pr}(\hat{\pi})\right) \cdot \left(r(\hat{\pi}, \Diamond T) + \mathsf{ER}_{\mathcal{M}}^{\hat{\pi}_n \to T}\right) \right)(\vec{u})\ \geq\ 0.$$

Table 3. Overview of the found ε-optimal values

Name	Instance	Direction	ε: 0.1 No heuristic	ε: 0.1 heuristic	ε: 0.05 no heuristic	ε: 0.05 heuristic
network_dp	(2,1,1)	Min	2.79e0	2.79e0	2.789e0	2.789e0
	(4,1,1)		1.28e0	1.28e0	1.276e0	1.276e0
	(2,1,1)	Max	4.77e0	4.77e0	4.769e0	4.769e0
	(4,1,1)		3.80e0	3.80e0	3.800e0	3.800e0
network_ps	(2,1,1)	min	2.61e-1	2.61e-1	2.610e-1	2.610e-1
	(2,5,1)		3.94e0	3.94e0	N/A	N/A
	(4,1,1)		1.77e0	1.77e0	1.766e0	1.766e0
	(2,1,1)	max	4.57e-1	4.57e-1	4.569e-1	4.569e-1
	(4,1,1)		2.03e1	2.03e1	2.029e1	2.029e1
Maze	(1)	Min	7.05e1	7.05e1	7.050e1	7.050e1
	(2)		7.06e1	7.06e1	7.059e1	7.059e1
	(1)	Max	1.54e6	1.54e6	1.535e6	1.535e6
	(2)		1.55e6	1.55e6	1.545e6	1.545e6
	(4)		1.86e6	1.86e6	1.861e6	1.861e6
	(8)		1.81e6	1.81e6	1.810e6	1.810e6
	(16)		2.06e6	2.06e6	2.062e6	2.062e6
	(32)		2.33e6	2.33e6	2.334e6	2.334e6
	(64)		2.17e6	2.17e6	2.171e6	2.171e6
samplerocks	(1,1)	Max	3.90e1	3.90e1	3.904e1	3.911e1
4 × 4 grid	(1)	Min	4.04e1	4.04e1	4.036e1	4.036e1
	(2)		4.03e1	4.03e1	4.026e1	4.026e1
	(4)		4.42e1	4.42e1	N/A	N/A
	(1)	Max	1.05e4	1.05e4	1.048e4	1.048e4
	(2)		1.07e4	1.07e4	1.068e4	1.069e4

For $n = 1$, Definition 8 corresponds to Definition 4. The claim that local monotonicity implies local monotonicity for n steps (for all n) can be proven by induction over n. The claim that global monotonicity follows from this can be shown similar to the proof of [39, Theorem 2, Equation 2], with as main difference that no state $s \in S$ exists for which $\mathrm{Pr}^{s \rightarrow T} = 0$, cf. Remark 1.

C.2   Proof of Lemma 7

Lemma 7. *For any state s with* $\mathsf{succ}(s) = \{s_1, s_2\}$, $f = \mathcal{P}(s, s_1)$, *region R and* $s_1 \preceq^{rew} s_2$: *if for all $\vec{u} \in R$*

1. $r(s) \geq f(\vec{u}) \cdot (U_R(s_2) - L_R(s_1))$ *then* $s_2 \preceq^{rew} s$, *and*
2. $r(s) \leq f(\vec{u}) \cdot (L_R(s_2) - U_R(s_1))$ *then* $s \preceq^{rew} s_2$.

Proof. Let $f' = \mathcal{P}(s, s_2) = 1 - f$. For the first case, we derive for all $\vec{u} \in R$:

$$
\begin{aligned}
r(s) &\geq f(\vec{u}) \cdot (U_R(s_2) - L_R(s_1)) \\
&\geq f(\vec{u}) \cdot \left(\mathsf{ER}^{s_2 \rightarrow T} - \mathsf{ER}^{s_1 \rightarrow T} \right) \\
&= f(\vec{u}) \cdot \mathsf{ER}^{s_2 \rightarrow T} - f(\vec{u}) \cdot \mathsf{ER}^{s_1 \rightarrow T} \\
&= (1 - f'(\vec{u})) \cdot \mathsf{ER}^{s_2 \rightarrow T} - f(\vec{u}) \cdot \mathsf{ER}^{s_1 \rightarrow T} \\
&= \mathsf{ER}^{s_2 \rightarrow T} - f'(\vec{u}) \cdot \mathsf{ER}^{s_2 \rightarrow T} - f(\vec{u}) \cdot \mathsf{ER}^{s_1 \rightarrow T}.
\end{aligned}
$$

From this. it immediately follows for all $\vec{u} \in R$:

$$
\underbrace{r(s) + f(\vec{u}) \cdot \mathsf{ER}^{s_1 \rightarrow T} + f'(\vec{u}) \cdot \mathsf{ER}^{s_2 \rightarrow T}}_{= \mathsf{ER}^{s \rightarrow T}} \geq \mathsf{ER}^{s_2 \rightarrow T},
$$

so $s_2 \preceq^{rew} s$.

For the second case, the proof follows in a similar way.

References

1. Amato, C., Bernstein, D.S., Zilberstein, S.: Optimizing fixed-size stochastic controllers for POMDPs and decentralized POMDPs. Auton. Agents Multi Agent Syst. **21**(3), 293–320 (2010)
2. Baier, C., Hensel, C., Hutschenreiter, L., Junges, S., Katoen, J.P., Klein, J.: Parametric Markov chains: PCTL complexity and fraction-free Gaussian elimination. Inf. Comput. **272** (2020)
3. Baier, C., Katoen, J.P.: Principles of Model Checking. MIT Press, Cambridge (2008)
4. Bonet, B.: Solving large POMDPs using real time dynamic programming. In: Proceedings of the AAAI 2021 Fall Symposium on POMDPs (1998)
5. Bonet, B., Geffner, H.: Solving POMDPs: RTDP-Bel vs. point-based algorithms. In: IJCAI, pp. 1641–1646 (2009)

6. Bork, A., Junges, S., Katoen, J.-P., Quatmann, T.: Verification of Indefinite-Horizon POMDPs. In: Hung, D.V., Sokolsky, O. (eds.) ATVA 2020. LNCS, vol. 12302, pp. 288–304. Springer, Cham (2020). https://doi.org/10.1007/978-3-030-59152-6_16
7. Bork, A., Katoen, J.-P., Quatmann, T.: Under-approximating expected total rewards in POMDPs. In: TACAS 2022. LNCS, vol. 13244, pp. 22–40. Springer, Cham (2022). https://doi.org/10.1007/978-3-030-99527-0_2
8. Braziunas, D., Boutilier, C.: Stochastic local search for POMDP controllers. In: AAAI, pp. 690–696. AAAI Press/The MIT Press, Menlo, Park (2004)
9. Carr, S., Jansen, N., Topcu, U.: Verifiable RNN-based policies for POMDPs under temporal logic constraints. In: IJCAI, pp. 4121–4127. ijcai.org (2020)
10. Carr, S., Jansen, N., Wimmer, R., Serban, A.C., Becker, B., Topcu, U.: Counterexample-guided strategy improvement for POMDPs using recurrent neural networks. In: IJCAI, pp. 5532–5539. ijcai.org (2019)
11. Chatterjee, K., Chmelik, M., Davies, J.: A symbolic SAT-based algorithm for almost-sure reachability with small strategies in POMDPs. In: AAAI. pp. 3225–3232. AAAI Press, Menlo, Park (2016)
12. Chen, T., Feng, Y., Rosenblum, D.S., Su, G.: Perturbation analysis in verification of discrete-time Markov chains. In: Baldan, P., Gorla, D. (eds.) CONCUR 2014. LNCS, vol. 8704, pp. 218–233. Springer, Heidelberg (2014). https://doi.org/10.1007/978-3-662-44584-6_16
13. Cleaveland, M., Ruchkin, I., Sokolsky, O., Lee, I.: Monotonic safety for scalable and data-efficient probabilistic safety analysis. In: ICCPS, pp. 92–103. ACM (2022)
14. Daws, C.: Symbolic and parametric model checking of discrete-time Markov chains. In: Liu, Z., Araki, K. (eds.) ICTAC 2004. LNCS, vol. 3407, pp. 280–294. Springer, Heidelberg (2005). https://doi.org/10.1007/978-3-540-31862-0_21
15. Doshi, F., Pineau, J., Roy, N.: Reinforcement learning with limited reinforcement: using Bayes risk for active learning in POMDPs. In: ICML. ACM International Conference Proceeding Series, vol. 307, pp. 256–263. ACM (2008)
16. Fang, X., Calinescu, R., Gerasimou, S., Alhwikem, F.: Fast parametric model checking through model fragmentation. In: ICSE. pp. 835–846. IEEE (2021)
17. Filieri, A., Ghezzi, C., Tamburrelli, G.: Run-time efficient probabilistic model checking. In: ICSE. ACM (2011)
18. Gainer, P., Hahn, E.M., Schewe, S.: Accelerated model checking of parametric Markov chains. In: Lahiri, S.K., Wang, C. (eds.) ATVA 2018. LNCS, vol. 11138, pp. 300–316. Springer, Cham (2018). https://doi.org/10.1007/978-3-030-01090-4_18
19. Gouberman, A., Siegle, M., Tati, B.: Markov chains with perturbed rates to absorption: theory and application to model repair. Perform. Eval. 130, 32–50 (2019)
20. Hahn, E.M., Han, T., Zhang, L.: Synthesis for PCTL in parametric Markov decision processes. In: Bobaru, M., Havelund, K., Holzmann, G.J., Joshi, R. (eds.) NFM 2011. LNCS, vol. 6617, pp. 146–161. Springer, Heidelberg (2011). https://doi.org/10.1007/978-3-642-20398-5_12
21. Hahn, E.M., Hermanns, H., Zhang, L.: Probabilistic reachability for parametric Markov models. Int. J. Softw. Tools Technol. Transfer 13, 319 (2011). https://doi.org/10.1007/s10009-010-0146-x
22. Hartmanns, A., Klauck, M., Parker, D., Quatmann, T., Ruijters, E.: The quantitative verification benchmark set. In: Vojnar, T., Zhang, L. (eds.) TACAS 2019. LNCS, vol. 11427, pp. 344–350. Springer, Cham (2019). https://doi.org/10.1007/978-3-030-17462-0_20
23. Hauskrecht, M.: Value-function approximations for partially observable Markov decision processes. J. Artif. Intell. Res. 13, 33–94 (2000)

24. Hensel, C., Junges, S., Katoen, J.-P., Quatmann, T., Volk, M.: The probabilistic model checker STORM. Int. J. Softw. Tools Technol. Transfer 1–22 (2021). https://doi.org/10.1007/s10009-021-00633-z

25. Hutschenreiter, L., Baier, C., Klein, J.: Parametric Markov chains: PCTL complexity and fraction-free Gaussian elimination. In: GandALF. EPTCS, vol. 256 (2017)

26. Jansen, N., Corzilius, F., Volk, M., Wimmer, R., Ábrahám, E., Katoen, J.-P., Becker, B.: Accelerating parametric probabilistic verification. In: Norman, G., Sanders, W. (eds.) QEST 2014. LNCS, vol. 8657, pp. 404–420. Springer, Cham (2014). https://doi.org/10.1007/978-3-319-10696-0_31

27. Junges, S.: Parameter synthesis in Markov models. Ph.D. thesis, RWTH Aachen University, Germany (2020)

28. Junges, S., et al.: Parameter synthesis for Markov models. CoRR abs/1903.07993 (2019)

29. Junges, S., et al.: Finite-state controllers of POMDPs using parameter synthesis. In: UAI. AUAI Press, Monterey,(2018)

30. Kaelbling, L.P., Littman, M.L., Cassandra, A.R.: Planning and acting in partially observable stochastic domains. Artif. Intell. **101**(1–2), 99–134 (1998)

31. Lanotte, R., Maggiolo-Schettini, A., Troina, A.: Parametric probabilistic transition systems for system design and analysis. Formal Aspects Comput. **19**(1), 93–109 (2007)

32. Littman, M.L., Cassandra, A.R., Kaelbling, L.P.: Learning policies for partially observable environments: Scaling up. In: ICML, pp. 362–370. Morgan Kaufmann (1995)

33. Lovejoy, W.S.: Computationally feasible bounds for partially observed Markov decision processes. Oper. Res. **39**(1), 162–175, 104504 (1991)

34. Pathak, S., Ábrahám, E., Jansen, N., Tacchella, A., Katoen, J.P.: A greedy approach for the efficient repair of stochastic models. In: NFM. LNCS, vol. 9058 (2015)

35. Quatmann, T., Dehnert, C., Jansen, N., Junges, S., Katoen, J.-P.: Parameter synthesis for Markov models: faster than ever. In: Artho, C., Legay, A., Peled, D. (eds.) ATVA 2016. LNCS, vol. 9938, pp. 50–67. Springer, Cham (2016). https://doi.org/10.1007/978-3-319-46520-3_4

36. Russell, S., Norvig, P.: Artificial Intelligence: A Modern Approach, 4th edn. Pearson, Boston (2020)

37. Silver, D., Veness, J.: Monte-Carlo planning in large POMDPs. In: NIPS, pp. 2164–2172. Curran Associates, Inc., Red Hook (2010)

38. Smith, T., Simmons, R.G.: Heuristic Search Value Iteration for POMDPs. In: UAI. pp. 520–527. AUAI Press (2004)

39. Spel, J., Junges, S., Katoen, J.-P.: Are parametric Markov chains monotonic? In: Chen, Y.-F., Cheng, C.-H., Esparza, J. (eds.) ATVA 2019. LNCS, vol. 11781, pp. 479–496. Springer, Cham (2019). https://doi.org/10.1007/978-3-030-31784-3_28

40. Spel, J., Junges, S., Katoen, J.P.: Are parametric Markov chains monotonic? CoRR abs/1907.08491 (2019), extended version

41. Spel, J., Junges, S., Katoen, J.-P.: Finding provably optimal Markov chains. In: TACAS 2021. LNCS, vol. 12651, pp. 173–190. Springer, Cham (2021). https://doi.org/10.1007/978-3-030-72016-2_10

42. Spel, J., Stein, S., Katoen, J.P.: POMDP controllers with optimal budget (artifact). Zenodo , 104504 (2022). https://doi.org/10.5281/zenodo.6793377

43. Wang, Y., Chaudhuri, S., Kavraki, L.E.: Bounded policy synthesis for POMDPs with safe-reachability objectives. In: International Foundation for Autonomous

Agents and Multiagent Systems Richland, AAMAS, pp. 238–246. SC, USA/ACM (2018)

44. Winkler, T., Junges, S., Pérez, G.A., Katoen, J.: On the complexity of reachability in parametric Markov decision processes. In: Proceedings of CONCUR. LIPIcs, vol. 140, pp. 14:1–14:17. Schloss Dagstuhl - Leibniz-Zentrum für Informatik (2019)

45. Winterer, L., et al.: Motion planning under partial observability using game-based abstraction. In: CDC, pp. 2201–2208. IEEE (2017)

46. Yang, L., Murugesan, S., Zhang, J.: Real-time scheduling over Markovian channels: when partial observability meets hard deadlines. In: GLOBECOM, pp. 1–5. IEEE (2011)

Markovian Agents and Population Models

A Logical Framework for Reasoning About Local and Global Properties of Collective Systems

Michele Loreti$^{(\boxtimes)}$ and Aniqa Rehman

University of Camerino, Camerino, Italy
`{michele.loreti,aniqa.rehman}@unicam.it`

Abstract. Collective adaptive systems (CAS) are composed of a large number of entities that interact with each other to reach local or global goals. Entities operate without any centralized control and should adapt their behavior to the changes in the environment where they operate. Due to the intricacies of these interactions and adaptation, it is difficult to predict the behavior of CAS. For this reason, formal tools are needed to specify and verify this behavior to ensure consistency, reliability, correctness, and safety properties. In this paper, we present a novel logical framework that permits specifying properties of CAS at both local and global levels: local properties refer to the behavior of individuals, while global properties refer to the whole system. An exact model checking algorithm, whose complexity is linear with the size of the formula and with the size of the model is also presented together with another one based on statistical model checking that permits handling systems composed by a large number of agents. Finally, a simple scenario is used to evaluate the advantages of the proposed approach.

Keywords: Temporal Logics · Multi-agent Systems · Local and Global properties · Statistical Model Checking

1 Introduction

Collective adaptive systems (CAS) are composed of a huge amount of components that interact with each other and with the enclosing environment to reach local and global goals. Each component in the system may exhibit autonomic behavior depending on its properties, objectives, and actions. Decision-making in such systems is complicated, and interaction between components may introduce new and sometimes unexpected behaviors. Due to the intricacies of these interactions and adaptation, it is difficult to predict the behavior of CAS. For this reason, formal tools are needed to specify and verify this behavior to ensure consistency, reliability, correctness, and safety properties. The adopted tools should

This research has been partially supported by Italian PRIN project "IT-MaTTerS" n, 2017FTXR7S, and by POR MARCHE FESR 2014–2020, project "MIRACLE", CUP B28I19000330007.

© Springer Nature Switzerland AG 2022
E. Ábrahám and M. Paolieri (Eds.): QEST 2022, LNCS 13479, pp. 133–149, 2022.
https://doi.org/10.1007/978-3-031-16336-4_7

encompass both functional and non-functional aspects of behavior. In particular, the design process should be supported by robust modelling techniques which are able to describe CAS and to reason about their behavior in both qualitative and quantitative terms [10].

Different models and formalism have been proposed to describe CAS. These are mainly based on *process specification languages* that permit describing the behaviour of the involved components, the enclosing environment and their interactions [2,4,11,14,16,18]. In these models, systems are mainly considered from a *global perspective* where the behavior of the single component somehow disappears in the multitude of the system state. However, sometimes we could be interested in a *local perspective* where we are not considering the whole state, but we focus on a specific component.

Let us consider a system composed of N agents that can be either *blue* or *red* coloured. The goal is to guarantee that without any centralized control agents evolve to a *balanced configuration* where the number of *blue agents* is *similar* to the number of *red agents*. In the system, each agent can autonomously change its colour according to what are its perceptions about the state of the other agents. Different approaches can be considered to *program* the behaviour of an agent. To compare the possible solutions, we have to define requirements that involve both *local* and *global properties*. From a *global* point of view we can require that *"Eventually, the system is able to reach a balanced configuration"* (**R1**). From a *local* point of view, one could be also interested in verifying that *"Agents are not continuously changing their state"* (**R2**).

The goal of this paper is to introduce a novel framework that permits specifying and verifying properties of CAS at both *global* and *local level*. We first introduce a variant of *Stochastic Processes* tailored to describe quantitative behaviour of CAS. For the sake of simplicity, in this paper we will focus on a variant of Discrete Time Markov Chain (DTMC), named *Multi-Agent Discrete Time Markov Chain* (MA-DTMC) that will be used to render the behavior of a system composed by a multitude of agents by providing both *global* and *local perspectives*. However, an extension to other classes of stochastic processes is straightforward. Hence, we will present *Global and Local Temporal Logic* (GLoTL), a temporal logic that permits specifying both *global* and *local* properties of multi-agent systems described as MA-DTMC models.

Moreover, an exact model checking algorithm is presented to check if a given GLoTL formula is verified or not. The proposed algorithm follows an on-the-fly approach [12]. Indeed, the state space is generated in a step wise fashion to compute the satisfaction probability of a given global formula Φ. We will show that the proposed approach is linear with the size of formula Φ. Unfortunately, we will observe that when the number of agents in the system is increasing, the number of states that one has to consider to compute the satisfaction probability is so big that an exact computation is almost impossible. For this reason, a methodology based on *statistical model-checking* [3] is also presented. In this paper, the *red/blue scenario* described above is used as a running example to evaluate the advantages of the proposed approach.

Related Work. Many recent publications have proposed temporal logic and tools to support the analysis of CAS or Multi-Agent Systems. We refer here the ones that we consider more relevant for this paper. In [12], Probabilistic Computation Tree Logic (PCTL) is used to specify the properties of CAS, and an on-the-fly model checking algorithm is proposed to check if a given model satisfies or not a formula. A similar approach has been proposed in [7] where Continuous Stochastic Logic (CSL) is used to specify the properties of interacting agents whose behavior is modelled via Deterministic Timed Automaton (DTA) with a single clock. Differently from [12], where only global formulas are taken into account, in [7] local properties are considered. However, these are verified by altering the behavior of the agents. Finally, in [1], an extension of Linear Temporal Logic (LTOL) is used to specify the properties of Multi-Agent Systems and to verify against Doubly-Labeled Transition System (DLTS), which are used to specify system behavior. The idea is that *local properties* is specified by means of predicates on agent states, while global behavior is rendered as temporal modalities on message passing. Differently from the approach presented in [1] the one presented in this paper focuses on *quantitative aspects* and, also due to the use of a simpler formalism, it is equipped with efficient model checking algorithms.

Structure of the Paper. The paper is structured as follows. In Sect. 2 *Multi-Agent Discrete Time Markov Chain* are introduced. In Sect. 3 syntax and semantics of GLoTL is presented. In Sect. 4 we will present the exact model checking algorithm with another algorithm based on *Statistical Model Checking* can be used to check the satisfaction of GLoTL. Section 5 concludes the paper.

2   Modelling Global and Local Behaviours

Collective adaptive systems (CAS) consist of a large number of entities, the agents, that interact with each other to reach local or global goals. To reason about behavior of CAS, models are needed to formally specify possible computations and, in the case of quantitative analysis, to provide tools that permit measuring the *probability* of the set of computations of our interest. These models should represent the behavior of a system at two different levels. A *global level*, where the whole state is considered, and a *local level*, where one can focus on the behavior of each single agent.

Let $\mathcal{S}$ be a set of *agent states*, a *Multi-Agent Stochastic Process* with size N consists of a set of random variables $\{X(t), t \in T\}$ that assume values in $\mathcal{S}^N$. Intuitively, $X(t)$ consists of a tuple of N *agent states* each of which represents the state of one of the N agents in the system. For the sake of simplicity, in this paper, we will focus on *Discrete Time Markovian Processes* and we will introduce *Multi-Agent Discrete Time Markov Chain* (MA-DTMC) that consist of a simple extension of standard Discrete Time Markov Chain (DTMC).

2.1   Multi-agent Discrete Time Markov Chain

A *Discrete Time Markov Chain* (DTMC) $\mathcal{D}$ is a tuple of $(Q, \mathbf{P})$, where Q is the finite set of states, $\mathbf{P} : Q \times Q \rightarrow [0, 1]$ is the transition probability matrix (where, for any $q \in Q$, $\sum_{q' \in Q} \mathbf{P}(q, q') = 1$). For any $q, q' \in Q$, $\mathbf{P}(q, q') = p$ indicates the probability to jump in the next time step from state q to state q'. When we use DTMC to model behaviour of CAS, each state in Q must describe the specific configuration of each agent operating in the system. Given a set $\mathcal{S}$ of *agent states*, a *Multi-Agent Discrete Time Markov Chain* (MA-DTMC) with size N, $\mathcal{M}^N$, is a DTMC $(\mathcal{S}^N, \mathbf{P}^N)$. Each state in $\mathcal{M}^N$ is a vector in $\mathcal{S}^N$ and represents the state of each of the N agents operating in the system.

Given a $\vec{q} \in \mathcal{S}^N$, we let $\vec{q}[i] \in \mathcal{S}$ denote the state of agent in position i. Moreover, for any $\vec{q} \in \mathcal{S}^N$ and $s \in \mathcal{S}$ we let $\#(\vec{q}, s)$ and $\%(\vec{q}, s)$ denote the number and the fraction of agents in the state s, respectively. Formally:

$$\#(\vec{q}, s) = |\{i | \vec{q}[i] = s\}| \qquad \%(\vec{q}, s) = \frac{\#(\vec{q}, s)}{N}$$

Example 1 (Red/Blue Scenario 1/5). Let us consider the *Red/Blue Scenario* introduced in Sect. 1. We can assume that each agent in the system can be either in the state B, indicating that the agent is *blue coloured*, or in the state R, when the agent is *red*. Our scenario can be modelled via a $MA - DTMC$ $(\mathcal{S}^N_{rb}, \mathbf{P}^N_{rb})$, where $\mathcal{S}_{rb} = \{\mathsf{B}, \mathsf{R}\}$ while $\mathbf{P}_{rb}$, that represents the probabilistic evolution of our system, depends on the agent behaviour.

A *global* path π over $\mathcal{M}^N$ is a non empty (infinite) sequence of states $\vec{q}_0 \vec{q}_1 \vec{q}_2 \cdots$ of states in $\mathcal{S}^N$ such that, for any i, $\mathbf{P}(\vec{q}_i, \vec{q}_{i+1}) > 0$. Similarly, a *finite path fragment* $\hat{\pi}$ consists of a finite sequence $\vec{q}_0 \cdots \vec{q}_k$ of states in $\mathcal{S}^N$ such that, for any $i < k$, $\mathbf{P}(\vec{q}_i, \vec{q}_{i+1}) > 0$.

Given a *global* path π we will use $\pi[i]$ to denote the state at position i, while $\pi[i..]$ indicates the path obtained from π by considering only the states from position i onward. Similar notations are used for a *finite path fragment* $\hat{\pi}$. We let $len(\hat{\pi})$ denote the number of elements in $\hat{\pi}$, while we say that $\hat{\pi}$ is a prefix of π, written $\hat{\pi} \prec \pi$, whenever for any $i \leq len(\hat{\pi})$ $\hat{\pi}[i] = \pi[i]$.

We will say that a path π (resp. $\hat{\pi}$) starts from state $\vec{q}$ whenever $\pi[0] = \vec{q}$ (resp. $\hat{\pi}[0] = \vec{q}$). Moreover, we will let $Paths_{\mathcal{M}^N}(\vec{q})$ and $FinPaths_{\mathcal{M}^N}(\vec{q})$ denote the set of paths and path fragments starting from $\vec{q}$ while $Paths_{\mathcal{M}^N}$ and $FinPaths_{\mathcal{M}^N}$ denote $\cup_{\vec{q} \in \mathcal{S}^N} Paths_{\mathcal{M}^N}(\vec{q})$ and $\cup_{\vec{q} \in \mathcal{S}^N} FinPaths_{\mathcal{M}^N}(\vec{q})$, respectively.

Any path $\pi \in Paths_{\mathcal{M}^N}$ represents a computation that can be experienced in the system described by $\mathcal{M}^N$. We can observe that, if we focus on the specific agent i in the sequence of states occurring in π, we have a *local* perspective of the computation executed by agent i. Given a *global* path π, we let $\pi \downarrow i$ denote the *local path* obtained from π by considering infinite sequence of agent states $s_0 s_1 \ldots (s_i \in \mathcal{S})$ such that for any j, $s_j = \pi[j][i]$. We say that $\pi_\ell = s_0 s_1 \ldots$ is a *local path* for agent i from the state $\vec{q} \in \mathcal{S}^N$ if and only if there exists $\pi \in Paths_{\mathcal{M}^N}(\vec{q})$ such that $\pi_\ell = \pi \downarrow i$. We let $Paths^i_{\mathcal{M}^N}(\vec{q})$ denote the set of local paths of agent i from $\vec{q}$, while $Paths^i_{\mathcal{M}^N} = \cup_{\vec{q} \in \mathcal{S}^N} Paths^i_{\mathcal{M}^N}(\vec{q})$.

GLOBAL FORMULAS

$$\Phi ::= \text{true} \mid \neg\Phi \mid \Phi_1 \wedge \Phi_2 \mid \%[\phi] \bowtie p \mid \mathcal{X}\,\Phi \mid \Phi_1\,\mathcal{U}^{\leq k}\,\Phi_2$$

LOCAL FORMULAS

$$\phi ::= \text{true} \mid \alpha \mid \neg\phi \mid \phi_1 \wedge \phi_2 \mid \mathcal{X}\,\phi \mid \phi_1\,\mathcal{U}^{\leq k}\,\phi_2$$

Fig. 1. Syntax of GLoTL formulas.

We say that a MA-DTMC $(\mathcal{S}^N, \mathbf{P}^N)$ is *decomposable* if and only if there exists a function $\mathfrak{F} : \mathcal{S}^\star \to \mathcal{S} \times \mathcal{S} \to [0,1]$ such that:

$$\mathbf{P}^N(\vec{q}_1, \vec{q}_2) = \prod_{i=1}^{N} \mathfrak{F}(\vec{q}_1)\,[\vec{q}_1[i], \vec{q}_2[i]] \tag{1}$$

Equation 1 states that when $\mathcal{M}^N$ is decomposable, its *global transition probability matrix* can be expressed in terms of function $\mathfrak{F}$ that describes the probabilistic behaviour of agent s when it is operating in a given system configuration. Indeed, given a tuple of states in $\vec{q} \in \mathcal{S}^\star = \bigcup_{N \in \mathbb{N}} \mathcal{S}^N$, $\mathfrak{F}(\vec{q})$ yields the *local probability matrix* of agents in $\mathcal{S}$.

Example 2 (Red/Blue Scenario 2/5). We can now define function $\mathfrak{F}_{rb}$ that describes the probabilistic behaviour of agents defined in Example 1. We assume that, at each computational step, an agent interacts with the other agents to decide if it has to change its colour or it can remain in the same state. In particular, an agent changes its state whenever it observes an agent having the same colour. Let α_m be the probability that in a computational step an agent is able to meet another agent, function $\mathfrak{F}_{rb}(\vec{q})$ can be defined as follows:

$$\mathfrak{F}(\vec{q})[\mathsf{R}, \mathsf{B}] = \alpha_m \cdot \%(\vec{q}, \mathsf{R}) \qquad \mathfrak{F}(\vec{q})[\mathsf{R}, \mathsf{R}] = 1 - \alpha_m \cdot \%(\vec{q}, \mathsf{R})$$

$$\mathfrak{F}(\vec{q})[\mathsf{B}, \mathsf{R}] = \alpha_m \cdot \%(\vec{q}, \mathsf{B}) \qquad \mathfrak{F}(\vec{q})[\mathsf{B}, \mathsf{B}] = 1 - \alpha_m \cdot \%(\vec{q}, \mathsf{B})$$

We can observe that the higher is the fraction of agents of a given colour, the higher is the probability that one of them will change it.

3   Global and Local Temporal Logic

In this section, we present *Global and Local Temporal Logic* (GLoTL) a temporal logic that permits specifying both *global* and *local* properties of multi-agent systems. Our formalism follows a *linear time* approach and it is inspired by *Signal Temporal Logic* (STL) [9]. We will first introduce GLoTL syntax and discuss the operators of our logic. After that, we will give the semantics of GLoTL in terms of MA-DTMCs. Finally, we will discuss how the interpretation of the proposed logical framework can extended to a wider class of models.

GLOBAL FORMULAS   LOCAL FORMULAS

false $= \neg$true   false $= \neg$true

$\Phi_1 \vee \Phi_2 = \neg(\neg\Phi_1 \wedge \neg\Phi_2)$ $\phi_1 \vee \phi_2 = \neg(\neg\phi_1 \wedge \neg\phi_2)$

$\Phi_1 \rightarrow \Phi_2 = \neg\Phi_1 \vee \Phi_2$ $\phi_1 \rightarrow \phi_2 = \neg\phi_1 \vee \phi_2$

$\Diamond^{\leq k}\Phi = $ true $\mathcal{U}^{\leq k} \Phi$ $\Diamond^{\leq k}\phi = $ true $\mathcal{U}^{\leq k} \phi$

$\Box^{\leq k}\Phi = \neg\Diamond^{\leq k}\neg\Phi$ $\Box^{\leq k}\phi = \neg\Diamond^{\leq k}\neg\phi$

$\%[\phi] \in [a,b] = \%[\phi] \geq a \wedge \%[\phi] \leq b$

$\%[\phi] \notin [a,b] = \neg\%[\phi] \in [a,b]$

Fig. 2. Derivable Logical Operators

3.1 Syntax

The syntax of GLoTL formulas is reported in Fig. 1. Two categories of formulas are considered: *global formulas* and *local formulas*.

A *global formula* Φ, which permits specifying properties of *global computations*, is built from standard *Boolean operators* (true, $\neg$ and $\wedge$), and *temporal operators* ($\mathcal{X} \Phi$ and $\Phi_1 \mathcal{U}^{\leq k} \Phi_2$). Finally, a novel operator $\%[\phi] \bowtie p$ is used to specify that, at a given point in the computation, the *fraction of agents* satisfying *local formula* ϕ is $\bowtie p$, where $\bowtie \in \{\leq, <, >, \geq\}$. A *local formula* ϕ is used to specify properties of the single agents. The syntax of ϕ is similar to the one already considered for *global formulas* and is built from *atomic proposition* $\alpha \in \mathcal{AP}$ via standard *Boolean operators* (true, $\neg$ and $\wedge$), and *temporal operators* ($\mathcal{X} \phi$ and $\phi_1 \mathcal{U}^{\leq k} \phi_2$). We will let Γ and Λ denote the set of global and local formulas, respectively.

Other logical operators can be derived as macros of the above defined ones. A list of derivable operators is reported in Fig. 2 for both *global* and *local* formulas: *disjunction*, $\vee$, and *implication*, $\rightarrow$, from $\wedge$ and $\neg$; *eventually*, $\Diamond^{\leq k}$, and *globally*, $\Box^{\leq k}$, from $\mathcal{U}^{\leq k}$.

Given a global formula Φ (resp. a local formula ϕ), the *horizon* of Φ (resp. ϕ) is the max number of time steps that are needed to guarantee its satisfaction. Function horizon is inductively defined in Fig. 3 where Ψ denote either a global or a local formula.

3.2 Semantic

Let $\mathcal{M}^N = (\mathcal{S}^N, \mathbf{P}^N)$ be a MA-DTMC and $\mathcal{L}: \mathcal{S} \rightarrow 2^{\mathcal{AP}}$ be a labelling function associating each *state* s in $\mathcal{S}$ with the set of atomic proposition $\mathcal{A} \in \mathcal{AP}$ satisfied by s. Satisfaction of *global* and *local formulas* are defined via the relations $\models^{\mathcal{M}^N,\mathcal{L}}$ and $\models_\ell^{\mathcal{M}^N,\mathcal{L}}$ defined in Fig. 4 and Fig. 5. Moreover, probabilistic semantics of *global formulas* Φ is defined according to a function μ (Eq. 2) associating each state $\vec{q} \in \mathcal{S}^N$ the probability that $\vec{q}$ satisfied Φ.

For the large part of operators, definition of $\models^{\mathcal{M}^N,\mathcal{L}}$ (Fig. 4) is straightforward. Any $\pi \in Paths_{\mathcal{M}^N}$ satisfies true, π satisfies $\Phi_1 \wedge \Phi_2$ if and only if both Φ_1

$$\text{horizon}(\text{true}) = 0$$

$$\text{horizon}(\alpha) = 0$$

$$\text{horizon}(\%[\phi] \bowtie p) = \text{horizon}(\phi)$$

$$\text{horizon}(\neg\Psi) = \text{horizon}(\Psi)$$

$$\text{horizon}(\Psi_1 \wedge \Psi_2) = \max\{\text{horizon}(\Psi_1), \text{horizon}(\Psi_2)\}$$

$$\text{horizon}(\mathcal{X}\,\Psi) = 1 + \text{horizon}(\Psi)$$

$$\text{horizon}(\Psi_1\,\mathcal{U}^{\leq k}\,\Psi_2) = k + \max\{\text{horizon}(\Psi_1) - 1, \text{horizon}(\Psi_2)\}$$

Fig. 3. Horizon of Local and Global Formulas

$$\pi \models^{\mathcal{M}^N,\mathcal{L}} \text{true}$$

$$\pi \models^{\mathcal{M}^N,\mathcal{L}} \neg\Phi \qquad \Longleftrightarrow \pi \not\models^{\mathcal{M}^N,\mathcal{L}} \Phi$$

$$\pi \models^{\mathcal{M}^N,\mathcal{L}} \Phi_1 \wedge \Phi_2 \qquad \Longleftrightarrow \pi \models^{\mathcal{M}^N,\mathcal{L}} \Phi_1 \wedge \pi \models^{\mathcal{M}^N,\mathcal{L}} \Phi_2$$

$$\pi \models^{\mathcal{M}^N,\mathcal{L}} \%[\phi] \bowtie p \qquad \Longleftrightarrow \frac{|\{i\,|\,\pi{\downarrow}i \models_{\ell}^{\mathcal{M}^N,\mathcal{L}}\phi\}|}{N} \bowtie p$$

$$\pi \models^{\mathcal{M}^N,\mathcal{L}} \mathcal{X}\,\Phi \qquad \Longleftrightarrow \pi[1..] \models^{\mathcal{M}^N,\mathcal{L}} \Phi$$

$$\pi \models^{\mathcal{M}^N,\mathcal{L}} \Phi_1\,\mathcal{U}^{\leq k}\,\Phi_2 \Longleftrightarrow \exists 0 \leq h \leq k.\,\pi[h..] \models^{\mathcal{M}^N,\mathcal{L}} \Phi_2 \wedge \forall 0 \leq i < h.\,\pi[i..] \models \Phi_1$$

Fig. 4. Global Formulas: Satisfaction relation

and Φ_2 are satisfied by π, while π satisfies $\neg\Phi$ if and only if it does not satisfy Φ. Temporal formula $\mathcal{X}\,\Phi$ is satisfied by π if the computation starting from step 1 satisfies Φ. Finally, π satisfies $\Phi_1\,\mathcal{U}^{\leq k}\,\Phi_2$ if and only if there exists an index h such that $\pi[h..]$ satisfies Φ_2 and for any index i less then h, $\pi[i..]$ satisfies Φ_1. The only interesting case is for $\%[\phi] \bowtie p$ that is satisfied by a global path π if and only if the fraction of *agents* in π having a *local computation* satisfying (local formula) ϕ is $\bowtie p$. Local computations of each single agent are obtained by considering $\pi \downarrow i$, the projection of π on the index i (for $0 \leq i < N$). Definition of $\models_\ell^{\mathcal{L}}$ is standard and is similar to what already discussed for the global case.

$$\pi_\ell \models_\ell^{\mathcal{M}^N,\mathcal{L}} \text{true}$$

$$\pi_\ell \models_\ell^{\mathcal{M}^N,\mathcal{L}} \alpha \qquad \Longleftrightarrow \alpha \in \mathcal{L}(\pi_\ell[0])$$

$$\pi_\ell \models_\ell^{\mathcal{M}^N,\mathcal{L}} \neg\phi \qquad \Longleftrightarrow \pi_\ell \not\models_\ell^{\mathcal{M}^N,\mathcal{L}} \phi$$

$$\pi_\ell \models_\ell^{\mathcal{M}^N,\mathcal{L}} \phi_1 \wedge \phi_2 \qquad \Longleftrightarrow \pi_\ell \models_\ell^{\mathcal{M}^N,\mathcal{L}} \phi_1 \wedge \pi_\ell \models_\ell^{\mathcal{M}^N,\mathcal{L}} \phi_2$$

$$\pi_\ell \models_\ell^{\mathcal{M}^N,\mathcal{L}} \mathcal{X}\,\phi \qquad \Longleftrightarrow \pi_\ell[1..] \models_\ell^{\mathcal{M}^N,\mathcal{L}} \phi$$

$$\pi_\ell \models_\ell^{\mathcal{M}^N,\mathcal{L}} \phi_1\,\mathcal{U}^{\leq k}\,\phi_2 \Longleftrightarrow \exists 0 \leq h \leq k.\,\pi_\ell[h..] \models_\ell^{\mathcal{M}^N,\mathcal{L}} \phi_2 \wedge \forall 0 \leq i < h.\,\pi_\ell[i..] \models \phi_1$$

Fig. 5. Local Formulas: Satisfaction relation

We can now define function μ amounting the probability that a given state $\vec{q}$ in a MA-DTMC $\mathcal{M}^N$ satisfies a global formula Φ, given a labelling function $\mathcal{L}$:

$$\mu(\mathcal{M}^N, \mathcal{L}, \vec{q}, \Phi) = Pr_{\mathcal{M}^N}\{\pi \in Paths_{\mathcal{M}^N}(\vec{q})|\pi \models^{\mathcal{M}^N, \mathcal{L}} \Phi\} \qquad (2)$$

Theorem 1. *For any MA-DTMC $\mathcal{M}^N$, time step $t \in \mathbb{N}$, formula Φ and labelling function $\mathcal{L}$, $\{\pi \in Paths_{\mathcal{M}^N}|\pi[t..] \models^{\mathcal{M}^N, \mathcal{L}} \Phi\}$ is measurable.*

Example 3 (Red/Blue Scenario 3/5). We can show now how we can use GLoTL to specify the requirements of our running example described in Sect. 1. First of all we let the global formula Φ_{bal} express that the system is *balanced*:

$$\Phi_{bal} = (\%[blue] \in [0.5 - \varepsilon, 0.5 + \varepsilon])$$

where the atomic proposition *blue* indicates a *blue agent*. Requirement **R1**, that is *"Within a given number of steps, the system is able to reach a balanced configuration."*, can be expressed by the following formula:

$$\Phi_1 = \Diamond^{\leq k_1}\Phi_{bal}$$

where k_1 indicates the number of required steps.

We can also require that the balanced configuration is preserved for the next k_2:

$$\Phi_2 = \Diamond^{\leq k_1}\Box^{\leq k_2}\Phi_{bal}$$

Both the above defined properties consider the system as a whole and do not say anything about the behaviour of the single agents. However, in our second requirement **(R2)** we are considering that *"When the system is balanced, agents are not continuously changing their state"*. We say that an agent is *locally stable* if whenever it changes its colour it remains on the same colour for at least k_3 steps:

$$\phi_{br} = blue \rightarrow \mathcal{X} (red \rightarrow \Box^{\leq k_3} red) \qquad \phi_{rb} = red \rightarrow \mathcal{X} (blue \rightarrow \Box^{\leq k_3} blue)$$

These properties can be used to specify requirements **R2**:

$$\Phi_3 = \Box^{\leq k_4} (\Phi_{bal} \rightarrow \%[\phi_{br} \vee \phi_{rb}] \geq .90)$$

This property states that in the next k_4 steps, if the system is balanced then it is also *locally stable*.

4   Model Checking GLoTL

In this section, we will introduce an exact model checking algorithm that permits computing the satisfaction probability of a given GLoTL formula. Moreover, to cope with the problem of state space explosion, another one based on statistical model checking [3] is also presented. Both these algorithms rely on an *operational semantics* of GLoTL formulas. The proposed approach is similar to the one presented in [8] where they define a structural operational semantics for a variant of LTL temporal logic.

$$\text{L-TRUE} \frac{}{\text{true} \xrightarrow{\mathcal{A}}_L \text{true}} \qquad \text{L-FALSE} \frac{}{\text{false} \xrightarrow{\mathcal{A}}_L \text{false}}$$

$$\text{L-ATOM1} \frac{\alpha \in \mathcal{A}}{\alpha \xrightarrow{\mathcal{A}}_L \text{true}} \qquad \text{L-ATOM2} \frac{\alpha \notin \mathcal{A}}{\alpha \xrightarrow{\mathcal{A}}_L \text{false}}$$

$$\text{L-AND} \frac{\phi_1 \xrightarrow{\mathcal{A}}_L \phi_1' \quad \phi_2 \xrightarrow{\mathcal{A}}_L \phi_2'}{\phi_1 \wedge \phi_2 \xrightarrow{\mathcal{A}}_L \phi_1' \wedge \phi_2'} \qquad \text{L-OR} \frac{\phi_1 \xrightarrow{\mathcal{A}}_L \phi_1' \quad \phi_2 \xrightarrow{\mathcal{A}}_L \phi_2'}{\phi_1 \vee \phi_2 \xrightarrow{\mathcal{A}}_L \phi_1' \vee \phi_2'}$$

$$\text{L-NEG} \frac{\phi \xrightarrow{\mathcal{A}}_L \phi'}{\neg \phi \xrightarrow{\mathcal{A}}_L \neg \phi'} \qquad \text{L-NEXT} \frac{}{\mathcal{X} \phi \xrightarrow{\mathcal{A}}_L \phi}$$

$$\text{L-UNTIL1} \frac{\phi_2 \xrightarrow{\mathcal{A}}_L \phi_2'}{\phi_1 \, \mathcal{U}^{\leq 0} \, \phi_2 \xrightarrow{\mathcal{A}}_L \phi_2'}$$

$$\text{L-UNTIL2} \frac{\phi_1 \xrightarrow{\mathcal{A}}_L \phi_1' \quad \phi_2 \xrightarrow{\mathcal{A}}_L \phi_2'}{\phi_1 \, \mathcal{U}^{\leq k+1} \, \phi_2 \xrightarrow{\mathcal{A}}_L \phi_1' \wedge (\phi_1 \, \mathcal{U}^{\leq k-1} \, \phi_2) \vee \phi_2'}$$

Fig. 6. Operational Semantics of Local Formulas

4.1 Operational Semantics of GLoTL Formulas

Let $\mathcal{AP}$ be a set of atomic propositions, we let $\rightarrow_L \subseteq \Lambda \times 2^{\mathcal{AP}} \times \Lambda$ be the transition relation defined in Fig. 6 where, to simplify the notation, derivable operators false and $\vee$ are used. Let $\mathcal{A} \subseteq \mathcal{AP}$, we will write $\phi_1 \xrightarrow{\mathcal{A}}_L \phi_2$ to denote that $(\phi_1, \mathcal{A}, \phi_2) \in \rightarrow_L$. The *local transition relation* $\rightarrow_L$, which is inductively defined on the syntax of local formulas, describes what should be satisfied after a step given the labelling of the current state. Namely, $\phi_1 \xrightarrow{\mathcal{A}}_L \phi_2$ indicates that a local computation starting from an agent state with labels $\mathcal{A} \subseteq \mathcal{AP}$ satisfies the formula ϕ_1 if and only if in the next step the formula ϕ_2 is satisfied.

Theorem 2. *Let* $\mathcal{M}^N = (\mathcal{S}^N, \mathbf{P}^N)$ *be a MA-DTMC and* $\mathcal{L} : \mathcal{S} \rightarrow 2^{\mathcal{AP}}$ *be a labelling function. For any* local *path* π_ℓ *of* $\mathcal{M}^N$, $\pi_\ell \models_\ell^{\mathcal{M}^N, \mathcal{L}} \phi_1$ *if and only if* $\phi_1 \xrightarrow{\mathcal{L}(\pi_\ell[0])} \phi_2$ *and* $\pi_\ell[1..] \models_\ell^{\mathcal{M}^N, \mathcal{L}} \phi_2$.

Transition relation of global formulas $\rightarrow_G \subseteq \Gamma \times 2^{\mathcal{AP}^\star} \times \Gamma$ is defined in Fig. 7. The rules are almost the same as the one considered for relation $\rightarrow_L$. However, while $\rightarrow_L$ is labelled with a set of atomic propositions, $\rightarrow_G$ is labelled with a tuple of set of atomic propositions. This because $\rightarrow_G$ can be thought of as a sort of *synchronous parallel composition* of local formulas. Indeed, transition at global level is obtained by considering all the transitions performed at local level. To manage the interplay between *local* and *global* formulas, an auxiliary operator is introduced. Indeed, we consider the operator $\langle \phi_1, \ldots, \phi_N \rangle \bowtie p$ whose semantics is the following:

$$\pi \models^{\mathcal{M}^N, \mathcal{L}} \langle \phi_1, \ldots, \phi_N \rangle \bowtie p \Leftrightarrow \frac{\left| \left\{ i \mid \pi \downarrow i \models_\ell^{\mathcal{M}^N, \mathcal{L}} \phi_i \right\} \right|}{N} \bowtie p$$

$$\text{G-True} \frac{}{\text{true} \xrightarrow{\vec{A}}_G \text{true}} \qquad \text{G-False} \frac{}{\text{false} \xrightarrow{\vec{A}}_G \text{false}}$$

$$\text{G-Frc} \frac{\forall i. \phi \xrightarrow{\vec{A}[i]}_L \phi'_i}{\%[\phi] \bowtie p \xrightarrow{\vec{A}}_G \langle \phi'_1, \ldots, \phi'_N \rangle \bowtie p}$$

$$\text{G-Vec} \frac{\forall i. \phi_i \xrightarrow{\vec{A}[i]}_L \phi'_i}{\langle \phi_1, \ldots, \phi_N \rangle \bowtie p \xrightarrow{\vec{A}}_G \langle \phi'_1, \ldots, \phi'_N \rangle \bowtie p}$$

$$\text{G-And} \frac{\Phi_1 \xrightarrow{\vec{A}}_G \Phi'_1 \quad \Phi_2 \xrightarrow{\vec{A}}_G \Phi'_2}{\Phi_1 \wedge \Phi_2 \xrightarrow{\vec{A}}_G \Phi'_1 \wedge \Phi'_2} \qquad \text{G-Or} \frac{\Phi_1 \xrightarrow{\vec{A}}_G \Phi'_1 \quad \Phi_2 \xrightarrow{\vec{A}}_G \Phi'_2}{\Phi_1 \vee \Phi_2 \xrightarrow{\vec{A}}_G \Phi'_1 \vee \Phi'_2}$$

$$\text{G-Neg} \frac{\Phi \xrightarrow{\vec{A}}_G \Phi'}{\neg \Phi \xrightarrow{\vec{A}}_G \neg \Phi'} \qquad \text{G-Next} \frac{}{\mathcal{X} \Phi \xrightarrow{\vec{A}}_G \Phi}$$

$$\text{G-Until1} \frac{\Phi_2 \xrightarrow{\vec{A}}_G \Phi'_2}{\Phi_1 \, \mathcal{U}^{\leq 0} \, \Phi_2 \xrightarrow{\vec{A}}_G \Phi'_2}$$

$$\text{G-Until2} \frac{\Phi_1 \xrightarrow{\vec{A}}_G \Phi'_1 \quad \Phi_2 \xrightarrow{\vec{A}}_G \Phi'_2}{\Phi_1 \, \mathcal{U}^{\leq k+1} \, \Phi_2 \xrightarrow{\vec{A}}_G \Phi'_1 \wedge (\Phi_1 \, \mathcal{U}^{\leq k-1} \, \Phi_2) \vee \Phi'_2}$$

Fig. 7. Operational Semantics of Global Formulas

This operator is used to assign each agent with a local formulas. The following theorem is the corresponding at global level of Theorem 2. With an abuse of notation, given a labelling function $\mathcal{L} : \mathcal{S} \to 2^{\mathcal{AP}}$ and a state $\vec{q} \in \mathcal{S}^N$, we let $\mathcal{L}(\vec{q})$ denote the set tuple $\vec{A} \in 2^{\mathcal{AP}^*}$ such that, for any i $\vec{A}[i] = \mathcal{L}(\vec{q})$.

Theorem 3. *Let* $\mathcal{M}^N = (\mathcal{S}^N, \mathbf{P}^N)$ *be a MA-DTMC and* $\mathcal{L} : \mathcal{S} \to 2^{\mathcal{AP}}$ *be a labelling function. For any* path π *of* $\mathcal{M}^N$, $\pi \models^{\mathcal{M}^N, \mathcal{L}} \Phi_1$ *if and only if* $\Phi_1 \xrightarrow{\mathcal{L}(\pi[0])}_G \Phi_2$ *and* $\pi[1..] \models^{\mathcal{M}^N, \mathcal{L}} \Phi_2$.

It is easy to see that both $\to_L$ and $\to_G$ are deterministic. For this reason, in what follows we will write $\text{after}(\Phi_1, \vec{A}) = \Phi_2$ if and only if $\Phi_1 \xrightarrow{\vec{A}}_G \Phi_2$.

4.2 Exact Model Checking Algorithm

Operational semantics of GLoTL formulas is used in this section to define an exact model checking algorithm. The algorithm is straightforward and is based on the idea that to compute the satisfaction probability of a formula Φ by a state $\vec{q}$ one has to perform an exploration of the state space driven by the operational semantics of Φ. The exploration terminates when either a formulas that

$$\frac{}{\text{true} \uparrow} \qquad \frac{}{\text{false} \downarrow} \qquad \frac{\varPsi_i \uparrow}{(\varPsi_1 \vee \varPsi_2) \uparrow} \qquad \frac{\varPsi_1 \downarrow \quad \varPsi_2 \downarrow}{(\varPsi_1 \vee \varPsi_2) \downarrow}$$

$$\frac{\varPsi_i \downarrow}{(\varPsi_1 \wedge \varPsi_2) \downarrow} \qquad \frac{\varPsi_1 \uparrow \quad \varPsi_2 \uparrow}{(\varPsi_1 \vee \varPsi_2) \uparrow} \qquad \frac{\varPsi \downarrow}{(\neg \varPsi) \uparrow} \qquad \frac{\varPsi \uparrow}{(\neg \varPsi) \downarrow}$$

$$\frac{\frac{|\{i | \phi_i \uparrow\}|}{N} \bowtie p}{\langle \phi_1, \dots, \phi_N \rangle \bowtie p \uparrow} \qquad \frac{\frac{|\{i | \phi_i \downarrow\}|}{N} \overline{\bowtie} 1 - p}{\langle \phi_1, \dots, \phi_N \rangle \bowtie p \downarrow}$$

Fig. 8. Acceptance and Rejection Criteria

Algorithm 1. Compute Probabilities of Global Formulas

1: **function** SATPROBABILITY($\mathcal{M}^N = (\mathcal{S}^N, \mathbf{P}^N), \mathcal{L}, \vec{q}, \varPhi$)
2:     **if** $\varPhi \uparrow$ **then**
3:         **return** 1.0
4:     **end if**
5:     **if** $\varPhi \downarrow$ **then**
6:         **return** 0.0
7:     **end if**
8:     after($\varPhi, \mathcal{L}(\vec{q})) = \varPhi'$
9:     **return** $\sum_{\vec{q}'} \mathbf{P}[\vec{q}, \vec{q}'] \cdot$ SATPROBABILITY($\mathcal{M}^N, \mathcal{L}, \vec{q}, \varPhi'$)
10: **end function**

---

is *accepting*, denoted by $\varPhi \uparrow$, or *rejecting*, denoted by $\varPhi \downarrow$, is reached. Accepting and rejecting conditions are defined Fig. 8 where we use $\varPsi$ to denote both a local and a global formula. All the rules are as expected and follows usual Boolean interpretation. The only interesting case is the one related to the *auxiliary* formula $\langle \phi_1, \dots, \phi_N \rangle \bowtie p$. Indeed, this formula is accepting whenever the fraction of local formulas ϕ_i that are *accepting* is $\bowtie p$. The same formula is *rejecting* when the fraction of local formulas ϕ_i that are rejecting is $\overline{\bowtie}(1 - p)$, where $\overline{\bowtie}$ indicates the opposite relation of $\bowtie$.

Given a MA-DTMC $\mathcal{M}^N$, a labelling function $\mathcal{A}$, a state $\vec{q}$ in $\mathcal{M}^N$ and a global formula $\varPhi$, function SATPROBABILITY($\mathcal{M}^N, \mathcal{L}, \vec{q}, \varPhi$) defined in Algorithm 1 can be used to compute recursively the probability $\mu(\mathcal{M}^N, \mathcal{L}, \vec{q}, \varPhi)$. If the formula $\varPhi$ is accepting function SATPROBABILITY just return 1.0, while 0.0 is returned when $\varPhi$ is rejecting. If $\varPhi$ is neither accepting nor rejecting, first the formula $\varPhi'$ obtained from $\varPhi$ *after* the labelling $\mathcal{L}(\vec{q})$ of $\vec{q}$ is computed, then the resulting probability is obtained by summing $\mathbf{P}[\vec{q}, \vec{q}'] \cdot$ SATPROBABILITY($\mathcal{M}^N, \mathcal{L}, \vec{q}', \varPhi'$), for any $\vec{q}'$ reachable from $\vec{q}$.

We can observe that in SATPROBABILITY the exploration of the state space reachable from $\vec{q}$ is made *on-the-fly*. This means that we do not need an explicit generation of all the state space. This is useful in particular when we one considers *decomposeable* models, where the probability matrix $\mathbf{P}$ can be defined in terms of a simpler (local) function $\mathfrak{F}$ like in Example 2.

Theorem 4. *Let* $\mathcal{M}^N = (\mathcal{S}^N, \mathbf{P}^N)$ *be a MA-DTMC and* $\mathcal{L} : \mathcal{S} \to 2^{\mathcal{AP}}$ *be a labelling function. For any state* $\vec{q}$ *of* $\mathcal{M}^N$ *and formula* $\varPhi$:

Algorithm 2. Checking Global Path Properties

1: **function** CHECKPATH($\mathcal{L}, \hat{\pi}, \Phi$)
2:     $i = 0$
3:     **while** $i < len(\hat{\pi})$ **do**
4:         after($\Phi, \mathcal{L}(\hat{\pi}[i])) = \Phi'$
5:         **if** $\Phi \uparrow$ **then**
6:             **return** 1
7:         **end if**
8:         **if** $\Phi \downarrow$ **then**
9:             **return** 0
10:        **end if**
11:        $i = i + 1$
12:        $\Phi = \Phi'$
13:    **end while**
14:    **return** 0
15: **end function**

$$\text{SATPROB}(\mathcal{M}^N, \mathcal{L}, \vec{q}, \Phi) = \mu(\mathcal{M}^N, \mathcal{L}, \vec{q}, \Phi)$$

Finally, we can observe that the number of steps needed to compute SATPROB linearly depends on the size of the model $\mathcal{M}$ and on the *horizon* of the formula Φ (see Fig. 3). Indeed, after at most horizon(Φ) reductions of formulas Φ either an accepting or a reject formula is reached. Unfortunately, when the number of agents operating in a system increases, the use of exact model checking is hard or even impossible. For instance, if we consider our running scenario with $N = 100$ agents, the size of the state space will be 2^{100}. To overcome this problem, in the forthcoming section a statistical based approach is proposed to compute the satisfaction probability of a formula.

4.3 Statistical Model Checking of GLoTL

In this section, an approach based on *Statistical Model-Checking* [3] is presented. In the proposed approach, given a time horizon $T \in \mathbb{N}$, first a number of computations of $\mathcal{M}^N$ are sampled. After that the satisfaction probabilities of a Φ are computed in the time interval $[0, T]$. We will show that the proposed approach is linear in the size of T and in the size of the formula Φ (defined in terms of its temporal horizon).

In the following, we first introduce the algorithms that can be used to check if a given *finite path fragment* $\hat{\pi}$ satisfies a given *global formula* Φ. After that, we will show how multiple invocations of this procedure permit estimating the satisfaction probabilities.

Function CHECKPATH, defined in Algorithm 2, takes as an input a labelling function $\mathcal{L}$, a fine path fragment $\hat{\pi}$ and a global formula Φ and returns 1 if the formula is satisfied by the given path fragment, 0 otherwise. The behaviour is straightforward: the algorithm iterate for all the elements i in the path and

---

Algorithm 3. Statistical Estimation of Satisfaction Probability

---

1: **function** ESTIMATE($\mathcal{M}^N, \mathcal{L}, K, \vec{q}, \Phi$)
2:   $T = \mathsf{horizon}(\Phi)$
3:   $sum = 0$
4:   **for** i **from** 0 **to** K-1 **do**
5:     $\pi = \text{SAMPLE}(\mathcal{M}^N, T)$
6:     $sum = sum + \text{CHECKPATH}(\mathcal{L}, \hat{\pi}, \Phi)$
7:   **end for**
8:   **return** $\frac{sum}{K}$
9: **end function**

---

returns true if an *accepting* formula can be reached by letting the formula evolving at each step by $\mathcal{L}(\hat{\pi}[i])$. The following statements guarantee that the value returned by CHECKPATH is coherent with the semantics of global formulas.

Theorem 5. *For any MA-DTMC $\mathcal{M}^N$, labelling function $\mathcal{L}$, local path fragment $\hat{\pi}$, such that $len(\hat{\pi}) \geq \mathsf{horizon}(\Phi)$, CHECKPATH$(\mathcal{L}, \hat{\pi}, \Phi) = 1$ if and only if for any $\pi \in Paths_{\mathcal{M}^N}$ such that $\hat{\pi} \prec \pi$, $\pi \models^{\mathcal{M}^N, \mathcal{L}} \Phi$*

To estimate the satisfaction probability of a *global formula* we rely on Statistical Model Checking. Statistical Model Checking (SMC) is a formal verification technique used to analyse stochastic systems and combines simulation and statistical methods [13,17]. SMC is based on the idea that satisfaction probability of a formula is statistically estimated via a sequence of simulations.

Given a formula Φ, to estimate the probability that a state $\vec{q}$ of $\mathcal{M}$ satisfies Φ, we *sample K* global paths from $\vec{q}$ in $\mathcal{M}$. For each sampled path π_i we add to variable *sum* 1 if Φ is satisfied, 0 otherwise. The probability that Φ is satisfied by $\vec{q}$ is estimated as $\frac{sum}{K}$. The procedure ESTIMATE is reported in Algorithm 3 where function SAMPLE$(\mathcal{M}^N, \vec{q}, T)$ is used to sample a path of length T starting from $\vec{q}$ in $\mathcal{M}^N$[1]. We can observe that the number of operations needed to estimate satisfaction probability of Φ is $O(K \cdot \mathsf{horizon}(\Phi) \cdot \Sigma)$, where Σ we assume that the cost of sampling a path of length T is $O(T)$.

We can observe that the estimated probability p^* is the result of a *random experiment*, the one associated with the sampling of K paths from $\mathcal{M}$. For this reason, we can speak about the probability that this value differs from the exact one, indicated as $\hat{p}$, more than a value ε. We have that the higher the number of sampled paths K, the smaller the probability that the obtained value is greater than ε. Indeed, by using the Chernoff-Hoeffding Bound we have that (see [17] for all the details) $Pr(|p^* - \hat{p}| > \varepsilon) \leq 2e^{-2K\varepsilon^2}$. For this reason, if we want to limit this probability of obtaining an error greater than ε by a threshold α, the number of paths to be sampled is $K = \frac{1}{2\varepsilon^2} \log\left(\frac{2}{\alpha}\right)$.

Example 4 (Red/Blue Scenario 4/4). The results of analysis of the formulas Φ_1, Φ_2 and Φ_3, described in Example 3, are reported in Fig. 9 and Fig. 10.

---

[1] We omit here the details of function SAMPLE that should be straightforward.

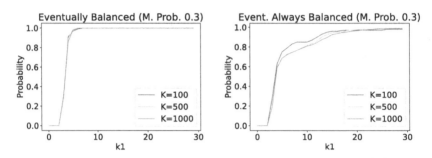

Fig. 9. Satisfaction probability of property Φ_1, on the left, and Φ_2, on the right ($N = 10^3$, $k_2 = 10$).

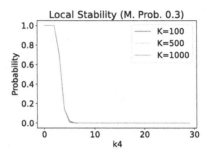

Fig. 10. Satisfaction probability of property Φ_3 ($N = 10^3$, $k_3 = 10$).

The analyses have been conducted via a prototype implementation of the described framework in the Java framework Sibilla[2]. In Fig. 9 the probability to see formulas Φ_1 (on the left) and Φ_2 (on the right) while the one of Φ_3 is plotted in Fig. 10. These probabilities are computed by considering a system composed by $N = 10^3$ agents, a meeting probability $\alpha_m = 0.3$ and a number of samplings $K \in \{100, 500, 10000\}$. The formulas are verified by considering where $\varepsilon = 0.025$, $k_2 = 10$ while the k_1 and k_4 range from 0 to 30.

We can observe that, if we evaluate our system from a global point of view, it works well. Indeed, all the properties are satisfied with an high probability a balanced configuration is reached after few steps. Moreover, the higher the number of samplings, the smaller the statistical error we have and the smoother is the plot we obtain.

However, if we observe the plot on the right side of Fig. 10 we can soon realise that from a *local point of view* the system is not working well since the agents are continuously changing their colour, even if the system is globally balanced. This is due to the fact that at each time step when the system is balanced, on the average, $\alpha_m \cdot 0.5$ agents will change their colour. This is evident in Fig. 11 where we show how the satisfaction probability of formulas Φ_1, Φ_2 and Φ_3 changes according to different values of α_m. We can observe that the higher the value of

---

[2] https://github.com/quasylab/sibilla.

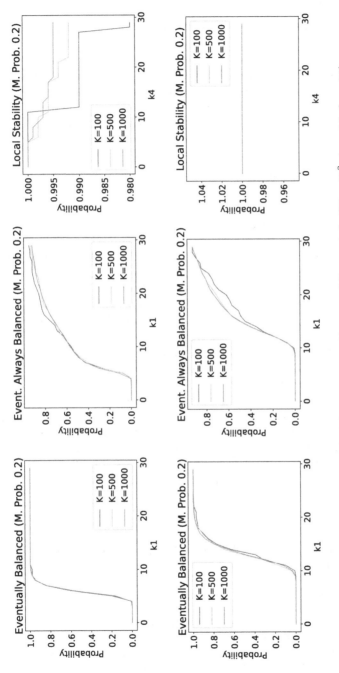

Fig. 11. Impact of Meeting probability α_m on satisfaction of Φ_{bal} and Φ_4 ($N = 10^3$, $k_2 = 10$, $k_3 = 10$).

α_m the shorter the time needed by our system to reach a balanced configuration. However, the higher is α_m the higher the fraction of agents that are not *locally stable*.

To guarantee *local stability* a more sophisticated solution, as the one presented in [6], should be considered. However, due to lack of space, we do not discuss it here. Nevertheless, we can observe that, thanks to the use of GLoTL formulas we have been able to point out a critical aspect of our system that is not evident if we limit our observations only at a global level.

All the experiments have been performed on a standard laptop with a 2, 3 GHz 8-Core Intel Core i9 and 32 Gb of RAM. The execution time ranged from 1.2 s for Φ_1 and $K = 100$ to 31.7 s for Φ_3 and $K = 1000$.

5   Concluding Remarks

In this paper, we have introduced a novel framework that permits specifying and verifying properties of CAS at both *global* and *local level*. First, we have introduced *Multi-Agent Discrete Time Markov Chain* (MA-DTMC), a variant of Discrete Time Markov Chain (DTMC) that can be used to describe the behavior of a system composed by a multitude of agents from both *global* and *local perspectives*. Moreover, we have introduced *Global and Local Temporal Logic* (GLoTL), a temporal logic that permits specifying both *global* and *local* properties of a multi-agent systems described as MA-DTMC models. Finally, Exact and Statistical Model Checking algorithms have been proposed to verify both local and global properties. A prototype implementation of framework has been implemented and integrated in Java framework, Sibilla.

For the sake of simplicity, in this paper we focused on MA-DTMC. However, the proposed approach can easily extended to take into account *Continuous Time Markov Chain* (CTMC). We will modify the exact model checking by applying some standard techniques while, the statistical model checking, in this case, remains the same, and the time will be continuous. As future work, we plan to extend the scalable analysis to the techniques based on fluid and mean-field approximation, such as the one considered in [5,12,15].

References

1. Alrahman, Y.A., Perelli, G., Piterman, N.: Reconfigurable interaction for MAS modelling. In: Proceedings of the 19th International Conference on Autonomous Agents and MultiAgent Systems, pp. 7–15 (2020)
2. Abolhasanzadeh, B., Jalili, S.: Towards modeling and runtime verification of self-organizing systems. Expert Syst. Appl. **44**, 230–244 (2016)
3. Agha, G., Palmskog, K.: A survey of statistical model checking. ACM Trans. Model. Comput. Simul. (TOMACS) **28**(1), 1–39 (2018)
4. Aldini, A.: Design and verification of trusted collective adaptive systems. ACM Trans. Model. Comput. Simul. (TOMACS) **28**(2), 1–27 (2018)
5. Bortolussi, L., Hillston, J., Latella, D., Massink, M.: Continuous approximation of collective system behaviour: a tutorial. Perform. Eval. **70**(5), 317–349 (2013)

6. Bortolussi, L., Hillston, J., Loreti, M.: Fluid approximation of broadcasting systems. Theor. Comput. Sci. **816**, 221–248 (2020)
7. Bortolussi, L., Lanciani, R., Nenzi, L.: Model checking Markov population models by stochastic approximations. Inf. Comput. **262**, 189–220 (2018)
8. Claessen, K., Mårtensson, J.: An operational semantics for weak PSL. In: Hu, A.J., Martin, A.K. (eds.) FMCAD 2004. LNCS, vol. 3312, pp. 337–351. Springer, Heidelberg (2004). https://doi.org/10.1007/978-3-540-30494-4_24
9. Donzé, A.: On signal temporal logic. In: Legay, A., Bensalem, S. (eds.) RV 2013. LNCS, vol. 8174, pp. 382–383. Springer, Heidelberg (2013). https://doi.org/10.1007/978-3-642-40787-1_27
10. Hillston, J.: Quantitative analysis of collective adaptive systems. In: Mazzara, M., Voronkov, A. (eds.) PSI 2015. LNCS, vol. 9609, pp. 1–5. Springer, Cham (2016). https://doi.org/10.1007/978-3-319-41579-6_1
11. Inverso, O., Trubiani, C., Tuosto, E.: Abstractions for collective adaptive systems. In: Margaria, T., Steffen, B. (eds.) ISoLA 2020. LNCS, vol. 12477, pp. 243–260. Springer, Cham (2020). https://doi.org/10.1007/978-3-030-61470-6_15
12. Latella, D., Loreti, M., Massink, M.: On-the-fly PCTL fast mean-field approximated model-checking for self-organising coordination. Sci. Comput. Program. **110**, 23–50 (2015)
13. Legay, A., Lukina, A., Traonouez, L.M., Yang, J., Smolka, S.A., Grosu, R.: Statistical model checking. In: Steffen, B., Woeginger, G. (eds.) Computing and Software Science. LNCS, vol. 10000, pp. 478–504. Springer, Cham (2019). https://doi.org/10.1007/978-3-319-91908-9_23
14. Loreti, M., Hillston, J.: Modelling and analysis of collective adaptive systems with CARMA and its tools. In: Bernardo, M., De Nicola, R., Hillston, J. (eds.) SFM 2016. LNCS, vol. 9700, pp. 83–119. Springer, Cham (2016). https://doi.org/10.1007/978-3-319-34096-8_4
15. Michaelides, M., Hillston, J., Sanguinetti, G.: Geometric fluid approximation for general continuous-time Markov chains. Proc. Roy. Soc. A **475**(2229), 20190100 (2019)
16. Nicola, R.D., Di Stefano, L., Inverso, O.: Multi-agent systems with virtual stigmergy. Sci. Comput. Program. **187**, 102345 (2020)
17. Reijsbergen, D., de Boer, P.-T., Scheinhardt, W., Haverkort, B.: On hypothesis testing for statistical model checking. Int. J. Softw. Tools Technol. Transfer **17**(4), 377–395 (2015). https://doi.org/10.1007/s10009-014-0350-1
18. Vissat, L.L., Hillston, J., Marion, G., Smith, M.J.: MELA: modelling in ecology with location attributes. In: Tribastone, M., Wiklicky, H. (eds.) Proceedings 14th International Workshop Quantitative Aspects of Programming Languages and Systems, QAPL 2016, Volume 227 of EPTCS, Eindhoven, The Netherlands, 2–3 April 2016, pp. 82–97 (2016)

Jump Longer to Jump Less: Improving Dynamic Boundary Projection with h-Scaling

Francesca Randone[1]($\boxtimes$), Luca Bortolussi[2] iD, and Mirco Tribastone[1] iD

[1] IMT School for Advanced Studies Lucca, Lucca, Italy
francesca.randone@imtlucca.it
[2] Department of Mathematics and Geosciences, Università degli Studi di Trieste, Trieste, Italy

Abstract. The master equation describes exactly the dynamics of a Markov Population Process (MPP) by associating one differential equation for each discrete state of the process. It is well known that MPPs are prone to suffer from the so-called *curse of dimensionality*, making the master equation intractable in most cases. We propose a novel approach, called h-scaling, that covers the state space of an MPP with a smaller number of states by an appropriate re-scaling of the MPP transition rate functions. When the original state space is bounded, this procedure may significantly reduce the number of the states while returning an approximate master equation that still retains good accuracy. We present h-scaling together with some theoretical results on asymptotic correctness and numerical examples taken from the performance evaluation literature. Moreover, we show that h-scaling can be combined with a recently proposed framework called dynamic boundary projection, which couples subsets of the master equation with mean-field approximations, to further reduce the number of equations without penalizing accuracy.

Keywords: Markov Population Processes · Master equation · Mean-field models · Approximation Methods

1 Introduction

Markov Population Processes (MPPs) are models used to describe systems of interacting agents. The underlying population is represented as a continuous-time Markov Chain (CTMC) in which each component of the state vector represents one class of agents. Transitions between states are modeled by transition classes, each consisting in a transition vector, modeling the change induced on the state vector, and a state-dependent rate function, modeling the frequency with which the transition happens, given that the system is in a certain state [4].

For their versatility and their ability to encode complex stochastic dynamics, they have been used to study the performance of computer-based systems such as queueing networks [2], virtualized environments [1], peer-to-peer networks [21], malware propagation dynamics [3] and allocation strategies [14, 16, 20, 22].

© Springer Nature Switzerland AG 2022
E. Ábrahám and M. Paolieri (Eds.): QEST 2022, LNCS 13479, pp. 150–170, 2022.
https://doi.org/10.1007/978-3-031-16336-4_8

In all these applications however, the analysis of the underlying stochastic model is hindered by the exponential growth of the state space, referred to as the *curse of dimensionality*. This makes a direct solution of the master equation (ME) intractable in many cases, and the application of stochastic simulation algorithms (e.g., [6]) computationally intensive. Different techniques have been proposed to solve this problem such as fluid (aka mean-field) approximations [8,11], truncation methods [15] and aggregation techniques [5].

Taking a different point of view, MPPs can be thought of as processes evolving on a multi-dimensional grid in which they perform jumps of a certain magnitude, with a certain rate. The state space is then the set of vertices in the grid reachable by the process. In this paper we propose an approximation based on a simple re-scaling: we consider a process that performs longer jumps at lower frequency, where the two quantities of interests, i.e. magnitude of the jumps and rate functions, are re-scaled proportionally. When the original process evolves on a finite state space, the advantage is immediately visible: the number of reachable states is reduced, or, to rephrase the idea in terms of multi-dimensional grids, we cover the original state space with a coarser grid, so that the process reaches fewer of its vertices.

To further reduce the number of equations needed in the approximation we propose to couple h-scaling with a recently proposed method called dynamic boundary projection (DBP) [18]. In DBP an approximate ME, describing the evolution of a subset of the state space, is coupled with a mean-field equation that shifts dynamically the observed subset over the state space. DBP has been proposed to refine the accuracy of the mean-field equation when used as an approximation of the average behavior of a finite-size MPP, a topic which has enjoyed considerable interest recently [10,11]. Here, we show that coupling h-scaling and DBP performs better than simply applying h-scaling. Furthermore, this allows us to reduce the number of equations used by DBP while keeping a high accuracy of the approximation.

From a theoretical point of view, we show that our scaling preserves the behaviour of the original system in the thermodynamic limit. Beside the mean-field limit, under additional hypotheses, it is possible to show that a sequence of processes of increasing size tends to a Gaussian process commonly referred to as the linear noise approximation (LNA) [19]. We show that, given a sequence of processes admitting a mean-field limit and a LNA, and considering a second sequence, obtained from the first by applying the proposed scaling with a fixed parameter, both sequences share the same limiting processes, up to an error due to possibly different initial conditions.

Finally we show some applications of our scaling coupled with DBP to systems taken from the performance evaluation literature, showing the advantages of this approach in terms of computational effort.

Structure of the Paper In Sect. 2 we introduce some background results used in the rest of the paper. In Sect. 3 we present our scaling and we explain how it can be used to approximate efficiently the mean dynamics of an MPP, either applying it directly (static h-scaling), or coupling it with DBP (scaled DBP).

In Sect. 4 we state the theoretical results on the preservation of the limit behaviour while in Sect. 5 we present a multi-scale extension of our method. Finally, Sect. 6 shows some applications of our methodology, while conclusions and future works are discussed in Sect. 7.

2   Background

Markov Population Processes. We consider an MPP $X(t)$ evolving on $\mathcal{S} \subseteq \mathbb{N}^m$. Suppose that $X(0) = x_0 \in \mathbb{N}^m$ with probability 1 and that there exists a finite set $\mathcal{L} \subset \mathbb{Z}^m$ such that for every $l \in \mathcal{L}$ the process X performs transitions

$$x \to x + l \quad \text{with rate} \quad \beta_l(x) \tag{1}$$

(we assume $\beta_l(x) = 0$ if x or $x + l$ are not in $\mathcal{S}$).

The exact dynamics of $X(t)$ can be described by its ME:

$$\frac{dP(x)}{dt} = \sum_{l \in \mathcal{L}} \beta_l(x - l)P(x - l; t) - \sum_{l \in \mathcal{L}} \beta_l(x)P(x; t) \quad \forall x \in \mathcal{S}. \tag{2}$$

The (exact) mean dynamics of $X(t)$ is given by $\mathbb{E}[X(t)] = \sum_{x \in \mathcal{S}} x P(x; t)$.

Mean-Field Approximation. Due to the quick growth of the state-space this method is rarely feasible, and it is common to resort to approximations. One of the most common approaches uses a classic limit results, first stated by Kurtz [12]. We refer to slightly more general formulation that can be found in [9].

We assume that the process X has size $\gamma_{\bar{N}}$ and there exists a sequence of processes $(X^N)_{N \geq N_0}$ such that $X^{\bar{N}} = X$ and each X^N has size γ_N with $\lim_{N \to \infty} \gamma_N = \infty$. In the most general case every X^N is defined on a state-space $\mathcal{S}^N$, has initial condition x_0^N and performs transitions $x \to x + l^N$ firing with rate $\beta_l^N(x)$ for each l in a given set $\mathcal{L}$.

A particular case is given when the original sequence of processes is *density-dependent*, whereby γ_N grows linearly with N, for each N and $l \in \mathcal{L}$ there exists a vector v_l such that $l^N = \frac{v_l}{\gamma_N}$ and for each N and $l \in \mathcal{L}$ there exists a (locally) Lipschitz continuous and (locally) bounded function $g_l : E \to \mathbb{R}_{\geq 0}$ such that $\beta_l^N(x) = \gamma_N g_l(x)$.

Given a sequence $(X^N)_{N \geq N_0}$, we consider the sequence of normalized processes $(\hat{X}^N)_{N \geq N_0}$, where each process has state vector $\hat{X}^N = \frac{1}{\gamma_N} X^N$, starts in $\hat{x}_0^N = \frac{x_0^N}{\gamma_N}$ and performs transitions $x \to x + \frac{l^N}{\gamma_N}$ with rate $\hat{\beta}_l^N(x) = \beta_l^N(\gamma_N x)$. We denote the normalized state space by $\hat{\mathcal{S}}^N$.

Finally, we define the *drift* $F^N : \hat{\mathcal{S}}^N \to \mathbb{R}^m$ as

$$F^N(x) = \sum_{l \in \mathcal{L}} \hat{l}^N \hat{\beta}_l^N(x).$$

Under these assumptions the following theorem holds.

Theorem 1 (Convergence to deterministic limit for MPPs [4]). *Let $E \subseteq \mathbb{R}^m$ be a closed set such that $\cup_N \hat{S}^N \subseteq E$. Suppose that there exists $x_0 \in E$ such that $\lim_N \hat{x}_0^N = x_0$ and a Lipschitz vector field $F : E \to \mathbb{R}^m$ such that*

$$\lim_N \sup_{x \in \hat{S}^N} \|F^N(x) - F(x)\| = 0.$$

Assuming the rates of convergence in Theorem 4.2 of [4] are verified, for any fixed time instant $T > 0$ and $\forall \epsilon \geq 0$

$$\lim_{N \to \infty} P\left(\sup_{0 \leq t \leq T} \|\hat{X}^N(t) - \hat{x}(t)\| > \epsilon \right) = 0$$

where $\hat{x}(t)$ is the solution to the initial value problem:

$$\begin{cases} \frac{d\hat{x}}{dt} = F(\hat{x}(t)) \\ \hat{x}(0) = x_0 \end{cases} \tag{3}$$

and $\hat{x}(t) \in E \, \forall t \geq 0$.

Observe that in the case of density-dependent processes the drift is independent of N and the hypotheses of the Theorem hold trivially.

For a general MPP defined as in (1) we define its *mean-field approximation* as the solution to the initial value problem:

$$\begin{cases} \frac{dx}{dt} = \sum_{l \in \mathcal{L}} l \beta_l(x) \\ x(0) = X(0). \end{cases}$$

Observe that if X is a density-dependent process admitting a deterministic limit defined by Eq. (3) we have $x(t) = \gamma_N \hat{x}(t)$. In particular, while $\hat{x}(t)$ approximates the normalized process, and therefore the average proportion of agents in a certain class, $N\hat{x}(t)$ approximates the mean number of agents in each class.

Linear Noise Approximation. Another classic limit result has been obtained by Van Kampen applying a "size expansion" to the ME [19]. It takes into account the stochastic fluctuations of the process around its deterministic limit. It can be proved that, in a first order approximation, such fluctuations behave as a Gaussian process with zero mean. The result is stated in the following theorem.

Theorem 2 (Convergence to Linear Noise Approximation for MPPs [19]). *Consider a sequence of density-dependent MPPs $\left(X^N \right)_{N \geq N_0}$, each starting in $X^N(0) = N\hat{x}_0$ and suppose that the drift $F(x)$ is continuously differentiable in E. Then, letting $\left(\hat{X}^N \right)_{N \geq N_0}$ denote the sequence of normalized processes and $\hat{x}(t)$ the solution of the initial value problem in (3) we have that*

$$\lim_{N \to \infty} \sqrt{N} \|\hat{X}^N - \hat{x}(t)\| = \xi(t) \tag{4}$$

where $\xi(t)$ is a Gaussian process identified by equations of the first two moments:

$$\frac{d\mu_i}{dt} = \sum_j \left(\sum_{l \in \mathcal{L}} \frac{\partial \beta_l}{\partial x_j}(\hat{x}(t)) \right) \mu_j(t) \tag{5}$$

$$\frac{d\Sigma_{ij}}{dt} = \sum_k \left(\sum_{l \in \mathcal{L}} l_i \frac{\partial \beta_l}{\partial x_k}(\hat{x}(t)) \right) \Sigma_{kj}(t) + \sum_k \left(\sum_{l \in \mathcal{L}} l_j \frac{\partial \beta_l}{\partial x_k}(\hat{x}(t)) \right) \Sigma_{ik}(t)$$
$$+ \sum_{l \in \mathcal{L}} l_i l_j \beta_l(\hat{x}(t)) \tag{6}$$

Dynamic Boundary Projection. In DBP, an hyper-rectangular subset of the state space, called a *truncation*, is shifted dynamically across the state space. This is achieved by coupling a truncated ME describing the evolution of the states in the subset with a mean-field approximation accounting for transitions outside the current truncation.

Let $n \in \mathbb{N}^m$ and let $\mathcal{S} \subset \mathbb{N}^m$. We define the *truncation of $\mathcal{S}$ of size n indexed by y* as the set:

$$\mathcal{T}(n, y) = \{x \in \mathcal{S} : y_i \leq x_i \leq y_i + n_i \, \forall i = 1, 2, \ldots, m\}. \tag{7}$$

Observe that $\mathcal{T}(n, y)$ has at most $\mathcal{N}(n) = \prod_{i=1}^m (n_i + 1)$ states. For each truncation a special role is played by states "on the border", i.e. those states from which X can perform a transition that takes it outside the current truncation. This leads us to define the sets:

$$\partial \mathcal{T}_l(n, y) = \{x \in \mathcal{T}(n, y) : x + l \notin \mathcal{T}(n, y)\}, \text{ for } l \in \mathcal{L},$$
$$\partial \mathcal{T}(n, y) = \bigcup_{l \in \mathcal{L}} \partial \mathcal{T}_l(n, y) = \{x \in \mathcal{T}(n, y) : \exists l \in \mathcal{L} \text{ s.t. } x + l \notin \mathcal{T}(n, y)\}$$

We consider an augumented approximation [13] called *boundary projection* (BP), in which every jump from $x \in \partial \mathcal{T}_l(n, y)$ to x' is redirected with same rate to a state x^* defined as:

$$x_i^* = \begin{cases} \min(y_i + n_i, x_i') & \text{if } x_i' > x_i \\ \max(y_i, x_i') & \text{if } x_i' < x_i \\ x_i & \text{if } x_i' = x_i. \end{cases} \tag{8}$$

After performing the augmentation, for each state $x \in \mathcal{T}(n, y)$ we have a set of jump vectors $\mathcal{L}^n(x)$ such that for every $l \in \mathcal{L}$ we can now define a new vector $l^{(n)}(x)$ given by:

$$l^{(n)}(x) = \begin{cases} l & \text{if } x \notin \partial \mathcal{T}_l(n, y), \\ x^* - x & \text{if } x \in \partial \mathcal{T}_l(n, y), \end{cases}$$

where x^* is the target state in which the transition $x \to x + l$ has been redirected and the associated transition rates are $\beta_{l^{(n)}(x)}(x) = \beta_l(x)$.

Let $X_y^{(n)}$ be the BP of X on $\mathcal{T}(n,y)$ and let $Q(n,y)$ be its transition rate matrix. It can be proved [18] that for every y the transition rate matrices $Q(n,y)$ have the same functional form. In particular they can be written as:

$$[Q^{(n)}(y)]_{x,x'} = \begin{cases} \sum_{l \in \mathcal{L}} \mathbb{I}_{\{x'+l^{(n)}(x')=x\}} \beta_l(x'+y) & \text{if } x \neq x' \\ -\sum_{l \in \mathcal{L}} \mathbb{I}_{\{l^{(n)}(x) \neq 0\}} \beta_l(x+y) & \text{if } x = x' \end{cases} \quad \text{for } x, x' \in \mathcal{T}(n,0).$$

(9)

This allows us to write the ME for the process $X_y^{(n)}$ as:

$$\frac{dP_y^{(n)}}{dt} = Q^{(n)}(y) P_y^{(n)}(\,\cdot\,;t)$$

where $P_y^{(n)}(\,\cdot\,;t)$ is an $\mathcal{N}(n)$-dimensional vector indexed by the states in $\mathcal{T}(n,0)$ and each component $P_y^{(n)}(x;t)$ is the probability of $X_y^{(n)}$ being in the state $x+y$.

To pass from BP to DBP, we need to write mean-field equation for y. In order to do so we define the functions:

$$\left[\Pi^{(n)}(x,y)\right]_i = \begin{cases} x_i & x_i < y_i \\ x_i - n_i & x_i > y_i + n_i \qquad \forall x, y \in \mathcal{S} \\ y_i & y_i \leq x \leq y_i + n_i. \end{cases}$$

$$\mathcal{Y}_l(n,x) = \Pi^{(n)}(x+l,0) \quad \forall l \in \mathcal{L}, \forall x \in \partial \mathcal{T}_l(n,0).$$

Then, the equation for DBP with parameter n are given by:

$$\frac{dY^{(n)}}{dt} = \sum_{l \in \mathcal{L}} \sum_{x \in \partial \mathcal{T}_l(n,0)} \mathcal{Y}_l(n,x) \beta_l(x+Y^{(n)}(t)) P^{(n)}(x;t)$$

(10)

$$\frac{dP^{(n)}}{dt} = Q^{(n)}(Y^{(n)}(t)) P^{(n)}(\,\cdot\,,t).$$

Supposing the original process X has initial condition $X(0) = x_0 \in \mathbb{N}^m$ with probability 1, we rewrite x_0 as $x_0^* + y_0$ with $y_0 = \Pi^{(n)}(x_0,0)$ and $x_0^* \in \mathcal{T}(n,0)$. Then, we set the initial condition for DBP to:

$$Y^{(n)}(0) = y_0 \qquad P^{(n)}(x;0) = \begin{cases} 1 & \text{if } x = \bar{x}_0^*, \\ 0 & \text{else.} \end{cases}$$

(11)

3 h-Scaling

Consider an MPP as the one defined in (1). Let $R(\mathcal{S})$ be a minimal (with respect to inclusion) hyper-rectangle in $\mathbb{N}^m$ containing $\mathcal{S}$ and let $v_R \in R(\mathcal{S})$ be such that $(v_R)_i = \min\{x_i | x \in R(\mathcal{S})\}$. We can imagine $\mathcal{S}$ as a subset of vertices of the m-dimensional grid covering $R(\mathcal{S})$ having edges of length 1, where v_R is a the vertex of $R(\mathcal{S})$ with minimal components. We want to cover $R(\mathcal{S})$ with a coarser grid, so that less vertices will be contained in the original hyper-volume.

To do this we fix a scalar parameter $h > 1$; in some cases h can be chosen to be a vector, extending all the present results, as will be discussed in Sect. 5. Now, let $H(\mathcal{D})$ be the convex hull in $\mathbb{R}^m$ of any discrete set $\mathcal{D}$. We define the state space of the scaled process as:

$$\mathcal{S}^h = \{x = v_R + h(k_1 e_1 + \ldots + k_m e_m) | k_i \in \mathbb{N} \, \forall i = 1, \ldots, m\} \cap \mathcal{H}(R(\mathcal{S}))$$

where e_i is the m-dimensional with 1 as i-th component and 0 else.

We now define a process X^h evolving on $\mathcal{S}^h$. We set

$$X^h(0) = v_R + h \left\lfloor \frac{x_0 - v_R}{h} \right\rfloor \in \mathcal{S}^h \tag{12}$$

and define the transitions of X^h as

$$x \to x + hl \quad \text{at rate} \quad \frac{1}{h} \beta_l(x) \mathbb{I}_{\{x + hl \in \mathcal{S}^h\}} \tag{13}$$

for all $l \in \mathcal{L}$ and with $\mathbb{I}_{\mathcal{C}}$ denoting the characteristic function of the set $\mathcal{C}$. We call the process X^h so defined the h-scaling of X.

Example 1. An $M/M/k$ queue can be seen as an MPP defined by the following transition classes, denoting respectively exogenous arrivals with Poisson rate λ and service with rate μ:

$$x \to x + 1 \quad \text{at rate } \lambda, \qquad x \to x - 1 \quad \text{at rate } \mu \min(x, k).$$

We assume that the queue starts with zero customers, i.e. $X(0) = 0$.

The original state space is $\mathcal{S} = \mathbb{N}$ and $\mathcal{H}(R(\mathcal{S})) = \mathbb{R}_{\geq 0}$ so we will have that $\mathcal{S}^h$ it is still infinite, with $|\mathcal{S}| = |\mathcal{S}^h|$, but it is different from $\mathcal{S}$:

$$\mathcal{S}^h = \{hn \, | \, n \in \mathbb{N}\}.$$

The h-scaling is X^h, such that $X^h(0) = 0$ and performing transitions:

$$x \to x + h \quad \text{at rate} \quad \frac{\lambda}{h}, \qquad x \to x - h \quad \text{at rate} \quad \frac{\mu}{h} \min(x, k). \tag{14}$$

A variation of the $M/M/k$ queue is the $M/M/k/N$ queue, in which no more than N jobs are accepted in the queue. In this case the transition classes are:

$$x \to x + 1 \quad \text{at rate } \lambda \mathbb{I}_{\{x < N\}}, \qquad x \to x - 1 \quad \text{at rate } \mu \min(x, k).$$

In this case, $\mathcal{S} = \{0, 1, \ldots, N\}$ and the h-scaling for the $M/M/k/N$ queue is defined by the transitions

$$x \to x + h \quad \text{at rate} \quad \frac{\lambda}{h} \mathbb{I}_{\{x < h \lfloor \frac{N}{h} \rfloor\}}, \qquad x \to x - h \quad \text{at rate} \quad \frac{\mu}{h} \min(x, k). \tag{15}$$

Therefore, in this case, $\mathcal{S}^h = \{0, h, \ldots, h \lfloor \frac{N}{h} \rfloor\}$ so we are indeed reducing the number of states as $|\mathcal{S}^h| = \lfloor \frac{N}{h} \rfloor + 1 \leq N + 1 = |\mathcal{S}|$.

Interpretation of h-Scaling. When h is integer it is possible to give a physical interpretation of the process X^h. For example, consider the $M/M/k$ of the example and set $h = 2$. X^2 is a process performing transitions $x \to x + 2$ with rate $\frac{\lambda}{2}$ and $x \to x - 2$ with rate $\frac{\mu}{2} \min(x, k)$. This means that in X^2 every event involves two perfectly synchronized agents that arrive and leave the queue together, and each event takes place after a time which is, on average, exactly two times the average time after which a single agent would perform that transition given the same initial conditions. The same can be said for any integer value of h. For $h \in \mathbb{Q}$, the state space of X^h takes values in the real space. However, since the components of the state vector of an MPP represent population counts, non-integer values escape physical intelligibility. Nevertheless, we will show that rational values of h are useful to tune the accuracy of the approximation.

3.1   Static Scaling

A first approximation can be obtained simply by solving the ME for X^h, which will yield a smaller number of equations than the one for X. This is possible when the state-space is finite (or when we consider a sufficiently large truncation of an infinite state-space to contain most of the probability mass [15]). We will call this approximation *static scaling*. Without loss of generality let us consider the finite state space $\mathcal{S} = \{0, 1, \ldots, N_1\} \times \ldots \times \{0, 1, \ldots, N_m\}$, thus $|\mathcal{S}| = \prod_{i=1}^{m}(N_i + 1)$ is the number of equations of the ME.

Applying the scaling in (13) for $h > 1$ gives as new state space $\mathcal{S}^h = \{0, h, \ldots, h\lceil \frac{N_1}{h} \rceil\} \times \ldots \times \{0, h, \ldots, h\lceil \frac{N_m}{h} \rceil\}$, with $|\mathcal{S}^h| = \prod_{i=1}^{m}\left(\lceil \frac{N_i}{h} + 1 \rceil + 1\right) \leq |\mathcal{S}|$. For all $x \in \mathcal{S}^h$ the ME for X^h can be written as

$$\frac{dP^h(x)}{dt} = \sum_{l \in \mathcal{L}} \frac{1}{h}\beta_l(x - hl)P^h(x - hl; t) - \sum_{l \in \mathcal{L}} \frac{1}{h}\beta_l(x)\mathbb{I}_{\{x+hl\in\mathcal{S}^h\}}P^h(x; t). \quad (16)$$

We can then approximate $\mathbb{E}[X]$ by solving (16) and computing

$$\mathbb{E}[X^h] = \sum_{x \in \mathcal{S}^h} x P^h(x).$$

Example 2. As we have seen, h-scaling applied to an $M/M/k/N$ queue yields (15). The ME for X^h is then

$$\frac{dP^h(x)}{dt} = \begin{cases} -\frac{\lambda}{h}P^h(0; t) + \frac{\mu}{h}\min(h, k)P^h(h; t) & x = 0 \\ -\left(\frac{\lambda}{h} + \frac{\mu}{h}\min(x, k)\right)P^h(x; t) + \frac{\lambda}{h}P^h(x - h; t) \\ \quad + \frac{\mu}{h}\min(x + h, k)P^h(x + h; t) & x \neq 0, h\lfloor \frac{N}{h} \rfloor \\ -\frac{\mu}{h}\min\left(h\lfloor \frac{N}{h} \rfloor, k\right)P^h\left(h\lfloor \frac{N}{h} \rfloor; t\right) \\ \quad + \frac{\lambda}{h}P^h\left(h\left(\lfloor \frac{N}{h} \rfloor - 1\right); t\right) & x = h\lfloor \frac{N}{h} \rfloor \end{cases}$$

In Fig. 1 we can see the results for $h = 1.2, 1.4, 1.6, 1.8, 2.0$ applied to an $M/M/k/N$ queue with parameters $k = 4, N = 50, \lambda = 3.95, \mu = 1$. We can see that the number of equations is progressively reduced up to 50% while the mean estimated using the h-scaling still keeps a low relative error with respect to the true mean (at steady-state no more than 4% of the true value).

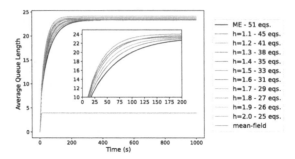

Fig. 1. Application of static h-scaling to the $M/M/k/N$ queue.

3.2 Scaled Dynamic Boundary Projection

Here we combine h-scaling with DBP [18]. We now assume that X evolves on a state space $\mathcal{S} \subseteq \mathbb{N}^m$ (not necessarily finite) and that DBP with parameter $n \in \mathbb{N}^m$ can be applied to X yielding system (10), consisting of $\mathcal{N}(n) + m$ equations. We are going to show how to apply DBP to the scaled process X^h and how this results in a reduction of the number of approximating equations.

In this case our idea is to cover the portion of the state space inside a truncation $\mathcal{T}(n, y)$ with a coarser grid. To do so we define the truncations:

$$\mathcal{T}^h(n, y) = \{x = y + h(k_1 e_1 + \ldots + k_m e_m) \mid k_i \in \mathbb{N} \, \forall i = 1, \ldots, m\}$$
$$\cap \mathcal{H}(\mathcal{T}(n, y)). \quad (17)$$

Observe that the states in $\mathcal{T}^h(n, y)$ are not necessarily in $\mathcal{S}^h$ (although they are if $y \in \mathcal{S}^h$). Moreover, $\mathcal{T}^h(n, y)$ has $\mathcal{N}^h(n) = \prod_{i=1}^{m} \left(\lfloor \frac{n_i}{h} \rfloor + 1 \right) \leq \mathcal{N}(n)$ states for any y. This implies that the number of equations in scaled DBP is reduced by a factor $\frac{1}{h^m}$. Once the definition of truncation in $\mathcal{S}^h$ is clarified the derivation of the equations for scaled DBP follows step by step the one for the original process and is reported in detail in the Appendix.

The equations for scaled DBP with parameters n and h are given by:

$$\frac{dY^{(n,h)}}{dt} = \sum_{l \in \mathcal{L}} \sum_{x \in \partial T_l^h(n,0)} \mathcal{Y}_l^h(n, x) \frac{1}{h} \beta_l(x + Y^{(n,h)}(t)) P^{(n,h)}(x; t)$$
$$\frac{dP^{(n,h)}}{dt} = Q^{(n,h)}(Y^{(n,h)}(t)) P^{(n,h)}(\cdot, t). \quad (18)$$

Example 3. We now consider the $M/M/k$ queue with $\lambda = 3.85, \mu = 1$ and $k = 4$. In principle, the model has an infinite state space, but it converges to its steady state distribution, so it is possible to select a finite truncation of the state space so that the probability mass outside it is arbitrarily small. To select the minimal truncation that we can take as ground truth, we start by considering the ME with 500 states and progressively reduce the number of states so the error introduced is less than 0.001% of the Average Queue Length (AQL) at steady state. We obtain that to correctly capture the dynamics of the queue we need 375 equations.

DBP can be applied to this system to reduce the number of equations needed to approximate the dynamics of the system and we see that using $n = 170$, the relative error between the DBP approximation and the solution of the ME at steady state is less than 1%.

We can further reduce the number of equations by coupling DBP with h-scaling (equations can be found in the Appendix), at the price of a bigger error. As can be seen from Fig. 2 and Table 1, for the same value of h, scaled DBP performs better than h-scaling while yielding an heavier reduction in the number of equations. We will see that this is the case also for more complex examples.

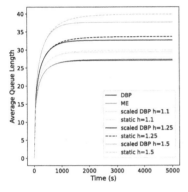

Fig. 2. h-scaling and scaled DBP applied to the $M/M/k$ queue.

Table 1. Approximated value of the AQL of the $M/M/k$ queue at steady state (t = 5000), with relative error, number of equation and reduction in the number of equations when h-scaling and scaled DBP are applied. h-scaling with $h = 1$ is considered the ground truth.

	h-scaling			scaled DBP		
h	err	# eqs.	red.	err	# eqs.	red.
1.0	-	376	-	0.91%	172	54.25%
1.1	9.25%	341	9.31%	7.54%	156	58.51%
1.25	22.78%	301	19.94%	19.29%	138	63.29%
1.5	45.50%	251	33.24%	37.4%	115	69.41%

4   Limit Behaviour

We now prove that for a fixed $h > 1$ the approximating process X^h shares the same limiting behaviour as X. To lighten notation we will assume $v_R = 0$.

Preservation of the Mean-Field Limit. Suppose that the original process X is part of a sequence $(X^N)_{N \geq N_0}$ satisfying the hypotheses of Theorem 1. Fix $h > 1$ and consider a new sequence $(X^{N,h})_{N \geq N_0}$ where each $X^{N,h}$ is obtained applying the h-scaling to X^N.

It is immediate to observe that the new sequence $(X^{N,h})_{N \geq N_0}$ still satisfies the hypotheses of the theorem, and, in particular, for every N

$$F^{N,h}(x) = \sum_{l \in \mathcal{L}} h \hat{l}^n \frac{1}{h} \hat{\beta}_l^N(x) = F^N(x),$$

which means that under the proposed scaling the drift function is preserved for every N and independent of h. This implies that the deterministic limit process

$\hat{x}(t)$ defined by (3) is exactly the same for both sequences (in fact, observe that $\lim_N h \lfloor \frac{x_0^N}{h} \rfloor = x_0$, so also the limiting initial condition is the same).

This can be summed up in the following theorem:

Theorem 3. *Consider a sequence of processes* $\left(X^N\right)_{N \geq N_0}$ *and consider the sequence of approximating processes* $\left(X^{N,h}\right)_{N \geq N_0}$ *obtained applying the scaling in* (13) *for a fixed* $h > 1$. *If the original sequence admits a deterministic limit* $\hat{x}$ *in the sense of Theorem 1, then the sequence of approximating processes admits the same limit.*

In the special case of density-dependent processes for all $l \in \mathcal{L}$ the scaled transitions become

$$x \to x + \frac{h}{\gamma_n} v_l \quad \text{at rate} \quad \frac{\gamma_N}{h} g_l(x).$$

This is equivalent to saying that the sequence $\left(X^{N,h}\right)_{N \geq N_0}$ is a density dependent family with respect to the parameter $\frac{\gamma_N}{h}$. This is the same observation that in [7] led to prove the limit behaviour for $h \to 0$ and it is easily explained by the fact that we are approximating systems of size γ_N with systems of size $\frac{\gamma_N}{h}$ scaling coherently both the magnitude of the jumps and the transition rates, which are the two quantities involved in the density-dependence assumption.

Extension to Other Mean-Field Limit Results. The fact the deterministic approximation is preserved with exactly the same limit drift F allows us to extend to $\left(X^{N,h}\right)_{N \geq N_0}$ other limit results, provided they hold for the original sequence. In particular, if E is compact and the ODE in (3) admits a globally asymptotically stable fixed point x^*, every sequence of invariant measures of X^N tends weakly to the Dirac distribution centered on x^* [3]. The same is true for any sequence of invariant measures of the sequence $\left(X^{N,h}\right)_{N \geq N_0}$. Analogously, we can straightforwardly extend to $\left(X^{N,h}\right)_{N \geq N_0}$ results related to mean-field independence (also called propagation of chaos) [3], both in transient and steady-state, and to fast simulation [9] (provided the original sequence satisfies the required assumptions).

Preservation of the LNA. Similarly to the mean-field limit, we assume that our original process X belongs to a sequence of processes $\left(X^N\right)_{N \geq N_0}$ satisfying the hypotheses of Theorem 2. Again, we fix $h > 1$ and consider a sequence $\left(X^{N,h}\right)_{N \geq N_0}$ obtained by applying the h-scaling to each process of the original sequence. Then, the following result hold:

Theorem 4. *Suppose that for the sequence* $\left(X^N\right)_{N \geq N_0}$ *the hypotheses of Theorem 2 are verified and, in addition:*

– *equation* (3) *admits a globally asymptotically stable equilibrium* x^*;
– *for each* N $X^N(0) = N\hat{x}_0$;
– *for each* N $\gamma_N = N$.

Then, letting $\mu(t)$ and $\Sigma(t)$ denote the mean and the covariance matrix of limiting Gaussian process for the original sequence, we have that the sequence of approximating processes $\left(X^{N,h}\right)_{N \geq N_0}$ admits a Gaussian limiting process with mean $\mu^h(t)$ and covariance matrix $\Sigma^h(t)$ such that:

$$\lim_{t \to \infty} \mu^h(t) = \lim_{t \to \infty} \mu(t) = 0 \text{ and } \Sigma^h(t) = \Sigma(t) \, \forall \, t \geq 0.$$

A proof of the Theorem can be found in the Appendix.

5   Multi-scale Approximation

In some cases, it can be useful to consider, instead of a single scalar parameter h, an m-dimensional vector $\underline{h} \in \mathbb{R}^m$, $h_i \geq 1 \, \forall \, i = 1, \dots, m$. In this case we will re-scale the jumps in different components using different values h_i. This can be desirable when one component evolves on a much larger space than the others. However, we cannot choose $\underline{h}$ arbitrarily, since we need to preserve the quantities transformed by the transitions. If in the original process a agents of class i transition into b agents of class j, the scaling parameters h_i and h_j must be chosen so that in the scaled process the corresponding transition preserves this conversion.

Formally, we can index the components of the original process with the set $I = \{1, \dots, m\}$. We say that component i transitions to j, written $i \leftrightarrow j$, if there exists a sequence of transitions that transforms an agent of class i into an agent of class j. We require that when choosing $\underline{h}$, if $i \leftrightarrow j$, then $h_i = h_j$. We call a vector $\underline{h}$ satisfying this assumption a *valid h-scaling vector*.

Observe that the constraints on $\underline{h}$ do not depend on the rate functions but only on the transition vectors l. The scalar case corresponds to $h_i = h_j \, \forall i, j \in I$.

All previous results can be extended straightforwardly to the multi-scale case, provided $\underline{h}$ is a valid vector. For an application see Sect. 6.1.

Example 4. A system with K classes of customers can be represented as an MPP with state $x = (x_{Q_1}, \dots, x_{Q_K})$, where each component identifies a class of customers in the queue. A common choice for the rate functions is to consider classes of customers subject to an *egalitarian* policy [16,17] where all customers are assigned the same weight; in particular, we use the rate functions adopted in [22]. The transition classes associated with the process are for $i = 1, \dots, K$:

$$a_i = +e_{Q_i}, \text{ at rate } N\lambda_i, \qquad d_i = -e_{Q_i}, \text{ at rate } \mu \frac{x_{Q_i}}{\sum_{j=1}^{K} x_{Q_j} + N}.$$

In this system agents cannot transition from one class into another one, so we can consider h-scaling vectors with $h_i \neq h_j$ for each $i, j \in I$.

Example 5. Consider a Malware Propagation model composed of N nodes, where each node can be dormant (D), active (A) or susceptible (S). Since the total number of nodes is constant, the model can be described by the state vector $x = (x_D, x_A)$ where the number of susceptible nodes at each state is given by $N - x_D - x_A$. Following [3,10] we consider the following transitions classes:

$$l_1 = -e_D + e_A \qquad \text{at rate} \quad \left(1 + \frac{10x_A}{x_D + 0.5N}\right) x_D,$$

$$l_2 = -e_A \qquad \text{at rate } 5x_A,$$

$$l_3 = +e_D \qquad \text{at rate} \quad \left(0.1 + \frac{10}{N}x_D\right)(N - x_D - x_A).$$

Since $D \leftrightarrow A$ we need to set $h_D = h_A$ for each valid h-scaling vector, i.e. we cannot use a multi-scale approach in this case.

For a valid h-scaling vector $\underline{h}$ we define the scaled state space $\mathcal{S}^{\underline{h}}$ as

$$\mathcal{S}^{\underline{h}} = \{x = v_R + (k_1 h_1 e_1 + \ldots + k_m h_m e_m)|k_i \in \mathbb{N} \forall i = 1, ..., m\} \cap \mathcal{H}(R(\mathcal{S})).$$

The scaled process $X^{\underline{h}}$ will have initial condition $X^{\underline{h}}(0)$ with components $X_i^{\underline{h}}(0) = (v_R)_i + h_i \lfloor \frac{(x_0)_i - (v_R)_i}{h_i} \rfloor \in \mathcal{S}^{\underline{h}}$.

Finally, we observe that, by definition of valid h-scaling vector, for each transition vector $l \in \mathcal{L}$, the scaling parameters h_i associated with non-null components of l, i.e. components i such that $l_i \neq 0$, are all equal to a certain value h_l. Then, for all $l \in \mathcal{L}$, we define the transitions of $X^{\underline{h}}$ as

$$x \to x + h_l \quad \text{at rate} \quad \frac{1}{h_l}\beta_l(x)\mathbb{I}_{\{x+h_l l \in \mathcal{S}^{\underline{h}}\}}.$$

6   Examples

We propose two examples previously analyzed in [18], representing systems in which the mean-field approximation performs a significant error with respect to the true dynamics of the system. While in [18] the average over a sufficient number of simulations was taken as ground truth, we now compare our results with the solution of the ME (truncated to a sufficient number of states when necessary). We show that while computing the mean from the original ME requires a prohibitive amount of time, this can be reduced significantly applying h-scaling and even more efficiently using DBP and its scaled version. All experiments were performed on a laptop equipped with a 2.8 GHz Intel i7 quad-core processor and 16 GB RAM.

6.1   Egalitarian Processor Sharing

First we look at the model proposed in Example 4 with two classes of customers, where we set the parameters to $N = 5, \lambda_1 = N \cdot 0.5, \lambda_2 = N \cdot 0.4, \mu = N$ and the initial condition to $x_{Q_1}(0) = x_{Q_2}(0) = 0$.

The considered model has an infinite state space. However, choosing a truncation of the type

$$\{(x_{Q_1}, x_{Q_2}) | x_{Q_1} \leq N_1, x_{Q_2} \leq N_2\}$$

for sufficiently large parameters N_1, N_2, it is possible to approximate the mean dynamics with arbitrary accuracy.

As we have done for Example 3, we start by setting $N_1 = 150, N_2 = 120$ and we progressively reduce the state space until the error introduced exceeds the 0.001% of the AQL at steady state. We find that to keep the error below the chosen threshold we need to set $N_1 = 135, N_2 = 108$. This truncation corresponds to an approximated ME with almost 15000 equations, so that the time needed to solve it is considerably high ($\approx$8 h). We can apply static h-scaling to reduce the number of equations, and consequently the computational time, but this comes at the cost of a relative error that can reach 27% of the AQL at steady state (see Table 2 and Fig. 3a).

For this example DBP already provide a significative advantage, by allowing us to achieve an error of less than 1% by choosing $n = (70, 56)$ that corresponds to roughly 4000 equations. This reduces the computational time significantly. Moreover using scaled DBP with the same h used for the static case, shows that a smaller relative error can be achieved with a smaller number of equations, by reducing the computational time to less than 2 min with a relative error of less than 10% (see Table 2 and Fig. 3b).

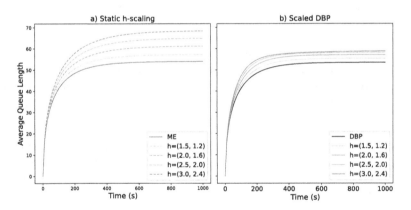

Fig. 3. Application of h-scaling and scaled DBP to Egalitarian Processor Sharing with two classes of customers.

6.2 Malware Propagation

We consider the Malware Propagation Model of Example 5 with $N = 80$ agents and initial conditions $x_D(0) = x_A(0) = 40$.

Table 2. Approximated value of the AQL of the system presented in Example 4 at steady state (t = 1000), with relative error, number of equations, reduction in the number of equations and computational time (in seconds) when h-scaling and scaled DBP are applied. h-scaling with $h = 1$ is considered the ground truth.

		h-scaling				scaled DBP			
h	AQL	err	# eqs.	red.	time	AQL	err	# eqs.	red.	time
(1.0, 1.0)	53.97	-	14824	-	3.09e4	53.46	0.94%	4047	72.70%	4.44e3
(1.5, 1.2)	57.23	6.04%	8281	44.14%	1.67e4	55.36	2.57%	2209	85.01%	1.73e3
(2.0, 1.6)	61.19	13.38%	4624	68.81%	5.43e3	57.12	5.83%	1296	91.26%	5.20e2
(2.5, 2.0)	64.95	20.35%	3025	79.59%	1.40e3	58.26	7.94%	841	94.32%	1.43e2
(3.0, 2.4)	68.39	26.71%	2116	85.72%	8.99e2	58.82	8.97%	576	96.11%	1.18e2

Since this system has a finite state-space given by $\{(x_D, x_A)|x_A + x_D \leq N\}$, its ME can be solved exactly. Taking into account the constraint, the ME yields $\frac{(N+1)\cdot(N+2)}{2}$ equations, instead of $(N + 1)^2$, and the same is true for static h-scaling $(\frac{1}{2}(\lfloor\frac{N}{h}\rfloor + 1)(\lfloor\frac{N}{h}\rfloor + 2)$ equations). Analogously, in DBP and scaled DBP the only states of the truncation with probability greater than 0 will be the states (x_A, x_B) such that $x_A + Y_A^{(n,h)}(t) + x_B + Y_B^{(n,h)}(t) \leq 80$. However since the component $Y^{(n,h)}$ varies in time we write the equations for all $(n+1)^2$ states.

We measure the error as the sum of the relative errors on the three different components S, A, D at steady state. Again, we choose n for DBP by requiring that the error is below 1%. This happens when we set $n = 50$ and reduces significantly the number of equations and the computational time. Moreover, using scaled DBP allows us to outperform static h-scaling in terms of accuracy, and further reduce the computational times while keeping the error below 10% (see Fig. 4 and Table 3, additional graphs are provided in the Appendix).

Table 3. Total relative error in the mean dynamics of Example 5 at steady state (t = 5), with number of equations, reduction in the number of equations and computational time (in seconds) when h-scaling and scaled DBP are applied. h-scaling with $h = 1$ is considered the ground truth

		h-scaling				scaled DBP		
h	err	# eqs.	red.	time	err	# eqs.	red.	time
1.0	-	3321	-	5.19e3	0.87%	2601	21.68%	2.38e3
1.25	2.92%	2145	35.41%	1.73e3	2.22%	1681	49.38%	1.00e3
1.5	5.91%	1485	55.28%	6.61e2	4.28%	1156	65.19%	6.18e2
1.75	7.68%	1081	67.45%	3.20e2	6.23%	841	74.68%	1.63e2
2.0	12.1%	861	74.07%	2.48e2	8.06%	676	79.64%	1.00e2

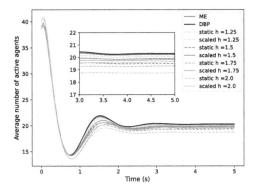

Fig. 4. Average number active agents in the Malware Propagation model computed using h-scaling and scaled DBP.

7   Conclusion and Related Works

In this paper we have proposed a method that lessens the curse of dimensionality in Markov population processes by studying the dynamics of a rescaled process. Seeing the state space of the original process as a multi-dimensional grid of size $h = 1$, our method covers it with a coarser grid with length $h > 1$, rescaling transition rates by a factor $\frac{1}{h}$. The resulting process has an exponentially smaller state space with respect to the number of dimensions.

A similar idea to the one proposed in this paper was proposed in [7] in the context of biochemical networks modeling using the process algebra Bio-PEPA. While the re-scaling of jump magnitudes and rate function is the same, our work substantially differs from the one in [7] for several reasons. First, we use the proposed scaling for the approximation of a given system, while in [7] the different discretizations are treated as systems on their own. Second, our scaling can be applied to a generic Markov Population Processes, abstracting away from a process algebraic description. Third, we propose the introduction of different re-scaling factors for different classes of agents, stating formally under which conditions this is possible.

Theoretically, in [7] it is proved that as the scaling parameter h tends to 0 the associated family of systems satisfies the density-dependent assumption and so the dynamics of the associated process tends to the mean-field limit. Here we study the case $h > 1$ and its convergence properties to a mean field or to an LNA, proving that these limits are preserved in the re-scaled sequence.

A natural question is whether it is possible to establish an error bound depending on h, or at least some convergence rate toward such limits. While this is not currently clear we leave these questions open for future works.

A Appendix

A.1   Derivation of Scaled DBP

Having defined the truncations for $\mathcal{S}^h$ as in Sect. 3.2 we proceed as in the derivation for DBP.

The border sets for the scaled truncations are defined as:

$$\partial T_l^h(n,y) = \left\{ x \in T^h(n,y) : x + hl \notin T^h(n,y) \right\}, \text{ for } l \in \mathcal{L},$$
$$\partial T^h(n,y) = \bigcup_{l \in \mathcal{L}} T_l^h(n,y) = \left\{ x \in T^h(n,y) : \exists l \in \mathcal{L} \text{ s.t. } x + hl \notin T^h(n,y) \right\}$$

We can then define the boundary projection of X^h on $T^h(n,y)$, in which every jump from $x \in \partial T_l(n,y)$ to x' is redirected with same rate to x^* defined as:

$$x_i^* = \begin{cases} \min(y_i + h\lfloor \frac{n_i}{h} \rfloor, x_i') & \text{if } x_i' > x_i \\ \max(y_i, x_i') & \text{if } x_i' < x_i \\ x_i & \text{if } x_i' = x_i. \end{cases}$$

After performing the augmentation we get the jump vectors $l^{(n,h)}(x)$ defined exactly as before. Then, letting $X_y^{(n,h)}$ be the boundary projection of X^h on $T^h(n,y)$, its transition matrix $Q^{(n,h)}(y)$ can be written for $x, x' \in T^h(n,0)$ as:

$$[Q^{(n,h)}(y)]_{x,x'} = \begin{cases} \sum_{l \in \mathcal{L}} \mathbb{I}_{\{x' + l^{(n,h)}(x') = x\}} \frac{1}{h} \beta_l(x' + y) & \text{if } x \neq x' \\ -\sum_{l \in \mathcal{L}} \mathbb{I}_{\{l^{(n,h)}(x) \neq 0\}} \frac{1}{h} \beta_l(x + y) & \text{if } x = x'. \end{cases}$$

So the ME for $X_y^{(n,h)}$ as:

$$\frac{dP_y^{(n,h)}}{dt} = Q^{(n,h)}(y) P_y^{(n,h)}(\,\cdot\,; t)$$

where $P_y^{(n,h)}(\,\cdot\,; t)$ is an $\mathcal{N}^h(n)$-dimensional vector.

Again, to pass to DBP, we need to define the functions:

$$\Pi_i^{(n,h)}(x,y) = \begin{cases} x_i & x_i < y_i \\ y_i + h\lceil x_i - (y_i + \lfloor \frac{n_i}{h} \rfloor) \rceil & x_i > y_i + n_i \qquad \forall\, x, y \in \mathcal{S}^h \\ y_i & y_i \le x \le y_i + n_i. \end{cases}$$

$$\mathcal{Y}_l^h(n,x) = \Pi^{(n,h)}(x + l, 0) \quad \forall\, l \in \mathcal{L}, \forall\, x \in \partial T_l^h(n,0).$$

Observe that the second case in the definition of $\Pi^{(n,h)}(x,y)$ is motivated by the fact that x may not be in the form $y + h(k_1 e_1 + \ldots + k_m e_m)$, and, to mirror what happens in classic DBP, we want the function to return the closes y' in this form so that $T^h(n,y')$ contains x.

Then the equations for scaled DBP with parameter n are given by:

$$\frac{dY^{(n,h)}}{dt} = \sum_{l \in \mathcal{L}} \sum_{x \in \partial T_l^h(n,0)} \mathcal{Y}_l^h(n,x) \frac{1}{h} \beta_l(x + Y^{(n,h)}(t)) P^{(n,h)}(x;t)$$

(19)

$$\frac{dP^{(n,h)}}{dt} = Q^{(n,h)}(Y^{(n,h)}(t)) P^{(n,h)}(\cdot,t).$$

Again, supposing $X(0) = x_0$ with probability 1, to define the initial condition we set:

$$\left[Y^{(n)}(0)\right]_i = \max\left(0, x_{0,i} - h\left\lfloor \frac{n_i}{h} \right\rfloor\right)$$

$$x_0^* = h\left\lfloor \frac{x_0 - Y^{(n)}(0)}{h} \right\rfloor$$

$$P^{(n)}(x;0) = \begin{cases} 1 & \text{if } x = x_0^*, \\ 0 & \text{else.} \end{cases}$$

A.2   Equations for Example 3

Equations for scaled DBP read:

$$\frac{dY^{(n,h)}}{dt} = -\frac{\mu}{h}\min\left(Y^{(n,h)}(t),k\right) P^{(n,h)}(0;t) + \frac{\lambda}{h} P^{(n,h)}\left(h\left(\left\lfloor \frac{N}{h} \right\rfloor - 1\right);t\right)$$

$$\frac{dP^{(n,h)}(x)}{dt} = \begin{cases} -\frac{\lambda}{h} P^{(n,h)}(0;t) + \frac{\mu}{h}\min\left(h + Y^{(n,h)}(t),k\right) P^{(n,h)}(h;t) & x = 0 \\ -\left(\frac{\lambda}{h} + \frac{\mu}{h}\min\left(x + Y^{(n,h)}(t),k\right)\right) P^{(n,h)}(x;t) & \\ \quad +\frac{\lambda}{h} P^{(n,h)}(x - h;t) & \\ \quad +\frac{\mu}{h}\min\left(x + h + Y^{(n,h)}(t),k\right) P^{(n,h)}(x + h;t) & x \neq 0, h\left\lfloor \frac{N}{h} \right\rfloor \\ -\frac{\mu}{h}\min\left(h\left\lfloor \frac{N}{h} \right\rfloor + Y^{(n,h)}(t),k\right) P^{(n,h)}\left(h\left\lfloor \frac{N}{h} \right\rfloor;t\right) & \\ \quad +\frac{\lambda}{h} P^{(n,h)}\left(h\left(\left\lfloor \frac{N}{h} \right\rfloor - 1\right);t\right) & x = h\left\lfloor \frac{N}{h} \right\rfloor \end{cases}$$

A.3   Proof of Theorem 4

Theorem 5. *Suppose that for the sequence* $\left(X^N\right)_{N \geq N_0}$ *the hypotheses of Theorem 2 are verified and, in addition:*

- *equation (3) admits a globally asymptotically stable equilibrium* x^*;
- *for each* N $X^N(0) = N\hat{x}_0$;
- *for each* N $\gamma_N = N$.

Then, letting $\mu(t)$ *and* $\Sigma(t)$ *denote the mean and the covariance matrix of limiting Gaussian process for the original sequence, we have that the sequence of approximating processes* $\left(X^{N,h}\right)_{N \geq N_0}$ *admits a Gaussian limiting process with mean* $\mu^h(t)$ *and covariance matrix* $\Sigma^h(t)$ *such that:*

$$\lim_{t \to \infty} \mu^h(t) = \lim_{t \to \infty} \mu(t) = 0 \text{ and } \Sigma^h(t) = \Sigma(t) \,\forall t \geq 0.$$

Proof. Theorem 2 guarantees that under the hypothesis $\mu(t)$ and $\Sigma(t)$ exist.

The rest of the proof is obtained by following the same derivation used in [19] with the ansatz:

$$\hat{X}^{N,h}(t) = \hat{x}(t) + \sqrt{\frac{h}{N}}\xi^h(t). \tag{20}$$

and verifying that $\xi^h(t)$ is a Gaussian Process whose mean $\mu^h(t)$ and covariance $\Sigma^h(t)$ satisfy exactly the same ODEs as $\mu(t)$ and $\Sigma(t)$, i.e. (5) and (6).

Furthermore, in the sequence of the approximated process $\left(X^{N,h}\right)_{N\geq0}$ we have redefined the initial conditions as $X^{N,h}(0) = h\lfloor\frac{N\hat{x}_0}{h}\rfloor$, while the initial condition for the deterministic process remains unchanged. Therefore, when setting the initial condition for $\mu^h(y)$ we need to take into account that for the ansatz to be valid at time $t = 0$ the Gaussian Limit Process possibly has non-zero mean, namely $\mu^h(0) = \sqrt{\frac{h}{N}}\left(\lfloor\frac{N\hat{x}_0}{h}\rfloor - N\hat{x}_0\right)$.

So, in general, $\xi^h(t)$, describing the fluctuations of $X^{N,h}$, is different from $\xi(t)$, describing the fluctuations of X^N, since $\mu^h(0) \neq \mu(0) = 0$ (observe that instead the covariance matrix is still the same, i.e. $\Sigma^h(t) = \Sigma(t)\,\forall t \geq 0$).

However, Eq. (5) is exactly the variational equation associated with the ODEs defining the deterministic limit (3), so, regardless of its initial condition, its solution must tend to 0 as $\hat{x}(t)$ tends to the equilibrium x^*. This implies $\lim_{t\to\infty}\mu^h(t) = \lim_{t\to\infty}\mu(t) = 0$.

Observe that all the introduced hypotheses are needed for the correct application of the ansatz: the differentiability of the drifts is needed to apply the Taylor expansion as in [19], while the presence of a globally asymptotically stable equilibrium ensures that the ansatz remains valid for $t \in [0, +\infty)$.

A.4   Additional Data on Malware Propagation Model

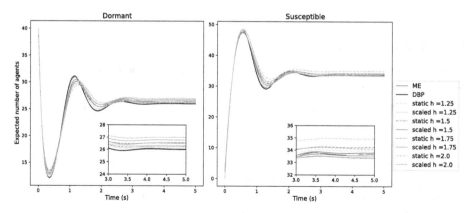

Fig. 5. Average number of dormant and susceptible agents in the Malware Propagation model computed using h-scaling and scaled DBP.

References

1. Anselmi, J., Verloop, I.M.: Energy-aware capacity scaling in virtualized environments with performance guarantees. Perform. Eval. **68**(11), 1207–1221 (2011)
2. Baskett, F., Chandy, K.M., Muntz, R.R., Palacios, F.G.: Open, closed, and mixed networks of queues with different classes of customers. J. ACM **22**(2), 248–260 (1975)
3. Benaim, M., Le Boudec, J.Y.: A class of mean field interaction models for computer and communication systems. Perform. Eeval. **65**(11–12), 823–838 (2008)
4. Bortolussi, L., Hillston, J., Latella, D., Massink, M.: Continuous approximation of collective system behaviour: a tutorial. Perf. Eeval. **70**(5), 317–349 (2013)
5. Buchholz, P.: Exact and ordinary lumpability in finite Markova chains. J. Appl. Probab. **31**(1), 59–75 (1994)
6. Cao, Y., Li, H., Petzold, L.: Efficient formulation of the stochastic simulation algorithm for chemically reacting systems. J. Chem. Phy. **121**(9), 4059–4067 (2004)
7. Ciocchetta, F., Degasperi, A., Hillston, J., Calder, M.: Some investigations concerning the CTMC and the ode model derived from bio-PEPA. Electr. Notes Theoret. Comput. Sci. **229**(1), 145–163 (2009)
8. Darling, R.: Fluid limits of pure jump Markov processes: a practical guide. arXiv preprint math/0210109 (2002)
9. Darling, R., Norris, J.R.: Differential equation approximations for markov chains. Probability surveys **5**, 37–79 (2008)
10. Gast, N., Bortolussi, L., Tribastone, M.: Size expansions of mean field approximation: transient and steady-state analysis. Perform. Eval. **129**, 60–80 (2019)
11. Gast, N., Van Houdt, B.: A refined mean field approximation. In: Proceedings of the ACM on Measurement and Analysis of Computing Systems, vol. 1, pp. 1–28 (2017)
12. Kurtz, T.G.: Solutions of ordinary differential equations as limits of pure jump Markov processes. J. Appl. Prob. **7**(1), 49–58 (1970)
13. Liu, Y., Li, W., Masuyama, H.: Error bounds for augmented truncation approximations of continuous-time Markova chains. Oper. Res. Lett. **46**(4), 409–413 (2018)
14. Minnebo, W., Van Houdt, B.: A fair comparison of pull and push strategies in large distributed networks. IEEE/ACM Trans. Netw. **22**(3), 996–1006 (2013)
15. Munsky, B., Khammash, M.: The finite state projection algorithm for the solution of the chemical master equation. J. Chem. Phy. **124**(4) (2006)
16. Parekh, A.K., Gallager, R.G.: A generalized processor sharing approach to flow control in integrated services networks: the single-node case. IEEE/ACM Trans. Netw. **3**, 344–357 (1993)
17. Parekh, A.K., Gallager, R.G.: A generalized processor sharing approach to flow control in integrated services networks: the multiple node case. IEEE/ACM Trans. Netw. **2**(2), 137–150 (1994)
18. Randone, F., Bortolussi, L., Tribastone, M.: Refining mean-field approximations by dynamic state truncation. Proc. ACM Measur. Anal. Comput. Syst. **5**(2), 1–30 (2021)
19. Van Kampen, N.G.: Stochastic Processes in Physics and Chemistry, vol. 1. Elsevier, New York (1992)

20. Xie, Q., Dong, X., Lu, Y., Srikant, R.: Power of d choices for large-scale bin packing: A loss model. ACM SIGMETRICS Perform. Eval. Rev. **43**(1), 321–334 (2015)
21. Yang, X., De Veciana, G.: Service capacity of peer to peer networks. In: IEEE INFOCOM 2004, vol. 4, pp. 2242–2252. IEEE (2004)
22. Zhu, L., Casale, G., Perez, I.: Fluid approximation of closed queueing networks with discriminatory processor sharing. Perform. Eval. **139** (2020)

Dynamical Systems

An Algorithm for the Formal Reduction of Differential Equations as Over-Approximations

Giuseppe Squillace[1]([✉]), Mirco Tribastone[1], Max Tschaikowski[2], and Andrea Vandin[3,4]

[1] IMT School for Advanced Studies Lucca, Lucca, Italy
`giuseppe.squillace@imtlucca.it`
[2] Aalborg University, Aalborg, Denmark
[3] Sant'Anna School of Advanced Studies, Pisa, Italy
[4] DTU Technical University of Denmark, Lyngby, Denmark

Abstract. Models of complex systems often consist of state variables with structurally similar dynamics that differ in the specific values of some parameters. Examples are multi-class epidemiological models, chemical reaction networks describing multiple components (e.g., binding sites) with equivalent functional behavior, and models of electric circuits with replicated designs. In these cases, the analysis may be expensive due to the model size. Here we consider models defined as systems of polynomial ordinary differential equations (ODEs) with positive solutions. We present an algorithm to reduce the computational cost by transforming the original ODE model into one for which we can compute an appropriate over-approximation on a smaller set of state variables. The algorithm is based on the theory of differential inequalities and consists of two steps. The first step computes a differential hull, which is an ODE system providing lower and upper bounds for each state variable. The hull is constructed such that variables with structurally similar dynamics but originally different parameters may now be represented by the same lower and upper bounds. Based on this, the second step exploits already developed notions of exact model reduction for ODEs to lump such variables. The algorithm is showcased on several case studies and its results are favourably compared against CORA, a well-known tool for reachability analysis of dynamical systems.

Keywords: Ordinary differential equations · Model reduction · Reachability analysis

1 Introduction

Ordinary differential equations (ODEs) are a fundamental model to describe the behavior of dynamical systems. In many cases, they represent classes of entities governed by structurally similar laws governed by different parameters. Such heterogeneous parameters may encode different dynamical behavior of the same

© Springer Nature Switzerland AG 2022
E. Ábrahám and M. Paolieri (Eds.): QEST 2022, LNCS 13479, pp. 173–191, 2022.
https://doi.org/10.1007/978-3-031-16336-4_9

underlying phenomenon, such as age- or location-dependent rates for the contagion of a disease [4], class-dependent service demands in a queuing system [3], and conformation-dependent binding affinities in protein interaction networks [13].

When there is a large degree of heterogeneity, intended as the number of classes used in the model, the analysis becomes increasingly more complex. This problem justifies the quest for reduction methods that simplify the description whilst retaining some formal relationship with the original models. Here we tackle this issue by designing an algorithm that aims to homogenize class-dependent behavior into representative equations that suitably summarize the difference in parameters. The idea builds on an earlier approach that yields so-called *differential hulls* for heterogeneous ODE systems [27]. Differential hulls provide lower- and upper-bounds on the original equations by relying on the theory of differential inequalities [23,24,26] which can be traced back to the seminal work of Müller [21]. Here, equations with different parameters are lower- and upper-bounded with the same differential inequality by taking appropriate minimum and maximum values across the parameters. Thus, the resulting system of differential inequalities is independent from the number of aforementioned classes and its solution give an envelope within which the original trajectories live, effectively constituting a formal over-approximation of the original model.

The main limitation of [27] is that no algorithm is given to build differential hulls; that is, the method requires a priori knowledge of the existence of "equivalent" dynamical equations up to the choice of the parameters. The main contribution of this paper is an algorithm that builds differential hulls for ODEs.

We focus on polynomial ODEs with positive solutions. This is already a general class to which other forms of nonlinearity can be reduced [19], and it essentially covers many dynamical models of systems where the state variables are physical quantities such as concentration of molecules and populations of agents. Our algorithm takes as input a tolerance ε that, informally, defines the amount of heterogeneity allowed in the model parameters. The procedure consists of two steps. First, the algorithm splits the set of parameters into blocks so long as the difference between the values of the minimum and the maximum elements in each block is less than ε. The differential hull is then built by doubling the number of variables in the system (e.g., from the original n to $2n$), replacing each variable with a pair of new ones representing its upper and lower bounds. This is obtained by appropriately substituting in the equations of each pair of new variables the original parameters with either of the extremal elements in the block it belongs to. If the original system consisted of structurally similar equations with different parameters (and these parameters are at most ε away), the intended output of this first step is to have replicated equations that have the same dynamical behavior due to the consistent choice of the parameter bounds. Since the algorithm is agnostic to the form of the original model, the second step performs an automated discovery of the replicated behavior. This is done with backward differential equivalence (BDE) [5,6,8], a reduction algorithm that exactly lumps ODE variables that have the same solution when starting from the same initial condition. Overall, the procedure returns a reduced differential hull that still bounds the original dynamics while using fewer than $2n$ variables.

We use case studies from engineering, biology, biochemistry, organic chemistry, and epidemiology to compare against our method against CORA [1], a state-of-the-art tool for over-approximation/reachability analysis. The comparison is justified by the fact that the proposed approach can be tied to over-approximation. Indeed, the first step of our algorithm splits the parameters into blocks where the difference between the maximum and the minimum is less than ε. Afterwards, the algorithm substitutes each parameter with the maximum or the minimum of the block it belongs to. These extremal elements of each block define the admissible parameter values that over-approximation techniques such as CORA take into account to compute the reachable sets. The investigation reveals that CORA computes in general tighter bounds than the proposed approach, but at higher time and space requirements.

Further Related Work. Many common over-approximation techniques rely on Lyapunov-like functions [12,20] known from stability theory of ODEs. However, the automatic computation of Lyapunov-like functions remains a challenging task in case of nonlinearity [14]. Instead, approaches such as CORA or Flow* approximate the nonlinear model by a multivariate polynomial or an affine system, see [2,10] and references therein. The research on approximate quotients of ODE systems, instead, can be traced to the 1960s [17] where the authors over-approximated the dynamics of mono-molecular reaction networks. Li and Rabitz extend this approximate lumping to general CRNs [18], but an explicit error bound was not given. In a similar vein, approximate quotients in ecology have been studied from the point of view of finding a reduced ODE system whose derivatives are as close as possible (in norm) to the derivatives of the original ODE system [15]. This is also exploited in ε-BDE [9], a reduction technique that is based on a partition-refinement algorithm of BDE [6] and aims to lump ODE variables with nearby trajectories, essentially by relaxing the requirement of exact symmetry imposed by the BDE approach used in this paper. Using a case study from [9], we show that our method can provide bounds for larger differences in the model parameters than ε-BDE.

2   Background

Backward Differential Equivalence. Let us consider a polynomial ODE system composed of a set of variables $\mathcal{V} = \{x_1, ..., x_n\}$. The dynamics of variable x_i is in the form $\dot{x}_i = q_i, 1 \leq x_i \leq n$, where q_i is a multivariate polynomial over $\mathcal{V}$. We say that q_i is in normal form when each monomial $x^\alpha \equiv \prod_{x_i \in \mathcal{V}} x_i^{\alpha_{x_i}}$, where $\alpha \in \mathbb{N}_0^{\mathcal{V}}$ is a multi-index, appears in q_i at most once. In this way, we can define $c(q_i, x^\alpha)$ as the coefficient of the monomial x^α in a normal form polynomial q_i.

The notion of BDE [6,9] relates variables that have the same solutions at all time points if they start from the same initial conditions. In the polynomial ODE systems that we consider, this technique makes pairwise comparisons between the coefficients of any two variables in the same equivalence class.

Definition 1 (Backward differential equivalence (BDE)). *Fix a polynomial ODE, a partition $\mathcal{H}$ of $\mathcal{V}$ and write $x_i \sim_{\mathcal{H}}^{B} x_j$ if all coefficients of the following polynomial are zero,*

$$q_{i,j}^{H} := (q_i - q_j)\left[x_{H',1}/x_{H'}, \ldots, x_{H',|H'|}/x_{H'} : H' \in \mathcal{H}\right]$$

i.e., when

$$\sum_{\alpha \in \mathbb{N}_0^V} |c(q_{i,j}^{H}, x^\alpha)| = 0. \tag{1}$$

A partition $\mathcal{H}$ is a BDE if $\mathcal{H} = \mathcal{V}/(\sim_{\mathcal{H}}^{B^*} \cap \sim_{\mathcal{H}})$.

Following the definition, a partition is a BDE partition if the differences between the coefficients on the same monomials are zero for any two variables in the same block.

Differential Hulls. We use the notation $x \leq y$ for the vectors $x = (x_1, \ldots, x_n)$ and $y = (y_1, \ldots, y_n)$ in $\mathbb{R}$ if and only if $x_i \leq y_i$ for all $1 \leq i \leq n$. The strict inequality, $x < y$, is defined similarly. The differential hull is a vector field with $2n$ variables that provides upper and lower bounds for the dynamics of the original ODE system defined on the set of variables $\mathcal{V} = \{x_1, \ldots, x_n\}$.

Definition 2 (Differential Hull [27]). *We call $(g_{\underline{1}}, \ldots, g_{\underline{n}}, g_{\overline{1}}, \ldots, g_{\overline{n}}) : \mathbb{R}_{>0}^{2n} \rightarrow \mathbb{R}^{2n}$ a differential hull of the polynomial ODE system $(q_1, \ldots, q_n) : \mathbb{R}_{>0}^{n} \rightarrow \mathbb{R}^n$ when, for all $1 \leq i \leq n$ $g_{\underline{i}}, g_{\overline{i}}$ are polynomials and for any $\underline{x} \leq x \leq \overline{x}$,*

$$\underline{x}_i = x_i \implies g_{\underline{i}}(\underline{x}, \overline{x}) \leq q_i(x) \qquad and \qquad x_i = \overline{x}_i \implies q_i(x) \leq g_{\overline{i}}(\underline{x}, \overline{x})$$

The previous definition is very general because the only condition a differential hull should satisfy is that it should over-approximate the dynamics of a polynomial vector field q.

Theorem 1. *Let g be a differential hull of q. Then, if the solution of the polynomial ODE system $(\underline{\dot{x}}, \overline{\dot{x}}) = g(\underline{x}, \overline{x})$ subject to $0 < \underline{x}(0) \leq x(0) \leq \overline{x}(0)$ exists and is positive on $[0; T]$, where $T > 0$, then the solution of $\dot{x} = q(x)$ exists on $[0; T]$ as well and satisfies $\underline{x}(t) \leq x(t) \leq \overline{x}(t)$ for all $0 \leq t \leq T$.*

3   Computing Differential Hulls

Algorithm 1 takes as input a tolerance value $\varepsilon > 0$ and a polynomial ODE system O, given by $\dot{x}_i = q_i(x)$ with $1 \leq i \leq n$. Line 2 sorts all coefficients $\{(i, \alpha, c(q_i, x^\alpha)) \in O \mid 1 \leq i \leq n, \alpha \in \mathbb{N}_0^n\}$ of O in increasing order and splits them into blocks whose members are within distance ε. More in detail, we start from the minimum parameter and add the next one in the same block until the difference between the first and last inserted is not greater than ε. Blocks are collected in the resulting partition, P. Lines 4–5 define two new equations $\overline{x}_i$ and $\underline{x}_i$, respectively the lower and upper bound of x_i. In lines 6–11, the algorithm

---

Algorithm 1. computeDifferentialHull

Require: An ODE system O, a tolerance ε .
1: $DHull = \{\}$
2: $P = \text{groupParameters}(O,\varepsilon)$
3: **for** each x_i in O **do**
4:      $\dot{\overline{x}}_i = []$
5:      $\dot{\underline{x}}_i = []$
6:      **for** each monomial M in O **do**
7:          $\overline{M} = \text{upperBound}(M,P,x_i)$
8:          $\underline{M} = \text{lowerBound}(M,P,x_i)$
9:          $\text{append}(\dot{\overline{x}}_i,\overline{M})$
10:         $\text{append}(\dot{\underline{x}}_i,\underline{M})$
11:     **end for**
12:     $\text{add}(DHull,\dot{\overline{x}}_i)$
13:     $\text{add}(DHull,\dot{\underline{x}}_i)$
14: **end for**
15: **return** $DHull$

---

considers the monomials M in equation x_i. It computes the lower and upper bound for each of them and appends these results to $\dot{\overline{x}}$ and $\dot{\underline{x}}$, respectively.

The procedure to compute the upper bound is shown in Algorithm 2. It requires a monomial M, the coefficients partition P already calculated by Algorithm 1, and variable x_i. In lines 2–3, the procedure retrieves the coefficient and the variables associated with the monomial M. In lines 4–8, the algorithm substitutes the original parameter of the monomial. If the coefficient of the monomial p is positive, the computation picks the maximum parameter in the block p belongs to (line 5), otherwise the minimum (line 7). In lines 9–15, the method takes care of the variables x_j. The idea is similar. The method picks the upper or lower bound of x_j depending on the value of p. The first condition in line 10 represents the case where the variable x_j is the same variable as $\dot{x}_i$. We are computing $\dot{\overline{x}}_i$ and we find x_j equals to x_i in q_i, in this case, since the variable defines itself, the algorithm will pick $\overline{x}_j$ no matter what is the value of p.

We omit the algorithm for the lower bound, called in line 9 of Algorithm 1, because it is similar to Algorithm 2. In lines 12–13 Algorithm 1 composes the new equations to the differential hull and returns it.

Theorem 2. *The time and space complexity of Algorithm 1 and Algorithm 2 is polynomial in the size of the ODE model.*

Running Example. Let us take the simple polynomial ODE system:

$$\dot{x}_1 = -k_2 x_1, \qquad \dot{x}_2 = k_1 x_1 - k_3 x_2, \qquad \dot{x}_3 = k_2 x_1 - k_3 x_3$$

with $k_1 = 1.0$, $k_2 = 1.1$, and $k_3 = 1.2$ and initial conditions all equal to 1.

We now consider the application of the Algorithm 1 with a tolerance parameter $\varepsilon = 0.2$. In the first step, the procedure splits the parameters in a single block

---

Algorithm 2. upperBound

Require: A monomial M, the parameters partition P, variable $\dot{x}_i$.
1: $\overline{M} = \{\}$
2: $(\cdot, \cdot, p) = \text{getParameter}(M)$
3: $V = \text{getVariables}(M)$
4: **if** $p > 0$ **then**
5:     add($\overline{M}$,getMax(p,P))
6: **else**
7:     add($\overline{M}$,getMin(p,P))
8: **end if**
9: **for** each x_j in V **do**
10:    **if** $x_j == x_i$ or $p > 0$ **then**
11:       add($\overline{M}$,$\overline{x}_j$)
12:    **else**
13:       add($\overline{M}$,$\underline{x}_j$)
14:    **end if**
15: **end for**
16: **return** $\overline{M}$

B_1 where the tolerance is exactly 0.2, corresponding to the difference $k_3 - k_1$. We now discuss the detailed process to compute the upper bound $\dot{\overline{x}}_2$. In line 6, Algorithm 1 considers every monomial in the dynamics $\dot{x}_2$ of ODE system O. For the first term $k_1 x_1$, since k_1 is positive, line 5 of Algorithm 2 picks k_3, the maximum parameter for this block. Similarly, in line 11, the maximum value that x_1 could assume is $\overline{x}_1$, which is the upper bound of x_1. In this way, the algorithm provides the first term $k_3 \overline{x}_1$ of $\dot{\overline{x}}_2$. The computation proceeds with the maximization of the second terms $-k_3 x_2$. Since $-k_3$ is negative, the algorithm takes the parameter k_1. Moreover, we fall in the case where the condition $x_j == x_i$ in line 10 is true; for this reason Algorithm 2 replaces x_2 with $\overline{x}_2$ rather than $\underline{x}_2$. Summing up all the steps, the algorithm computes the upper bound of $\dot{x}_2$ with the equation $\dot{\overline{x}}_2 = k_3 \overline{x}_1 - k_1 \overline{x}_2$. The lower bound is computed similarly and, for this reason, is omitted.

Overall, the differential hull for the system reads:

$$\dot{\underline{x}}_1 = -k_3 \underline{x}_1 \qquad \dot{\underline{x}}_2 = k_1 \underline{x}_1 - k_3 \underline{x}_2 \qquad \dot{\underline{x}}_3 = k_1 \underline{x}_1 - k_3 \underline{x}_3$$
$$\dot{\overline{x}}_1 = -k_1 \overline{x}_1 \qquad \dot{\overline{x}}_2 = k_3 \overline{x}_1 - k_1 \overline{x}_2 \qquad \dot{\overline{x}}_3 = k_3 \overline{x}_1 - k_1 \overline{x}_3$$

In Fig. 1 (left), we plot both the dynamics of the differential hull and the original system when all initial conditions are equal to 1. Every trajectory x_i falls in a band bounded by the two equations $\underline{x}_i$ and $\overline{x}_i$. Importantly, we notice that, due to the choice of initial conditions, the solutions for $\underline{x}_2$ and $\underline{x}_3$ coincide, and so do the solutions for $\overline{x}_2$ and $\overline{x}_3$. This is due to the fact that the partition of variables $\{\{\underline{x}_1\}, \{\overline{x}_1\}, \{\underline{x}_2, \underline{x}_3\}, \{\overline{x}_2, \overline{x}_3\}\}$ satisfies the BDE criterion in Eq. 1. This gives the following BDE-reduced differential hull where variables $\underline{x}_2$ and $\overline{x}_2$ are taken as the representatives of their respective blocks.

$$\dot{\underline{x}}_1 = -k_3 \underline{x}_1, \quad \dot{\overline{x}}_1 = -k_1 \overline{x}_1, \quad \dot{\underline{x}}_2 = k_1 \underline{x}_1 - k_3 \underline{x}_2, \quad \dot{\overline{x}}_2 = k_3 \overline{x}_1 - k_1 \overline{x}_2.$$

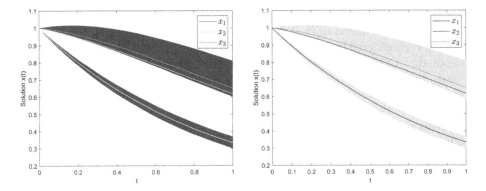

Fig. 1. (left) Over-approximation by means of differential hulls for the running example. (right) CORA over-approximation of the running example.

It is important to notice that the bounds computed over-approximate not only the dynamics for the parameters under study. Indeed, any set of parameters giving rise to the same differential hull will be over-approximated by the hull. Specifically, the following can be shown.

Theorem 3. *Let O be an ODE system over variables $x_1, \ldots, x_n$ and let P be the partition as computed by Algorithm 1 and Algorithm 2. Assume that all blocks of P have common signs (i.e., for any $B \in P$ and $(\cdot, \cdot, p_1), (\cdot, \cdot, p_2) \in B$, it holds that $p_1 \cdot p_2 \geq 0$). Then, an ODE system O' over $x_1, \ldots, x_n$ gives rise to the same differential hull as O when*

– *O' has no more monomials than O, that is, if $(j, \beta, \cdot) \notin B$ for each $B \in P$, then $c(q'_j, x^\beta) = 0$ and;*
– *the parameters of O' yield the same minima and maxima over partition P, i.e., for all $(j, \beta, \cdot) \in B$ and all $B \in P$ we have that*

$$\min\{c(q_i, x^\alpha) \mid (i, \alpha, \cdot) \in B\} \leq c(q'_j, x^\beta) \leq \max\{c(q_i, x^\alpha) \mid (i, \alpha, \cdot) \in B\},$$

where $c(q'_j, x^\beta)$ denotes the coefficient of monomial x^β in q'_j of O'.

Remark 1. The assumption on P having blocks with common signs can be always enforced by means of a prepartitioning. This being said, we wish to point out that all models considered in the evaluation section did not require a prepartitioning, i.e., Theorem 3 could be applied directly.

The foregoing result ties differential hulls to reachability analysis, where an amount of perturbation is considered among the grouped parameter. This justifies the comparison against CORA in the next section. For completeness, we next show the application of CORA to our running example.

CORA requires choosing how to represent the reachability set and the amount of perturbation in the parameters. In this case, we decided to represent the sets with the zonotopes. We set up the parameters to their average values, that is 1.1, allowing an amount of perturbation equal to 0.1. In this way, we consider the following range of uncertainty [1.0; 1.2], that represent the set of all the possible parameters considered by the differential hull. In Fig. 1 (right) we show the bounds computed by CORA. In this example, the two techniques provide almost the same bounds. In the next section, we will present several models from different fields to compare the bounds provided by our approach and the ones by CORA. It can be noted that the two techniques provide almost identical bounds. We will see in the next section that CORA tends to give better bounds compared to our approach, while requiring significantly more time and space.

4   Case Studies

In this section, we consider a number of case studies. The CORA implementation was carried out in Matlab, while the BDE reductions of Algorithm 1 were performed by invoking ERODE [7].

4.1   SIR Model

The SIR model describes the spread of an infection in a population composed of three main actors: infected (I), susceptible (S), and the recovered individuals (R) [16]. The infected individuals are the ones that could infect the susceptibles; the recovered obtained a permanent immunization from infection because they already got the disease. The model has two types of parameters: β, the infection rate, and γ, the recovery rate. In this context, we consider the following multiclass SIR model of individuals with class-specific infection and recovery rates:

$$\dot{S}_i = \sum_{j=1}^{N} -S_i\beta_{i,j}I_j, \qquad \dot{I}_i = -\gamma_i I_i + \sum_{j=1}^{N} S_i\beta_{i,j}I_j, \qquad \dot{R}_i = \gamma_i I_i,$$

where the parameters $\beta_{i,j}$ represent cross-class infection rates. For consistency across all number of classes, the parameters were chosen using the same level of heterogeneity, as follows:

$$\theta_{\text{SIR}} = |\max_{i,j} \beta_{i,j} - \min_{i,j} \beta_{i,j}| + |\max_i \gamma_i - \min_i \gamma_i| = 0.2$$

All parameter values and the initial conditions are provided in the Appendix.

We computed the differential hull running our algorithm with the tolerance ε equal to θ_{SIR}, then we reduced it with BDE. The reduced differential hull is an SIR model where all the lower and the upper bounds for each class collapse into one, so that the reduction achieved by BDE is: $\{\{\underline{S}_1, ..., \underline{S}_N\}, \{\overline{S}_1, ..., \overline{S}_N\},$ $\{\underline{I}_1, ..., \underline{I}_N\}, \{\overline{I}_1, ..., \overline{I}_N\}, \{\underline{R}_1, ..., \underline{R}_N\}, \{\overline{R}_1, ..., \overline{R}_N\}\}$.

In Fig. 2, we show the comparison between CORA and differential hulls for the SIR model with two different classes; the bounds computed considering an higher number of classes are similar. CORA has tighter bounds, but it is more time consuming. Indeed, Table 1, which lists the CORA runtimes, shows a fast increase with respect to the number of classes, issuing out of memory errors for 8. Our algorithm instead required less than 1 s in all cases. This is an expected result because, as stated in Theorem 2, the cost of the algorithm is polynomial and is based on the substitution of parameters and variables.

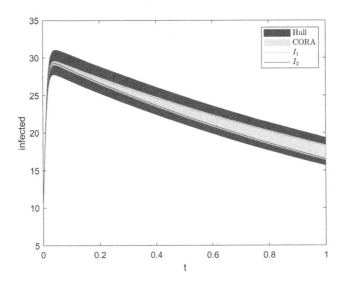

Fig. 2. Bounds of the infected individuals computed by our algorithm against CORA.

Table 1. CORA running times for the SIR model.

Number of classes	2	4	6	8
CORA runtime	12.98 s	43.43 s	162.96 s	Out of memory

4.2 Polymerization

In chemistry, polymerization is the process by which monomers react to form longer chains. We consider next the polymerization model presented in [28] which describes the formation of polycyclic aromatic hydrocarbons in flame combustion. The underlying system of polynomial ODEs is induced by the law of mass action [29]. Let us consider, for instance, the reaction $A_i + H \xrightarrow{\alpha_i} A_{\tilde{i}} + H_2$. The terms on the left side are called reagents, while those on the right are called products. An instance of each reagent is *consumed* when the reaction occurs, and one of each product is *produced*. The kinetic reaction rate is α_i, instead. The reaction occurs at speed $\alpha_i A_i H$, where the variables denote the current concentration (the current amount) of the corresponding species. Consequently, the monomial $\alpha_i A_i H$ will appear with negative sign in the ODEs of the reagents (A_i and H), and with positive sign in those of the products ($A_{\tilde{i}}$ and H_2).

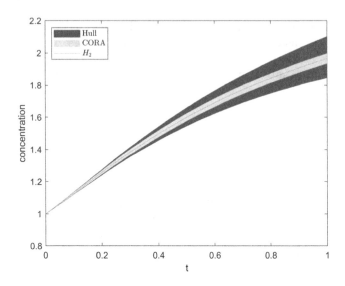

Fig. 3. Bounds of the molecule H_2 computed by Algorithm 1 against CORA.

$$A_i + H \xrightarrow{\alpha_i} A_{\tilde{i}} + H_2 \qquad (i,1) \qquad\qquad (2)$$

$$A_{\tilde{i}} + H_2 \xrightarrow{\overline{\alpha}_i} A_i + H \qquad (i,\overline{1})$$

$$A_{\tilde{i}} + C_2H_2 \xrightarrow{\beta_i} A_iCHCH^{\tilde{}} \qquad (i,2) \qquad\qquad (3)$$

$$A_iCHCH^{\tilde{}} \xrightarrow{\overline{\beta}_i} A_{\tilde{i}} + C_2H_2 \qquad (i,\overline{2})$$

$$A_iCHCH^{\tilde{}} + C_2H_2 \xrightarrow{\gamma_i} A_{i+1} + H \qquad (i,3) \qquad \ldots \qquad (i+1,1)$$

Here $A_{\tilde{i}}$ is an aromatic radical formed by H abstraction from A_i, and $A_iCHCH^{\tilde{}}$ is a radical formed by adding C_2H_2 to $A_{\tilde{i}}$. We enumerate the reactions $(i,1)$ and their reverse versions $(i,\overline{1})$. The reverse reaction is a reaction where the products became the reagents and vice versa. Since the reaction network is infinite we restrict our analysis to truncated version of this model, where we consider the dynamics of polymers up to length N (i.e., with $i \in \{1,...,N\}$). To do this we redirect the flux to A_{i+1}, when $i + 1 > N$ to A_1 in order to mimic the fact that polymers longer than N are unstable. Similarly to the previous case study, let us define the following level of heterogeneity:

$$\theta_{\mathrm{Poly}} = |\max_i \alpha_i - \min_i \alpha_i| + |\max_i \beta_i - \min_i \beta_i| + |\max_i \gamma_i - \min_i \gamma_i|$$

For the omitted parameters, the difference between the maximum and the minimum is zero. This keeps a level of heterogeneity equal to 0.2 for each model. For simplicity, only a part of the parameters was subject to perturbation; the respective values and the initial conditions can be found in the Appendix. We ran Algorithm 1 with $\varepsilon = 0.2$, obtaining the reduced differential hull through BDE. The variables are lumped according to the following partition:

$$\{\{\underline{A}_1, ..., \underline{A}_N\}, \{\overline{A}_1, ..., \overline{A}_N\}, \{\underline{A}_{\hat{1}}, ..., \underline{A}_N\}, \{\overline{A}_{\hat{1}}, ..., \overline{A}_N\}, \{\underline{H}\}, \{\overline{H}\}, \{\underline{H}_2\}, \{\overline{H}_2\},$$
$$\{\underline{C}_2\underline{H}_2\}, \{\overline{C}_2\overline{H}_2\}, \{\underline{A}_1\underline{CHCH}, ..., \underline{A}_N\underline{CHCH}\}, \{\overline{A}_1\overline{CHCH}, ..., \overline{A}_N\overline{CHCH}\}\}.$$

It can be noted the lower and upper bounds of each molecule-family were lumped together. Figure 3 shows the over-approximations of H_2 obtained by CORA and differential hulls. Also in this case study, the plot show the results only for $N = 2$, but the results are similar also for bigger models. As shown in Table 2, CORA provides tighter over-approximations but becomes computationally challenging as the number of molecules grows.

4.3   Protein Interaction Network

We next consider a prototypical model from systems biology where molecule A has multiple binding sites to which a molecule B can bind reversibly [11]. Since the number of reactions grows exponentially with the number of the binding sites, we only show the case for two binding sites. We indicate with A_{10} and A_{01} the complex obtained when A and B are bound via the first or second binding site, respectively. We denote with A_{11} when A is bounded with two molecules of B. The following reaction network describes this model:

$$A + B \xrightarrow{k_{b_1}} A_{10} \qquad\qquad A_{10} \xrightarrow{k_{u_1}} A + B$$
$$A + B \xrightarrow{k_{b_2}} A_{01} \qquad\qquad A_{01} \xrightarrow{k_{u_2}} A + B$$
$$A_{01} + B \xrightarrow{k_{b_1}} A_{11} \qquad\qquad A_{11} \xrightarrow{k_{u_1}} B + A_{01}$$
$$A_{10} + B \xrightarrow{k_{b_2}} A_{11} \qquad\qquad A_{11} \xrightarrow{k_{u_2}} B + A_{10}$$

The parameters k_b and k_u represent, respectively, the rate for binding and unbinding of molecules B to/from A. We define the level of heterogeneity as $\theta_{Protein} = |k_{b1} - k_{b2}|$. Without loss of generality, the heterogeneity was only applied to the binding parameters, with the specific parameters being reported in the Appendix. We applied our algorithm with a tolerance equal to 0.2 and computed the reduced differential hull. The reduction computed by BDE was

$$\{\{\underline{A}\}, \{\overline{A}\}, \{\underline{B}\}, \{\overline{B}\}, \{\underline{A}_{01}, \underline{A}_{10}\}, \{\overline{A}_{01}, \overline{A}_{10}\}, \{\underline{A}_{11}\}, \{\overline{A}_{11}\}\}$$

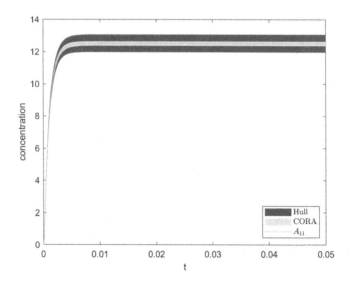

Fig. 4. Bounds of the molecule A_{11}.

Table 3. CORA running times of the protein interaction network.

N	2	4	6
CORA runtime	12.51 s	376.77 s	Out of memory

It can be noted that all molecules with the same amount of occupied binding site were lumped together. This yields an exponential reduction because the size of the original model increases exponentially in N (i.e., $2^N + 1$), while that of the reduced one polynomially (i.e., $N + 2$). We report in Fig. 4 the bounds computed with our technique and CORA; instead, Table 3 reports the computation times of CORA.

4.4   Electrical Network

We consider a simplified (inductance free) version of a power distribution electrical network enjoying a so-called H-tree topology [25]. In this setting, let us denote with N the depth of the tree and let the resistance and the capacitance at depth i be denoted by $R_{i,k}$ and $C_{i,k}$, respectively. We consider a constant source voltage v_s equal to $2.0\,$V. Denoting the voltage at $C_{i,k}$ by $v_{i,k}$, we then obtain the following affine ODE system

$$\dot{v}_{1,1} = \frac{v_S - v_{1,1}}{R_{1,1}C_{1,1}} - \frac{v_{1,1} - v_{2,1}}{R_{2,1}C_{1,1}} - \frac{v_{1,1} - v_{2,2}}{R_{2,2}C_{1,1}}, \qquad \dot{v}_{i,k} = \frac{v_{i-1,l} - v_{i,k}}{R_{i,k}C_{i,k}},$$

where $1 \leq i \leq N$, $k = 1, ..., 2^{i-1}$, and $l = \lceil k/2 \rceil$, with $\lceil \cdot \rceil$ denoting the ceil function. As a baseline, we considered a network with depth $N = 2$. For the sake of

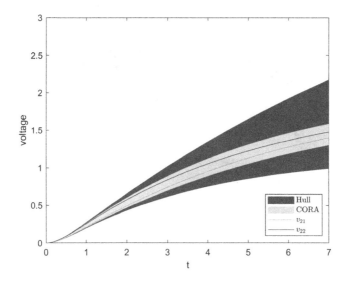

Fig. 5. Bounds of the voltages in the second level of the H-tree.

Table 4. CORA running times of the H-tree circuit model.

N	2	4	6
CORA runtime	53.99 s	231.56 s	Out of memory

simplicity, we define the associated ODE system with the following set of parameters $\mathcal{P} = \{b_2 = 1/(R_{2,1}C_{1,1}), b_3 = 1/(R_{2,2}C_{1,1}), a_{1,1} = 1/(R_{1,1}C_{1,1}), a_{2,1} = 1/(R_{2,1}C_{2,1}), a_{2,2} = 1/(R_{2,2}C_{2,2})\}$. We defined the following level of heterogeneity by

$$\theta_{Htree} = |b_2 - b_3| + |a_{2,1} - a_{2,2}|.$$

Similarly to the foregoing case studies, the differential hull was computed through Algorithm 1 and reduced afterwards via the BDE technique. The values of parameter and initial conditions can be found in the Appendix. The following variables were lumped: $\{\{\underline{v}_{1,1}\}, \{\overline{v}_{1,1}\}, \{\underline{v}_{2,1}, \underline{v}_{2,2}\}, \{\overline{v}_{2,1}, \overline{v}_{2,2}\}\}$. As expected, the voltages of the same level are lumped together. The bounds for the voltages at the second level in case of a heterogeneity equal to 0.2 can be found in Fig. 5. We considered larger models by increasing the height N of the H-Tree. Table 4 reports the computational times required to calculate the respective over-approximations.

Remark 2 (ε-BDE). This model was already studied in [9], where it was reduced through ε-BDE, an approximate version of the BDE reduction. As anticipated earlier, we next discuss the bounds computed by the differential hull with one guaranteed by ε-BDE. Indeed in [9] a theorem states that, under certain conditions, ε-BDE assures a formal bound error between the original model and

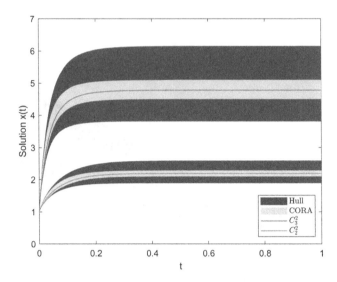

Fig. 6. Two largest over-approximations in the n-Hexane model (these of C_3^2 and C_2^2, respectively). CORA provided tighter bounds but required around 10 s, while the proposed technique less than one second.

the reduced one. Unfortunately, the applicability of the aforementioned theorem hinges on restrictive assumptions and allows only for small heterogeneity in the parameters in practice. Instead, the differential hull always succeeds in computing error bounds for approximate lumpable trajectories. This case study is an example where the hetoregeneity expressed by the parameters is too large to apply the theorem. Instead, as we can in Fig. 5, the differential hull approach is able to provide formal bounds.

4.5 Conversion of Light Alkanes over H-ZSM-5

Catalytic conversions of light alkanes into industrial chemicals, such as olefins, aromatics, oxygenates, and organic nitrides, are promising candidates for traditional petroleum-based or coal-based producing routes. We consider the conversion of n-alkanes over H-ZSM5, which is commonly used in converting methanol to gasoline and diesel. In [22], the authors considered three n-alkanes: the n-Butane, the n-Pentane, and the n-Hexane. They investigated the three different conversions reporting the entire reaction networks for each n-alkanes.

We applied our framework to the n-Hexane conversion of H-ZSM5 for the original parameters from [22]. The heterogeneity parameter was set to $\varepsilon = 15$, while the reactions were

$$C_6H_{14} \xrightarrow{k_1} C_1 + C_5^{2-} \qquad\qquad C_6H_{14} \xrightarrow{k_2} C_2 + C_4^{2-}$$

$$C_6H_{14} \xrightarrow{k_3} C_3 + C_3^{2-} \qquad\qquad C_6H_{14} \xrightarrow{k_4} C_4 + C_2^{2-}$$

$$C_6H_{14} \xrightarrow{k_5} H_2 + C_6^{2-} \qquad\qquad C_6^{2-} \xrightarrow{k_6} C_3^{2-} + C_3^{2-}$$

$$C_5^{2-} \xrightarrow{k_7} C_2^{2-} + C_3^{2-}$$

Similarly to before, the parameters and the initial conditions are reported in the Appendix. Likewise, the BDE algorithm was used to reduced the differential hull, giving rise to the following partition of the variables:

$$\{\{\underline{C}_6\underline{H}_{14}\}, \{\overline{C}_6\overline{H}_{14}\}, \{\underline{C}_1, \underline{C}_4\}, \{\overline{C}_1\overline{C}_4\}, \{\underline{C}_5^2\}, \{\overline{C}_5^2\}, \{\underline{C}_2, \underline{C}_4^2\},$$

$$\{\overline{C}_2, \overline{C}_4^2\}, \{\underline{C}_3, \underline{H}_2\}, \{\overline{C}_3, \overline{H}_2\}, \{\underline{C}_3^2\}, \{\overline{C}_3^2\}, \{\underline{C}_2^2\}, \{\overline{C}_2^2\}, \{\underline{C}_6^2\}, \{\overline{C}_6^2\}\}$$

We compare our approach against CORA. In Fig. 6, we show the bounds computed for the molecules with the largest differential hull bounds, C_3^2 and C_2^2. The CORA bounds are tighter, as expected. At the same time, CORA's running time is around 10 s, while our approach remains under 1 s. Unlike to the other case studies, the computational advantage of differential hulls cannot be exploited on larger models instances.

5   Conclusion

Despite major efforts, the over-approximation of nonlinear models given in terms of ordinary differential equations (ODEs) remains computationally challenging. This work proposes an efficient algorithmic approach for the over-approximation of nonlinear ODE models by combining results from the theory of differential inequalities and nonlinear model reduction. More specifically, by enforcing a homogeneity across model parameters in dependence on a given numerical threshold parameter, the algorithm constructs a system of differential inequalities that a) is guaranteed to over-approximate the original ODE system in presence of uncertain/noisy parameters and; b) can be often reduced, thanks to homogeneous parameters, while preserving the aforementioned over-approximation. The applicability of the approach was demonstrated by complementing the established over-approximation tool CORA on models from epidemiology, (bio)chemistry and electrical engineering. Future work will integrate the approach into the software tool ERODE [7].

Acknowledgments. This work was supported in part by DFF project REDUCTO 9040-00224B, the Poul Due Jensen Grant 883901, the Villum Investigator Grant S4OS, and the PRIN project SEDUCE 2017TWRCNB.

A   Appendix

A.1   Proofs

Proof (Theorem 2). Trivial.

Proof (Theorem 3). The only nontrivial fact to be aware about is that a parameter block with different signs will give rise a different differential hull because of the if-statements in algorithms `upperBound` and `lowerBound` will be evaluated differently.

A.2   Experiments

We next report the parameter values and the initial conditions.

SIR. Here we provide the parameters and runtimes for the SIR model considered in Sect. 4.1 (Tables 5 and 6).

Table 5. Parameters of the SIR model.

Parameters	$\beta_{1,1}$	$\beta_{1,2}$	$\beta_{2,1}$	$\beta_{2,2}$	γ_1	γ_2
Actual values	2.46	2.45	2.53	2.55	0.5	0.6

Table 6. Initial conditions of the SIR model.

Variables	S_1	S_2	I_1	I_2	R_1	R_2
Initial conditions	20	20	10	10	0	0

Polymerization. Here we provide the parameters and runtimes for the polymerization model considered in Sect. 4.2 (Tables 7 and 8).

Table 7. Parameters of the Polymerization model.

Parameters	α_1	α_2	$\overline{\alpha}_1$	$\overline{\alpha}_2$	β_1	β_2	$\overline{\beta}_2$	$\overline{\beta}_2$	γ_2	γ_2
Actual values	0.55	0.60	1.95	2.00	1.5	1.6	0.01	0.01	0.25	0.25

Table 8. Initial conditions of the Polymerization model.

Variables	A_1	A_2	$A_1^\sim$	$A_2^\sim$	H	H_2	C_2H_2	$A_1CHCH^\sim$	$A_2CHCH^\sim$
Initial conditions	1	1	1	1	1	1	1	1	1

Protein Interaction Network. Here we provide the parameters and runtimes for the model considered in Sect. 4.3 (Tables 9 and 10).

Table 9. Parameters of the Protein interaction network.

Parameters	k_{b_1}	k_{b_2}	k_{u_1}	k_{u_2}
Actual values	20.10	19.90	0.1	0.1

Table 10. Initial conditions of the Protein interaction network.

Variables	A	B	A_{10}	A_{01}	A_{11}
Initial conditions	50	50	0	0	0

Electrical Network. Here we provide the parameters and runtimes for the model considered in Sect. 4.4 (Tables 11 and 12).

Table 11. Parameters of the Electrical network.

Parameters	b_2	b_3	$a_{1,1}$	$a_{2,1}$	$a_{2,2}$
Actual values	0.56	0.66	1.12	0.40	0.50

Table 12. Initial conditions of the Electrical network.

Variables	$v_{1,1}$	$v_{2,1}$	$v_{2,2}$
Initial conditions	0.56	0.66	1.12

n-Hexane Model. Here we provide the parameters and runtimes for the model considered in Sect. 4.5 (Tables 13 and 14).

Table 13. Parameters of the n-Hexane model.

Parameters	k_1	k_2	k_3	k_4	k_5	k_6	k_7
Actual values	17	54	42	13	32	32	14

Table 14. Initial conditions of the n-Hexane model.

Variables	C_6H_{14}	C_1	C_5^2	C_2	C_4^2	C_3	C_3^2	C_4	C_2^2	H_2	C_6^2
Initial conditions	1	1	1	1	1	1	1	1	1	1	1

References

1. Althoff, M., Kochdumper, N.: CORA 2016 manual. TU Munich 85748 (2016)
2. Asarin, E., Dang, T., Girard, A.: Reachability analysis of nonlinear systems using conservative approximation. In: HSCC (2003)
3. Bortolussi, L., Gast, N.: Mean field approximation of uncertain stochastic models. In: DSN, pp. 287–298 (2016)
4. Cardelli, L., et al.: Exact maximal reduction of stochastic reaction networks by species lumping. Bioinformatics **37**(15), 2175–2182 (2021)
5. Cardelli, L., Tribastone, M., Tschaikowski, M., Vandin, A.: Forward and backward bisimulations for chemical reaction networks. In: CONCUR, pp. 226–239 (2015)
6. Cardelli, L., Tribastone, M., Tschaikowski, M., Vandin, A.: Symbolic computation of differential equivalences. In: POPL, pp. 137–150 (2016)
7. Cardelli, L., Tribastone, M., Tschaikowski, M., Vandin, A.: ERODE: a tool for the evaluation and reduction of ordinary differential equations. In: TACAS (2017)
8. Cardelli, L., Tribastone, M., Tschaikowski, M., Vandin, A.: Maximal aggregation of polynomial dynamical systems. Proc. Natl. Acad. Sci. **114**(38), 10029–10034 (2017)
9. Cardelli, L., Tribastone, M., Tschaikowski, M., Vandin, A.: Guaranteed error bounds on approximate model abstractions through reachability analysis. In: QEST, pp. 104–121 (2018)
10. Chen, X., Ábrahám, E., Sankaranarayanan, S.: Flow*: an analyzer for non-linear hybrid systems. In: CAV, pp. 258–263 (2013)
11. Conzelmann, H., Fey, D., Gilles, E.D.: Exact model reduction of combinatorial reaction networks. BMC Syst. Biol. **2**(1), 1–25 (2008)
12. Duggirala, P.S., Mitra, S., Viswanathan, M.: Verification of annotated models from executions. In: EMSOFT, pp. 26:1–26:10. IEEE Press (2013)
13. Feret, J., Danos, V., Krivine, J., Harmer, R., Fontana, W.: Internal coarse-graining of molecular systems. Proc. Natl. Acad. Sci. **106**(16), 6453–6458 (2009)
14. Girard, A., Pappas, G.J.: Approximate bisimulations for nonlinear dynamical systems. In: IEEE Conference on Decision and Control and European Control Conference (2005)
15. Iwasa, Y., Levin, S.A., Andreasen, V.: Aggregation in model ecosystems II. Approximate aggregation. Math. Med. Biol. **6**(1), 1–23 (1989)
16. Kermack, W.O., McKendrick, A.G.: A contribution to the mathematical theory of epidemics. Proc. R. Soc. London. Ser. A Contain. Papers. Math. Phys. Character. **115**(772), 700–721 (1927)
17. Kuo, J.C.W., Wei, J.: Lumping analysis in monomolecular reaction systems. Analysis of approximately lumpable system. Indus. Eng. Chem. Fundam. **8**(1), 124–133 (1969)
18. Li, G., Rabitz, H.: A general analysis of approximate lumping in chemical kinetics. Chem. Eng. Sci. **45**(4), 977–1002 (1990)
19. Liu, J., Zhan, N., Zhao, H., Zou, L.: Abstraction of elementary hybrid systems by variable transformation. In: Bjørner, N., de Boer, F. (eds.) FM 2015. LNCS, vol. 9109, pp. 360–377. Springer, Cham (2015). https://doi.org/10.1007/978-3-319-19249-9_23
20. Majumdar, R., Zamani, M.: Approximately bisimilar symbolic models for digital control systems. In: CAV, pp. 362–377 (2012)
21. Müller, M.: Über das Fundamentaltheorem in der Theorie der gewöhnlichen Differentialgleichungen. Mathematische Zeitschrift **26**, 619–645 (1927)

22. Narbeshuber, T., Vinek, H., Lercher, J.: Monomolecular conversion of light alkanes over H-ZSM-5. J. Catal. **157**(2), 388–395 (1995)
23. Ramdani, N., Meslem, N., Candau, Y.: Reachability of uncertain nonlinear systems using a nonlinear hybridization. In: HSCC, pp. 415–428 (2008)
24. Ramdani, N., Meslem, N., Candau, Y.: Computing reachable sets for uncertain nonlinear monotone systems. Nonlinear Anal. Hybrid Syst. **4**(2), 263–278 (2010)
25. Rosenfeld, J., Friedman, E.G.: Design methodology for global resonant H-tree clock distribution networks. IEEE Trans. VLSI Syst. **15**(2), 135–148 (2007)
26. Scott, J.K., Barton, P.I.: Bounds on the reachable sets of nonlinear control systems. Automatica **49**(1), 93–100 (2013)
27. Tschaikowski, M., Tribastone, M.: Approximate reduction of heterogenous nonlinear models with differential hulls. IEEE Trans. Autom. Contr. **61**(4), 1099–1104 (2016)
28. Turányi, T., Tomlin, A.S.: Reduction of reaction mechanisms. In: Analysis of Kinetic Reaction Mechanisms, pp. 183–312. Springer, Heidelberg (2014). https://doi.org/10.1007/978-3-662-44562-4_7
29. Voit, E.O.: Biochemical systems theory: a review. ISRN Biomath. **2013**, 53 (2013)

Stability Analysis of Planar Probabilistic Piecewise Constant Derivative Systems

Spandan Das$^{(\boxtimes)}$ ⓘ and Pavithra Prabhakar

Kansas State University, Manhattan, USA
{spandan,pprabhakar}@ksu.edu

Abstract. In this paper, we study the probabilistic stability analysis of a subclass of stochastic hybrid systems, called the *Planar Probabilistic Piecewise Constant Derivative Systems (Planar PPCD)*, where the continuous dynamics is deterministic, constant rate and planar, the discrete switching between the modes is probabilistic and happens at boundary of the invariant regions, and the continuous states are not reset during switching. These aptly model piecewise linear behaviors of planar robots. Our main result is an exact algorithm for deciding *absolute* and *almost sure stability* of Planar PPCD under some mild assumptions on mutual reachability between the states and the presence of non-zero probability self-loops. Our main idea is to reduce the stability problems on planar PPCD into corresponding problems on Discrete-time Markov Chains with edge weights.

Keywords: Stability · Probabilistic Piecewise Constant Derivative Systems · Discrete-time Markov Chain · Convergence

1 Introduction

Stability of Stochastic Hybrid Systems (SHS) [28] is a desirable property, as it guarantees eventual convergence of executions to a point of equilibrium, even in the presence of random errors. In this paper, we investigate the stability of a certain kind of SHS where the continuous state space is planar and dynamics has constant rate, where the rates are discrete and chosen probabilistically. More precisely, we study *Probabilistic Piecewise Constant Derivative Systems (PPCD)*, that consist of a finite number of discrete states representing different modes of operation each associated with a constant rate dynamics, and probabilistic mode switches enabled at certain polyhedral boundaries. Such systems can aptly model piecewise linear behaviour of planar robots.

Safety analysis of SHS has been extensively studied in the context of both non-stochastic as well as stochastic hybrid systems [1,8,17,18,26]; stability on the other hand is relatively less explored, especially, from a computational point of view. It is well-known that even for non-stochastic hybrid systems decidability (existence of exact algorithms) for safety is achievable only under restrictions on

This work was partially supported by NSF CAREER Grant No. 1552668 and NSF Grant No. 2008957.

E. Ábrahám and M. Paolieri (Eds.): QEST 2022, LNCS 13479, pp. 192–213, 2022.
https://doi.org/10.1007/978-3-031-16336-4_10

the dynamics and the dimension [14]. More recently, decidability of stability of hybrid systems has been explored in the non-stochastic setting [24]. The main contribution of this paper is the identification of a practically useful subclass of stochastic hybrid systems for which stability is decidable along with an exact stability analysis algorithm.

The classical stability analysis techniques build on the notion of Lyapunov functions that provide a certificate of stability. While the notion of Lyapunov functions have been extended to the hybrid system setting, computing them is a challenge. Typically, they require solving certain complex optimization problems, for instance, to deduce coefficients of polynomial templates, and more importantly, need the exploration of increasingly complex templates. In this paper, we take an alternate route where we present graph theory based reductions to show the decidability of stability analysis.

Our broad approach is to reduce a planar PPCD, that is a potentially infinite state probabilistic system, to that of a Finite State Discrete-time Markov Chain such that the stability of the planar PPCD can be deduced exactly by algorithmically checking certain properties of the reduced system. We study two notions of stability, namely, absolute stability and almost sure stability. In the former, we seek to ensure that every execution converges, while in the latter, we require that the probability of the set of system executions that converge be 1. Absolute convergence ignores the probabilities associated with the transitions, and hence, can be solved using previous results on stability analysis of Piecewise Constant Derivative systems [23], where one checks for certain diverging transitions and cycles. Checking almost sure convergence is much more challenging. We show that almost sure convergence can be characterized by certain constraints based on the stationary distribution of the reduced system. For this result to hold, we need mild conditions on the PPCD that ensure the existence of this stationary distribution. The proof relies on several insights, including the properties of planar dynamics, and convergence results on infinite sequences of random variables.

The rest of the paper is organized as follows. In Sect. 2, we discuss related works. In Sect. 3, we model motion of a planar robot with faulty angle actuator using PPCD. In Sect. 4, we define important definitions and notations related to Markov Chains. In Sect. 5, we develop algorithms for analyzing convergence of Markov Chains. We analyze stability of general and planar PPCDs in Sect. 6. Finally, we conclude in Sect. 7.

2   Related Work

Stability is a well studied problem in classical control theory, where Lyapunov function based methods have been extensively developed. They have been extended to hybrid systems using multiple and common Lyapunov functions [4,9,19,30]. However, constructing Lyapunov functions is computationally challenging, hence, alternate approximate methods have been explored. For example, in one approach the state space is divided into certain regions and shown that the system inevitably ends up in a certain region, thus ensuring stability [12,13,20,21]. Another approach is based on abstraction, where a simplified model (known as the abstract model) is created based on the original model and stability analysis on the simplified model is mapped back to the original one [1,2,5,8,10,22,23,25].

While stability has been extensively studied in non-probabilistic setting, investigations of stability for probabilistic systems are limited. Sufficient conditions for stability of Stochastic Hybrid Systems via Lyapunov functions is discussed in the survey [29]. Almost sure exponential stability [6,7,11,15] and asymptotic stability in distribution [31,32] for Stochastic Hybrid Systems have also been studied. Most of these works on probabilistic stability analysis provide approximate mehtods for analysis. We provide a simple class of Stochastic Hybrid Systems that have practical application in modeling planar robots, and an exact decidable algorithm for probabilistic stability analysis.

3   Case Study: Planar Robot with a Faulty Actuator

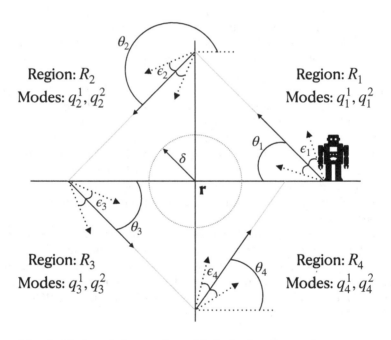

Fig. 1. Motion of planar robot with faulty heading angle actuator

Consider a robot navigating in a 2D plane at some constant speed v as shown in Fig. 1. The plane is divided into four regions R_1, R_2, R_3, R_4 corresponding to the four quadrants, and the robot has a unique direction θ_i (mode of operation) in which it moves while in the region R_i, and changes its mode of operation at the boundary of the regions. Due to faulty actuator, the robot heading angle may deviate from θ_i by an amount ϵ_i. We model this as probabilistically choosing one of the k_i uniformly distanced angles $\theta_i^1, \cdots, \theta_i^{k_i}$ in the interval $[\theta_i - \epsilon_i, \theta_i + \epsilon_i]$ with probabilities $p_i^1, \cdots, p_i^{k_i}$, respectively. The whole system can be modelled as a planar PPCD with $\sum_{i=1}^4 k_i$ modes, where for every i and $1 \leq j \leq k_i$, the mode q_i^j corresponds to the robot traversing with heading angle θ_i^j with speed v in the region R_i. The mode switching is possible between R_i and R_j if they are neighbors, that is, they share a common boundary. For instance, we can switch between quadrants 1 and 2 or 4 and 1 but not 1 and 3. We can move to any mode corresponding to a neighbor q_i^j with probability p_i^j.

The objective of the navigation is to reach a target point r on the 2D plane arbitrarily closely. More precisely, we want to check whether the robot reaches within a $\delta > 0$ ball around r for any arbitrarily small δ. We want to check if all executions of the robot have this property, i.e., if the planar PPCD is absolutely stable, as well as if the probability of convergence is 1, i.e., the planar PPCD is almost surely stable.

4   Preliminaries

In this section, we will discuss important concepts related to Discrete-time Markov Chain (DTMC), Weighted Discrete-time Markov Chain (WDTMC) and convergence of WDTMC.

4.1   Discrete-Time Markov Chain

Let $Dist(S)$ denote the set of all probability distributions on the set S. Let us define Discrete-time Markov Chain (DTMC) on the set of states S.

Definition 1 (Discrete-time Markov Chain). *The Discrete-time Markov Chain (DTMC) is defined as the tuple $\mathcal{M} = (S, P)$ where*

- *S is a set of states.*
- *$P : S \mapsto Dist(S)$ is a function from the set of states S to the set of all probability distributions over S, $Dist(S)$.*

We use $\mathrm{P}(s_1, s_2)$ to denote $\mathrm{P}(s_1)(s_2)$ and $\mathrm{P}^n(s_1, s_2)$ to denote the probability of going from s_1 to s_2 in n-steps.

A path of a DTMC $\mathcal{M}$ is a sequence of states $\sigma = s_1, s_2, \ldots$ such that for all $i < |\sigma|$, $\mathrm{P}(s_i, s_{i+1}) > 0$, where $|\sigma|$ is the length of the sequence. A path of length 2 is called an edge and the set of all edges is denoted as $\mathcal{E}$. The ith state of the path σ is denoted by σ_i and the last state of σ is denoted as σ_{end}. $\sigma[i : j]$ denotes the subsequence $\sigma_i, \sigma_{i+1}, \ldots, \sigma_j$. We say s_2 is reachable from s_1 (denoted $s_1 \rightsquigarrow s_2$) if there is a path σ on $\mathcal{M}$ such that $\sigma_1 = s_1$ and $\sigma_{end} = s_2$. The set of all finite paths of a DTMC $\mathcal{M}$ is denoted as $Paths_{fin}(\mathcal{M})$ and the set of all infinite paths is denoted as $Paths(\mathcal{M})$.

The probability of a finite path σ, denoted $\mathrm{P}(\sigma)$, is the product of the probabilities of each of its edges, $\mathrm{P}(\sigma) := \prod_{i<|\sigma|} \mathrm{P}(\sigma_i, \sigma_{i+1})$. The probability of σ with respect to a distribution ρ, denoted $P_\rho(\sigma)$ is the product of $\mathrm{P}(\sigma)$ and the probability of σ_1 under ρ, i.e., $P_\rho(\sigma) := \rho(\sigma_1) \cdot P(\sigma)$. We can associate a probability measure Pr to the set of infinite paths $Paths(\mathcal{M})$ of a DTMC $\mathcal{M}$ using probability of the cylinder sets of the finite paths as discussed in [3]. A path property $\mathbb{P}$ is said to be *almost surely* satisfied if the set of all paths having property $\mathbb{P}$ has probability 1, i.e., $Pr\{\sigma \mid \sigma \text{ has } \mathbb{P}\} = 1$.

Next we define some subclasses of DTMC and show that it has some nice convergence properties.

Definition 2 (Irreducibility). *A DTMC $\mathcal{M}$ is called irreducible if for any $s_1, s_2 \in S$, $s_1 \rightsquigarrow s_2$ and $s_2 \rightsquigarrow s_1$.*

Definition 3 (Periodicity). *A state $s \in S$ in a DTMC $\mathcal{M}$ is called periodic if there is a natural number $n > 1$ such that, for any path σ starting and ending at s, $|\sigma|$ is a multiple of n. A DTMC $\mathcal{M}$ is called aperiodic if none of its states is periodic.*

We say a probability distribution is stationary for a DTMC $\mathcal{M}$ if the next step distribution remains unchanged.

Definition 4 (Stationary Distribution). *A distribution $\rho^* \in Dist(S)$ is called the stationary distribution of DTMC $\mathcal{M}$ if,*

$$\rho^*(s) = \sum_{s' \in S} \rho^*(s')P(s', s), \quad \forall s \in S.$$

For finite, irreducible DTMC, the stationary distribution is unique. The following theorem guarantees existence of limiting distribution for finite, irreducible and aperiodic DTMC and associates it with the stationary distribution of the DTMC (see [27]).

Theorem 1. *For a finite, irreducible and aperiodic DTMC $\lim_{n \to \infty} P^n(s_1, s_2)$ exists for all $s_1, s_2 \in S$ and $\lim_{n \to \infty} P^n(s_1, s_2) = \rho^*(s_2)$ where $\rho^* \in Dist(S)$ is the unique stationary distribution of $\mathcal{M}$.*

Note that, $P^n(s_1, s_2)$ does not depend on s_1 as $n \to \infty$.

4.2   Weighted Discrete-Time Markov Chain

Let us now define Weighted Discrete-time Markov Chain (WDTMC) that extend DTMC with weighted edges. Basically, a WDTMC can be observed as a Markov Reward Process where rewards are associated to individual transitions rather than nodes.

Definition 5 (Weighted DTMC). *The weighted DTMC (WDTMC) $\mathcal{M}_W = (S, P, W)$ is a tuple such that (S, P) is a DTMC and $W : \mathcal{E} \mapsto \mathbb{R}$ is a weight function where $\mathcal{E}$ is the set of all possible edges of $\mathcal{M}_W$.*

We also define disjoint union of two WDTMC $\mathcal{M}_W^1$ and $\mathcal{M}_W^2$ as a WDTMC $\mathcal{M}_W^1 \sqcup \mathcal{M}_W^2$ whose states and edges are disjoint unions of states and edges of $\mathcal{M}_W^1$ and $\mathcal{M}_W^2$ respectively. With the weight function W defined, it is possible to associate weights to individual paths of $\mathcal{M}_W$.

Definition 6 (Weight of a path). *The weight of a path σ of WDTMC $\mathcal{M}_W$, denoted $W(\sigma)$, is defined as,*

$$W(\sigma) := \sum_{i < |\sigma|} W(\sigma_i, \sigma_{i+1})$$

For $\sigma \in Paths(\mathcal{M}_W)$, the quantity $\lim_{n \to \infty} \sum_{i=1}^{n} W(\sigma_i, \sigma_{i+1})$ is denoted by $W(\sigma[1 : \infty])$. It is easy to observe that, $W(\sigma) = W(\sigma[1 : \infty])$. A simple path is a path without state repetition and a simple cycle is a path where only the starting and the ending states are same. We use the notation $\mathcal{SP}(\mathcal{M}_W)$ for the set of all simple paths and the notation $\mathcal{SC}(\mathcal{M}_W)$ for the set of all simple cycles of a WDTMC $\mathcal{M}_W$.

4.3   Convergence of Weighted Discrete-Time Markov Chain

Let us define the notions of absolute and probabilistic convergence of WDTMC. A WDTMC is said to be absolutely convergent if the weight of every infinite path diverges to $-\infty$.

Definition 7 (Absolute Convergence of WDTMC). *A WDTMC $\mathcal{M}_W$ is said to be absolutely convergent if for all infinite path $\sigma \in Paths(\mathcal{M}_W)$, $W(\sigma)$ diverges to $-\infty$, i.e.,*

$$W(\sigma[1 : \infty]) = -\infty.$$

Further, a WDTMC is said to be almost surely convergent if the weight of an infinite path diverges to $-\infty$ with probability 1.

Definition 8 (Almost Sure Convergence of WDTMC). *We say that a WDTMC $\mathcal{M}_W$ is almost surely convergent if for any path σ of $\mathcal{M}_W$, $W(\sigma)$ diverges to $-\infty$ with probability 1. In other words,*

$$Pr\{\sigma \in Paths(\mathcal{M}_W) : W(\sigma[1 : \infty]) = -\infty\} = 1.$$

Remark 1. Let us explain the reason behind defining such a strange notion of convergence. For reasons that will be clarified later, we actually want to check for an infinite path σ of $\mathcal{M}_W$, if the product of weights of the edges converge to 0, i.e., $\lim_{n \to \infty} \prod_{i=1}^{n} W(\sigma_i, \sigma_{i+1}) = 0$, provided $0 < W(\sigma_i, \sigma_{i+1}) < \infty$ for all $i \in \mathbb{N}$. This condition is equivalent to $\lim_{n \to \infty} \sum_{i=1}^{n} \log(W(\sigma_i, \sigma_{i+1})) = -\infty$. Hence for convenience, we consider log of original weights as weights of individual edges, and check if sum of weights of edges of an infinite path diverge to $-\infty$.

4.4   Probabilistic Bisimulation

Probabilistic bisimulation [3] on a WDTMC is an equivalence relation on its set of states such that probabilities of corresponding edges agree for two related states.

Definition 9 (Probabilistic Bisimulation). *A probabilistic bisimulation on a WDTMC* $\mathcal{M}_W$ *is an equivalence relation $\sim$ on S such that for any $s_1, s_2 \in S$ with $s_1 \sim s_2$, $P(s_1, T) = P(s_2, T)$ for each equivalence class T of $\sim$.*

Note that, $P(s, T) = \sum_{t \in T} P(s, t)$ for $s \in S$. Let us now use probabilistic bisimulation to relate infinite paths of a WDTMC.

Definition 10 (Bisimulation-Equivalent Paths). *Given a probabilistic bisimulation $\sim$ on a WDTMC $\mathcal{M}_W$, two infinite paths $\pi = \pi_1, \pi_2, \ldots$ and $\tilde{\pi} = \tilde{\pi}_1, \tilde{\pi}_2, \ldots$ are said to be bisimulation equivalent, denoted $\pi \sim \tilde{\pi}$, if they are statewise related by $\sim$, i.e.,*

$$\pi \sim \tilde{\pi} \text{ iff } \pi_i \sim \tilde{\pi}_i \text{ for all } i \geq 1$$

A set of infinite paths is $\sim$ bisimulation-closed for some probabilistic bisimulation $\sim$, if for any path in the set, all its bisimulation-equivalent paths are also in the set. In other words, $\Pi \subseteq Paths(\mathcal{M}_W)$ is $\sim$ bisimulation-closed if for any $\pi \in \Pi$ and any $\tilde{\pi} \sim \pi$, $\tilde{\pi} \in \Pi$. Let us denote by $Pr_s(\Pi)$ the set of all paths in Π that start from $s \in S$. The following lemma [3] equates the probability of two sets of paths that start from $\sim$ related states and are subset of the same $\sim$ bisimulation-closed set.

Lemma 1. *Let $\sim$ be a probabilistic bisimulation on a WDTMC $\mathcal{M}_W$. For all states s_1, s_2 of $\mathcal{M}_W$, $s_1 \sim s_2$ implies $Pr_{s_1}(\Pi) = Pr_{s_2}(\Pi)$, for all $\sim$ bisimulation-closed events $\Pi \subseteq Paths(\mathcal{M}_W)$.*

4.5   Polyhedral Sets

We denote the set of all polyhedral subsets of $\mathbb{R}^n$ by $Poly(n)$. The facets of a polyhedral subset A are the largest polyhedral subsets of the boundary of A. We denote the boundary of a polyhedral subset A by $\partial(A)$ and the set of all facets of A by $\mathbb{F}(A)$. We say a polyhedral subset P is positive scaling invariant if for all $x \in P$ and $\alpha > 0$, $\alpha x \in P$.

5   Analyzing Convergence of Weighted Discrete-Time Markov Chains

In this section, we discuss necessary and sufficient conditions for absolute and almost sure convergence of WDTMC. For our analysis, we will assume all paths of the WDTMC start from a single state called the *initialization point* (denoted s_{init}) of the WDTMC. In other words we restrict our attention to the set of paths $\Sigma' := \{\sigma \in Paths(\mathcal{M}_W) \mid \sigma_1 = s_{init}\}$. Consequently, we consider only those edges $\mathcal{E}' = \Sigma' \cap \mathcal{E}$, which are reachable from s_{init}. We abuse notation and use Σ for Σ' and $\mathcal{E}$ for $\mathcal{E}'$ for the rest of the section.

5.1   Analyzing Absolute Convergence of Weighted DTMC

Here we provide a necessary and sufficient condition for analyzing absolute convergence of a WDTMC. We begin with the following proposition (proved in Appendix A.1) which states that for any finite path $\sigma \in Paths_{fin}(\mathcal{M}_W)$, we can get one simple path and a set of simple cycles such that their total weight equals the weight of σ.

Proposition 1. *For any finite path σ of $\mathcal{M}_W$ there exist a simple path $\sigma_s \in SP(\mathcal{M}_W)$ and a set of simple cycles $SC_\sigma \subseteq SC(\mathcal{M}_W)$ such that $W(\sigma) = W(\sigma_s) + \sum_{\mathcal{C} \in SC_\sigma} W(\mathcal{C})$.*

We use Proposition 1 to prove the following main theorem which states that, a WDTMC is absolutely convergent iff there is no edge of infinite weight and no cycle of weight greater or equal to 0 reachable from the initial point.

Theorem 2. *The WDTMC $\mathcal{M}_W$ is absolutely convergent iff,*

1. *There does not exist an edge $e \in \mathcal{E}$ reachable from s_{init} such that $W(e) = \infty$.*
2. *For any simple cycle $\mathcal{C}$ reachable from s_{init}, $W(\mathcal{C}) < 0$.*

Proof. ($\Rightarrow$) To show that the conditions 1 and 2 are necessary, we have to prove that if either of them is negated then $\mathcal{M}_W$ is not absolutely convergent. If condition 1 is false then there is an edge $e = (s_1, s_2)$ with $W(s_1, s_2) = \infty$ such that for some finite path σ starting from s_{init}, $\sigma_{|\sigma|-1} = s_1$ and $\sigma_{|\sigma|} = s_2$. But that implies $W(\sigma) = \sum_{i=1}^{|\sigma|-1} W(\sigma_i, \sigma_{i+1}) = \infty$. So for any infinite path σ' with prefix σ, $W(\sigma') = \infty$. Thus $\mathcal{M}_W$ is not absolutely convergent. On the other hand if we suppose condition 2 is false then there is a simple cycle $\mathcal{C} \in SC(\mathcal{M}_W)$ with $W(\mathcal{C}) \geq 0$ such that for some finite path σ starting from s_{init}, there exists an index j such that $\mathcal{C} = \sigma[j : |\sigma|]$. Now we can easily construct the following infinite path $\sigma_\infty = \sigma \cdot \mathcal{C} \cdot \mathcal{C} \ldots$ by concatenating $\mathcal{C}$ infinite times to σ. Clearly, σ_∞ starts at s_{init} since σ starts at s_{init} and $W(\sigma_\infty) = W(\sigma) + \sum_{n \in \mathbb{N}} W(\mathcal{C}) \geq W(\sigma)$. Since for any finite path σ, $W(\sigma)$ is also finite, $W(\sigma_\infty)$ is bounded below by some finite quantity and cannot diverge to $-\infty$. Thus, $\mathcal{M}_W$ is not absolutely convergent.

($\Leftarrow$) Conversely, suppose both conditions 1 and 2 hold. Now, let σ be an arbitrary infinite path starting from s_{init} and $\sigma[1 : i]$ be its finite prefix of length $i \in \mathbb{N}$. By Proposition 1, there exist a simple path $\sigma[1 : i]_s$ and a set of simple cycles $SC_{\sigma[1:i]}$ such that $W(\sigma[1 : i]) = W(\sigma[1 : i]_s) + \sum_{\mathcal{C} \in SC_{\sigma[1:i]}} W(\mathcal{C})$. Now, for any $i \in \mathbb{N}$, $W(\sigma[1 : i]_s)$ is at most $\sum_{(s_1,s_2) \in \mathcal{E}} \max\{W(s_1, s_2) \mid (s_1, s_2) \in \mathcal{E}\} < \infty$. Also, $SC_{\sigma[1:i]}$ is a set of simple cycles where each cycle has weight at most $\max_{\mathcal{C} \in SC(\mathcal{M}_W)} W(\mathcal{C}) < 0$ (here we abuse notation and denote the set of all simple cycles reachable from s_{init} as $SC(\mathcal{M}_W)$). Thus, for all $K \in \mathbb{R}$, there exists $i \in \mathbb{N}$

such that

$$\sum_{(s_1,s_2)\in\mathcal{E}} \max\{W(s_1,s_2) \mid (s_1,s_2)\in\mathcal{E}\} + \sum_{\mathcal{C}\in\mathcal{SC}_{\sigma[1:i]}} W(\mathcal{C}) < K$$

$$\Rightarrow W(\sigma[1:i]_s) + \sum_{\mathcal{C}\in\mathcal{SC}_{\sigma[1:i]}} W(\mathcal{C}) < K$$

$$\Rightarrow W(\sigma[1:i]) < K.$$

But this implies $W(\sigma) = \lim_{i\to\infty} W(\sigma[1:i]) = -\infty$ for any infinite path σ starting from s_{init}, i.e., $\mathcal{M}_W$ is absolutely convergent.   □

5.2   Analyzing Almost Sure Convergence of Weighted DTMC

In this subsection, we will provide a necessary and sufficient condition for almost sure convergence of a WDTMC. We assume a WDTMC $\mathcal{M}_W$ is finite, irreducible and aperiodic and thus has the limiting distribution equal to its stationary distribution ρ^* (Theorem 1).

Given a WDTMC $\mathcal{M}_W$, we begin by defining random variables $\{X_j^e \mid e \in \mathcal{E}; j\in\mathbb{N}\}$ on the set of infinite paths $Paths(\mathcal{M}_W)$, that captures the information of whether an edge $e\in\mathcal{E}$ appears on the jth step of an infinite path σ. More precisely,

$$X_j^e(\sigma) = \begin{cases} 1 \text{ if } (\sigma_j,\sigma_{j+1}) = e \\ 0 \text{ else.} \end{cases}$$

Note that for some $e\in\mathcal{E}$ and $\sigma\in Paths(\mathcal{M}_W)$, $\sum_{j=1}^n X_j^e(\sigma)$ gives the number of times e appears on $\sigma[1:n+1]$. Now, the following lemma (proved in Appendix A.3) gives that, for any edge $e\in\mathcal{E}$, the average of $\{X_j^e \mid j\in\mathbb{N}\}$ almost surely converges to $P_{\rho^*}(e)$, which is the probability of e with respect to the stationary distribution ρ^*.

Lemma 2. *For any edge $e\in\mathcal{E}$ of a WDTMC $\mathcal{M}_W$,*

$$Pr\left\{\sigma\in Paths(\mathcal{M}_W) : \lim_{n\to\infty} \frac{\sum_{j=1}^n X_j^e(\sigma)}{n} = P_{\rho^*}(e)\right\} = 1.$$

Next, we define partial average weight upto n for an infinite path σ as

$$\frac{(S_\sigma)_n}{n} := \frac{\sum_{i=1}^n W(\sigma_i,\sigma_{i+1})}{n},$$

and note that,

$$\frac{(S_\sigma)_n}{n} = \frac{\sum_{e\in\mathcal{E}}(\# \text{ times } e \text{ appears on } \sigma[1:n+1])\cdot W(e)}{n}$$

$$= \frac{\sum_{e\in\mathcal{E}}\left(\sum_{j=1}^n X_j^e(\sigma)\right)\cdot W(e)}{n} \tag{1}$$

We now state the main lemma of this subsection which essentially states that, the average weight of an infinite path almost surely converges to a quantity that depends only on the weights and probabilities of the edges.

Lemma 3. *For a WDTMC* $\mathcal{M}_W$,

$$Pr\left\{\sigma \in Paths(\mathcal{M}_W) : \lim_{n\to\infty} \frac{(S_\sigma)_n}{n} = \sum_{e\in\mathcal{E}} P_{\rho^*}(e)\,W(e)\right\} = 1.$$

Proof. We have already established that,

$$\frac{(S_\sigma)_n}{n} = \frac{\sum_{e\in\mathcal{E}}\left(\sum_{j=1}^{n} X_j^e(\sigma)\right)\cdot W(e)}{n} \quad \text{[Eq. 1]}$$

$$\text{Thus, } \lim_{n\to\infty}\frac{(S_\sigma)_n}{n} = \lim_{n\to\infty}\frac{\sum_{e\in\mathcal{E}}\left(\sum_{j=1}^{n} X_j^e(\sigma)\right)\cdot W(e)}{n}$$

$$\Rightarrow \lim_{n\to\infty}\frac{(S_\sigma)_n}{n} = \sum_{e\in\mathcal{E}} P_{\rho^*}(e)\cdot W(e) \text{ almost surely} \quad \text{[by Lemma 2]}$$

$\square$

We say $\sum_{e\in\mathcal{E}} P_{\rho^*}(e)W(e)$ is the effective weight of the WDTMC $\mathcal{M}_W$ and denote it as $W_\mathcal{E}$. The main theorem basically states that a WDTMC is almost surely convergent iff its effective weight is strictly less than 0.

Theorem 3. *A WDTMC* $\mathcal{M}_W$ *is almost surely convergent iff* $W_\mathcal{E} < 0$, *where* $W_\mathcal{E} = \sum_{e\in\mathcal{E}} P_{\rho^*}(e)\,W(e)$ *is the effective weight of* $\mathcal{M}_W$.

Proof. Observe that, weight of an infinite path σ, $W(\sigma)$, can be written as $\lim_{n\to\infty} n \cdot ((S_\sigma)_n/n)$, where $((S_\sigma)_n/n)$ is the partial average weight upto n for the infinite path σ. Since,

$$\lim_{n\to\infty}\left(\frac{(S_\sigma)_n}{n}\right) = \sum_{e\in\mathcal{E}} P_{\rho^*}(e)W(e) \text{ almost surely} \quad \text{[by Lemma 3]}$$

$$\Rightarrow W(\sigma) = \lim_{n\to\infty} n\cdot\left(\frac{(S_\sigma)_n}{n}\right) = \lim_{n\to\infty} n\cdot\left(\sum_{e\in\mathcal{E}} P_{\rho^*}(e)W(e)\right) \text{ almost surely}$$

Thus, $W(\sigma)$ diverges to $-\infty$ almost surely if and only if $\sum_{e\in\mathcal{E}} P_{\rho^*}(e)W(e) < 0$. In other words, $\mathcal{M}_W$ is almost surely convergent iff $W_\mathcal{E} < 0$. $\square$

5.3 Computability

Based on Theorems 2 and 3 we present two algorithms here (Appendix A.2) for checking absolute and almost sure convergence of a WDTMC. For the first algorithm, assuming the WDTMC is finite, we first check for existence of an infinite weight edge by Breadth First Search (BFS) [16] and then for a cycle with non-negative weight using a variant of the Bellman-Ford algorithm [16]. If neither of them is found then the WDTMC is deemed absolutely convergent by Theorem 2. Since BFS takes time linear to the size of its input and Bellman-Ford takes time quadratic to the size of its input, the time complexity of this algorithm is $O(|S|^2)$, where S is the set of states of $\mathcal{M}_W$.

For the second algorithm, assuming the WDTMC is finite, irreducible and aperiodic, existence of an infinite weight edge is checked by Breadth First Search (BFS). If such an edge exists then the WDTMC is deemed not almost surely convergent (by Theorem 3). Otherwise, the stationary distribution ρ^* of the WDTMC is calculated by solving a set of linear equations mentioned in Definition 4. The value $\sum_{e \in \mathcal{E}} P_{\rho^*}(e) \mathrm{W}(e)$ is then calculated (where $\mathcal{E}$ is the set of transitions of the WDTMC) and compared to 0. The WDTMC is deemed almost surely convergent only if $\sum_{e \in \mathcal{E}} P_{\rho^*}(e) \mathrm{W}(e) < 0$. Since BFS takes time linear to its input size and solving a set of linear equations takes time at most cubic in the number of variables, the time complexity of this algorithm is $O(|S|^3)$, where S is the set of states of $\mathcal{M}_W$.

6   Probabilistic Piecewise Constant Derivative Systems

In this section, we present the details of the Probabilistic Piecewise Constant Derivative Systems (PPCD) and provide a characterization of absolute and almost sure stability by a reduction to that of DTMCs.

6.1   Formal Definition of PPCD

We model PPCDs as consisting of a discrete set of modes, each associated with an invariant and probabilistic transitions between modes that are enabled at the boundaries of the invariants.

Definition 11 (PPCD). *The Probabilistic Piecewise Constant Derivative System (PPCD) is defined as the tuple $\mathcal{H} := (Q, \mathcal{X}, Inv, Flow, Edges)$ where*

- *Q is the set of discrete locations,*
- *$\mathcal{X} = \mathbb{R}^n$ is the continuous state space for some $n \in \mathbb{N}$,*
- *$Inv : Q \rightarrow Poly(n)$ is the invariant function which assigns a positive scaling invariant polyhedral subset of the state space to each location $q \in Q$,*
- *$Flow : Q \rightarrow \mathcal{X}$ is the Flow function which assigns a flow vector, say $Flow(q) \in \mathcal{X}$, to each location $q \in Q$,*
- *$Edges \subseteq Q \times (\cup_{q \in Q} \mathbb{F}(Inv(q))) \times Dist(Q)$ is the probabilistic edge relation such that $(q, f, \rho) \in Edges$ where for every (q, f), there is a at most one ρ such that $(q, f, \rho) \in Edges$ and $f \in \mathbb{F}(Inv(q))$. f is called a Guard of the location q.*

Next, we discuss the semantics of the PPCD. An execution starts from a location $q_0 \in Q$ and some continuous state $x_0 \in \mathcal{X}$ and evolves continuously for some time T according to the dynamics of q_0 until it reaches a facet f_0 of the invariant of q_0. Then a probabilistic discrete transition is taken if there is an edge

(q_0, f_0, ρ_0) and the state q_0 is probabilistically changed to q_1 with probability $\rho_0(q_1)$. The execution (tree) continues with alternating continuous and discrete transitions.

Formally, for any two continuous states $x_1, x_2 \in \mathcal{X}$ and $q \in Q$, we say that there is a *continuous transition* from x_1 to x_2 with respect to q if $x_1, x_2 \in Inv(q)$, there exists $T \geq 0$ such that $x_2 = x_1 + Flow(q) \cdot T$, $x_1 + Flow(q) \cdot t \notin \partial(Inv(q_0))$ for any $0 \leq t < T$ and $x_2 \in \partial(Inv(q_0))$. We note that there is a unique continuous transition from any state (q, x) since it requires the state to evolve until it reaches the boundary for the first time, which corresponds to a unique time of evolution T. Further, if for all $t \geq 0$, $x_1 + Flow(q) \cdot t \in Inv(q)$ then we say x_1 has an infinite edge with respect to q. For two locations $q_1, q_2 \in Q$, we say there is a *discrete transition* from q_1 to q_2 with probability p via $\rho \in Dist(Q)$ and $f \in \mathbb{F}(q_1)$ if $f \subseteq Inv(q_2)$, $(q_1, f, \rho) \in Edges$ and $p = \rho(q_2)$.

We capture the semantics of a PPCD using a WDTMC, wherein we combine a continuous transition and a discrete transition to represent a probabilistic transition of the DTMC. In addition, to reason about convergence, we also need to capture the relative distance of the states from the equilibrium point, which is captured using edge weights. Let us fix 0 as the equilibrium point for the rest of the section. The weight on a transition from (q_1, x_1) to (q_2, x_2) captures the logarithm of the relative distance of x_1 and x_2 from 0, that is, it is $(\|x_2\|/\|x_1\|)$, where $\|x\|$ captures the distance of state x from 0.

Definition 12 (Semantics of PPCD). *Given a PPCD $\mathcal{H}$, we can construct the WDTMC $\mathcal{M}_\mathcal{H} := (S_\mathcal{H}, P_\mathcal{H}, W_\mathcal{H})$ where,*

- *$S_\mathcal{H} = Q \times \mathcal{X}$*
- *$P_\mathcal{H}$ and $W_\mathcal{H}$ are defined as follows for any (q_1, x_1) and (q_2, x_2):*
  - *If there is a continuous transition from x_1 to x_2 with respect to q_1 and there is a discrete transition from q_1 to q_2 with probability p via some $\rho \in Dist(Q)$ and $f \in \mathbb{F}(q_1)$, and $x_2 \in f$, then $P_\mathcal{H}((q_1, x_1), (q_2, x_2)) = p$ and $W_\mathcal{H}((q_1, x_1), (q_2, x_2)) = \log(\|x_2\|/\|x_1\|)$*
  - *If x_1 has an infinite edge with respect to q_1, then $P_\mathcal{H}((q_1, x_1), (q_2, x_2)) = 1$ if $(q_1, x_1) = (q_2, x_2)$ and 0, otherwise, and $W_\mathcal{H}((q_1, x_1), (q_1, x_1)) = \infty$.*
  - *Otherwise, $P_\mathcal{H}((q_1, x_1), (q_2, x_2)) = W_\mathcal{H}((q_1, x_1), (q_2, x_2)) = 0$.*

Since all executions of the PPCD $\mathcal{H}$ start from location q_0 and state x_0, we consider only those paths of the semantics $\mathcal{M}_\mathcal{H}$ which start from (q_0, x_0) and denote them as $Paths(\mathcal{M}_\mathcal{H})$. We say a path $\sigma = (q_0, x_0), (q_1, x_1), \ldots$ converges to 0 if norm of the corresponding state sequence $\|x_0\|, \|x_1\|, \ldots$ converges to 0. Stability of a PPCD $\mathcal{H}$ is defined in terms of convergence of paths of its semantics $\mathcal{M}_\mathcal{H}$ as follows,

Definition 13 (Stability of PPCD). *A PPCD $\mathcal{H}$ is called absolutely stable if every path of $\mathcal{M}_\mathcal{H}$ converges to 0. Analogously, $\mathcal{H}$ is called almost surely stable if any path of $\mathcal{M}_\mathcal{H}$ converges to 0 with probability 1, i.e.,*

$$Pr\{\sigma \in Paths(\mathcal{M}_\mathcal{H}) : \sigma \text{ converges to } 0\} = 1.$$

We now characterize stability of a PPCD $\mathcal{H}$ in terms of its semantics $\mathcal{M}_{\mathcal{H}}$. Basically we state that, $\mathcal{H}$ is absolutely (almost surely) stable iff $\mathcal{M}_{\mathcal{H}}$ is absolutely (almost surely) convergent.

Theorem 4 (Characterization of Stability). *A PPCD $\mathcal{H}$ is absolutely stable iff its semantics $\mathcal{M}_{\mathcal{H}}$ is absolutely convergent and it is almost surely stable iff $\mathcal{M}_{\mathcal{H}}$ is almost surely convergent.*

Proof. Note that, a path σ of $\mathcal{M}_{\mathcal{H}}$ converges to 0 iff $W(\sigma)$ diverges to $-\infty$. To observe this, let $\sigma = (q_0, x_0), (q_1, x_1), \ldots$. Then, $\|x_0\|, \|x_1\|, \ldots$ converge to 0 iff,

$$\lim_{n \to \infty} \frac{\|x_n\|}{\|x_0\|} = 0 \quad [\text{since } \|x_0\| \neq 0]$$

$$\Longleftrightarrow \lim_{n \to \infty} \frac{\|x_1\|}{\|x_0\|} \cdot \frac{\|x_2\|}{\|x_1\|} \cdots \frac{\|x_n\|}{\|x_{n-1}\|} = 0$$

$$[\text{since } \|x_i\| \neq 0 \text{ for all } i = 1, \ldots, n-1 \text{ if } \sigma \text{ is infinite}]$$

$$\Longleftrightarrow \lim_{n \to \infty} \log \left(\frac{\|x_1\|}{\|x_0\|} \cdot \frac{\|x_2\|}{\|x_1\|} \cdots \frac{\|x_n\|}{\|x_{n-1}\|} \right) = -\infty$$

$$\Longleftrightarrow \lim_{n \to \infty} \log \left(\frac{\|x_1\|}{\|x_0\|} \right) + \log \left(\frac{\|x_2\|}{\|x_1\|} \right) + \cdots \log \left(\frac{\|x_n\|}{\|x_{n-1}\|} \right) = -\infty$$

$$\Longleftrightarrow \lim_{n \to \infty} W(\sigma[1 : n]) = -\infty$$

Thus, every infinite path of $\mathcal{M}_{\mathcal{H}}$ converges to 0 iff weight of every infinite path diverges to $-\infty$ and the set of infinite paths of $\mathcal{M}_{\mathcal{H}}$ converging to 0 has probability 1 iff the set of infinite paths having weight diverging to $-\infty$ has probability 1. In other words, $\mathcal{H}$ is absolutely (almost surely) stable iff $\mathcal{M}_{\mathcal{H}}$ is absolutely (almost surely) convergent. □

6.2 Stability of Planar PPCD

In general, semantics of a PPCD has infinite number of states and thus the algorithms developed in Sect. 5.3 cannot be applied to decide absolute (almost sure) convergence of the semantics. However, if the continuous state space of a PPCD $\mathcal{H}$ is $\mathbb{R}^2$, then we can reduce $\mathcal{M}_{\mathcal{H}}$ to a finite WDTMC that provides an exact characterization of $\mathcal{M}_{\mathcal{H}}$. A PPCD with $\mathcal{X} = \mathbb{R}^2$ is called a planar PPCD. Since for each location q, $Inv(q)$ is positively scaled, the facets of $Inv(q)$ are rays emanating from origin. Given constant flow for each location q, a continuous transition starting at a point of some facet $f_1 \in \cup_{q \in Q} \mathbb{F}(Inv(q))$ ends up at a unique point of a unique facet $f_2 \in \cup_{q \in Q} \mathbb{F}(Inv(q))$. This property is not observed if the continuous state space is of three or higher dimensions (Fig. 2). Also, if two continuous transitions start from different points x_1, x_1' of the same facet f_1, they end up in unique points x_2, x_2' (respectively) of a unique facet f_2 such that $\|x_2\|/\|x_1\| = \|x_2'\|/\|x_1'\|$. This gives us the following lemma,

Lemma 4. *Let $e = ((q_1, x_1), (q_2, x_2))$, $e' = ((q_1, x_1'), (q_2, x_2'))$ be two edges of $\mathcal{M}_{\mathcal{H}}$ (where $\mathcal{H}$ is a planar PPCD) such that, $P_{\mathcal{H}}(e), P_{\mathcal{H}}(e') > 0$, and $x_1, x_1' \in f_1$ where $f_1 \in \cup_{q \in Q} \mathbb{F}(Inv(q))$. Then $P_{\mathcal{H}}(e) = P_{\mathcal{H}}(e')$ and $W_{\mathcal{H}}(e) = W_{\mathcal{H}}(e')$.*

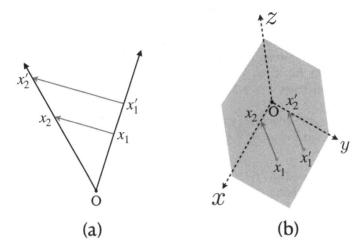

Fig. 2. (a) In $\mathbb{R}^2$, continuous transition with constant rate starting from any point in a facet leads to a unique point in a unique facet. (b) In $\mathbb{R}^3$, even with constant rate, continuous transitions starting from different points in the same facet may end up in different facets.

A proof of Lemma 4 is provided in Appendix A.4. For the rest of the section, we will assume all paths of the semantics $\mathcal{M}_\mathcal{H}$ of a planar PPCD $\mathcal{H}$ start at (q_0, x_0) and $x_0 \in f_0$, where f_0 is a facet in $\cup_{q \in Q} \mathbb{F}(Inv(q))$.

We now define the quotient of a planar PPCD $\mathcal{H}$, which is a finite WDTMC having the same convergence properties as $\mathcal{M}_\mathcal{H}$. Here we consider the set of states as $Q \times \cup_{q \in Q} \mathbb{F}(Inv(q))$ instead of $Q \times \mathcal{X}$ and use Lemma 4 to define the probabilistic edges and their weights.

Definition 14 (Quotient of PPCD). *Let $\mathcal{H}$ be a planar PPCD and $\mathcal{M}_\mathcal{H}$ be its semantics. We define the WDTMC $\mathcal{H}^{red} = (S^{red}, P^{red}, W^{red})$ as follows,*

- $S^{red} = Q \times \bigcup_{q \in Q} \mathbb{F}(Inv(q))$
- $P^{red}((q_1, f_1), (q_2, f_2)) = P_\mathcal{H}((q_1, x_1), (q_2, x_2))$ *for some $x_1 \in f_1$ and $x_2 \in f_2$ such that $P_\mathcal{H}((q_1, x_1), (q_2, x_2)) > 0$, and 0 otherwise.*
- $W^{red}((q_1, f_1), (q_2, f_2)) = W_\mathcal{H}((q_1, x_1), (q_2, x_2))$ *for some $x_1 \in f_1$ and $x_2 \in f_2$ such that $P_\mathcal{H}((q_1, x_1), (q_2, x_2)) > 0$, and 0 otherwise.*

The above definition is well-defined, that is, the choice of x_1 and x_2 do not matter due to Lemma 4.

We will eventually prove that a planar PPCD $\mathcal{H}$ is absolutely (almost surely) stable if and only if its quotient WDTMC $\mathcal{H}^{red}$ is absolutely (almost surely) convergent. First, let us show that for every infinite path σ of $\mathcal{M}_\mathcal{H}$, there is a path π in $\mathcal{H}^{red}$ with same weight and vice versa.

Lemma 5 (Conservation of weight). *For every infinite path σ of $\mathcal{M}_\mathcal{H}$, there is a path π in $\mathcal{H}^{red}$ such that $W(\sigma) = W(\pi)$ and vice versa.*

Proof. ($\Rightarrow$) Let $\sigma = \sigma_1, \sigma_2, \ldots$ be an infinite path of $\mathcal{M}_{\mathcal{H}}$. By assumption, $\sigma_i \in f_i$ where $f_i \in \bigcup_{q \in Q} \mathbb{F}(q)$ is a facet, for each $i \in \mathbb{N}$. Suppose for each i, $\sigma_i = (q_i, x_i)$. Since for each i, there is an edge between (q_i, x_i) and (q_{i+1}, x_{i+1}) in $\mathcal{M}_{\mathcal{H}}$, there should be an edge between (q_i, f_i) and (q_{i+1}, f_{i+1}) in $\mathcal{H}^{red}$. Using Lemma 4 we can conclude that for all i, $W((q_i, x_i), (q_{i+1}, x_{i+1})) = W((q_i, f_i), (q_{i+1}, f_{i+1}))$. Thus we can construct the infinite path $\pi = ((q_1, f_1), (q_2, f_2), \ldots)$ such that $W(\pi) = W(\sigma)$.

($\Leftarrow$) To prove the converse, we show by induction that for any $n \in \mathbb{N}$ if there is a path π of length n in $\mathcal{H}^{red}$ then there is a path σ of length n in $\mathcal{M}_{\mathcal{H}}$ with same weight as π.

Base Case: Suppose $((q_1, f_1), (q_2, f_2))$ is an edge of $\mathcal{H}^{red}$. Then there exist $x_1 \in f_1$ and $x_2 \in f_2$ such that $x_2 = x_1 + Flow(q_1) \cdot t$, for some $t \geq 0$, i.e., there is an edge between (q_1, x_1) and (q_2, x_2) in $\mathcal{M}_{\mathcal{H}}$. Also by Lemma 4, $W((q_1, x_1), (q_2, x_2)) = W((q_1, f_1), (q_2, f_2))$. Hence base case is proved.

Now suppose $((q_1, f_1), \ldots, (q_n, f_n), (q_{n+1}, f_{n+1}))$ is a path of $\mathcal{H}^{red}$ and by induction hypothesis we have a path $((q_1, x_1), \ldots, (q_n, x_n))$ in $\mathcal{M}_{\mathcal{H}}$ such that $W((q_1, f_1), \ldots, (q_n, f_n)) = W((q_1, x_1), \ldots, (q_n, x_n))$. Since there is an edge between (q_n, f_n) and (q_{n+1}, f_{n+1}), there exist $x'_n \in f_n$ and $x'_{n+1} \in f_{n+1}$ such that

$$x'_{n+1} = x'_n + Flow(q_n) \cdot t \tag{2}$$

for some $t \geq 0$. Since $\mathcal{X} = \mathbb{R}^2$, f_n and f_{n+1} are rays. By Eq. 2, there is a straight line of slope $Flow(q_n)$ that intersects both of them. But then any straight line with slope $Flow(q_n)$ intersecting f_n will also intersect f_{n+1}, in fact, if we take the straight line with slope $Flow(q_n)$ passing through x_n, it will intersect f_{n+1}. That means there exists $t \geq 0$ and $x_{n+1} \in f_{n+1}$ such that $x_{n+1} = x_n + Flow(q_n) \cdot t$. This is because for t to be negative, f_n and f_{n+1} must intersect and x_n and x'_n must lie on opposite sides of this intersection point on f_n. But this is impossible since f_n and f_{n+1} intersect only at 0 and both of them get terminated at 0. Thus there exist $x_{n+1} \in f_{n+1}$ such that $((q_n, x_n), (q_{n+1}, x_{n+1}))$ is an edge of $\mathcal{M}_{\mathcal{H}}$. By Lemma 4, $W((q_n, x_n), (q_{n+1}, x_{n+1})) = W((q_n, f_n), (q_{n+1}, f_{n+1}))$. Hence our claim is proved for all $n \in \mathbb{N}$, i.e., it holds for infinite paths of $\mathcal{H}^{red}$ as well. $\square$

Using Lemma 5, we now prove the main theorem which states that a PPCD is absolutely (almost surely) stable if and only if its quotient WDTMC is absolutely (almost surely) stable.

Theorem 5. *A planar PPCD $\mathcal{H}$ is absolutely (almost surely) stable iff its quotient WDTMC $\mathcal{H}^{red}$ is absolutely (almost surely) convergent.*

Proof. A PPCD $\mathcal{H}$ is absolutely stable iff $\mathcal{M}_{\mathcal{H}}$ is absolutely convergent (Theorem 4). By Lemma 5, it is easy to observe that every infinite path of $\mathcal{M}_{\mathcal{H}}$ diverge to $-\infty$ if and only if every infinite path of $\mathcal{H}^{red}$ diverge to $-\infty$. Thus, we can conclude that $\mathcal{H}$ is absolutely stable if and only if $\mathcal{H}^{red}$ is absolutely stable.

On the other hand, a PPCD $\mathcal{H}$ is almost surely stable iff $\mathcal{M}_{\mathcal{H}}$ is almost surely convergent (Theorem 4). Let us show that $\mathcal{M}_{\mathcal{H}}$ is almost surely convergent iff $\mathcal{H}^{red}$ is almost surely convergent. Since we have assumed that all paths of $\mathcal{M}_{\mathcal{H}}$

start from (q_0, x_0), all paths of $\mathcal{H}^{red}$ will start from (q_0, f_0), where f_0 is the facet containing x_0. Let us define the equivalence relation $\sim$ on the set of states of the WDTMC $\mathcal{M}_\mathcal{H} \sqcup \mathcal{H}^{red}$ as,

$$(q_i, x_i) \sim (q_j, x_j) \text{ if } q_i = q_j \text{ and } x_i, x_j \text{ belong to the same facet}$$
$$(q_i, x_i) \sim (q_j, f_j) \text{ if } q_i = q_j \text{ and } x_i \in f_j$$
$$(q_i, f_i) \sim (q_j, f_j) \text{ if } q_i = q_j \text{ and } f_i = f_j,$$

where $q_i, q_j \in Q$, $x_i, x_j \in \mathcal{X}$ and $f_i, f_j \in \cup_{q \in Q} \mathbb{F}(Inv(q))$. Note that, the set of equivalence classes of $\sim$ is given by $\{(q, f) \mid q \in Q, f \in \mathbb{F}(Inv(q))\}$. Now by Lemma 4, we can easily deduce that $\sim$ is a probabilistic bisimulation on $\mathcal{M}_\mathcal{H} \sqcup \mathcal{H}^{red}$. Observe that, the set

$$\Pi = \{\pi \in Paths(\mathcal{M}_\mathcal{H} \sqcup \mathcal{H}^{red}) : W(\pi[1 : \infty]) = -\infty\}$$

is $\sim$ bisimulation-closed. To see this, take any $\pi \in \Pi$ and $\tilde{\pi} \sim \pi$. By Lemma 4, $W(\pi_i, \pi_{i+1}) = W(\tilde{\pi}_i, \tilde{\pi}_{i+1})$ for all i. Thus, $W(\tilde{\pi}[1 : \infty]) = -\infty$ as well, i.e., $\tilde{\pi} \in \Pi$. Now, we have $Pr_{(q_0, x_0)}(\Pi) = Pr_{(q_0, f_0)}(\Pi)$ as a direct consequence of Lemma 1, i.e.,

$$Pr\{\sigma \in Paths(\mathcal{M}_\mathcal{H}) \mid \sigma_1 = (q_0, x_0) \text{ and } W(\sigma) \text{ diverges to } -\infty\}$$
$$= Pr\{\pi \in Paths(\mathcal{H}^{red}) \mid \pi_1 = (q_0, f_0) \text{ and } W(\pi) \text{ diverges to } -\infty\}.$$

Hence, $Pr\{\sigma \in Paths(\mathcal{M}_\mathcal{H}) \mid \sigma_1 = (q_0, x_0) \text{ and } W(\sigma) \text{ diverges to } -\infty\} = 1$ if and only if $Pr\{\pi \in Paths(\mathcal{H}^{red}) \mid \pi_1 = (q_0, f_0) \text{ and } W(\pi) \text{ diverges to } -\infty\} = 1$, i.e., $\mathcal{M}_\mathcal{H}$ is almost surely convergent iff $\mathcal{H}^{red}$ is almost surely convergent. Thus, $\mathcal{H}$ is almost surely stable iff $\mathcal{H}^{red}$ is almost surely convergent. $\square$

Since $\mathcal{H}^{red}$ is finite, we can use the algorithms developed in Sect. 5.3 to decide its absolute (almost sure) convergence. This in turn decides absolute (almost sure) stability of $\mathcal{H}$ by Theorem 5.

7 Conclusion

In this paper, we showed the decidability of absolute and almost sure convergence of Planar Probabilistic Piecewise Constant Derivative Systems (PPCD), that are a practically useful subclass of stochastic hybrid systems and can model motion of planar robots with faulty actuators. We give a computable characterization of absolute and almost sure convergence through a reduction to a finite state DTMC. In the future, we plan to extend these ideas to analyze higher dimensions PPCD and SHS with more complex dynamics. In particular, the idea of reduction can be applied to higher dimensional PPCD but we will need to extend our analysis to a Markov Decision Process that will appear as the reduced system.

A   Appendix

A.1   Proof of Proposition 1

Here we provide a detailed proof of Proposition 1 which states that,
For any finite path σ of $\mathcal{M}_W$ there exist a simple path $\sigma_s \in \mathcal{SP}(\mathcal{M}_W)$ and a set of simple cycles $\mathcal{SC}_\sigma \subseteq \mathcal{SC}(\mathcal{M}_W)$ such that $W(\sigma) = W(\sigma_s) + \sum_{\mathcal{C} \in \mathcal{SC}_\sigma} W(\mathcal{C})$.

Proof. We traverse σ and whenever a cycle $\mathcal{C}$ is encountered, remove its edges from σ and add the cycle to the set $\mathcal{SC}_\sigma$. This process is repeated until $\mathcal{SC}_\sigma$ contains only simple cycles and the remaining edges of σ form a simple path $\sigma_s = \sigma - (\cup\{\mathcal{C} \mid \mathcal{C} \in \mathcal{SC}_\sigma\})$. Let $\mathcal{E}^{\sigma_s}$ denote the set of edges of σ_s and for each $\mathcal{C} \in \mathcal{SC}_\sigma$, $\mathcal{E}^{\mathcal{C}}$ denote the set of edges of $\mathcal{C}$. Clearly, $\{\mathcal{E}^{\sigma_s}\} \cup \{\mathcal{E}^{\mathcal{C}} \mid \mathcal{C} \in \mathcal{SC}_\sigma\}$ is a partition of the set of edges of σ. Thus $W(\sigma) = W(\sigma_s) + \sum_{\mathcal{C} \in \mathcal{SC}_\sigma} W(\mathcal{C})$. Hence, our claim is proved.   □

A.2   Algorithms from Sect. 5.3

Based on the discussions of Sect. 5.3, we provide pseudocodes for algorithms for checking absolute (almost sure) convergence of a finite (finite, irreducible and aperiodic) WDTMC.

Algorithm 1. Checking absolute convergence of WDTMC

Input: A WDTMC $\mathcal{M}_W := (S, P, W)$
Output: Yes/No
1: Convert $\mathcal{M}_W$ to a weighted graph $G = (V, E, W')$ where,
   $V = S$, $E = \{(s_1, s_2) \in S \times S \mid P(s_1, s_2) > 0\}$,
   and $W' : E \to \mathbb{R}$ defined as $W'(e) := -W(e)$
2: Run BFS on G to check existence of edge with weight $-\infty$
3: **if** (edge with $-\infty$ weight exists) **then**
4:     Return No
5: **end if**
6: Run Bellman-Ford algorithm on G
7: **if** (cycle with negative weight is found) **then**
8:     Return No
9: **else**
10:     Let $d : V \to \mathbb{R}_{\geq 0}$ define the shortest distance of each $v \in V$ from s_{init}
11:     Mark in E all edges (u, v) such that $d(v) = d(u) + W'(u, v)$
12:     Delete from G all unmarked edges
13:     Run DFS on G (with unmarked edges deleted) to check for a cycle
14:     **if** (a cycle is found) **then**
15:         Return No
16:     **else**
17:         Return Yes
18:     **end if**
19: **end if**

Algorithm 2. Checking almost sure convergence of WDTMC

Input: A WDTMC $\mathcal{M}_W := (S, \mathrm{P}, \mathrm{W})$
Output: Yes/No

1: Convert $\mathcal{M}_W$ to a weighted graph $G = (V, E, W')$ where,
   $V = S$, $E = \{(s_1, s_2) \in S \times S \mid \mathrm{P}(s_1, s_2) > 0\}$,
   and $W' : E \to \mathbb{R}$ defined as $W'(e) := \mathrm{W}(e)$
2: Run BFS on G to check existence of edge with weight ∞
3: **if** (edge with ∞ weight exists) **then**
4:     Return No
5: **end if**
6: Calculate stationary distribution ρ^* of $\mathcal{M}_W$ by solving the set of linear equations,

$$\rho^*(s) = \sum_{s' \in S} \rho^*(s') \mathrm{P}(s', s), \quad \forall s \in S$$

$$\sum_{s \in S} \rho^*(s) = 1$$

7: $asWeight \leftarrow 0$
8: **for** $e \in E$ **do**
9:     $asWeight = asWeight + P_{\rho^*}(e)W'(e)$
10: **end for**
11: **if** $asWeight < 0$ **then**
12:     Return Yes
13: **else**
14:     Return No
15: **end if**

A.3   Proof of Lemma 2

We prove Lemma 2 here which essentially states that,
For any edge $e \in \mathcal{E}$ of a WDTMC $\mathcal{M}_W$,

$$Pr\left\{ \sigma \in Paths(\mathcal{M}_W) : \lim_{n \to \infty} \frac{\sum_{j=1}^{n} X_j^e}{n} = P_{\rho^*}(e) \right\} = 1,$$

Proof. Construct the DTMC $\mathcal{M}' = (S', \mathrm{P}')$ from $\mathcal{M}_W$, where $S' = S \cup \mathcal{E}$ and for each $e = (s, s') \in \mathcal{E}$, $(s, e), (e, s') \in \mathcal{E}'$ with $\mathrm{P}'(s, e) = \mathrm{P}(s, s')$ and $\mathrm{P}'(e, s') = 1$ ($\mathcal{E}'$ is the set of edges of $\mathcal{M}'$). Note that, there is a one to one correspondence between $Paths(\mathcal{M}_W)$ and $Paths(\mathcal{M}')$, where each edge $e = (s, s')$ in $\sigma \in Paths(\mathcal{M}_W)$ is replaced by consecutive edges (s, e) and (e, s') in the corresponding path $\sigma' \in Paths(\mathcal{M}')$. Thus, $(\sigma_j, \sigma_{j+1}) = e$ if and only if $\sigma'(2j) = e$, where σ' is the corresponding path of σ. Now, let us define random variables $\{Y_j^x \mid x \in S'; j \in \mathbb{N}\}$ as,

$$Y_j^x = \begin{cases} 1 & \text{if } \sigma_j' = x \\ 0 & \text{else} \end{cases}$$

for $\sigma' \in Paths(\mathcal{M}')$. Then, it is easy to observe that, $\sum_{j=1}^{n} X_j^e = \sum_{j=1}^{2n} Y_{2j}^e$. Note that, $\mathcal{M}'$ is finite and irreducible. Hence, by strong law of large numbers for any

$x \in S'$ [27],

$$\lim_{n \to \infty} \frac{\sum_{j=1}^{2n} Y_j^x}{2n} = \rho^{*\prime}(x) \text{ almost surely,}$$

where $\rho^{*\prime}$ is the stationary distribution of $\mathcal{M}'$. Since for any $x \in S'$,

$$\lim_{n \to \infty} \frac{\sum_{j=1}^{2n} Y_{2j}^x}{2n} = \lim_{n \to \infty} \frac{\sum_{j=1}^{2n} Y_j^x}{2n}$$

$$\text{Thus, } \lim_{n \to \infty} \frac{\sum_{j=1}^{n} X_j^e}{n} = 2 \left(\lim_{n \to \infty} \frac{\sum_{j=1}^{2n} Y_{2j}^e}{2n} \right) = 2\rho^{*\prime}(e) \text{ almost surely.} \quad (3)$$

Consider $\rho : S' \to [0, 1]$ as

$$\rho(x) = \begin{cases} \frac{\rho^*(x)}{2} & \text{if } x \in S \\ \frac{P(x)\rho^*(s)}{2} & \text{if } x = (s, s') \in \mathcal{E}. \end{cases}$$

where ρ^* is the stationary distribution of $\mathcal{M}_W$. Let us observe that, $\sum_{x \in S'} \rho(x) = \sum_{s \in S} \rho^*(s)/2 + \sum_{s \in S} \sum_{(s,s') \in \mathcal{E}} \rho^*(s)P(s, s')/2 = 1$, i.e., ρ is a probability distribution.

Note that, for any $x \in S$,

$$\sum_{x' \in S'} \rho(x')P'(x', x) = \sum \{\rho(e)P'(e, x) : e = (s', x) \in \mathcal{E}\}$$

$$= \sum \left\{ \frac{P(e)\rho^*(x)}{2} : e = (s', x) \in \mathcal{E} \right\}$$

$$= \frac{\rho^*(x)}{2} = \rho(x).$$

And for any $x = (s, s') \in \mathcal{E}$,

$$\sum_{x' \in S'} \rho(x')P'(x', x) = \rho(s)P'(s, x) = \frac{\rho^*(s)}{2} \cdot P(x) = \rho(x).$$

Thus, for all $x \in S'$, $\rho(x) = \sum_{x' \in S'} \rho(x')P'(x', x)$, i.e., ρ is a stationary distribution for $\mathcal{M}'$. Since $\mathcal{M}'$ is finite and irreducible, it has a unique stationary distribution. Thus, $\rho = \rho^{*\prime}$, which ultimately provides for any $e = (s, s') \in \mathcal{E}$,

$$\lim_{n \to \infty} \frac{\sum_{j=1}^{n} X_j^e}{n} = 2 \left(\frac{P(e)\rho^*(s)}{2} \right) \text{ almost surely} \quad [\text{by Eq. 3}]$$

$$= \rho^*(s)P(e) \text{ almost surely}$$

$$= P_{\rho^*}(e) \text{ almost surely,}$$

This proves Lemma 2.   □

A.4   Proof of Lemma 4

We prove Lemma 4 here which states the following,
Let $e = ((q_1, x_1), (q_2, x_2))$, $e' = ((q_1, x_1'), (q_2, x_2'))$ be two edges of $\mathcal{M}_\mathcal{H}$ (where $\mathcal{H}$ is a planar PPCD) such that, $P_\mathcal{H}(e), P_\mathcal{H}(e') > 0$, and $x_1, x_1' \in f_1$ where $f_1 \in \bigcup_{q \in Q} \mathbb{F}(Inv(q))$. Then $P_\mathcal{H}(e) = P_\mathcal{H}(e')$ and $W_\mathcal{H}(e) = W_\mathcal{H}(e')$.

Proof. Since continuous state space of $\mathcal{H}$ is $\mathbb{R}^2$, there is a unique facet f_2 for f_1 such that $x_2, x_2' \in f_2$ (assuming $W_\mathcal{H}(e), W_\mathcal{H}(e') \neq \infty$). Now, since $P_\mathcal{H}(e)$ and $P_\mathcal{H}(e')$ depend only on q_1 and f_2, $P_\mathcal{H}(e) = P_\mathcal{H}(e')$. Since any facet is a ray emanating from the origin, it can be depicted by the formula $y = kx$, where $k \in \mathbb{R}$. Let $x_1 = (x_1[1], x_1[2])$ and $x_2 = (x_2[1], x_2[2])$. By property of PPCD, $x_2 = x_1 + Flow(q_1) \cdot T$ for some $T \geq 0$. Thus,

$$(x_2[1], x_2[2]) = (x_1[1], x_1[2]) + (Flow(q_1)[1]), Flow(q_1)[2])T \tag{4}$$

Let $f_1 : y = k_1 x$ and $f_2 : y = k_2 x$. So,

$$x_2[2] = k_2 \cdot x_2[1] \tag{5}$$
$$x_1[2] = k_1 \cdot x_1[1] \tag{6}$$

Using Eqs. 4, 5, 6 we can write $x_2[1] = c \cdot x_1[1]$ where c depends on k_1, k_2, $Flow(q_1)[1]$ and $Flow(q_1)[2]$. Thus $\frac{\|x_2\|}{\|x_1\|}$ can also be written in terms of k_1, k_2, $Flow(q_1)[1]$ and $Flow(q_1)[2]$ since $\frac{\|x_2\|}{\|x_1\|}$ is equal to either $|x_2[2]|/|x_1[2]|$ or $|x_2[2]|/|x_1[1]|$ or $|x_2[1]|/|x_1[2]|$ or $|x_2[1]|/|x_1[1]|$ and x_1 and x_2 dependent terms on numerator and denomenator always cancel off each other. Same is true for e' as well. Thus, $W_\mathcal{H}(e), W_\mathcal{H}(e')$ depend only on q, f_1 and f_2 and not on the points x_1, x_1', x_2, x_2'. Hence, they must be equal.   □

References

1. Alur, R., Dang, T., Ivanci, F.: Counter-example guided predicate abstraction of hybrid systems. In: Garavel, H., Hatcliff, J. (eds.) TACAS 2003. LNCS, vol. 2619, pp. 208–223. Springer, Heidelberg (2003). https://doi.org/10.1007/3-540-36577-X_15
2. Asarin, E., Dang, T., Girard, A.: Hybridization methods for the analysis of non-linear systems. Acta Inform. **43**, 451–476 (2007). https://doi.org/10.1007/s00236-006-0035-7
3. Baier, C., Katoen, J.P.: Principles of Model Checking (Representation and Mind Series). The MIT Press, Cambridge (2008)
4. Branicky, M.: Multiple Lyapunov functions and other analysis tools for switched and hybrid systems. IEEE Trans. Autom. Control **43**(4), 475–482 (1998). https://doi.org/10.1109/9.664150
5. Chen, X., Ábrahám, E., Sankaranarayanan, S.: Taylor model flowpipe construction for non-linear hybrid systems. In: 2012 IEEE 33rd Real-Time Systems Symposium, pp. 183–192 (2012). https://doi.org/10.1109/RTSS.2012.70

6. Cheng, P., Deng, F.: Almost sure exponential stability of linear impulsive stochastic differential systems. In: Proceedings of the 31st Chinese Control Conference, pp. 1553–1557. IEEE (2012)

7. Cheng, P., Deng, F., Yao, F.: Almost sure exponential stability and stochastic stabilization of stochastic differential systems with impulsive effects. Nonlinear Anal. Hybrid Syst. **30**, 106–117 (2018)

8. Clarke, E., et al.: Abstraction and counterexample-guided refinement in model checking of hybrid systems. Int. J. Found. Comput. Sci. **14**(04), 583–604 (2003)

9. Davrazos, G., Koussoulas, N.: A review of stability results for switched and hybrid systems. In: Mediterranean Conference on Control and Automation. Citeseer (2001)

10. Dierks, H., Kupferschmid, S., Larsen, K.G.: Automatic abstraction refinement for timed automata. In: Raskin, J.-F., Thiagarajan, P.S. (eds.) FORMATS 2007. LNCS, vol. 4763, pp. 114–129. Springer, Heidelberg (2007). https://doi.org/10.1007/978-3-540-75454-1_10

11. Do, K.D., Nguyen, H.: Almost sure exponential stability of dynamical systems driven by lévy processes and its application to control design for magnetic bearings. Int. J. Control **93**(3), 599–610 (2020)

12. Duggirala, P.S., Mitra, S.: Abstraction refinement for stability. In: 2011 IEEE/ACM Second International Conference on Cyber-Physical Systems, pp. 22–31. IEEE (2011)

13. Duggirala, P.S., Mitra, S.: Lyapunov abstractions for inevitability of hybrid systems. In: Proceedings of the 15th ACM international conference on Hybrid Systems: Computation and Control, pp. 115–124 (2012)

14. Henzinger, T.A., Kopke, P.W., Puri, A., Varaiya, P.: What's decidable about hybrid automata? J. Comput. Syst. Sci. **57**(1), 94–124 (1998). https://doi.org/10.1006/jcss.1998.1581, https://www.sciencedirect.com/science/article/pii/S0022000098915811

15. Hu, L., Mao, X.: Almost sure exponential stabilisation of stochastic systems by state-feedback control. Automatica **44**(2), 465–471 (2008)

16. Kleinberg, J., Tardos, E.: Algorithm Design. Addison-Wesley, Boston (2006)

17. Lal, R., Prabhakar, P.: Hierarchical abstractions for reachability analysis of probabilistic hybrid systems. In: 2018 56th Annual Allerton Conference on Communication, Control, and Computing (Allerton), pp. 848–855. IEEE (2018)

18. Lal, R., Prabhakar, P.: Counterexample guided abstraction refinement for polyhedral probabilistic hybrid systems. ACM Trans. Embedded Comput. Syst. **18**(5s), 1–23 (2019)

19. Liberzon, D.: Switching in Systems and control. Springer, Boston (2003). https://doi.org/10.1007/978-1-4612-0017-8

20. Podelski, A., Wagner, S.: Model checking of hybrid systems: from reachability towards stability. In: Hespanha, J.P., Tiwari, A. (eds.) HSCC 2006. LNCS, vol. 3927, pp. 507–521. Springer, Heidelberg (2006). https://doi.org/10.1007/11730637_38

21. Podelski, A., Wagner, S.: A sound and complete proof rule for region stability of hybrid systems. In: Bemporad, A., Bicchi, A., Buttazzo, G. (eds.) HSCC 2007. LNCS, vol. 4416, pp. 750–753. Springer, Heidelberg (2007). https://doi.org/10.1007/978-3-540-71493-4_76

22. Prabhakar, P., Duggirala, S., Mitra, S., Viswanathan, M.: Hybrid automata-based CEGAR for rectangular hybrid automata. Citeseer (2013). http://www.its.caltech.edu/pavithra/Papers/rtss2012tr.pdf

23. Prabhakar, P., Garcia Soto, M.: Abstraction based model-checking of stability of hybrid systems. In: Sharygina, N., Veith, H. (eds.) CAV 2013. LNCS, vol. 8044, pp. 280–295. Springer, Heidelberg (2013). https://doi.org/10.1007/978-3-642-39799-8_20

24. Prabhakar, P., Viswanathan, M.: On the decidability of stability of hybrid systems. In: Proceedings of the 16th International Conference on Hybrid Systems: Computation and Control, pp. 53–62 (2013)

25. Prabhakar, P., Vladimerou, V., Viswanathan, M., Dullerud, G.E.: Verifying tolerant systems using polynomial approximations. In: 2009 30th IEEE Real-Time Systems Symposium, pp. 181–190 (2009). https://doi.org/10.1109/RTSS.2009.28

26. Prajna, S., Jadbabaie, A.: Safety verification of hybrid systems using barrier certificates. In: Alur, R., Pappas, G.J. (eds.) HSCC 2004. LNCS, vol. 2993, pp. 477–492. Springer, Heidelberg (2004). https://doi.org/10.1007/978-3-540-24743-2_32

27. Ross, S.M.: Chapter 4 - Markov chains. In: Ross, S.M. (ed.) Introduction to Probability Models, 10th edn., pp. 191–290. Academic Press, Boston (2010). https://doi.org/10.1016/B978-0-12-375686-2.00009-1

28. Rutten, J.J., Kwiatkowska, M., Norman, G., Parker, D.: Mathematical Techniques for Analyzing Concurrent and Probabilistic Systems (Monograph series. No. 23), American Mathematical Society (2004)

29. Teel, A.R., Subbaraman, A., Sferlazza, A.: Stability analysis for stochastic hybrid systems: a survey. Automatica 50(10), 2435–2456 (2014). https://doi.org/10.1016/j.automatica.2014.08.006 https://www.sciencedirect.com/science/article/pii/S0005109814003070

30. Van Der Schaft, A.J., Schumacher, J.M.: An introduction to hybrid dynamical systems, vol. 251. Springer, London (2000). https://doi.org/10.1007/BFb0109998

31. Wang, B., Zhu, Q.: Asymptotic stability in distribution of stochastic systems with semi-Markovian switching. Int. J. Control 92(6), 1314–1324 (2019)

32. Yuan, C., Mao, X.: Asymptotic stability in distribution of stochastic differential equations with Markovian switching. Stoch. Process. Their Appl. 103(2), 277–291 (2003)

Tools

LCRL: Certified Policy Synthesis via Logically-Constrained Reinforcement Learning

Mohammadhosein Hasanbeig[1]($\boxtimes$), Daniel Kroening[2], and Alessandro Abate[1]

[1] Computer Science Department, University of Oxford,
Oxford, UK
{hosein.hasanbeig,alessandro.abate}@cs.ox.ac.uk
[2] Amazon Inc., London, UK
daniel.kroening@magd.ox.ac.uk

Abstract. LCRL is a software tool that implements model-free Reinforcement Learning (RL) algorithms over unknown Markov Decision Processes (MDPs), synthesising policies that satisfy a given linear temporal specification with maximal probability. LCRL leverages partially deterministic finite-state machines known as Limit Deterministic Büchi Automata (LDBA) to express a given linear temporal specification. A reward function for the RL algorithm is shaped on-the-fly, based on the structure of the LDBA. Theoretical guarantees under proper assumptions ensure the convergence of the RL algorithm to an optimal policy that maximises the satisfaction probability. We present case studies to demonstrate the applicability, ease of use, scalability and performance of LCRL. Owing to the LDBA-guided exploration and LCRL model-free architecture, we observe robust performance, which also scales well when compared to standard RL approaches (whenever applicable to LTL specifications). Full instructions on how to execute all the case studies in this paper are provided on a GitHub page that accompanies the LCRL distribution www.github.com/grockious/lcrl.

Keywords: Model-Free Reinforcement Learning · Policy Synthesis · Linear Temporal Logic · Limit Deterministic Büchi Automata

1 Introduction

Markov Decision Processes (MDPs) are extensively used for problems in which an agent needs to control a process by selecting actions that are allowed at the process' states and that affect state transitions. Decision making problems in MDPs are equivalent to resolving action non-determinism, and result in policy synthesis problems. Policies are synthesised to maximise expected long-term rewards

This work is in part supported by the HiClass project (113213), a partnership between the Aerospace Technology Institute (ATI), Department for Business, Energy and Industrial Strategy (BEIS) and Innovate UK.

obtained from the process. This paper introduces a new software tool, LCRL, which performs policy synthesis for unknown MDPs when the goal is that of maximising the probability to abide by a task (or constraint) that is specified using Linear Temporal Logic (LTL). LTL is a formal, high-level, and intuitive language to describe complex tasks [9]. In particular, unlike static (space-dependent) rewards, LTL can describe time-dependent and complex non-Markovian tasks that can be derived from natural languages [16,36,46]. Any LTL specification can be translated efficiently into a Limit-Deterministic Büchi Automaton (LDBA), which allows LCRL to automatically shape a reward function for the task that is later employed by the RL learner for optimal policy synthesis. LCRL is implemented in Python, the *de facto* standard programming language for machine learning applications.

1.1   Related Work

There exists a few tools that solve control (policy) synthesis in a model-free fashion, but not under full LTL specifications. One exception is the work in [6] which proposes an interleaved reward and discounting mechanism. However, the reward shaping dependence on the discounting mechanism can make the reward sparse and small, which might negatively affect convergence. The work in [17] puts forward a tool for an average-reward scheme based on earlier theoretical work. Other model-free approaches with available code-bases are either (1) focused on fragments of LTL and classes of regular languages (namely finite-horizon specs) or (2) cannot deal with unknown black-box MDPs. The proposed approach in [29,30] presents a model-free RL solution but for regular specifications that are expressed as deterministic finite-state machines. The work in [10,11] takes a set of LTL$_f$/LDL$_f$ formulae interpreted over finite traces as constraints, and then finds a policy that maximises an external reward function. The VSRL software tool [12–14,28] solves a control synthesis problem whilst maintaining a set of safety constraints during learning.

1.2   Contributions

The LCRL software tool has the architecture presented in Fig. 1, and presents the following features:

- LCRL leverages **model-free RL algorithms**, employing only traces of the system (assumed to be an unknown MDP) to formally synthesise optimal policies that satisfy a given LTL specification with maximal probability. LCRL finds such policies by learning over a set of traces extracted from the MDP under LTL-guided exploration. This efficient, guided exploration is owed to reward shaping based on the automaton [18–21,23,26]. The guided exploration enables the algorithm to focus only on relevant parts of the state/action spaces, as opposed to traditional Dynamic Programming (DP) solutions, where the Bellman iteration is exhaustively applied over the whole state/action spaces [5]. Under standard RL convergence assumptions, the LCRL output is an optimal policy whose traces satisfy the given LTL specification with **maximal** probability.

- LCRL is **scalable** owing to LTL-guided exploration, which allows LCRL to cope and perform efficiently with MDPs whose state and action spaces are significantly large. There exist a few LDBA construction algorithms for LTL, but not all of resulting automata can be employed for quantitative model-checking and probabilistic synthesis [31]. The succinctness of the construction proposed in [39], which is used in LCRL, is another contributing factor to LCRL scalability. The scalability of LCRL is evaluated in an array of numerical examples and benchmarks including high-dimensional Atari 2600 games [3,7].
- LCRL is the first RL synthesis method for LTL specifications in **continuous state/action** MDPs. So far no tool is available to enable RL, whether model-based or model-free, to synthesise policies for LTL on continuous-state/action MDPs. Alternative approaches for continuous-space MDPs [1,34,41,44] discretise the model into a finite-state MDP, or alternatively propose a DP-based method with value function approximation [15].
- LCRL displays **robustness** features to hyper-parameter tuning. Specifically, we observed that LCRL results, although problem-specific, are not significantly affected when hyper-parameters are not tuned with care.

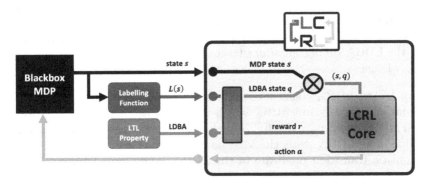

Fig. 1. The LCRL architecture: the inputs to the tool are the environment (MDP) - in particular its states s and corresponding labels $L(s)$ - as well as the LDBA generated from the user-defined LTL specification. The MDP (with state s) and the LDBA (with state q) are synchronised on-the-fly, thus generating the pair (s,q). A reward r is then automatically generated by LCRL, based on the environment label $L(s)$ and the automaton state q; actions a are selected accordingly and passed back to the environment (MDP). (Color figure online)

2   Logically-Constrained Reinforcement Learning (LCRL)

We assume the unknown environment is encompassed by an MDP, which in this work is a discrete-time stochastic control processes defined as the tuple $\mathfrak{M} = (\mathcal{S}, \mathcal{A}, s_0, P)$ over a set of continuous states $\mathcal{S} = \mathbb{R}^n$, and where $\mathcal{A} = \mathbb{R}^m$ is a set of continuous actions, and $s_0 \in \mathcal{S}$ is an initial state. $P : \mathcal{B}(\mathcal{S}) \times \mathcal{S} \times \mathcal{A} \to [0,1]$ is a conditional transition kernel which assigns to any pair comprising a state $s \in S$ and an action $a \in A$ a probability measure $P(\cdot|s,a)$ on properly measurable sets on S [4]. A finite state/action MDP is a special case in which $|S| < \infty$, $|A| < \infty$, and $P : S \times A \times S \to [0,1]$ is a transition probability matrix assigning a conditional probability to enter sets of states in S. A variable $R(s,a) \sim \Upsilon(\cdot|s,a) \in \mathcal{P}(\mathbb{R}^+)$ is defined over the MDP $\mathfrak{M}$, representing the reward obtained when action a is taken at a given state s, sampled from the reward distribution Υ defined over the set of probability distributions $\mathcal{P}(\mathbb{R}^+)$ on subsets of $\mathbb{R}^+$.

LCRL is a policy synthesis architecture for tasks that are expressed as specifications in LTL [18–26]. The LCRL Core in Fig. 1 is compatible with any general RL scheme that conforms with the environment state and action spaces. Inside the LCRL module the MDP and LDBA states are synchronised, resulting in an on-the-fly product MDP. Intuitively, the product MDP encompasses the extra dimension of the LDBA states, which is added to the state space of the original MDP. The role of the added dimension is to track the sequence of labels that have been read across episodes, and thus to act as a memory register for the given task. This allows to evaluate the (partial) satisfaction of the corresponding temporal property. More importantly, this synchronisation breaks down the non-Markovian LTL specification into a set of Markovian reach-avoid components, which facilitates the RL convergence to a policy whose traces satisfy the overall LTL specification. In practice, no product between the MDP and LDBA is computed: the LDBA simply *monitors* traces executed by the agents as episodes of the RL scheme.

Remark 1. The LDBA construction inherently introduces limited form of non-determinism, called ε-transitions, which is treated as an extra action over the original MDP action space [39,40]. Namely, when there exists a non-deterministic transition in an LDBA state, the MDP action space is augmented with the non-deterministic transition predicate of the LDBA. These non-deterministic transitions are automatically handled by LCRL during the learning and appropriate on-the-fly modifications are carried out, so that the RL agent can learn to deal with those non-deterministic transitions in order to reach the accepting conditions of the LDBA. We emphasise that the underlying assumption in LCRL is that the MDP model is unknown (Fig. 1), and thus a single state is obtained as output when given a state and an action as input.   □

LCRL defines a reward function R for the RL Core, whose objective is to maximise the expected discounted return [42]:

$$\mathbb{E}^\pi[\sum_{n=0}^{\infty} \gamma^n R(s_n, a_n)|s_0 = s], \tag{1}$$

where $\mathbb{E}^\pi[\cdot]$ denotes the expected value given that the agent follows the policy $\pi : S \times A \to [0,1]$ from state s; parameter $\gamma \in [0,1]$ is a discount factor; and

$s_0, a_0, s_1, a_1...$ is the sequence of state/action pairs, initialised at $s_0 = s$. This reward is intrinsically defined over the product MDP, namely it is a function of the MDP state (describing where the agent is in the environment) and the sate of the automaton (encompassing partial task satisfaction). For further details on the LCRL reward shaping, please refer to [18–21, 23, 26].

The discount factor γ affects the optimality of the synthesised policy and has to be tuned with care. There is standard work in RL on state-dependent discount factors [6, 35, 37, 45, 47], which is shown to preserve convergence and optimality guarantees. For LCRL the learner discounts the received reward whenever it's positive, and leaves it un-discounted otherwise:

$$\gamma(s) = \begin{cases} \eta & \text{if } R(s,a) > 0, \\ 1 & \text{otherwise,} \end{cases} \tag{2}$$

where $0 < \eta < 1$ is a constant [20, 47]. Hence, (1) reduces to an expected return that is bounded, namely

$$\mathbb{E}^\pi[\sum_{n=0}^{\infty} \gamma(s_n)^{N(s_n)} R(s_n, \pi(s_n))|s_0 = s], \ 0 < \gamma(s) \le 1, \tag{3}$$

where $N(s_n)$ is the number of times a positive reward has been observed at state s_n.

For any state $s \in \mathcal{S}$ and any action $a \in \mathcal{A}$, LCRL assigns a quantitative action-value $Q : \mathcal{S} \times \mathcal{A} \to \mathbb{R}$, which is initialised with an arbitrary and finite value over all state-action pairs. As the agent starts learning, the action-value $Q(s, a)$ is updated by a linear combination between the current $Q(s, a)$ and the target value:

$$R(s,a) + \gamma \max_{a' \in \mathcal{A}} Q(s', a'),$$

with the weight factors $1 - \mu$ and μ respectively, where μ is the learning rate.

An optimal stationary Markov policy synthesised by LCRL on the product MDP that maximises the expected return, is guaranteed to induce a finite-memory policy on the original MDP that maximises the probability of satisfying the given LTL specification [20]. Of course, in finite-state and -action MDPs, the set of stationary deterministic policies is finite and thus after a finite number of learning steps RL converges to an optimal policy. However, when function approximators are used in RL to tackle extensive or even infinite-state (or -action) MDPs, such theoretical guarantees are valid only asymptotically [21, 24].

2.1 Installation

LCRL can be set up by the `pip` package manager as easy as:

```
pip install lcrl
```

This allows to readily import LCRL as a package into any **Python** project

```
>>> import lcrl
```

Table 1. List of hyper-parameters and features that can be externally selected

Hyper-parameter	Default Value	Description
algorithm	'ql'	RL algorithm underlying LCRL Core, selected between (cf. Table 2): - 'ql': Q-learning, - 'nfq': Neural Fitted Q-iteration, - 'ddpg': Deep Deterministic Policy Gradient
episode_num	2500	number of learning episodes
iteration_num_max	4000	max number of iterations/steps within each episode
discount_factor	0.95	discounting coefficient η as in (2)
learning_rate	0.9	learning rate parameter μ
epsilon	0.1	value for epsilon-greedy exploration (= 0 for fully greedy)
test	true	run of closed-loop simulations to test the generated policy
save_dir	'./results'	directory address for saving the results
average_window	-1	number of episodes for moving-average window for plots (default value -1 for 30% of episode_num)

and employ its modules. Alternatively, the provided setup file found within the distribution package will automatically install all the required dependencies. The installation setup has been tested successfully on Ubuntu 18.04.1, macOS 11.6.5, and Windows 11.

2.2 Input Interface

LCRL training module lcrl.src.train inputs two main objects (cf. Fig. 1): an **MDP** black-box object that generates training episodes; and an **LDBA** object; as well as learning hyper-parameters[1] that are listed in Table 1.

MDP: An MDP is an object with internal attributes that are a priori unknown to the agent, namely the state space, the transition kernel, and the labelling function (respectively denoted by $\mathcal{S}$, P, and L). The states and their labels are observable upon reaching. To formalise the agent-MDP interface we adopt a scheme that is widely accepted in the RL literature [7]. In this scheme the learning agent can invoke the following methods from any state of the MDP:

- reset(): this resets the MDP to its initial state. This allows the agent to start a new learning episode whenever necessary.
- step(action): the MDP step function takes an action (the yellow signal in Fig. 1) as input, and outputs a new state, i.e. the black signal in Fig. 1.

A number of well-known MDP environments (e.g., the stochastic grid-world) are embedded as classes within LCRL, and can be found within the module lcrl.src.environments. Most of these classes can easily set up an MDP object. However, note that the state signal output by the step function needs to be fed

---

[1] These parameters are called hyper-parameters since their values are used to control the learning process. This is unlike other parameters, such as weights and biases in neural networks, which are set and updated automatically during the learning phase.

to a labelling function state_label(state), which outputs a list of labels (in string format) for its input state (in Fig. 1, the black output signal from the MDP is fed to the blue box, or labelling function, which outputs the set of label). For example, state_label(state) = ['safe', 'room1']. The labelling function state_label(state) can then be positioned outside of the MDP class, or it can be an internal method in the MDP class. The built-in MDP classes in lcrl.src.environments module have an empty state_label(state) method that are ready to be overridden at the instance level:

```
1   # create a SlipperyGrid object
2   gridworld_1 = SlipperyGrid()
3
4   # "state_label" function outputs the label of input state
5   # (input: state, output: string label)
6   def state_label(self, state):
7       # defines the labelling image
8       labels = np.empty([gridworld_1.shape[0], gridworld_1.shape[1]], dtype=object)
9       labels[0:40, 0:40] = 'safe'
10      labels[25:33, 7:15] = 'unsafe'
11      labels[7:15, 25:33] = 'unsafe'
12      labels[15:25, 15:25] = 'goal1'
13      labels[33:40, 0:7] = 'goal2'
14      # returns the label associated with input state
15      return labels[state[0], state[1]]
16
17  # now override the step function
18  SlipperyGrid.state_label = state_label.__get__(gridworld_1, SlipperyGrid)
```

Listing 1.1. Example of state_label(state) specification in the MDP object lcrl.src.environments.gridworld_1.

LDBA: An LDBA object is an instance of the lcrl.src.automata.ldba class. This class is structured according to the automaton construction in [39], and it encompasses modifications dealing with non-determinism, as per Remark 1. The LDBA initial state is numbered as 0, or can alternatively be specified using the class attribute initial_automaton_state once an LDBA object is created. The LDBA non-accepting sink state is numbered as -1. Finally, the set of accepting sets, on which we elaborate further below, has to be specified at the instance level by configuring accepting_sets (Listing 1.2 line 1). The key interface methods for the LDBA object are:

- accepting_frontier_function(state): this automatically updates an internal attribute of an LDBA class called accepting_sets. This is a list of accepting sets of the LDBA, e.g. $\mathcal{F} = \{F_1, ..., F_f\}$. For instance, if the set of LDBA accepting sets is $\mathcal{F} = \{\{3,4\}, \{5,6\}\}$ then this attribute is a list of corresponding state numbers accepting_sets = [[3,4],[5,6]]. As discussed above, the accepting_sets has to be specified once the LDBA class is instanced (Listing 1.2 line 1). The main role of the accepting frontier function is to determine if an accepting set can be reached, so that a corresponding reward is given to the agent (cf. red signal in Fig. 1). Once an accepting set is visited it will be temporarily removed from the accepting_sets until the agent visits all the accepting sets within accepting_sets. After that, accepting_sets is reset to the original list. To set up an LDBA class in LCRL the user needs to specify accepting_sets for the LDBA. LCRL then automatically shapes the reward function and calls the accepting_frontier_function whenever necessary. Further details on the accepting_frontier_function and the accepting_sets can be found in [18–21,23,26].

– step(label): LDBA step function takes a label set, i.e. the blue signal in Fig. 1, as input and outputs a new LDBA state. The label set is delivered to the step function by LCRL. The step method is empty by default and has to be specified manually after the LDBA class is instanced (Listing 1.2 line 5).

– reset(): this method resets the state and accepting_sets to their initial assignments. This corresponds to the agent starting a new learning episode.

```
1    goal1_or_goal2 = LDBA(accepting_sets=[[1, 2]])
2
3    # "step" function for the automaton transitions
4    # (input: label, output: automaton_state, non-accepting sink state is "-1")
5    def step(self, label):
6    # state 0
7    if self.automaton_state == 0:
8    if 'epsilon_1' in label:
9    self.automaton_state = 1
10   elif 'epsilon_2' in label:
11   self.automaton_state = 2
12   elif 'unsafe' in label:
13   self.automaton_state = -1  # non-accepting sink state
14   else:
15   self.automaton_state = 0
16   # state 1
17   elif self.automaton_state == 1:
18   if 'goal1' in label and 'unsafe' not in label:
19   self.automaton_state = 1
20   else:
21   self.automaton_state = -1  # non-accepting sink state
22   # state 2
23   elif self.automaton_state == 2:
24   if 'goal2' in label and 'unsafe' not in label:
25   self.automaton_state = 2
26   else:
27   self.automaton_state = -1  # non-accepting sink state
28   # step function returns the new automaton state
29   return self.automaton_state
30
31
32   # now override the step function
33   LDBA.step = step.__get__(goal1_or_goal2, LDBA)
```

Listing 1.2. Example of the specification of the step(label) method in the LDBA object lcrl.automata.goal1_or_goal2 for the LTL specification $(\Diamond \Box goal1 \vee \Diamond \Box goal2) \wedge \Box \neg unsafe$. The non-accepting sink state is numbered as -1.

If the automaton happens to have ε-transitions, e.g. Fig. 2, they have to distinguishable, e.g. numbered. For instance, there exist two ε-transitions in the LDBA in Fig. 2 and each is marked by an integer. Furthermore, the LDBA class has an attribute called epsilon_transitions, which is a dictionary to specify which states in the automaton contain ε-transitions. In Fig. 2, only state 0

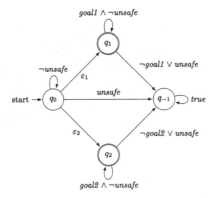

Fig. 2. LDBA for the LTL specification $(\Diamond\Box goal1 \vee \Diamond\Box goal2) \wedge \Box\neg unsafe$.

has outgoing ε-transitions and thus, the attribute `epsilon_transitions` in the LDBA object `goal1_or_goal2` has to be set to

`goal1_or_goal2.epsilon_transitions = {0:['epsilon_0', 'epsilon_1']}`

2.3 Output Interface

LCRL provides the results of learning and testing as `.pkl` files. Tests are closed-loop simulations where we apply the learned policy over the MDP and observe the results. For any selected learning algorithm, the learned model is saved as `learned_model.pkl` and test results as `test_results.pkl`. The instruction on how to load these files is also displayed at the end of training for ease of re-loading data and for post-processing. Depending on the chosen learning algorithm, LCRL generates a number of plots to visualise the learning progress and the testing results. These plots are saved in the `save_dir` directory. The user has the additional option to export a generated animation of the testing progress, LCRL prompts this option to the user following the completion of the test. During the learning phase, LCRL displays the progress in real-time and allows the user to stop the learning task (in an any-time fashion) and save the generated outcomes.

3 Experimental Evaluation

We apply LCRL on a number of case studies highlighting its features, performance and robustness across various environment domains and tasks. All the experiments are run on a standard machine, with an Intel Core i5 CPU at 2.5 GHz and with 20 GB of RAM. The experiments are listed in Table 2 and discussed next.

The `minecraft` environment [2] requires solving challenging low-level control tasks (`minecraft-tX`), and features many sequential goals. For instance, in

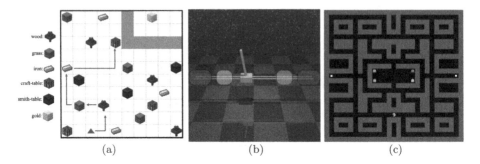

(a)   (b)   (c)

Fig. 3. (a) Synthesised policy by LCRL in `minecraft-t3`; (b) `cart-pole` experiment [43]; (c) `pacman-lrg` - the white square on the left is labelled as food 1 (f_1) and the one on the right as food 2 (f_2), the state of being caught by a ghost is labelled as (g) and the rest of the state space is labelled as neutral (n). (Color figure online)

`minecraft-t3` (Fig. 3a) the agent is tasked with collecting three items sequentially and to reach a final checkpoint, which is encoded as the following LTL specification: $\Diamond(wood \wedge \Diamond(grass \wedge \Diamond(iron \wedge \Diamond(craft\_table))))$, where $\Diamond$ is the known *eventually* temporal operator.

The `mars-rover` problems are realistic robotic benchmarks taken from [21], where the environment features continuous state and action spaces.

The known `cart-pole` experiment (Fig. 3b) [8,24,43] has a task that is expressed by the LTL specification $\Box\Diamond y \wedge \Box\Diamond g \wedge \Box\neg u$, namely, starting the pole in upright position, the goal is to prevent it from falling over ($\Box\neg u$, namely *always not u*) by moving the cart, whilst in particular alternating between the yellow (y) and green (g) regions ($\Box\Diamond y \wedge \Box\Diamond g$), while avoiding the red (unsafe) parts of the track ($\Box\neg u$).

The `robot-surve` example [38] has the task to repeatedly visit two regions (A and B) in sequence, while avoiding multiple obstacles (C) on the way: $\Box\Diamond A \wedge \Box\Diamond B \wedge \Box\neg C$.

Environments `slp-easy` and `slp-hard` are inspired by the widely used stochastic MDPs in [42, Chapter 6]: the goal in `slp-easy` is to reach a particular region of the state space, whereas the goal in `slp-hard` is to visit four distinct regions sequentially in a given order.

The `frozen-lake` benchmarks are adopted from the OpenAI Gym [7]: the first three are reachability problems, whereas the last three require sequential visits of four regions, in the presence of unsafe regions to be always avoided.

Finally, `pacman-sml` and `pacman-lrg` are inspired by the well-known Atari game Pacman, and are initialised in a tricky configuration (`pacman-lrg` as in Fig. 3c), which is likely for the agent to be caught: in order to win the game, the agent has to collect the available tokens (food sources) without being caught by moving ghosts. Formally, the agent is required to choose between one of the two available foods and then find the other one ($\Diamond[(f_1 \wedge \Diamond f_2) \vee (f_2 \wedge \Diamond f_1)]$), while

Table 2. Learning results with **LCRL**. MDP state and action space cardinalities are $|\mathcal{S}|$ and $|\mathcal{A}|$, the number of automaton states in LDBA is denoted by $|\mathcal{Q}|$, the optimal action value function in the initial state is denoted by "**LCRL** $\max_a Q(s_0, a)$", which represents the **LCRL** estimation of the maximum satisfaction probability. For each experiment, the reported result includes the mean and the standard error of ten learning trials with **LCRL**. This probability is also calculated by the PRISM model checker [33] and, whenever the MDP model can be processed by PRISM, it is reported in column "**max sat. prob. at** s_0". The closer "**LCRL** $\max_a Q(s_0, a)$" and "**max sat. prob. at** s_0" the better. Note that for continuous-state-action MDPs the maximum satisfaction probability cannot be precisely computed by model checking tools, unless abstraction approximation techniques are applied, hence "n/a". Furthermore, if the MDP state (or action) space is large enough, e.g. **pacman**, the model checkers tools cannot parse the model and the model checking process times out, i.e. "t/o". The column "**LCRL conv. ep.**" presents the episode number in which **LCRL** converged. Finally, "wall_clocktime" presents the average elapsed real time needed for **LCRL** to converge on a test machine. The rest of the columns provide the values of the hyper-parameters, as described in Table 1.

experiment	MDP $|\mathcal{S}|, |\mathcal{A}|$	LDBA $|\mathcal{Q}|$	LCRL $\max_a$ $Q(s_0, a)$	max sat. prob. at s_0	alg.	episode_ num	iteration_ num_max	discount_ factor*	learning_ rate†	wall_clock time*(min)
minecraft-t1	100, 5	3	0.991 ± 0.009	1	'ql'	500	4000	0.95	0.9	0.1
minecraft-t2	100, 5	3	0.991 ± 0.009	1	'ql'	500	4000	0.95	0.9	0.1
minecraft-t3	100, 5	5	0.993 ± 0.007	1	'ql'	1500	4000	0.95	0.9	0.25
minecraft-t4	100, 5	3	0.991 ± 0.009	1	'ql'	500	4000	0.95	0.9	0.1
minecraft-t5	100, 5	3	0.995 ± 0.005	1	'ql'	500	4000	0.95	0.9	0.1
minecraft-t6	100, 5	4	0.995 ± 0.005	1	'ql'	1500	4000	0.95	0.9	0.25
minecraft-t7	100, 5	5	0.993 ± 0.007	1	'ql'	1500	4000	0.95	0.9	0.5
mars-rover-1	∞, 5	3	0.991 ± 0.002	n/a	'nfq'	50	3000	0.9	0.01	550
mars-rover-2	∞, 5	3	0.992 ± 0.006	n/a	'nfq'	50	3000	0.9	0.01	540
mars-rover-3	∞, ∞	3	n/a	n/a	'ddpg'	1000	18000	0.99	0.05	14
mars-rover-4	∞, ∞	3	n/a	n/a	'ddpg'	1000	18000	0.99	0.05	12
cart-pole	∞, ∞	4	n/a	n/a	'ddpg'	100	10000	0.99	0.02	1
robot-surve	25, 4	3	0.994 ± 0.006	1	'ql'	500	1000	0.95	0.9	0.1
slp-easy-sml	120, 4	2	0.974 ± 0.026	1	'ql'	300	1000	0.99	0.9	0.1
slp-easy-med	400, 4	2	0.990 ± 0.010	1	'ql'	1500	1000	0.99	0.9	0.25
slp-easy-lrg	1600, 4	2	0.970 ± 0.030	1	'ql'	2000	1000	0.99	0.9	2
slp-hard-sml	120, 4	5	0.947 ± 0.039	1	'ql'	500	1000	0.99	0.9	1
slp-hard-med	400, 4	5	0.989 ± 0.010	1	'ql'	4000	2100	0.99	0.9	5
slp-hard-lrg	1600, 4	5	0.980 ± 0.016	1	'ql'	6000	3500	0.99	0.9	9
frozen-lake-1	120, 4	3	0.949 ± 0.050	0.9983	'ql'	400	2000	0.99	0.9	0.1
frozen-lake-2	400, 4	3	0.971 ± 0.024	0.9982	'ql'	2000	2000	0.99	0.9	0.5
frozen-lake-3	1600, 4	3	0.969 ± 0.019	0.9720	'ql'	5000	4000	0.99	0.9	1
frozen-lake-4	120, 4	6	0.846 ± 0.135	0.9728	'ql'	2000	2000	0.99	0.9	1
frozen-lake-5	400, 4	6	0.735 ± 0.235	0.9722	'ql'	7000	4000	0.99	0.9	2.5
frozen-lake-6	1600, 4	6	0.947 ± 0.011	0.9467	'ql'	5000	5000	0.99	0.9	9
pacman-sml	729,000, 5	6	0.290 ± 0.035	t/o‡	'ql'	80e3	4000	0.95	0.9	1600
pacman-lrg	4,251,000, 5	6	0.282 ± 0.049	t/o‡	'ql'	180e3	4000	0.95	0.9	3700

* coefficient η in (2)   † learning rate μ ‡ timed out: too large for model-checking tools   * on a machine running macOS 11.6.5 with Intel Core i5 CPU at 2.5 GHz and with 20 GB of RAM

avoiding the ghosts ($\Box \neg g$). We thus feed to the agent a conjunction of these associations, as the following LTL specification: $\Diamond[(f_1 \wedge \Diamond f_2) \vee (f_2 \wedge \Diamond f_1)] \wedge \Box \neg g$. Standard QL fails to find a policy generating satisfying traces for this experiment.

Table 3. Robustness of LCRL performance against hyper-parameter tuning, for the frozen-lake-1 experiment. Maximum probability of satisfaction is 99.83% as calculated by PRISM (cf. Table 2). The reported values are the percentages of times that execution of LCRL final policy produced traces that satisfied the LTL property. Statistics are taken over 10 trainings and 100 testing for each training, namely 1000 trials for each hyper-parameter configuration.

μ / η	0.2	0.4	0.6	0.8	0.99
0.2	$92.5 \pm 7.5\%$	$96.7 \pm 3.2\%$	$91.3 \pm 8.7\%$	$98.8 \pm 1.1\%$	$94.7 \pm 5.29\%$
0.4	$98.6 \pm 1.4\%$	$89.5 \pm 10.5\%$	$94.5 \pm 5.5\%$	$94.5 \pm 5.5\%$	$99.2 \pm 0.74\%$
0.6	$99.0 \pm 0.83\%$	$94.5 \pm 5.5\%$	$93.3 \pm 6.7\%$	$96.4 \pm 3.59\%$	$93.3 \pm 6.7\%$
0.8	$95.8 \pm 4.2\%$	$99.5 \pm 0.49\%$	$99.5 \pm 0.49\%$	$96.9 \pm 3.09\%$	$97.7 \pm 2.2\%$
0.99	$88.9 \pm 11.09\%$	$98.4 \pm 1.55\%$	$97.1 \pm 2.31\%$	$96.1 \pm 3.73\%$	$95.2 \pm 4.79\%$
overall avg.	**$95.676 \pm 4.268\%$**				

We emphasise that the two tasks in cart-pole and robot-surve are not co-safe, namely require possibly infinite traces as witnesses.

Additionally, we have evaluated the LCRL robustness to RL key hyper-parameter tuning, i.e. discount factor η and learning rate μ, by training the LCRL agent for 10 times and testing its final policy for 100 times. The evaluation results and an overall rate of satisfying the given LTL specifications are reported for the frozen-lake-1 experiments in Table 3. The statistics are taken across 10×100 tests, which results in 1000 trials for each hyper-parameter configuration.

4   Conclusions and Extensions

This paper presented LCRL, a new software tool for policy synthesis with RL under LTL and omega-regular specifications. There is a plethora of extensions that we are planning to explore. In the short term, we intend to: (1) directly interface LCRL with automata synthesis tools such as OWL [32]; (2) link LCRL with other model checking tools such as PRISM [33] and Storm [27]; and (3) embed more RL algorithms for policy synthesis, so that we can tackle policy synthesis problems for more challenging environments. In the longer term, we plan to extend LCRL such that (1) it will be able to handle other forms of temporal logic, e.g., signal temporal logic; and (2) it will have a graphical user-interface for the ease of interaction.

References

1. Abate, A., Prandini, M., Lygeros, J., Sastry, S.: Probabilistic reachability and safety for controlled discrete time stochastic hybrid systems. Automatica **44**(11), 2724–2734 (2008)

2. Andreas, J., Klein, D., Levine, S.: Modular multitask reinforcement learning with policy sketches. In: ICML, vol. 70, pp. 166–175 (2017)
3. Bellemare, M.G., Naddaf, Y., Veness, J., Bowling, M.: The Arcade learning environment: an evaluation platform for general agents. JAIR **47**, 253–279 (2013)
4. Bertsekas, D.P., Shreve, S.: Stochastic Optimal Control: The Discrete-Time Case. Athena Scientific (2004)
5. Bertsekas, D.P., Tsitsiklis, J.N.: Neuro-dynamic Programming, vol. 1. Athena Scientific (1996)
6. Bozkurt, A.K., Wang, Y., Zavlanos, M.M., Pajic, M.: Control synthesis from linear temporal logic specifications using model-free reinforcement learning. arXiv preprint:1909.07299 (2019)
7. Brockman, G., et al.: OpenAI gym. arXiv preprint:1606.01540 (2016)
8. Cai, M., Hasanbeig, M., Xiao, S., Abate, A., Kan, Z.: Modular deep reinforcement learning for continuous motion planning with temporal logic. IEEE Robot. Aut. Lett. **6**(4), 7973–7980 (2021). https://doi.org/10.1109/LRA.2021.3101544
9. Clarke Jr, E.M., Grumberg, O., Kroening, D., Peled, D., Veith, H.: Model Checking. MIT Press, London (2018)
10. De Giacomo, G., Iocchi, L., Favorito, M., Patrizi, F.: Foundations for restraining bolts: Reinforcement learning with LTLf/LDLf restraining specifications. In: ICAPS, vol. 29, pp. 128–136 (2019)
11. Favorito, M.: Reinforcement learning framework for temporal goals. https://github.com/whitemech/temprl (2020)
12. Fulton, N.: Verifiably safe autonomy for cyber-physical systems. Ph.D. thesis, Carnegie Mellon University Pittsburgh (2018)
13. Fulton, N., Platzer, A.: Safe reinforcement learning via formal methods: Toward safe control through proof and learning. In: Proceedings of the AAAI Conference on Artificial Intelligence (2018)
14. Fulton, N., Platzer, A.: Verifiably safe off-model reinforcement learning. In: TACAS, pp. 413–430 (2019)
15. Gordon, G.J.: Stable function approximation in dynamic programming. In: Proceedings of the Twelfth International Conference on Machine Learning, pp. 261–268. Elsevier (1995)
16. Gunter, E.: From natural language to linear temporal logic: Aspects of specifying embedded systems in LTL. In: Workshop on Software Engineering for Embedded Systems: From Requirements to Implementation (2003)
17. Hahn, E.M., Perez, M., Schewe, S., Somenzi, F., Trivedi, A., Wojtczak, D.: Mungojerrie: reinforcement learning of linear-time objectives. arXiv preprint arXiv:2106.09161 (2021)
18. Hasanbeig, M.: Safe and certified reinforcement learning with logical constraints. Ph.D. thesis, University of Oxford (2020)
19. Hasanbeig, M., Abate, A., Kroening, D.: Logically-constrained reinforcement learning. arXiv preprint:1801.08099 (2018)
20. Hasanbeig, M., Abate, A., Kroening, D.: Certified reinforcement learning with logic guidance. arXiv preprint:1902.00778 (2019)
21. Hasanbeig, M., Abate, A., Kroening, D.: Logically-constrained neural fitted Q-iteration. In: AAMAS. pp. 2012–2014. International Foundation for Autonomous Agents and Multiagent Systems (2019)
22. Hasanbeig, M., Abate, A., Kroening, D.: Cautious reinforcement learning with logical constraints. In: AAMAS. International Foundation for Autonomous Agents and Multiagent Systems (2020)

23. Hasanbeig, M., Kantaros, Y., Abate, A., Kroening, D., Pappas, G.J., Lee, I.: Reinforcement learning for temporal logic control synthesis with probabilistic satisfaction guarantees. In: Proceedings of the 58th Conference on Decision and Control, pp. 5338–5343. IEEE (2019)
24. Hasanbeig, M., Kroening, D., Abate, A.: Deep reinforcement learning with temporal logics. In: Bertrand, N., Jansen, N. (eds.) FORMATS 2020. LNCS, vol. 12288, pp. 1–22. Springer, Cham (2020). https://doi.org/10.1007/978-3-030-57628-8_1
25. Hasanbeig, M., Kroening, D., Abate, A.: Towards verifiable and safe model-free reinforcement learning. In: Proceedings of Workshop on Artificial Intelligence and Formal Verification, Logics, Automata and Synthesis (OVERLAY), pp. 1–10. Italian Association for Artificial Intelligence (2020)
26. Hasanbeig, M., Yogananda Jeppu, N., Abate, A., Melham, T., Kroening, D.: DeepSynth: Program synthesis for automatic task segmentation in deep reinforcement learning. In: AAAI Conference on Artificial Intelligence. Association for the Advancement of Artificial Intelligence (2021)
27. Hensel, C., Junges, S., Katoen, J.-P., Quatmann, T., Volk, M.: The probabilistic model checker STORM. Int. J. Softw. Tools Technol. Transfer **22**, 1–22 (2021). https://doi.org/10.1007/s10009-021-00633-z
28. Hunt, N., Fulton, N., Magliacane, S., Hoang, N., Das, S., Solar-Lezama, A.: Verifiably safe exploration for end-to-end reinforcement learning. arXiv preprint arXiv:2007.01223 (2020)
29. Icarte, R.T., Klassen, T., Valenzano, R., McIlraith, S.: Using reward machines for high-level task specification and decomposition in reinforcement learning. In: ICML, pp. 2107–2116 (2018)
30. Jothimurugan, K., Alur, R., Bastani, O.: A composable specification language for reinforcement learning tasks. In: NeurIPS, pp. 13041–13051 (2019)
31. Kini, D., Viswanathan, M.: Optimal translation of LTL to limit deterministic automata. In: Legay, A., Margaria, T. (eds.) TACAS 2017. LNCS, vol. 10206, pp. 113–129. Springer, Heidelberg (2017). https://doi.org/10.1007/978-3-662-54580-5_7
32. Křetínský, J., Meggendorfer, T., Sickert, S.: Owl: a library for ω-words, automata, and LTL. In: Lahiri, S.K., Wang, C. (eds.) ATVA 2018. LNCS, vol. 11138, pp. 543–550. Springer, Cham (2018). https://doi.org/10.1007/978-3-030-01090-4_34
33. Kwiatkowska, M., Norman, G., Parker, D.: PRISM 4.0: a. In: Gopalakrishnan, G., Qadeer, S. (eds.) CAV 2011. LNCS, vol. 6806, pp. 585–591. Springer, Heidelberg (2011). https://doi.org/10.1007/978-3-642-22110-1_47
34. Lee, I.S., Lau, H.Y.: Adaptive state space partitioning for reinforcement learning. Eng. Appl. Artif. Intell. **17**(6), 577–588 (2004)
35. Newell, R.G., Pizer, W.A.: Discounting the distant future: how much do uncertain rates increase valuations? J. Environ. Econ. Manag **46**(1), 52–71 (2003)
36. Nikora, A.P., Balcom, G.: Automated identification of LTL patterns in natural language requirements. In: ISSRE, pp. 185–194. IEEE (2009)
37. Pitis, S.: Rethinking the discount factor in reinforcement learning: a decision theoretic approach. arXiv preprint:1902.02893 (2019)
38. Sadigh, D., Kim, E.S., Coogan, S., Sastry, S.S., Seshia, S.A.: A learning based approach to control synthesis of Markov decision processes for linear temporal logic specifications. In: CDC, pp. 1091–1096. IEEE (2014)
39. Sickert, S., Esparza, J., Jaax, S., Křetínský, J.: Limit-deterministic Büchi automata for linear temporal logic. In: Chaudhuri, S., Farzan, A. (eds.) CAV 2016. LNCS, vol. 9780, pp. 312–332. Springer, Cham (2016). https://doi.org/10.1007/978-3-319-41540-6_17

40. Sickert, S., Křetínský, J.: MoChiBA: probabilistic LTL model checking using limit-deterministic Büchi automata. In: Artho, C., Legay, A., Peled, D. (eds.) ATVA 2016. LNCS, vol. 9938, pp. 130–137. Springer, Cham (2016). https://doi.org/10.1007/978-3-319-46520-3_9

41. Soudjani, S.E.Z., Gevaerts, C., Abate, A.: FAUST^2: Formal Abstractions of Uncountable-STate STochastic processes. In: Baier, C., Tinelli, C. (eds.) TACAS 2015. LNCS, vol. 9035, pp. 272–286. Springer, Heidelberg (2015). https://doi.org/10.1007/978-3-662-46681-0_23

42. Sutton, R.S., Barto, A.G.: Reinforcement Learning: An Introduction, vol. 1. MIT Press, Cambridge (1998)

43. Tassa, Y., et al.: Deepmind control suite. arXiv preprint:1801.00690 (2018)

44. Voronoi, G.: Nouvelles applications des paramètres continus à la théorie des formes quadratiques. Deuxième mémoire. Recherches sur les parallélloèdres primitifs. Journal für die reine und angewandte Mathematik **134**, 198–287 (1908)

45. Wei, Q., Guo, X.: Markov decision processes with state-dependent discount factors and unbounded rewards/costs. Oper. Res. Lett. **39**(5), 369–374 (2011)

46. Yan, R., Cheng, C.H., Chai, Y.: Formal consistency checking over specifications in natural languages. In: Proceedings of the 2015 Design, Automation & Test in Europe Conference & Exhibition, pp. 1677–1682. EDA Consortium (2015)

47. Yoshida, N., Uchibe, E., Doya, K.: Reinforcement learning with state-dependent discount factor. In: ICDL, pp. 1–6. IEEE (2013)

LN: A Meta-solver for Layered Queueing Network Analysis

Giuliano Casale^(✉), Yicheng Gao, Zifeng Niu, and Lulai Zhu

Department of Computing, Imperial College London,
London, UK
{g.casale,y.gao20,zifeng.niu19,
lulai.zhu15}@imperial.ac.uk

Abstract. We overview LN, a novel solver introduced in the LINE software package to analyze layered queueing network (LQN) models. The novelty of the LN solver lies in its capability to analyze LQNs with a user-defined combination of solution paradigms, including discrete-event and stochastic simulation, continuous-time Markov chain analysis (CTMC), normalizing constant evaluation (NC), matrix analytic methods (MAM), mean-field approximations (FLUID), and mean-value analysis (MVA). Being parametric in the solver used for each LQN layer, LN as a whole enables the efficient computation of advanced performance metrics such as marginal and joint state probabilities, response and passage time distributions, and transient measures, leveraging individual strengths of the supported solution paradigms. We discuss in particular recent developments added to NC, the default layer solver of LN, which significantly improve the solution of queueing network models obtained using loose layering of the LQN.

Keywords: Layered queueing networks · Computational algorithms · Class switching · Performance measures

1 Introduction

LINE[1] is an open-source software package for analyzing extended queueing network models [9]. The package implements several tens of solution algorithms grouped into solvers, each embodying a specific paradigm for queueing analysis, either simulation-based or analytical (CTMC, NC, MAM, FLUID, MVA). In this paper, we present the LN solver available within the LINE suite version 2.0, which adds a capability to analyze LQNs, a class of extended queueing networks featuring simultaneous resource possession. LN is the first LQN meta-solver, i.e., it offers the flexibility to parametrically choose any of the aforementioned paradigms to evaluate individual layers that compose an LQN. This feature greatly extends the scope of the original LQN solver available in the first version of LINE [28], which was supporting the solution of each layer based on mean-field

---

[1] http://line-solver.sf.net/.

E. Ábrahám and M. Paolieri (Eds.): QEST 2022, LNCS 13479, pp. 232–254, 2022.
https://doi.org/10.1007/978-3-031-16336-4_12

approximations only. The present paper is the first one to review LQN analysis methods available in the LINE 2.0.x releases.

LINE is open sourced under a permissive BSD-3-Clause license. It is mainly developed in MATLAB, with a few components coded in Java for computational efficiency. A royalty-free Docker image built on the MATLAB compiler runtime is also made available so that end users can run the tool as a service without licensing costs. Some solution methods are implemented based on external solvers that include JMT [2], LQNS [15], BuTools [19], KPC-Toolbox [12], and Q-MAM [3]. In essence, LINE acts as an integration point for multiple queueing analysis tools, providing a common model specification language for their joint use along with its native solvers. For example, an LQN may be analyzed via LQNS for a fast solution, and the result then verified using a slower simulation-based trace-driven execution of the LN solver. This is especially useful in research studies, to detect bugs and compare efficiency of alternative solution methods. JMT is used to visualize models through automated model-to-model transformation and for simulation-based analysis [9]. Besides, LINE is complementary to other efforts to broaden the availability of queueing network algorithms for performance evaluation educators and practitioners, such as Octave queueing [23] and PDQ [18]. Compared to these, LINE adds several advanced algorithms not available in existing software packages.

Related Work. State-of-the-art solvers specific to LQNs include for example LQNS [15] and DiffLQN [33,34]. LQNS is an established solver with a long record of application to real-world software engineering case studies. At heart, the tool applies to the LQN layers approximate mean-value analysis for extended queueing network models [15]. LQN analysis using GreatSPN [1] is also supported via its petrirsrvn tool.

DiffLQN is instead a solver that is based on the mean-field approximation theory developed in the context of PEPA models to scalably analyze LQNs [33,34]. The mean-field fluid paradigm is particularly suited to the solution of large models, as it becomes asymptotically exact in layers with multi-server FCFS stations once the number of jobs and servers grows large in a fixed ratio. Subsequent work on mean-field approximations has further generalized the fluid solutions to processor sharing (PS) stations, class switching, random environments and response time percentiles [28], differentiated service weights [38], multi-class FCFS approximations [9], and mixed models [30].

LN is built around the experience of these LQN solvers, integrating many of the approximate MVA and fluid methods proven effective in the above studies. In addition, it enables the analysis of LQNs using solution paradigms that are uncommon for LQNs, such as continuous-time Markov chains, matrix analytic methods, and normalizing constant evaluation techniques, all of which are not available in existing solvers. As we show later, these paradigms are helpful in computing several LQN metrics that are difficult or impossible to obtain with mean-value analysis or mean-field fluid approximations.

Theoretical Contributions. In developing the LN solver, we have advanced the theory of product-form queueing networks for models consisting of a single infinite-server node and m replicated queueing stations (i.e., having identical service demands in every class). Each station offers multi-class service, according to per-class service time distributions that are possibly load-dependent. We shall refer to such models as *homogeneous layers*, since they naturally arise from a certain LQN decomposition style known as loose layering [16, §3.2], that can be used to analyze arbitrary LQNs. The main theoretical contributions of this paper, are as follows:

- We develop a Gaussian quadrature method for approximating the normalizing constants of homogeneous layers, which leads to a fast computation of their associated performance metrics. By controlling the order of the quadrature, these methods can trade accuracy for speed, while retaining linear worst-case complexity in the total population size.
- We propose a method of moments algorithm for *exactly* solving homogeneous layers in linear time with respect to the total population size when the queueing stations have a single server. We show its ability to handle multi-class models with thousands of jobs in a few milliseconds. As opposed to existing methods such as MoM [7] and CoMoM [6] that can *theoretically* achieve linear complexity, the proposed technique is the first one that realizes this in concrete implementations by avoiding the use of exact arithmetic, which introduces overheads up to about log-linear in the total population size [6,7]. Moreover, it does so without solving systems of linear equations usually appearing in methods of moment algorithms. The result provides efficient approximations for more complex networks with multi-server stations [11,31] and for non-product-form models.
- We derive a related method for marginal probability computations in homogeneous layers featuring quadratic complexity in the total population size. This method also does not require solving systems of linear equations.

We illustrate the application of the above methods to LN and exemplify the other features of the solvers through case studies. In particular, we demonstrate the ability of LN, as a meta-solver, to study performance metrics that cannot be easily analyzed with other LQN solvers, for example integrated models of queueing and caching.

The rest of the paper is organized as follows. Section 2 describes the modeling formalism supported by the LN solver. The overall solution approach and advanced features unique to LN are discussed in Sect. 3. Section 4 elaborates novel solution algorithms offered by the solver. Some case studies are presented in Sect. 5 to illustrate the distinguishing features of LN. Finally, Sect. 6 is dedicated to the conclusions. Proofs of the solution algorithms are given in the Appendix together with an overview of the software architecture of LN.

2   LQN Formalism

LINE offers exact, approximate, asymptotic and simulation-based analysis of open, closed, and mixed multi-class queueing network models. In these models,

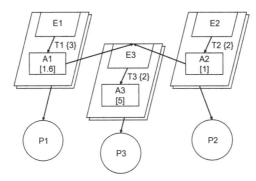

Fig. 1. Example of an LQN model.

jobs are probabilistically routed across a set of nodes, usually queueing sta-tions, where they receive service, typically subject to contention by other jobs. Each job belongs to a class, i.e., a type that defines its service, routing, and arrival characteristics at each node. LINE also supports extensions commonly required in applications such as class switching, non-exponential service times, load dependence, and priorities. The problem is to obtain station and system per-formance measures such as average queue lengths, utilizations, response times, and throughputs/arrival rates.

Among the most feature-rich extended queueing network models is the class of *layered queueing networks* (LQNs), which has found broad application in soft-ware performance engineering [15]. We point the reader to [35] for a comprehen-sive introduction to LQNs and discuss here only the essential concepts. In an LQN, job visits to the system are modeled as directed acyclic graphs that invoke entries exposed by tasks running on host processors. A workflow of one or more activities (i.e., service phases) is executed at each invocation of an entry. This workflow is called the *activity graph* bound to the entry. Within it, an activity may issue a synchronous call to an entry, while keeping a server in the task blocked, leading to simultaneous resource possession. Asynchronous calls are also possible, which behave similarly to job movements in ordinary queueing networks.

Figure 1 shows an example that contains all the basic elements of LQN mod-els: *tasks*, *host processors*, *entries*, and *activities*. Tasks are depicted as stacked parallelograms and their multiplicities are indicated within curly brackets. A task runs on a single processor (e.g., P1), which is represented by a circle. Specific services provided by a task are called entries and drawn as smaller parallelo-grams inside that task. Each rectangle denotes a particular activity performed during the execution of an entry. The number between square brackets specifies the service demand of the activity (e.g., 1.6 for activity A1). Workloads in LQNs are generated by a special task termed the reference task, e.g., tasks T1 and T2 in Fig. 1 which model two classes of users with 3 and 2 jobs each and both call entry E3 of task T3.

Extensive prior work in the area has shown that an LQN can be accurately solved by iteratively evaluating a collection of ordinary mixed queueing networks, obtained via decomposition, until reaching a fixed-point solution. Each decomposed sub-model conceptually represents a layer of the client-server system being modeled [29]. These models interact, in the sense that the outputs parameters of one model (e.g., its response times) may form the input parameters of another model (e.g., its service times). In LINE, a collection of interactive models is referred to as an *ensemble*, which is not restricted to LQNs and can encompass other formalisms such as caching models [17]. For this reason, LN may be seen as a general-purpose layered stochastic network solver.

3   LQN Decomposition and Iterative Solution

This section describes the algorithmic methods underpinning the LN solver. We particularly focus on the strategy to divide a given LQN model into multiple layers and the default, though customizable, solution paradigm applied to analyzing each resultant layer.

3.1   Layering Strategy

Prior art has extensively investigated layering strategies, i.e., methods for generating a decomposition of an LQN model into an ensemble of ordinary queueing networks on which solution algorithms can be instantiated. In the current release, the LN solver adopts loose layering [16, §3.2], which ensures that each layer includes replicated queueing stations (i.e., an LQN task or host processor) coupled with an infinite-server node to model the inter-request times of clients. An exception to this rule is that identical replicas of the queueing station are also generated by LN in the same layer. For example, the model in Fig. 1 features under the loose layering style 4 layers: T1→ P1, T2→P2, T3→P3, and (T1,T2)→T3, where → indicates a client-server relationship.

The rationale for choosing loose layering as the default strategy is that, for a total of m queueing stations, many queueing network analysis methods are computationally more efficient in solving m small models with a single (possibly replicated) queueing station rather than a monolithic model comprising m queueing stations. For example, a CTMC solver may be fairly scalable for single queueing systems, but easily incurs exponential state space explosion for queueing networks. A drawback of loose layering is that heterogeneous load balancing or fork/join sections are challenging to model as the participating queueing stations may be scattered across different layers.

Solutions of interactive models in an ensemble are reconciled through fixed-point iterations until performance metrics across the layers are consistent within a predefined numerical tolerance. To this end, the LQNS solver adopts an "elevator" algorithm whereby the graph that describes the client-server relationships is traversed top down and bottom up in an alternate fashion after topological sorting, thus cyclically inverting the order in which the layers are analyzed [16]. The same algorithm is implemented in LN to iterate over all the models within an ensemble.

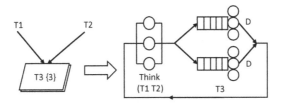

Fig. 2. A layer of the LQN model with replication $m = 2$ and multiplicity $c = 3$.

3.2 Homogeneous Layers

Let us introduce the notation for individual LQN layers obtained by LN through loose layering. These are closed queueing networks with m identical c-server queues and a delay (i.e., infinite-server) node. Jobs in class r have service demand D_r at the multi-server station and think time Z_r at the infinite-server station.

Service distributions are assumed to be of phase type in LN. They include special cases such as the *Disabled* distribution, which allows users to forbid the routing of a class to a station for debugging purposes, and the *Immediate* distribution, which characterizes negligible processing that takes zero time. The handling of the latter is solver-dependent. For example, LINE's CTMC solver applies stochastic complementation to remove the corresponding transitions in an exact fashion [25].

We now outline the mapping between LQN abstractions and product-form models. Under loose layering, a layer l consists of a queueing network with m identical queueing stations, modeling servers in that layer, and a single infinite-server station, modeling the clients. Thus, m denotes the replication factor of the server. We shall refer to such a queueing network as a *homogeneous* model. The number of servers in the queueing stations is equal to the multiplicity c of the task or host processor acting as servers in that layer. Figure 2 gives the model of a layer for the LQN where T3 acts as server.

Clients issuing synchronous calls to layer l are represented as jobs initialized in a *reference class* at the delay. Subsequently, these jobs cycle between the delay and the queues, switching class to represent the specific tasks, entries, and activities that the clients visit (or invoke) during execution. Parameters such as think times and service demands are iteratively updated as per the method of layers [29]. For example, if certain sections of the client workflow require access to another layer $l' \neq l$, the corresponding residence times are modeled as think times that already incorporate the queueing contention in layer l'. Moreover, the probability of a client executing a particular entry is set proportional to the last throughput of this entry. Unlike LQNS, LN updates routing probabilities at each iteration, because not all solution paradigms are visit-based.

Clients that send asynchronous calls to layer l are instead represented as open Poisson arrival streams. Coexistence of open and closed classes therefore gives rise to mixed models, which can be reduced to closed ones by demand scaling [5].

Service classes are mapped to a set of R chains, obtained by computing the strongly connected components of the routing matrix. Each chain j represents a client task to the layer, and has an associated number of jobs N_j that are initialized at the infinite-server station, starting in the chain reference class. Note that this does not loses information as it is possible to exactly recover the per-class performance metrics from the per-chain ones [5,36].

For ease of presentation, since a multi-chain model can always be reduced to a corresponding multi-class model with R classes, one per chain, we shall use the terms "chain" and "class" interchangeably.

3.3 Performance Metrics

Performance metric computation is solver-dependent. We focus here on the default solver used by LN, which is LINE's normalizing constant (NC) solver. We assume for simplicity that scheduling leads to product-form models and single-server stations ($c = 1$). Approximations to handle other cases, such as multi-class FCFS, are discussed later in the paper.

Let $N = (N_1, \ldots, N_R)$ be the population vector for a layer, $|N| = \sum_{r=1}^{R} N_r$, and recall that $G(m, N)$ is the normalizing constant of the state probabilities for the associated product-form model, which consists of m identical queueing stations and an infinite-server node. Denote by 1_r a row vector of all zeros with a one in the r-th dimension. We may exploit the following relations for the mean class-r throughput $X_r(N)$ and for the mean class-r queue length $Q_r(N)$ at any of the identical queueing stations:

$$X_r(N) = \frac{G(m, N - 1_r)}{G(m, N)} \qquad Q_r(N) = D_r \frac{G(m + 1, N - 1_r)}{G(m, N)}$$

The system throughput $X_r(N)$ is assumed to be computed at a reference station for which we set the mean number of visits of class r to unity. Little's law may then be combined with the previous relations to obtain other metrics such as mean response times and resource utilizations [14,22].

Before discussing the novel algorithms integrated in LN to compute these metrics, we remark that specific simplifications arise in evaluating normalizing constants for homogeneous layers due to the structure of the product-form solutions. At first, if either $Z_r = 0 \wedge D_r = 0$ or $N_r = 0$ holds for a class r, then this class can be removed from the model as it does not contribute to the normalizing constant. Define $\mathcal{R}_D$ as the set of remaining classes for which $D_r = 0$. We note that the contribution of such classes to $G(m, N)$ is given exactly by a factor $\prod_{s \in \mathcal{R}_D} Z_s^{N_s}/N_s!$. Hence, every model can be reduced without loss of generality to one where all classes have $D_r > 0$, which we will assume throughout.

3.4 Advanced LN Features

We now briefly overview advanced features and LQN extensions supported by LN, which are to the best of our knowledge unique to this solver.

Caching Layers. We have made an extension to the LQN formalism, enabling the inclusion of *cache* nodes. When visiting a cache, a job reads an item according to some probability, resulting in either a cache hit or a cache miss. The subsequent processing activities can depend on whether the read outcome was a hit or miss. Cache reads also activate a replacement policy (e.g., random replacement, FIFO, LRU) to evict infrequently used items. The LN solver features the ability to define caches in an LQN using specialized tasks and entries named CacheTask and ItemEntry, which capture the data access requirements of jobs traversing the LQN. More details can be found in [17] and in Sect. 5.

Multi-chain Joint and Marginal Probabilities. The LN solver allows the computation of joint and marginal state probabilities in each layer, leveraging the ability of the NC solver to evaluate normalizing constants. This makes it possible to obtain probabilistic measures, which may be useful for example in parameter inference and buffer overflow analysis.

Multi-chain Transient Analysis. With the FLUID solver, one can compute transient metrics and passage time distributions in each layer. This solver performs mean-field approximations for PS nodes based on the theory presented in [28]. As mentioned before, FCFS stations are also treated as PS nodes with service demands corrected through a hybrid $M/G/k$-diffusion approximation [9].

Response Time Distributions. Recently, we have demonstrated the possibility to couple LN with mixture density networks (MDNs) for response time distribution analysis [27]. The MDN-based approach considerably increases the precision of computing response time percentiles for LQNs compared to analytical approximations, which are notoriously difficult for multi-chain networks.

4   Novel Algorithms

LN's default layer solver, NC, implements state-of-the-art exact and approximate methods for normalizing constant analysis of mixed queueing networks. Historically, such methods were replaced in the early years of performance evaluation by exact and approximate MVA algorithms to overcome intrinsic numerical instabilities arising from the use of normalizing constants. However, recently developed techniques for computing normalizing constants exhibit superior complexity to their corresponding MVA counterparts, prompting a reconsideration of these methods, as we discuss throughout.

Particularly in the context of loose layering, we show that the normalizing constant numerical instabilities can be circumvented through appropriate scalings or log-sum-exp approximations [4], and propose several exact and asymptotic solution methods for queueing networks, that are not available in the traditional MVA framework.

4.1   Solving Homogeneous Layers with the CoMoM Algorithm

Recall that mixed queueing network models can be mapped with suitable transformation to a model consisting only of closed classes [5, §8.2.3]. On this basis,

LN analyzes layers by default using the NC solver, which implements the Class-oriented Method of Moments (CoMoM) algorithm proposed in [6]. For a model with N jobs belonging to a fixed number of R classes, CoMoM implementations require approximately log-quadratic time and log-linear space in $|N|$ to obtain an *exact* solution, thus being more scalable than the exact MVA algorithms. The latter have a time and space complexity of $O(|N|^R)$. Moreover, contrary to other moment-based methods, CoMoM can avoid degeneracies when the model consists of one or more replicated queueing stations, as in the case of loose layering. An equivalent result is not currently available for MVA. Among the complications, it is worth noting that MVA expressions such as the celebrated arrival theorem are bi-linear in their defining terms, mean queue lengths and mean throughputs, yielding systems of non-linear equations that are not as tractable as CoMoM's linear matrix recurrence relation.

Enhancements. LN evolves CoMoM by developing explicit solutions to its system of linear equations for homogeneous models. Such solutions are applicable to models with an arbitrary number of classes R. For a vector v with d dimensions, let $|v| = \sum_{i=1}^{d} v_i$ and define $\mathrm{diag}(v)$ as a diagonal matrix with the elements of v placed on the main diagonal. We give the following exact result.

Theorem 1. *Consider a product-form queueing network model with R classes, having m identical single-server queueing stations with service demand $D_r > 0$ in class r and an infinite-server node with think time Z_r in class r. Define the collection of normalizing constants*

$$g(m, N) = \begin{bmatrix} G(m, N) & G(m, N - 1_1) & \cdots & G(m, N - 1_{R-1}) \end{bmatrix}.$$

and the basis

$$\Lambda(N) = \begin{bmatrix} g(m+1, N) & g(m, N) \end{bmatrix}^T$$

Then the following matrix recurrence relation holds

$$\Lambda(N) = (\boldsymbol{F}_{1,R} + N_R^{-1} \boldsymbol{F}_{2,R}) \Lambda(N - 1_R) \tag{1}$$

in which

$$\boldsymbol{F}_{1,R} = \begin{bmatrix} D_R \boldsymbol{E}_{1,1} & \boldsymbol{0} \\ \boldsymbol{0} & \boldsymbol{0} \end{bmatrix} \quad \boldsymbol{F}_{2,R} = \begin{bmatrix} m D_R \boldsymbol{S} & Z_R \boldsymbol{S} \\ m D_R \boldsymbol{I} & Z_R \boldsymbol{I} \end{bmatrix}$$

$$\boldsymbol{S} = -m^{-1} \begin{bmatrix} -|\tilde{N}| - m & \tilde{Z}^T \\ -\mathrm{diag}(\tilde{D})^{-1} \tilde{N} & \mathrm{diag}(\tilde{D})^{-1} \mathrm{diag}(\tilde{Z}) \end{bmatrix}$$

where $\boldsymbol{E}_{1,1}$ is of order R with a single nonzero entry in position $(1,1)$, $\tilde{N} = (N_1, \ldots, N_{R-1})^T$, $\tilde{D} = (D_1, \ldots, D_{R-1})^T$, $\tilde{Z} = (Z_1, \ldots, Z_{R-1})^T$, $\boldsymbol{I}$ is the identity matrix of order R, and $\boldsymbol{0}$ is the zero matrix of order R.

A proof of the theorem is in the Appendix A. Note in particular that the knowledge of $\Lambda(N)$ and $\Lambda(N - 1_R)$ is sufficient to determine the expression of all the mean performance metrics introduced in Sect. 3.3.

Termination conditions for the matrix recurrence relation (1) are obtained noting that $G(\cdot, 0) = 1$ and, whenever any element of N is negative, $G(\cdot, N) = 0$.

Table 1. Relative error and runtime upon computing $\log G(m, N)$ exactly.

Classes	Total jobs	Method	Runtime [s]
8	40	Convolution	0.0033
8	40	CoMoM (original)	0.0047
8	40	CoMoM (enhanced)	0.0014
8	400	Convolution	1.4201
8	400	CoMoM (original)	1.1433
8	400	CoMoM (enhanced)	0.0016
8	4000	Convolution	Memory exhausted
8	4000	CoMoM (original)	Timeout
8	4000	CoMoM (enhanced)	0.0017
8	10^6	Convolution	Memory exhausted
8	10^6	CoMoM (original)	Timeout
8	10^6	CoMoM (enhanced)	0.2591

Numerical Stabilization. In principle, to prevent floating-point range exceptions, the proposed solution can either be computed using exact or multi-precision arithmetic. In practice, we have observed that scaling at each step the vector $\Lambda(N)$ so that $|\Lambda(N)| = 1$, and removing the effect of such scaling only in the final result, is sufficient to sanitize numerical problems in practical uses, except for negligible numerical fluctuations. This makes the theoretical and implementation complexity identical and, empirically, much faster than using exact arithmetic.

To illustrate this, we consider a challenging model with $R = 8$ classes, where $m = 1$, $Z_r = r$ and $D_r = 10^{-r}$, $r = 1, \ldots, R$. Jobs are split equally across the classes. The exponential spacing of the demands and the large population make the analysis numerically challenging. We set a timeout of 10 s to solve a model. The original CoMoM leverages exact arithmetic in Java, whereas the enhanced method implements in MATLAB the recursion we have proposed in Theorem 1 using standard floating-point arithmetic. Table 1 shows numerical results, which corroborate the high scalability of the enhanced CoMoM for homogeneous models. Results are obtained on an AMD Ryzen 7 2700X Processor with 64 GB of RAM. Note that at population $|N| = 4000$ convolution becomes unviable due to excessive memory requirements, but since the normalizing constant reaches order 10^{-602} it would have anyway exceeded the floating-point range during execution. Scaling methods for Convolution have been proposed in [21], however it is not difficult in our experience to generate examples of large models where this technique still cannot prevent floating-point range exceptions. Instead, the enhanced CoMoM can also solve the largest model with 10^6 jobs, agreeing within the first 11 digits of $\log G(m, N)$ with the results of the logistic expansion (LE) proposed in [8], which is asymptotically exact. We have also observed in all cases that at least the first 6-digits of the mean per-class throughputs computed by the enhanced CoMoM were identical to the ones obtained by the AQL approximate MVA method [37]. As no other exact method can reach this model scale, it is difficult to rigorously verify exactness, yet the result suggests no, or at least negligible, presence of error accumulation.

4.2 Solving Homogeneous Layers with Gaussian Quadratures

As illustrated in the last numerical example, the CoMoM method has slightly increasing time requirements to analyze a single layer as the population grows. This also occurs as R increases, since the CoMoM basis has $2R$ elements. While tens or hundreds of milliseconds may be negligible for a single model, LQNs are solved iteratively and can feature many layers, hence solution times compound quickly. In large models, it is therefore useful to trade accuracy for speed. LN uses to this aim quadrature methods for integral forms of the normalizing constant.

A simple expression for the normalizing constant is given by the McKenna-Mitra integral form [24]. This is in general a multidimensional integral, with one dimension for each queueing station in the model. Thus, in a homogeneous model for a layer one would expect a m-dimensional integral. We show however that in homogeneous models this integral form takes the following simpler expression.

Theorem 2. *Under the same assumptions of Theorem 1, the normalizing constant of state probabilities for the queueing network model admits the following integral form*

$$G(m, N) = \frac{1}{(m-1)! \prod_{r=1}^{R} N_r!} \int_{u=0}^{+\infty} e^{-u} u^{m-1} \prod_{r=1}^{R} (Z_r + D_r u)^{N_r} du \qquad (2)$$

A proof is given in Appendix C. The main difficulty associated with evaluating $G(m, N)$ directly is that (2) is prone to numerical difficulties. This is because quadratures do not operate directly in the log domain and are therefore numerically sensitive to the magnitude of the factors under the integration sign, one being an exponentially decaying function (e^{-u}), the other being a polynomial of high order $|N|+m-1$. A novel strategy developed in the NC solver to evaluate (2) is to use Gaussian quadrature methods coupled with the log-sum-exp trick [4]. We have implemented both Gauss-Legendre and Gauss-Laguerre quadratures for (2), finding them empirically better suited at evaluating normalizing constants than MATLAB's default integral method and overall the best evaluation methods unless job populations are asymptotically large. We point to Appendix B for a brief introduction of both Gauss-Legendre and Gauss-Laguerre quadratures.

Generally, Gauss-Laguerre quadrature enables increasingly precise evaluations of (2) for growing values of its order K, however it also faces numerical difficulties for large number of jobs N, for which the quadrature weights and the integrand can display vastly different magnitudes. In such cases, we evaluate instead $\log G(m, N)$ in the quadrature summation by applying to the expression the log-sum-exp method, using in particular the implementation described in [4].

Numerical Example. We consider the same models considered for CoMoM and numerically evaluate the integral form (2). MATLAB's integral method is run with an absolute tolerance of 10^{-12}. Node and weights for the Gaussian quadratures are precomputed offline: due to numerical instability we can reach for the Gauss-Laguerre method up to order 300, while for Gauss-Legendre we could precompute weights up to order 20000 in the range $u \in [0, 10^6]$. We also include in the test the LE asymptotic expansion implemented in NC, which is a scalable method for models with few stations and many classes. The method applies a Laplace's approximation to the simplex integral form for the normalizing constant in [8]. Asymptotically the LE results are tight to the exact solutions.

Table 2. Relative error and runtime upon approximating $\log G(m, N)$.

Classes	Total jobs	Method	Rel. error [%]	Runtime [s]
8	40	MATLAB's integral	0.0000	0.0006
8	40	Gauss-Legendre	0.0000	0.0004
8	40	Gauss-Laguerre	0.0000	0.0010
8	40	Logistic expansion	-0.1249	0.0012
8	400	MATLAB's integral	0.0144	0.0005
8	400	Gauss-Legendre	-0.0001	0.0006
8	400	Gauss-Laguerre	-0.0001	0.0010
8	400	Logistic expansion	0.0033	0.0013
8	4000	MATLAB's integral	Unstable	0.0008
8	4000	Gauss-Legendre	-0.0006	0.0021
8	4000	Gauss-Laguerre	-0.0006	0.0010
8	4000	Logistic expansion	0.0003	0.0013
8	10^6	MATLAB's integral	Unstable	0.0008
8	10^6	Gauss-Legendre	-0.0592	0.0095
8	10^6	Gauss-Laguerre	0.2508	0.0011
8	10^6	Logistic expansion	0.0000	0.0013

The results are given in Table 2. Overall, we see that Gauss-Legendre is typically sufficient except in large asymptotic models where LE solutions are closer to optimal. The lower performance of Gauss-Laguerre is interpreted as being due to the restriction of using up to 300 nodes and weights in the interpolation due to numerical instability in their computation. Since Gauss-Legendre quadratures of order $K = 2n - 1$ are exact for polynomials up to order n, and the normalizing constant is itself the integral of a polynomial of order $n = |N| + m - 1$, it is possible to use the $K = 2|N| + 2m - 3$ order as a threshold for when the quadrature will cease to be exact and switch afterwards to LE. For example, with Gaussian integration of order $n = 300$ and $m = 1$ the solution would switch to LE for $|N| \geq 6000$. Note that, on top of this, approximation errors are incurred in

Table 2 by the log-sum-exp trick used for numerical stabilization, which explains why small errors are incurred by Gauss-Legendre and Gauss-Laguerre also in cases where the quadrature should be exact. Another source of errors is that Gauss-Legendre requires a finite interval and has therefore been truncated to the range $u \in [0, 10^6]$, whereas the normalizing constant integral is defined in the range $u \in [0, \infty]$.

Summarizing, the numerical analysis reveals that a combination of Gauss-Legendre quadrature, for models with tens or hundreds of jobs, and LE, for larger models, provides an effective way to approximate homogeneous layers.

4.3   Computing Marginal Probabilities in Homogeneous Layers

Using normalizing constants instead of MVA simplifies the calculation of probabilistic measures on each layer, as illustrated in this section. Thanks to loose layering, specialized results can be derived to allow simple computation of marginal probabilities in a layer. We focus here in particular on the marginal probability $\pi_N(m, n)$ that n jobs are queueing or receiving services at any of the m identical queueing stations. This is also equal to the probability that $\pi_N(m, |N| - n)$ jobs are waiting at the infinite server station.

Computing $\pi_N(m, n)$ is in general a difficult problem, since with R classes there is a combinatorially-large number of job mixes that result in the same total job population n at the m queueing nodes. In this case, NC leverages a novel result, developed in the next theorem, which obtains marginal probabilities in $O(|N|^2)$ time and $O(|N|)$ space in homogeneous layers. For $m = 1$, this improves over MVA methods that require instead $O(|N|^R)$ time and space, while matching the complexity of CoMoM's extension to marginal probabilities [6, §VII], but without requiring the solution of a system of linear equations as CoMoM does for marginal probabilities. Another novelty is that, unlike CoMoM, the expression below applies also to homogeneous models with $m > 1$.

Theorem 3. *Under the same assumptions of Theorem 1, let $\pi_N(m, k)$ be the marginal probability that the m queueing stations have k resident jobs in total, $k = 0, \ldots, |N|$. Define the following basis of unnormalized probabilities:*

$$\Pi(N) = G(m, N) \left[\pi_N(m, |N|), \ldots, \pi_N(m, 0) \right]^T$$

with $\Pi(0) = (0, \ldots, 0, 1)^T$. Then $G(m, N) = |\Pi(N)|$ and the following exact recurrence relation holds

$$\Pi(N) = N_R^{-1} \boldsymbol{T}_R \Pi(N - 1_R) \tag{3}$$

where

$$\boldsymbol{T}_R = \begin{bmatrix} Z_R \left(|N| + m - 1 \right) D_r & 0 & \cdots & 0 \\ 0 & Z_R & (|N| + m - 2) D_r & \cdots & 0 \\ \vdots & \vdots & \vdots & \ddots & \vdots \\ 0 & 0 & 0 & Z_R & (m-1) D_r \\ 0 & 0 & 0 & 0 & Z_R \end{bmatrix}$$

A proof of the result is given in Appendix D. While the result is exact, this little says about its numerical stability. We have verified with numerical examples, using the load-dependent convolution algorithm, that the formulas in Theorem 3 match numerically the exact solutions, while avoiding exponential time and space requirements as the number of classes grows. In the tests we observe that the method is applicable using floating-point arithmetic only to models with up to, approximately, $|N| = 500$ jobs, provided that, without loss of generality [21], demands are rescaled beforehand to $D_r = 1, r = 1, \ldots, R$. Larger models require instead the use of exact or multi-precision arithmetic to prevent floating-point range exceptions, which heightens complexity by a log-linear factor in both time and space [7].

4.4 Multi-server Nodes, Load-Dependence and Multi-class FCFS

We here briefly discuss other strategies used in LN to accelerate the evaluation and cope with extended features. In cases where the scheduling policy does not yield a product-form, suitable approximations are coupled with the proposed algorithms to approximate the solution. In particular, first-come first-served (FCFS) multi-class stations are mapped to PS stations with demands iteratively adjusted with an interpolation that depends on a hybrid $M/G/k$-diffusion approximation, as proposed in [9]. We point to the original paper for results showing high accuracy.

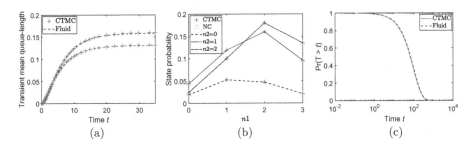

Fig. 3. Some meta-solver capabilities of LINE in analyzing LQN models.

Seidmann's approximation is used by default in NC to approximate FCFS and PS stations with multiple servers [31]. This is a simple method that replaces a c-server station with demands D_r with a sub-network consisting of a single-server queueing station having demands D_r/c and infinite server station having demands $D_r(c - 1)/c$. Under this transformation, as in the original system, a class-r job can traverse the two stations in D_r time overall when these are found both empty upon arrival. Moreover, the new infinite server station delay can be exactly aggregated within the pre-existing infinite server think time, so that the model retains overall the same number of stations.

Load-dependent modeling methods are also available in LINE to evaluate individual layers, which rely on the exact normalizing constant methods recently proposed in [11]. In essence, the latter factorize the normalizing constant of a load-dependent model into solving a single-server queueing network and scaling the resultant mean performance metrics by the normalizing constant of a related load-dependent model defined on a reduced state space. The work shows that this can be done either exactly or approximately, based on mean-value analysis (RD method) or a novel integral form of the normalizing constant (Norlund-Rice form). Although these methods may also be applicable to multi-server station analysis, Seidmann's approximation often suffices to achieve good accuracy while retaining the benefit of reducing the problem to a simple single-server model on which the CoMoM and Gaussian integration methods both apply.

5   Case Studies

5.1   Meta-solver Capabilities

The most distinctive feature of the LN solver is the meta-solver capability. Different solution paradigms can be used throughout individual iterations and across layers. Moreover, once the iteration has reached a fixed-point, multiple paradigms can be applied to obtain the metrics of interest. We here focus on the latter case.

We illustrate this feature on the example shown in Fig. 1, which describes a scenario where two job classes T1 and T2 require services from a server T3. There are 3 and 2 jobs for $class1$ and $class2$ respectively, and T3 is a FCFS service station with multiplicity $c = 2$. We consider transient analysis, for which LINE provides multiple solver options. Figure 3a shows the transient average queue length for the two job classes, representing T1 and T2 as clients, for the layer where T3 is modeled as a server. The plot shows a tight matching between the figures given by both CTMC and FLUID solvers. We can observe that the system reaches steady-state at around $t = 16$. In the figure, FLUID solver looks very accurate for this non-saturated single queue scenario but the accuracy depends on load and number of servers.

We also show meta-solver capabilities on steady-state probabilities. Figure 3b displays the joint steady-state probabilities calculated by NC and CTMC solvers for the whole state space at T3, given the service demand of A3 as 50. The results from both solvers are almost the same, but the speed of NC solver is much faster. In this example, the calculation of each probability takes CTMC solver around 4 s while the NC solver takes less than 50 ms.

Beyond joint probabilities, LINE allows us to compute response time percentiles with the FLUID and CTMC solvers. In the model, the operations of T3 are executed by the processor P3, here we use both solvers the obtain the response time distribution at the layer where T3 acts as client to P3. Results are shown in Fig. 3c, demonstrating agreement of the solutions.

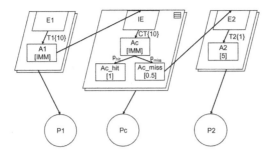

Fig. 4. Example of a multi-formalism model containing a caching layer (middle task)

5.2 Multi-formalism Capabilities

LINE can analyze, analytically or via simulation, models with integrated queueing and caching formalisms. LN can therefore also solve LQNs with caching, such as the three-layer model in Fig. 4. The host processors P1, Pc, P2 adopt the processor sharing scheduling policy. The number of users is represented by the multiplicity of the task T1 and the number of jobs is represented by the multiplicity of the task T2 and the cache task CT. As per Sect. 3.4, a cache task is a novel LQN element introduced by LN to describe data item reads from a cache, with different activities occurring based on whether a cache hit or a cache miss occurred. This is modeled by means of state-dependent class-switching.

In the example under study, items access probabilities obey a discrete uniform distribution and the cache is configured with a random replacement (RR) strategy. Arbitrary access probability distributions may be configured in LN and replacement strategies such as FIFO or LRU are supported. Jobs requested from T1 retrieve the items in the cache task CT. If the required items are cached, jobs will be processed by the *hit* activity with a probability of p_{hit}. Otherwise, jobs will be transformed to the *miss* activity with a probability of p_{miss} and be further processed by the task T2.

To solve an LQN model containing caching, LN first decomposes the entire model into a group of layers, as illustrated in Sect. 3. For the layer without cache nodes, the solutions are given in Sect. 4. On the other hand, for the layer

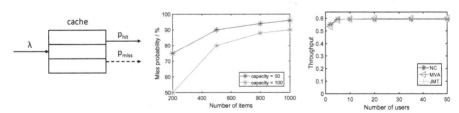

(a) Cache upper sub-model (b) Miss probability analysis (c) MVA and NC solutions

Fig. 5. Capabilities of LINE in analyzing multi-formalism LQN models

that involves a cache node, LN additionally divides the layer into two sub-models. In the upper sub-model, the cache node is isolated in an open model with Poisson arrivals, as shown in Fig. 5a. In the lower sub-model, the delay and the queueing station are contained in a closed queueing network with routing probabilities dynamically obtained from the p_{hit} and p_{miss} values obtained at the last iteration on the upper sub-model. More details can be found in [17].

For this model that combines both the queueing and caching stochastic formalisms, numerical results given by LINE are shown in Fig. 5. Figure 5b demonstrates the miss probabilities against different number of items, which decrease with the improvement of the cache capacity. Figure 5c compares the accuracy of the throughput for cache solved by MVA and NC solver respectively. MVA analyzes caches by the fixed-point iteration method (FPI) proposed in [10], whereas NC implements the normalizing constant asymptotics proposed in the same paper. The solvers are both compared against an equivalent model constructed with a JMT model using both queueing and Petri net formalisms, similar to the validation model used in [17], but adapted to the example at hand. The results indicate high accuracy of both MVA and NC in capturing the cache layer throughput, which is in general a function of the cache hit ratio.

6   Conclusion

We have presented the LN solver, the first meta-solver for LQNs, introducing new analysis methods for loose layering, in particular Gaussian integrals and a stabilized version of the exact CoMoM [6] to efficiently analyze layers in milliseconds. Case studies have shown the ability of the tool of combining several formalisms and solution methods in LQN analysis.

Future work will focus on extending LN to broaden the support for extended queueing models, such as fork-join networks and state-dependent queues.

Acknowledgments. LINE has been partially funded by the European Commission grants FP7-318484 (MODAClouds), H2020-644869 (DICE), H2020-825040 (RADON), and by the EPSRC grant EP/M009211/1 (OptiMAM).

A   Proof of Theorem 1

For a homogeneous model, the CoMoM recurrence relation may be written as

$$A\Lambda(N) = B\Lambda(N - 1_R)$$

where

$$A = \begin{bmatrix} A_{1,1} & A_{1,2} \\ 0 & A_{2,2} \end{bmatrix} \quad B = \begin{bmatrix} B_{1,1} & 0 \\ B_{2,1} & B_{2,2} \end{bmatrix}$$

Let $0_{I,J}$ indicate a block of zeros of size $I \times J$. Defining $\tilde{D} = (D_1, \ldots, D_R)^T$, we have

$$A_{1,1} = \begin{bmatrix} 1 & -\tilde{D}^T \\ 0_{R-1,1} & -m \operatorname{diag}(\tilde{D}) \end{bmatrix} \quad A_{1,2} = \begin{bmatrix} -1 & 0_{1,R-1} \\ \tilde{N} & -\operatorname{diag}(\tilde{Z}) \end{bmatrix} \quad A_{2,2} = N_R I$$

$$B_{1,1} = D_R E_{1,1} \quad B_{2,1} = m D_R I \quad B_{2,2} = Z_R I$$

The inverse of the block upper triagonal matrix A is now computed as

$$A^{-1} = \begin{bmatrix} A_{1,1}^{-1} & S A_{2,2}^{-1} \\ 0_{R-1,1} & A_{2,2}^{-1} \end{bmatrix}$$

with $S = -A_{1,1}^{-1} A_{1,2}$. Observe first that

$$A_{1,1}^{-1} = \begin{bmatrix} 1 & -m^{-1} e^T \\ 0_{R-1,1} & -m^{-1} \operatorname{diag}(\tilde{D})^{-1} \end{bmatrix} \quad A_{2,2}^{-1} = N_R^{-1} I$$

where $e^T = \tilde{D}^T \operatorname{diag}(\tilde{D})^{-1} = (1, \ldots, 1)$. Thus

$$S = -\begin{bmatrix} 1 & -m^{-1} e^T \\ 0_{R-1,1} & -m^{-1} \operatorname{diag}(\tilde{D})^{-1} \end{bmatrix} \begin{bmatrix} -1 & 0_{1,R-1} \\ \tilde{N} & -\operatorname{diag}(\tilde{Z}) \end{bmatrix}$$

$$= -m^{-1} \begin{bmatrix} -|\tilde{N}| - m & \tilde{Z}^T \\ -\operatorname{diag}(\tilde{D})^{-1} \tilde{N} & \operatorname{diag}(\tilde{D})^{-1} \operatorname{diag}(\tilde{Z}) \end{bmatrix}$$

Note that $A_{2,2}$ is the only block that depends on N_R. We can therefore write

$$A^{-1} B = F_{1,R} + N_R^{-1} F_{2,R}$$

where

$$F_{1,R} = \begin{bmatrix} A_{1,1}^{-1} & 0 \\ 0 & 0 \end{bmatrix} \begin{bmatrix} D_R E_{1,1} & 0 \\ m D_R I & Z_R I \end{bmatrix} = \begin{bmatrix} D_R E_{1,1} & 0 \\ 0 & 0 \end{bmatrix}$$

$$F_{2,R} = \begin{bmatrix} 0 & S \\ 0 & I \end{bmatrix} \begin{bmatrix} D_R E_{1,1} & 0 \\ m D_R I & Z_R I \end{bmatrix} = \begin{bmatrix} m D_R S & Z_R S \\ m D_R I & Z_R I \end{bmatrix}$$

B   Gaussian Quadratures

A Gauss-Laguerre quadrature of order K evaluates exponentially-weighted integrals by means of the approximation

$$\int_{x=0}^{\infty} e^{-x} f(x) dx \approx \sum_{k=1}^{K} w_k f(x_k) \tag{4}$$

where x_k denotes the k-th root of the Laguerre polynomial

$$L_K(x) = \sum_{j=0}^{K} \binom{K}{i} \frac{(-1)^j}{j!} x^j$$

and with weights $w_k = x_k \left((k+1)^2 \left[L_{k+1}(x_k) \right]^2 \right)^{-1}$.

Gauss-Legendre methods are similar but applicable to finite ranges $[a, b]$. Setting $a = 0$ and large b they can also help evaluating the normalizing constant. Their main benefit is that nodes and weights do not incur the same floating-point range exceptions as observed instead for Gauss-Laguerre quadratures of large order. We point to [20] for further details on Gauss-Legendre methods.

C   Proof of Theorem 2

For a homogeneous model with m identical single-server stations, the McKenna-Mitra integral takes the form

$$G(m, N) = \frac{1}{\prod_{r=1}^{R} N_r!} \int_{u_1=0}^{+\infty} \cdots \int_{u_m=0}^{+\infty} e^{-(u_1+\ldots+u_m)} h(u_1 + \ldots + u_m) du_1 \cdots du_m$$

where $h(u) = \prod_{r=1}^{R} (Z_r + D_r u)^{N_r}$. We note that the multidimensional integral may be interpreted as computing $E[h(U_1 + \ldots + U_m)]$ for the i.i.d. exponential random variables $U_i \sim Exp(1)$. The result then readily follows after noting that $U_1 + \ldots + U_m$ is Erlang-m distributed with density $f(u) = \frac{1}{(m-1)!} u^{m-1} e^{-u}$.

D   Proof of Theorem 3

Let the entries of $\Pi(N)$ be indicated with $\widetilde{\pi}_N(n)$, $n = |N|, \ldots, 0$. A probabilistic population constraint holds for homogeneous models with $m = 1$ [6, Thm. 6]

$$N_R \widetilde{\pi}_N(n) = Z_R \widetilde{\pi}_{N-1_R}(n) + n D_R \widetilde{\pi}_{N-1_R}(n-1)$$

for all $n = 1, \ldots, |N|$ and where $\widetilde{\pi}_{N-1_R}(n-1) = 0$ if $n = 0$. With a load-dependent queueing station $(m = 1)$, the derivation in [6] generalizes with similar passages to the following form

$$N_R \widetilde{\pi}_N(n) = Z_R \widetilde{\pi}_{N-1_R}(n) + n \frac{D_R}{\mu(n)} \widetilde{\pi}_{N-1_R}(n-1) \tag{5}$$

where $\mu(n)$ is the load-dependent scaling at the queueing station. Organizing (5) in matrix form, we get (3)

$$\boldsymbol{T}_R = \begin{bmatrix} Z_R |N| \dfrac{D_R}{\mu(|N|)} & 0 & \cdots & 0 \\ 0 & Z_R & (|N|-1) \dfrac{D_R}{\mu(|N|-1)} & \cdots & 0 \\ \vdots & \vdots & \vdots & \ddots & \vdots \\ 0 & 0 & 0 & & Z_R \dfrac{D_R}{\mu(1)} \\ 0 & 0 & 0 & & 0 \quad Z_R \end{bmatrix}$$

As assumed, consider now an homogeneous layer, where there are m identical *load-independent* single-server stations. The proof follows by noting that, if $m > 1$, the m queueing stations can be exactly replaced by a flow-equivalent server station with identical $D_1, \ldots, D_R$ and [26]

$$\mu(n) = \frac{n}{(n + m - 1)}$$

The final expression for $\boldsymbol{T}_R$ follows after plugging the above expression for $\mu(n)$.

E   Software Architecture Design

Figure 6 illustrates the key architectural elements of LINE, including the NetworkStruct data structure, and the Network, NetworkSolver, LayeredNetwork, EnsembleSolver and LayeredNetworkSolver classes.

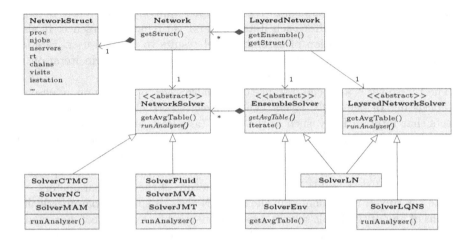

Fig. 6. Key architectural elements of LINE.

The Network object summarizes the model characteristics and acts as its persistence layer. The object is generated by the user either through a domain-specific language offered by LINE [9] or via model-to-model transformations from other formats (e.g., JMT's XML [2], PMIF [32]). Besides the model specification, a Network object can cache the model state space, its initial state, and retain information needed for the traffic equations in state-dependent models.

Each Network object is able to synthesize via the getStruct method a NetworkStruct data structure. The latter includes key model parameters, such as representations of service and arrival processes, job populations, and number of servers, among others. In addition, the data structure includes the routing table, the associated chains, and the average number of visits that each class pays to each node. NetworkStruct also offers indexing functions, that allow for example to differentiate between *stations*, where jobs can reside, and *nodes*, which are elements of the network traversed with zero service time (e.g., a fork).

The NetworkSolver object encodes a solver type, of which the aforementioned six LINE solvers are specific instances. The main role of this class is to ensure consistent computation of performance results, adopting identical conventions for reporting per-class and per-chain results to the end-user. Operational relationships are also applied by this object to derive certain performance metrics from the ones returned by the solvers, e.g. arrival rates from throughputs [14].

Each NetworkSolver object is equipped with a getAvgTable method that returns mean performance metrics for the model in a tabular format. The method invokes via the runAnalyzer method one of the solution methods offered by that solver, which operates solely on the NetworkStruct data structure. Model transformations that alter the model topology are conducted within runAnalyzer. An example is tagging a job class, which is used in response time distribution analysis.

The EnsembleSolver specifies the life-cycle for an iterative solution method that works on an ensemble of Network objects. This class allows to bind a Network-Solver to each particular Network in the ensemble, applying consistently actions before, in-between, and after each iteration, and verifying convergence. It also harmonizes the presentation of ensemble-level results to the end-user. Besides LN, the Env solver is another example of EnsembleSolver, wherein iteration is used to analyze random environments [13].

The LN solver is a special instance of EnsembleSolver, operating on an ensemble consisting of the LQN layers. The LayeredNetwork class encompasses the objects that form an LQN, such as the Entry, Task, Host, and Activity classes.

The LayeredNetwork class offers a getEnsemble method that generates, and stores within the LayeredNetwork object, the ensemble of Network models, each mapping to a distinct LQN layer. Similarly to Network, this class also exposes a getStruct method that builds a static data structure of the LQN parameters.

The LN solver, implemented in the SolverLN class, is specified parametrically in terms of any of the LINE solvers, or a custom combination thereof. For example, the end-user may require to use LINE's simulators on layers that include non-Markovian service distributions (e.g., Pareto) and MVA otherwise.

References

1. Amparore, E.G., Balbo, G., Beccuti, M., Donatelli, S., Franceschinis, G.: 30 Years of GreatSPN. In: Fiondella, L., Puliafito, A. (eds.) Principles of Performance and Reliability Modeling and Evaluation. SSRE, pp. 227–254. Springer, Cham (2016). https://doi.org/10.1007/978-3-319-30599-8_9
2. Bertoli, M., Casale, G., Serazzi, G.: The JMT simulator for performance evaluation of non-product-form queueing networks. In: Proceedings of ANSS, pp. 3–10. IEEE (2007)
3. Bini, D., Meini, B., Steffé, S., Pérez, J.F., Van Houdt, B.: SMCSolver and Q-MAM: tools for matrix-analytic methods. ACM SIGMETRICS Perform. Eval. Rev. **39**(4), 46–46 (2012)
4. Blanchard, P., Higham, D.J., Higham, N.J.: Accurately computing the log-sum-exp and softmax functions. IMA J. Numer. Anal. **41**(4), 2311–2330 (2021)
5. Bolch, G., Greiner, S., de Meer, H., Trivedi, K.S.: Queueing Networks and Markov Chains: Modeling and Performance Evaluation with Computer Science Applications. John Wiley & Sons, Hoboken (2006)
6. Casale, G.: CoMoM: efficient class-oriented evaluation of multiclass performance models. IEEE Trans. Softw. Eng. **35**(2), 162–177 (2009)
7. Casale, G.: Exact analysis of performance models by the method of moments. Perform. Eval. **68**(6), 487–506 (2011)
8. Casale, G.: Accelerating performance inference over closed systems by asymptotic methods. In: Proceedings of ACM SIGMETRICS, pp. 8:1–8:25. ACM (2017)

9. Casale, G.: Integrated performance evaluation of extended queueing network models with LINE. In: Proceedings of WSC, pp. 2377–2388. IEEE (2020)
10. Casale, G., Gast, N.: Performance analysis methods for list-based caches with nonuniform access. IEEE/ACM Trans. Netw. **29**(2), 651–664 (2021)
11. Casale, G., Harrison, P.G., Ong, W.H.: Facilitating load-dependent queueing analysis through factorization. Perform. Eval. **152**, 102241 (2021)
12. Casale, G., Muntz, R.R., Serazzi, G.: Geometric bounds: a noniterative analysis technique for closed queueing networks. IEEE Trans. Comput. **57**(6), 780–794 (2008)
13. Casale, G., Tribastone, M., Harrison, P.G.: Blending randomness in closed queueing network models. Perform. Eval. **82**, 15–38 (2014)
14. Denning, P.J., Buzen, J.P.: The operational analysis of queueing network models. ACM Comput. Surv. **10**(3), 225–261 (1978)
15. Franks, G., Al-Omari, T., Woodside, M., Das, O., Derisavi, S.: Enhanced modeling and solution of layered queueing networks. IEEE Trans. Softw. Eng. **35**(2), 148–161 (2009)
16. Franks, R.G.: Performance Analysis of Distributed Server Systems. Ph.D. thesis, Carleton University (2000)
17. Gao, Y., Casale, G.: JCSP: Joint caching and service placement for edge computing systems. In: Proceedings of IEEE/ACM IWQoS. IEEE (2022)
18. Gunther, N.J.: Analyzing Computer System Performance with Perl::PDQ. Springer, Heidelberg (2011). https://doi.org/10.1007/978-3-642-22583-3
19. Horváth, G., Telek, M.: BuTools 2: a rich toolbox for Markovian performance evaluation. In: Proceedings of VALUETOOLS, pp. 137–142. ICST (2017)
20. Johansson, F., Mezzarobba, M.: Fast and rigorous arbitrary-precision computation of Gauss-Legendre quadrature nodes and weights. SIAM J. Sci. Comput. **40**(6), C726–C747 (2018)
21. Lam, S.S.: Dynamic scaling and growth behavior of queueing network normalization constants. J. ACM **29**(2), 492–513 (1982)
22. Lazowska, E.D., Zahorjan, J., Graham, G.S., Sevcik, K.C.: Quantitative System Performance: Computer System Analysis Using Queueing Network Models. Prentice Hall, Hoboken (1984)
23. Marzolla, M.: The octave queueing package. In: Norman, G., Sanders, W. (eds.) QEST 2014. LNCS, vol. 8657, pp. 174–177. Springer, Cham (2014). https://doi.org/10.1007/978-3-319-10696-0_14
24. McKenna, J., Mitra, D.: Asymptotic expansions and integral representations of moments of queue lengths in closed Markovian networks. J. ACM **31**(2), 346–360 (1984)
25. Meyer, C.D.: Stochastic complementation, uncoupling Markov chains, and the theory of nearly reducible systems. SIAM Rev. **31**(2), 240–272 (1989)
26. Mitra, D., McKenna, J.: Asymptotic expansions for closed Markovian networks with state-dependent service rates. J. ACM **33**(3), 568–592 (1986)
27. Niu, Z., Casale, G.: A mixture density network approach to predicting response times in layered systems. In: Proceedings of MASCOTS, pp. 1–8. IEEE (2021)
28. Pérez, J.F., Casale, G.: LINE: evaluating software applications in unreliable environments. IEEE Trans. Reliab. **66**(3), 837–853 (2017)
29. Rolia, J.A., Sevcik, K.C.: The method of layers. IEEE Trans. Softw. Eng. **21**(8), 689–700 (1995)
30. Ruuskanen, J., Berner, T., Årzén, K.E., Cervin, A.: Improving the mean-field fluid model of processor sharing queueing networks for dynamic performance models in cloud computing. Perform. Eval. **151**, 102231 (2021)

31. Seidmann, A., Schweitzer, P.J., Shalev-Oren, S.: Computerized closed queueing network models of flexible manufacturing systems: a comparative evaluation. Large Scale Syst. **12**, 91–107 (1987)
32. Smith, C.U., Lladó, C.M., Puigjaner, R.: Performance model interchange format (PMIF 2): a comprehensive approach to queueing network model interoperability. Perform. Eval. **67**(7), 548–568 (2010)
33. Tribastone, M.: A fluid model for layered queueing networks. IEEE Trans. Softw. Eng. **39**(6), 744–756 (2013)
34. Waizmann, T., Tribastone, M.: DiffLQN: differential equation analysis of layered queuing networks. In: Companion of ICPE, pp. 63–68. ACM (2016)
35. Woodside, M.: Tutorial introduction to layered modeling of software performance (2013).   http://www.sce.carleton.ca/rads/lqns/lqn-documentation/tutorialh.pdf Accessed 08 May 2022
36. Zahorjan, J.: An exact solution method for the general class of closed separable queueing networks, vol. 8, pp. 107–112, New York, NY, USA (1979)
37. Zahorjan, J., Eager, D.L., Sweillam, H.M.: Accuracy, speed, and convergence of approximate mean value analysis. Perform. Eval. **8**(4), 255–270 (1988)
38. Zhu, L., Casale, G., Perez, I.: Fluid approximation of closed queueing networks with discriminatory processor sharing. Perform. Eval. **139**, 102094 (2020)

Eulero: A Tool for Quantitative Modeling and Evaluation of Complex Workflows

Laura Carnevali, Riccardo Reali$^{(\boxtimes)}$, and Enrico Vicario

Department of Information Engineering,
University of Florence, Florence, Italy
{laura.carnevali,riccardo.reali,
enrico.vicario}@unifi.it

Abstract. We present Eulero, a novel Java library enabling modeling of complex workflows and evaluation of their end-to-end response time Probability Density Function (PDF). Workflows consist of activities with general (i.e., non-exponential) duration with bounded support, composed through sequence, choice/merge, and split/join blocks, with unbalanced split and join constructs that break the structure of well-formed nesting. Eulero supports specification of workflows through structure trees, a hierarchical representation enabling the workflow decomposition into sub-workflows that can be efficiently analyzed in isolation. Eulero implements composition of the analysis results of these sub-workflows to provide a stochastically ordered approximation of the response time PDF of the overall workflow. The library supports random generation of workflow models controlling the main factors of computational complexity. Eulero exploits the SIRIO Library of the ORIS tool to represent monovariate PDFs and to model and analyze sub-workflows, and it is designed to facilitate usability, maintainability, and extensibility.

Keywords: stochastic workflows · response time probability density function · structured model · compositional evaluation · stochastic ordering · model random generation · software tools and libraries

1 Introduction

Workflows are a collection of activities orchestrated according to precedence constraints and control-flow constructs. In particular, workflows can be built using basic patterns such as *sequence, split/join, choice/merge* [8], and more complex patterns that break the well-formed nesting of basic patterns, such as *Directed Acyclic Graphs* (DAGs) and *loops* [21]. This expressivity enables workflows to represent processes of many application contexts, such as supply chain management [16], composition of web services [6], and cloud function as a service [24]. When activities are enriched with stochastic durations, quantitative analysis of the workflow *end-to-end response time* provides relevant metrics that can guide choices during system design, or be exploited to define *soft deadlines*

© Springer Nature Switzerland AG 2022
E. Ábrahám and M. Paolieri (Eds.): QEST 2022, LNCS 13479, pp. 255–272, 2022.
https://doi.org/10.1007/978-3-031-16336-4_13

or *penalty functions* [14,20], e.g. to define *Service Level Agreements* (SLA) of a composite web service, or to select contractors who maximize the profit of a manufacturing production. In both cases, summary statistics, such as mean value or variance, are not sufficient to characterize the considered metrics, and the *Probability Density Function* (PDF) of the workflow response time is required.

Depending on the type of duration distributions of activities, workflows underlie different classes of stochastic models, which rely on different analysis techniques and, hence, on different tools. In case of exponential durations, the underlying stochastic process of a workflow is a *Continuous Time Markov Chain* (CTMC), for which analysis is performed by exploiting the Markov property (i.e., the future state of the system depends on the current state only through the discrete logical location). For this class of models, evaluation can leverage several consolidated tools such as GreatSPN [2], PRISM [18], MRMC [15], SMART [5], and Storm [7]. However, in many application contexts, workflows include activities with deterministic or bounded or stochastic non-exponential duration. In this case, the Markov property does not hold in every state, and the underlying stochastic process is a *non-Markovian process*. Here, the end-to-end response time can be evaluated numerically exploiting the presence of *regenerations*, that are states where the Markov property holds true again. In particular, when the model is such that a regenerative state is eventually reached with probability 1 (w.p.1), the model belongs to the class of *Markov Regenerative Processes* (MRPs) [17], whose evaluation is performed by computing a *global kernel* characterizing behavior until the first regeneration is reached, and a *local kernel* characterizing sequencing and timing of visits to subsequent regenerations. Kernel-based evaluation can be performed with different techniques, implemented in consolidated tools. SHARPE [23], TimeNET [28], and Great-SPN [2] enable evaluation of models that satisfy the *enabling restriction*, i.e., the condition for which only one duration with general (GEN), i.e., non-exponential, distribution is enabled in each state [12]. ORIS [19] implements the *method of stochastic state classes*, which enables evaluation of MRPs beyond the enabling restriction [13]. Nevertheless, when the concurrency degree of activities with general distribution increases, the complexity of the analysis of the workflow model cannot be afforded. In this case, workflows are decomposed into sub-workflows that can be evaluated efficiently in isolation, and results are recombined to obtain an approximation of the actual end-to-end response time [3,4].

In this paper, we present the *Eulero* Java library, which implements the approach of [3,4] for efficient and accurate evaluation of the response time PDF of complex workflows, consisting of activities with GEN duration with bounded support, composed through sequence, choice/merge, and split/join blocks, with unbalanced split and join constructs that break the structure of well-formed nesting. The library (Sect. 2) is organized in three packages, supporting workflow modeling (Sect. 3), workflow evaluation (Sect. 4), and random generation of workflow models (Sect. 5). A workflow is modeled as a hierarchy of sub-workflows using structure trees, providing not only ease of modeling but also efficiency of analysis by facilitating the identification of sub-workflows that can be separately analyzed. Eulero uses the SIRIO Library of the ORIS tool [19] to represent monovariate PDFs, to model sub-workflows as Stochastic Time Petri Nets (STPNs) [26], and to perform their transient analysis to derive the sub-workflow response time PDF. The library is designed with the aim of facilitating usability, maintainability, and extensibility, by exploiting consolidated programming design patterns [11]. Possible extensions of the library are discussed in the conclusions of the paper (Sect. 6).

2   The Eulero Library

The Eulero library is available at https://github.com/oris-tool/eulero under the AGPL licence. The UML use case diagram of Fig. 1a shows the main functionalities, in Fig. 1b:

- The package **modeling** supports modeling of workflows affording various aspects of complexity (i.e., bounded generally distributed durations, high concurrency degree, unbalanced split and join constructs breaking the structure of well-formed nesting) in terms of structure trees, a hierarchical representation enabling accurate and efficient evaluation of the response time PDF.
- The package **evaluation** includes the sub-package **heuristics**, implementing compositional methods for the evaluation of the workflow response time PDF, and the sub-package **approximation**, supporting the derivation of the analytical form of stochastically ordered approximants of numerical PDFs.
- The package **modelgeneration** provides functionalities to randomly generate models according to the specification of the package **modeling**.

Eulero uses the SIRIO library for the representation of the analytical form of monovariate PDFs, for the derivation of their response time PDF of sub-workflows, and for the estimation of complexity of this evaluation.

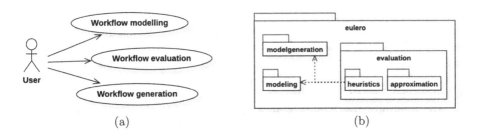

(a)   (b)

Fig. 1. Eulero library: (a) UML use case diagram and (b) package diagram.

3   Workflow Modeling

The package **modeling** implements the structure tree specification for the representation of stochastic workflows. In particular, we illustrate the formalism (Sects. 3.1 and 3.2) and how to use the package to instantiate the structure tree of a workflow (Sect. 3.3).

3.1 Structure Tree and STPN Blocks

We specify a class of stochastic workflows that can represent a wide set of control flow patterns [21,27]. In particular, workflows are built by recursively composing Single Entry Single Exit (SESE) blocks, defined as STPN blocks with a single starting place and a single ending place. A block execution starts when a token reaches its starting place, and eventually terminates with probability 1 (w.p.1) when a token reaches its final place. Blocks can be designed to model concurrent behaviors (fork-join structures, termed AND blocks), sequential behaviors (sequence and choice/merge structures, termed SEQ and XOR blocks, respectively), and acyclic behaviors where well-formed nesting is broken by an unbalanced composition of fork and join operators (termed DAG blocks).

Blocks are combined together in a *structure tree* $S = \langle N, E, n_0 \rangle$, where N is the set of nodes (i.e., blocks), E is the set of directed edges connecting each block with its component blocks, and n_0 is the root node (i.e., the overall workflow). Figure 2b shows an example of the structure tree of the workflow STPN of Fig. 2a, where a block is depicted as a box labeled with the block name and with either the activity name (for ACT blocks) or the block type (for SEQ, AND, and XOR blocks). The box of a DAG block also contains places and transitions connecting their component blocks. Since every block is defined as an STPN, then each workflow can in turn be mapped to a unique STPN. Conversely, the composition of blocks does not cover all the expressivity of STPNs.

3.2 Complexity Measures

The complexity of workflow evaluation is estimated by exploiting the state class graph provided by nondeterministic analysis of the Timed Petri Net (TPN) [1,25] underlying the workflow STPN (which can be obtained by exploring the structure tree). Since workflows are SESEs blocks, the state class graph has a single final state class, where all model transitions are disabled. Based on this, complexity is measured in terms of *concurrency degree c*, representing the maximum number of concurrent GEN transitions, and *sequencing degree q*, representing the maximum number of firings of GEN transitions in any path of the state class graph.

If the model is too complex, enumeration of state classes may require a non-negligible amount of time, making nondeterministic analysis unfeasible. In this case, the structure tree provides an abstraction to identify the workflow *unexpanded TPN*, which is the underlying TPN whose inner blocks are replaced with single activities having the same durations. Then, workflow complexity measures can be derived by exploring the structure tree through a bottom-up approach, combining nondeterministic analysis of the unexpanded TPN with the complexity measures of inner blocks, obtaining upper bounds C and $\bar{C}$ on the concurrency degree of the workflow TPN and unexpanded TPN, respectively, as well as sequencing degrees q and $\bar{q}$ of the workflow TPN and unexpanded TPN, respectively. Given two thresholds Θ_c and Θ_q, a workflow is termed *complex* if $C > \Theta_c \vee q > \Theta_q$, and *internally complex* if $\bar{C} > \Theta_c \vee \bar{q} > \Theta_q$.

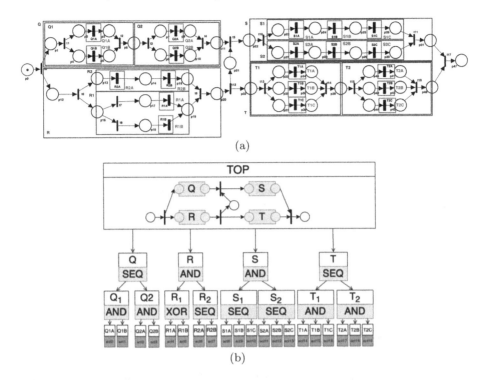

(a)

(b)

Fig. 2. (a) STPN model of a workflow: blocks are highlighted by boxes, blue for composite blocks and red for activity blocks. (b) Structure tree of the workflow of Fig. 2a: composite blocks are filled with light blue. (Color figure online)

3.3 Package Description

The package `modeling` is responsible for building stochastic workflows as Java objects. The related UML class diagram is shown in Fig. 3 and a code snippet illustrating the programmatic specification of the model of Fig. 2b is provided in Listing 1.1. Workflow modeling is implemented through the design pattern Composite [11], where the abstract class `Activity` defines a common interface, which is implemented by subtype classes `Simple`, `XOR`, `DAG`, `AND`, and `SEQ`. In particular, `Activity` defines block types common attributes, such as the minimum `min` and maximum `max` response time of the block, the degrees of concurrency `C` and `simplifiedC` corresponding to C and $\bar{C}$ respectively, the degrees of sequencing `Q` and `simplifiedQ` corresponding to q and $\bar{q}$ respectively, and the abstract methods `buildSTPN()` and `buildTPN()`, which are overridden by sub-typing classes to define the transformation of the block structure tree into the block STPN and its underlying TPN, respectively.

The class `Simple` enables the instantiation of simple activities whose piecewise duration PDF is encoded in the field `features` as a list of `StochasticTransitionFeature`. The latter is a class of the SIRIO library defining the support and analytical form of a PDF. Each feature is assumed to have unit

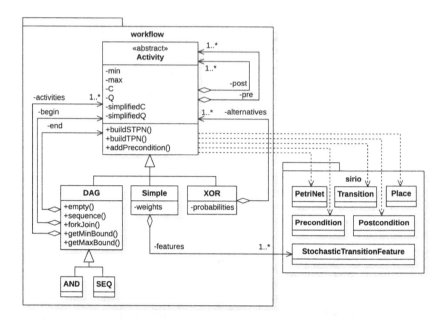

Fig. 3. UML class diagram of the package `workflow`.

measure, and therefore it is weighted by a value that guarantees unit measure over the whole PDF. These values are encoded by the attribute `weights`.

The class `XOR` enables the instantiation of XOR blocks having different alternative branches. Branches are referenced through the field `alternatives` and associated with a probability value encoded in the field `probabilities`.

The class `DAG` enables the instantiation of DAG blocks. `DAG` objects have three references to class `Activity`. The reference `activities` consists of a list referencing all the activities nested in a DAG block. The references `begin` and `end` refer to the initial and final fictitious activities, respectively, that are required to define a DAG block as a SESE block. The structure of a DAG is defined by adding preconditions to its internal activities through the method `addPrecondition()`, which updates the list attributes `pre` and `post` of the DAG activities. The `DAG` class also exposes three static methods `empty()`, `sequence()` and `forkJoin()`. The method `empty()` enables the generation of empty DAGs that can then be built by adding activities and preconditions, evaluating the response time bounds, and referencing the added activities. The methods `sequence()` and `forkjoin()` enable the generation of SEQ and AND blocks, respectively, which are implemented as derivation classes of `DAG`. Although `AND` and `SEQ` classes do not extend `DAG` functionalities, they have been treated as separate classes in order to handle them as a specific case in the analysis algorithm.

In Listing 1.1, blocks Q, R, S and T are created using the static methods of `DAG` and the constructors of `XOR` and `Simple` (lines 3 to 52). Then, the top node is created as a DAG block through the static method `empty()`, and populated connecting inner blocks by adding preconditions (lines 54 to 59). In doing so, it is mandatory that final blocks be preconditions of the node **end** (line 59), and that all initial nodes have the node **begin** as precondition (lines 55 and 56). Since DAGs are built step-by-step adding preconditions, then **min** and **max** are estimated and set using the related setter methods (lines 60 and 61), and the added **activities** are also referenced using the related setter method (line 62).

```
 1  StochasticTransitionFeature feature = StochasticTransitionFeature.
        newUniformInstance("0", "1");
 2
 3  Activity Q = DAG.sequence("Q",
 4      DAG.forkJoin("Q1",
 5          new Simple("Q1A", feature),
 6          new Simple("Q1B", feature)
 7
 8      ),
 9      DAG.forkJoin("Q2",
10          new Simple("Q2A", feature),
11          new Simple("Q2B", feature)
12      )
13  );
14
15  Activity R = DAG.forkJoin("R",
16      new XOR("R1",
17          List.of(
18              new Simple("R1A", feature),
19              new Simple("R1b", feature)
20          ),
21          List.of(0.3, 0.7)),
22      DAG.sequence("R2",
23          new Simple("R2A", feature),
24          new Simple("R2B", feature)
25      )
26  );
27
28  Activity S = DAG.forkJoin("S",
29      DAG.sequence("S1",
30          new Simple("S1A", feature),
31          new Simple("S1B", feature),
32          new Simple("S1C", feature)
33      ),
34      DAG.sequence("S2",
35          new Simple("S2A", feature),
36          new Simple("S2B", feature),
37          new Simple("S2C", feature)
38      )
39  );
40
41  DAG T = DAG.sequence("T",
42      DAG.forkJoin("T1",
43          new Simple("T1A", feature),
44          new Simple("T1B", feature),
45          new Simple("T1C", feature)
46      ),
47      DAG.forkJoin("T2",
48          new Simple("T2A", feature),
49          new Simple("T2B", feature),
50          new Simple("T2C", feature)
51      )
52  );
53
54  DAG top = DAG.empty("TOP");
55  Q.addPrecondition(top.begin());
56  R.addPrecondition(top.begin());
57  T.addPrecondition(R);
58  S.addPrecondition(R, Q);
```

```
59  top.end().addPrecondition(T, S);
60  top.setMin(top.getMinBound(top.end()));
61  top.setMax(top.getMaxBound(top.end()));
62  top.setActivities(Lists.newArrayList(Q, R, S, T));
```

Listing 1.1. Construction of the workflow structure tree of Fig. 2b.

The method `buildSTPN()` is an abstract method responsible for the construction of the workflow STPN. Construction is realized recursively by exploring the workflow structure tree, and adding `Place`, `Transition`, `Precondition` and `Postcondition` objects to a `PetriNet` object depending on the tree topology (the latter 5 classes belong to the SIRIO library). The method `buildSTPN()` of `DAG`, `AND`, and `XOR` merely adds places and immediate transitions, representing fork, join, choice, and merge structures specifying the nesting of inner blocks, before the method is called recursively by the inner blocks. The method `buildSTPN()` of `SEQ` simply chains the recursive calls made for inner activities belonging to the considered sequence. The method `buildSTPN()` of `Simple` adds a random switch of transitions whose distributions are encoded by the fields `features` and `weights`. The `buildTPN()` method is the abstract method that enables the construction of the underlying TPN. Similarly to `buildSTPN()`, it assigns temporal and complexity features to the workflow activities, so that these information can be exploited during the block complexity analysis.

4   Workflow Evaluation

The package `evaluation` implements a compositional technique [3,4] for the evaluation of the response time PDF of stochastic workflows. We provide an overview of the analysis heuristics (Sect. 4.1) and illustrate the use of the package (Sect. 4.2).

4.1   Analysis Heuristics

The end-to-end response time PDF of a workflow is evaluated by composing the results of separate analyses of sub-workflows. In turn, the sub-workflows are identified by a recursive exploration of the structure tree, which selects the most appropriate action to analyze a block, based on its type or complexity measures and according to the provided heuristics of analysis. Four actions are considered:

- *Numerical analysis* combines the numerical response time PDFs of the components of well-nested blocks, providing the overall response time PDF.
- *Forward transient analysis* is suitable to evaluate not well-nested blocks with limited complexity, providing the numerical form of their response time PDF. The analysis requires each activity to be associated with the analytical form of its duration distribution. If not, the analytical form of a stochastic upper bound PDF is derived for each numerical PDF.
- *Inner block analysis* can be applied to blocks with high complexity measures. It selects an internal block, evaluates its response time PDF with some action and replaces it with a new activity block.

– *Inner block replication* can be applied to complex not well-nested blocks. It identifies two sub-workflows sharing activities, separates them by replicating shared activities, and recombines them as children of an AND block, guaranteeing that the resulting response time PDF is a stochastic upper bound of the exact one [3]. For example, in Fig. 7, activities S and T share the same predecessor R. Applying inner block replication, two sub-workflows would be identified, consisting of activities Q, R and S, and activities T and R, respectively, where every instance of R is a replication of the original activity. The two sub-workflows would then be combined in an AND block, as shown in Fig. 4. If one or both sub-workflows had become well-nested, they would subsequently be evaluated through *numerical analysis*.

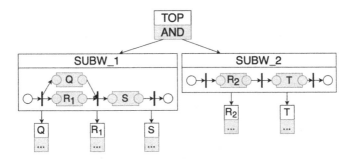

Fig. 4. Result of application of inner block replication to the DAG block of Fig. 7.

Heuristics can be defined to explore the structure tree and evaluate the workflow response time. In [3,4], two heuristics are defined, which both use *numerical analysis* for well-nested blocks and *forward transient analysis* for simple DAGs, as illustrated by Algorithm 1, but differ in the way complex DAGs are evaluated, by favouring either *inner block replication* or *inner block analysis*, as shown in Algorithms 2 and 3. Specifically, considering a workflow block b:

– If b is an activity block, then its exact response time PDF is its duration PDF for both heuristics (line 2 and 3).
– If b is, or can be reduced to, a well-nested composition of independent sub-workflows, then a numerical analysis is recursively applied to get the exact response time PDF, for both heuristics (lines 4 to 12).
– If b is a DAG block, *inner block analysis*, *inner block replication* or *forward transient analysis* are applied according to the complexity measures of the block, and depending on the considered heuristics (see Algorithms 2 and 3).

Algorithm 1: Evaluation of the response time PDF of a workflow block

---

CompositionalAnalysis(b, Θ_c, Θ_q, h)

    input : block b, concurrency degree threshold Θ_c, sequencing degree
               threshold Θ_q, heuristics h

    output: response time PDF $\phi(t)$ of b

1 **if** b *is an activity block* **then**
2 | **return** the duration PDF of b
3 **if** b *is a* SEQ *block or an* XOR *block or an* AND *block* **then**
4 | **foreach** *block* b_i *of* b **do**
5 | | $\phi_i(t) \leftarrow$ CompositionalAnalysis(b_i, Θ_c, Θ_q, h)
6 | **return** NumericalAnalysis($\phi_0(t), ..., \phi_n(t)$)
7 **if** b *is a* DAG *block* **then**
8 | **return** h(b)

Algorithm 2: Evaluation of the response time PDF of a DAG block

---

Heuristics1(b)

    input : workflow block b

    output: response time PDF $\phi(t)$ of b

1 **if** b *is internally complex to analyze* **then**
2 | **return** InnerBlockReplication(b)
3 **if** b *is complex to analyze* **then**
4 | **return** InnerBlockAnalysis(b)
5 **return** the PDF of b computed through forward transient analysis

---

Algorithm 3: Evaluation of the response time PDF of a DAG block

---

Heuristics2(b)

    input : workflow block b, concurrency degree threshold Θ_c, sequencing
               degree threshold Θ_q

    output: response time PDF $\phi(t)$ of b

1 **if** b *contains at least one composite block and is complex to analyze* **then**
2 | **return** InnerBlockAnalysis(b)
3 **if** b *is internally complex to analyze* **then**
4 | **return** InnerBlockReplication(b)
5 **return** the PDF of b computed through forward transient analysis

---

4.2 Package Description

The package evaluation implements the compositional evaluation method of [3, 4], recalled in Sect. 4.1. The related UML class diagram is provided in Fig. 5. The package contains the inner packages heuristics and approximator, both of which implement the design pattern Strategy [11].

In the package heuristics, the abstract class AnalysisHeuristicsStrategy implements the methods numericalXOR(), numericalAND(), numericalSEQ(), for numerical analysis operations on well-nested block types XOR, AND, SEQ, respectively, and forwardAnalysis(), innerBlockAnalysis(), and innerBlock-Replication(), which implement the remaining actions. Each of these methods makes one or more recursive calls to the abstract method analyze(), enabling the top-down evaluation of any structure tree. The order of execution of the actions is specified by overriding the analyze() method, using the fields CThreshold and QThreshold as the thresholds Θ_c and Θ_q for the complexity measures, respectively. The heuristics of Algorithms 2 and 3 are implemented by the classes AnalysisHeuristics1 and AnalysisHeuristics2, respectively. Note that the Strategy Pattern facilitates the addition of new heuristics strategies by extending the abstract class AnalysisHeuristicsStrategy with a new concrete class. Moreover, it is trivial to implement new actions as non-abstract methods of the class AnalysisHeuristicsStrategy.

The package approximation implements different approximation methods for numerical monovariate PDFs. The abstract class Approximator provides the abstract method getApproximationStochasticTransitionFeatures(), which processes a numerical monovariate PDF and returns an approximant piecewise PDF represented as a list of weights and StochasticTransitionFeature objects. Concrete approximators define the approximation logic by overriding this method. Again, the Strategy pattern enables an easy extension of the package. The class AnalysisHeuristicsStrategy uses the class Approximator in the method innerBlockAnalysis(): when this method is called, the most complex inner block of a DAG is picked up and evaluated by recursive call of analyze(), then the result is approximated by the selected Approximator, and finally, the obtained features and weights are passed to the constructor of the class Simple to generate the simple activity that replaces the complex inner block.

Listing 1.2 shows how to evaluate the model built in Listing 1.1. Considering values tC and tQ for complexity thresholds Θ_c and Θ_q, respectively, analysis horizon timeLimit, numerical precision step, and a specific approximator (lines 1 to 5), the analysis is performed by creating the heuristics (line 6) and calling analyze() (line 8). Figure 6 shows the evaluated response time PDF and CDF.

```
1 BigInteger tC = BigInteger.valueOf(3);
2 BigInteger tQ = BigInteger.valueOf(7);
3 BigDecimal timeLimit = model.max();
4 BigDecimal step = BigDecimal.valueOf(0.01);
5 Approximator approximator = new EXPMixtureApproximation();
6 AnalysisHeuristicStrategy strategy = new AnalysisHeuristics1(tC, tQ,
       approximator);
7 double[] cdf = strategy.analyze(model, timeLimit, step);
```

Listing 1.2. Evaluation of a workflow structure tree.

The analysis algorithm scales even on more complex models. In fact, analysis heuristics have been tested for structure trees having depth up to 6, concurrency and sequencing degree up to 64, and whose corresponding STPNs contain up to 7000 transitions. In all cases, analysis heuristics always took less than 1 s.

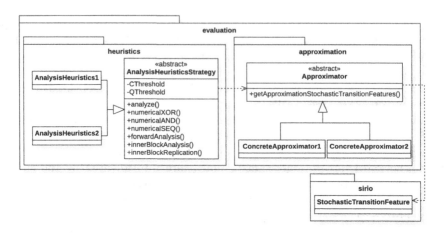

Fig. 5. UML class diagram of the package `evaluation`.

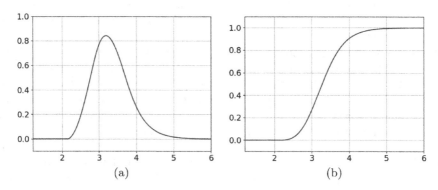

Fig. 6. PDF (a) and CDF (b) evaluated for the workflow of Fig. 2.

5 Workflow Generation

Many problems can be modeled as workflows, and experimention enabled by tools such as TGFF [9], Apache Airflow [10] or Luigi [22] results to be an added value. However, there is often a lack of models on which to conduct experiments, and when there are, they are often not complex enough to justify the use of compositional methods. In this section, we present a generation strategy to randomly build workflows that can be exploited to test quantitative evaluation methods, stressing different complexity factors, such as the degree of parallelism and sequencing. In particular, we illustrate how a model compliant with the structure tree specification is generated (Sect. 5.1), and describe its Java implementation through an illustrative example (Sect. 5.2).

5.1 Random Generation

Algorithm 4 implements the proposed strategy for random generation of work-flow structure trees. Specifically, given the depth of the structure tree, random generation of a workflow is carried out through a recursive procedure, which builds well-nested or DAG blocks in different ways, depending on the type of block that is drawn (lines 2 and 12). If a well-nested type (i.e., SEQ, AND, XOR) is selected, the number of its children is drawn too, and these are generated by the recursive call of the generation procedure (lines 3 to 5). Then, the block of the drawn type is created, assigning the created children to it (lines 6 to 11).

Algorithm 4: Random generation of a workflow

GenerateBlock(d, *settings*)

    input : depth level d, generation settings *settings*

    output: workflow b

1 **if** $d > 0$ **then**

2   **if** *type is* SEQ *or* AND *or* XOR **then**

3         $children \leftarrow \{\}$

4   **foreach** $i \in \{2, ..., \text{RandomMaximumChildren}()\}$ **do**

5 | $children \leftarrow children \cup \text{GenerateBlock}(d-1, \textit{settings})$

6   **if** *type is* SEQ **then**

7 | **return** GenerateSEQBlock(*children*)

8   **if** *type is* AND **then**

9 | **return** GenerateANDBlock(*children*)

10   **if** *type is* XOR **then**

11 | **return** GenerateXORBlock(*children*)

12   **if** *type is* DAG **then**

13 | **return** GenerateDAGBlock(*parameters*)

14 **return** GenerateSimpleBlock()

A DAG block consists of a set of nodes sorted according to a topological order in several consecutive levels, such that every node can have predecessor nodes in previous levels (see Fig. 7). In particular, a node belongs to the i-th level if, for all paths from the initial fictitious node to the considered one (both not included), the longest paths contain exactly i nodes. For instance, node E in Fig. 7 belongs to level L2 because, among all the paths from the initial node in to node E, the longest paths contain 2 nodes. Since DAG blocks break the well-formed nesting of concurrent blocks, yet ensuring that execution ends w.p.1, both conditions must be fulfilled to generate a not well-nested DAG. The first condition holds when there is at least one level where i) at least two nodes share at least one predecessor and ii) their predecessor sets do not coincide (e.g., C and D share A as predecessor, but D also has B) or coincide but contain at least 2 predecessors (e.g., E and F share all the predecessors, which are more then 1; otherwise, they would have been well-nested). The second condition holds if every node has at least one predecessor and at least one successor.

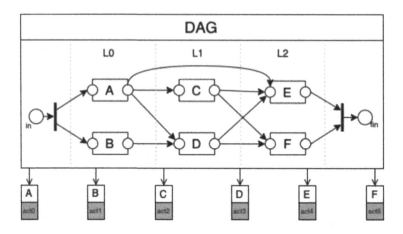

Fig. 7. A DAG block with nodes sorted according to a topological order.

Hence, random generation of a DAG (lines 12 and 13 of Algorithm 4) is achieved by drawing the number of levels, the number of nodes for each level, and randomly connecting nodes of different levels, ensuring that the two conditions described above are met. To guarantee not well-formed nesting, it is sufficient to randomly select two nodes from some level $i > 0$, and connect both of them to the same predecessor node, and one of them to a different predecessor node, randomly drawn from the $(i-1)$-th level. To guarantee that a block is a SESE, every node of a level is connected to at least one predecessor and one successor, except for the first-level nodes, which share the same and only predecessor node (the initial fictitious node in), and the last-level nodes, which share the same and only successor node (the final fictitious node fin). Finally, at the lowest level of the workflow structure tree, a simple block is created (line 14 of Algorithm 4).

5.2   Package Description

The package modelgeneration implements the approach of Sect. 5.1 for random generation of workflow structure trees. The UML class diagram of the package is shown in Fig. 8. The class RandomGenerator defines the recursive method generateBlock(), which implements the generation logic based on the depth treeDepth of the structure tree and on some setting parameters referenced by list settings. Each item of the list refers to a certain level of depth of the workflow, and collects a set of BlockTypeSetting objects, through which generation is driven. In particular, BlockTypeSetting is an abstract class that specifies a type and a probability of being drawn. For every item of the list settings, the sum of BlockTypeSetting probability fields must be equal to 1. BlockTypeSetting is extended to define block type related parameters, such as the minimum and maximum number of children for class WellNestedBlockSetting, or the minimum and maximum number of levels, nodes per level, connections between nodes and distance between nodes belonging to not consecutive levels, for class DAGBlockSetting. Note that XORBlockSet-ting, ANDBlockSetting, and SEQBlockSetting do not add functionalities to WellNestedBlockSetting, but are required to generate XOR, AND, and SEQ blocks in the method generateBlock(), respectively. In addition, it is possible to constrain the structure of a DAG by varying its generation parameters.

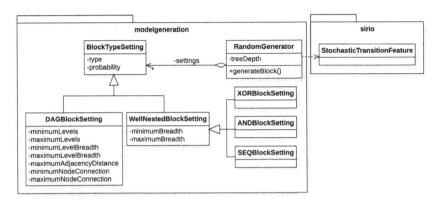

Fig. 8. UML class diagram of the package modelgeneration.

Listing 1.3 shows how to use the package modelgenerator to randomly generate workflow structure trees. For each level of the tree, at least one BlockTypeSetting object must be created and added to that level setting variable, specifying the occurrence probability of that block type and the related parameters (e.g., lines 4 to 8 and lines 10 to 12 for levels 1 and 2, respectively). Every level setting is added to a global setting variable (lines 14 to 16) which is then passed to a RandomGenerator object (line 19). Finally, the model is created invoking the method generateBlock() of the RandomGenerator object.

```
 1 int concurrencyDegree, sequenceFactor;
 2 concurrencyDegree = sequenceFactor = 3;
 3
 4 Set<BlockTypeSetting> level1Settings = new HashSet<>();
 5 BlockTypeSetting AND = new ANDBlockSetting(0.5, concurrencyDegree);
 6 BlockTypeSetting SEQ = new SEQBlockSetting(0.5, sequenceFactor);
 7 level1Settings.add(AND);
 8 level1Settings.add(SEQ);
 9
10 Set<BlockTypeSetting> level2Settings = new HashSet<>();
11 BlockTypeSetting DAG = new DAGBlockSetting(1.);
12 level2Settings.add(DAG);
13
14 ArrayList<Set<BlockTypeSetting>> settings = new ArrayList<>();
15 settings.add(level1Settings);
16 settings.add(level2Settings);
17
18 StochasticTransitionFeature feature = StochasticTransitionFeature.
        newUniformInstance("0", "1");
19 RandomGenerator randomGenerator = new RandomGenerator(feature, settings);
20 Activity model = randomGenerator.generateBlock(settings.size());
```

Listing 1.3. Random generation of a workflow structure tree.

6   Conclusions

Eulero is a Java library for efficient and accurate compositional evaluation of the response time PDF of complex stochastic workflows, where the presence of a large number of concurrent activities having GEN distribution with bounded support makes the analysis of the underlying stochastic process unfeasible. This issue is exacerbated by the presence of unbalanced split and join constructs breaking the structure of the well-formed nesting, which further increases the concurrency degree of activities with GEN distribution and requires a compositional approach to afford the workflow analysis. Specifically, the library supports the representation of workflows as structure trees, a hierarchical formalism that facilitates not only the workflow instantiation but also its evaluation by supporting the decomposition into sub-workflows that can be efficiently analyzed in isolation. Eulero implements evaluation heuristics to decompose the structure tree, exploiting various analysis techniques to evaluate the sub-workflows. Last but not least, the library also provides a package to instantiate randomly generated workflows, exposing parameters that impose structural constraints on the generated structure trees and thus allows the user to control the different factors of computational complexity in the model generation.

Eulero is designed to facilitate usability, maintainability and extensibility. In the perspective of modeling, new basic workflow patterns and even structures compliant with other formalisms can be defined, notably including the Business Process Model and Notation (BPMN). In the perspective of evaluation, the systematic use of the Strategy design pattern enables the definition of new analysis heuristics and new approximation methods for monovariate PDFs.

It is planned to support structured cycles that execute a body and then repeat it with constant probability $p > 0$ or terminate with probability $1 - p$. Future work also includes exploiting the library in specific application domains such as management of manufacturing processes and composition of web services. In these application contexts, Eulero methods will provide added-value features, such as predicting completion times of a production workflow or selecting SLA compliant services of a composite web service.

Data Availability Statement. An artifact of the Eulero library is available at https://doi.org/10.5281/zenodo.6841108. The artifact consists of the jar archive of the library, two Java classes containing the code snippets provided with the paper, a suite of models randomly generated through the procedure described in Sect. 5.2, a script to generate and evaluate a new suite of models, and a script to replicate the generation and the evaluation of a suite of models.

References

1. Alur, R., Dill, D.L.: A theory of timed automata. Theoret. Comput. Sci. **126**(2), 183–235 (1994)
2. Amparore, E.G., Balbo, G., Beccuti, M., Donatelli, S., Franceschinis, G.: 30 years of GreatSPN. Principles of performance and reliability modeling and evaluation, pp. 227–254 (2016)

3. Carnevali, L., Paolieri, M., Reali, R., Vicario, E.: Compositional safe approximation of response time distribution of complex workflows. In: Abate, A., Marin, A. (eds.) QEST 2021. LNCS, vol. 12846, pp. 83–104. Springer, Cham (2021). https://doi.org/10.1007/978-3-030-85172-9_5

4. Carnevali, L., Reali, R., Vicario, E.: Compositional evaluation of stochastic workflows for response time analysis of composite web services. In: Proceedings of the ACM/SPEC International Conference on Performance Engineering, pp. 177–188 (2021)

5. Ciardo, G., Jones, R.L., Miner, A.S., Siminiceanu, R.: Logical and stochastic modeling with SMART. In: Kemper, P., Sanders, W.H. (eds.) TOOLS 2003. LNCS, vol. 2794, pp. 78–97. Springer, Heidelberg (2003). https://doi.org/10.1007/978-3-540-45232-4_6

6. Curbera, F., Goland, Y., Klein, J., Leymann, F., Roller, D., Thatte, S., Weerawarana, S.: Business process execution language for web services (2002)

7. Dehnert, C., Junges, S., Katoen, J.-P., Volk, M.: A Storm is coming: a modern probabilistic model checker. In: Majumdar, R., Kunčak, V. (eds.) CAV 2017. LNCS, vol. 10427, pp. 592–600. Springer, Cham (2017). https://doi.org/10.1007/978-3-319-63390-9_31

8. van Der Aalst, W.M., Ter Hofstede, A.H., Kiepuszewski, B., Barros, A.P.: Workflow patterns. Dist. Paral. Datab. **14**(1), 5–51 (2003)

9. Dick, R.P., Rhodes, D.L., Wolf, W.: TGFF: task graphs for free. In: Proceedings of the Sixth International Workshop on Hardware/Software Codesign. (CODES/CASHE 1998), pp. 97–101. IEEE (1998)

10. Foundation, A.S.: Apache airflow

11. Gamma, E., Helm, R., Johnson, R., Vlissides, J.M.: Design Patterns: Elements of Reusable Object-Oriented Software. Addison-Wesley Professional, 1 edn. (1994)

12. German, R., Lindemann, C.: Analysis of stochastic Petri nets by the method of supplementary variables. Perform. Eval. **20**(1–3), 317–335 (1994)

13. Horváth, A., Paolieri, M., Ridi, L., Vicario, E.: Transient analysis of non-Markovian models using stochastic state classes. Perform. Eval. **69**(7–8), 315–335 (2012)

14. Jensen, E.D., Locke, C.D., Tokuda, H.: A time-driven scheduling model for real-time operating systems. In: RTSS, vol. 85, pp. 112–122 (1985)

15. Katoen, J.P., Zapreev, I.S., Hahn, E.M., Hermanns, H., Jansen, D.N.: The ins and outs of the probabilistic model checker MRMC. Perform. Eval. **68**(2), 90–104 (2011)

16. de Kok, T.G., Fransoo, J.C.: Planning supply chain operations: definition and comparison of planning concepts. Handbooks Oper. Res. Manage. Sci. **11**, 597–675 (2003)

17. Kulkarni, V.G.: Modeling and Analysis of Stochastic Systems. Chapman and Hall/CRC, Boca Raton (2016)

18. Kwiatkowska, M., Norman, G., Parker, D.: PRISM: probabilistic symbolic model checker. In: Field, T., Harrison, P.G., Bradley, J., Harder, U. (eds.) TOOLS 2002. LNCS, vol. 2324, pp. 200–204. Springer, Heidelberg (2002). https://doi.org/10.1007/3-540-46029-2_13

19. Paolieri, M., Biagi, M., Carnevali, L., Vicario, E.: The ORIS tool: quantitative evaluation of non-Markovian systems. IEEE Trans. Softw. Eng. **47**(6), 1211–1225 (2019)

20. Rahman, J., Lama, P.: Predicting the end-to-end tail latency of containerized microservices in the cloud. In: 2019 IEEE International Conference on Cloud Engineering (IC2E), pp. 200–210. IEEE (2019)

21. Russell, N., Ter Hofstede, A.H., Van Der Aalst, W.M., Mulyar, N.: Workflow control-flow patterns: a revised view. BPM Center Report BPM-06-22, BPMcenter. org, pp. 06–22 (2006)
22. SA, S.T.: Luigi
23. Trivedi, K.S., Sahner, R.: SHARPE at the age of twenty two. ACM SIGMETRICS Perform. Eval. Rev. $\mathbf{36}(4)$, 52–57 (2009)
24. Van Eyk, E., Iosup, A., Abad, C.L., Grohmann, J., Eismann, S.: A SPEC RG cloud group's vision on the performance challenges of FaaS cloud architectures. In: Companion of the 2018 ACM/SPEC International Conference on Performance Engineering, pp. 21–24 (2018)
25. Vicario, E.: Static analysis and dynamic steering of time-dependent systems. IEEE Trans. Softw. Eng. $\mathbf{27}(8)$, 728–748 (2001)
26. Vicario, E., Sassoli, L., Carnevali, L.: Using stochastic state classes in quantitative evaluation of dense-time reactive systems. IEEE Trans. Softw. Eng. $\mathbf{35}(5)$, 703–719 (2009)
27. Zheng, Z., Trivedi, K.S., Qiu, K., Xia, R.: Semi-Markov models of composite web services for their performance, reliability and bottlenecks. IEEE Trans. Serv. Comput. $\mathbf{10}(3)$, 448–460 (2015)
28. Zimmermann, A.: Modelling and performance evaluation with TimeNET 4.4. In: Bertrand, N., Bortolussi, L. (eds.) QEST 2017. LNCS, vol. 10503, pp. 300–303. Springer, Cham (2017). https://doi.org/10.1007/978-3-319-66335-7_19

Applications

Preference-Aware Computation Offloading for IoT in Multi-access Edge Computing Using Probabilistic Model Checking

Kaustabha Ray[✉] and Ansuman Banerjee

Advanced Computing and Microelectronics Unit, Indian Statistical Institute, Kolkata, India
kaustabharay@gmail.com, ansuman@isical.ac.in

Abstract. A key cornerstone of Multi-Access Edge Computing (MEC) is an offloading policy utilized to determine whether to execute computation tasks on IoT devices or to offload the tasks to MEC servers for processing. In this work, we propose a Probabilistic Model Checking based offloading policy catering to device user preferences. We model the interactions between the various components of the MEC environment using a Turn-Based Stochastic Multi-Player Game (SMG). We present experiments on practical scenarios on data gathered from a test-bed setup with benchmark applications to show the benefits of an adaptive preference-aware approach over conventional approaches in MEC offloading.

Keywords: Multi-Access Edge Computing · Probabilistic Model Checking

1 Introduction

In recent times, Multi-Access Edge Computing (MEC) is showing much promise as the preferred application service provisioning model to facilitate convenient access to services for Internet-of-Things (IoT) device users [17,24,27,38]. The central idea of MEC is to have service providers deploy their application services on MEC servers located near mobile base stations. Computationally intensive tasks from mobile IoT devices are either executed locally on the devices or routed to, and served from nearby MEC servers on their route as they move around, with improved computation running times. This provisioning model is increasingly being acknowledged as a near-user low latency convenient alternative to traditional cloud computing for several classes of real-time latency driven applications. Driven by new innovations in MEC, the number of application services (e.g. object recognition, obstacle identification, navigation, maps, games, e-commerce etc.) hosted at servers to be used by IoT users (e.g. autonomous vehicles, drones, users on the move) is also growing at a considerable pace.

Motivation and Objectives of This Work: A key challenge in MEC is to devise an offloading policy which decides whether to execute the task locally on

© Springer Nature Switzerland AG 2022
E. Ábrahám and M. Paolieri (Eds.): QEST 2022, LNCS 13479, pp. 275–297, 2022.
https://doi.org/10.1007/978-3-031-16336-4_14

the IoT devices or on a MEC server. While on one hand, executing tasks on MEC servers enables low computation running times, on the other hand, the data transfer associated with offloading the task to the MEC server can lead to high bandwidth usage and in turn high battery consumption. This can occur since the data transmitter is often a higher energy consuming component than the CPU [4]. The users of these IoT devices often have several preferences such as long Battery Cycle Life (BCL) and high quality of experience (QoE). However, the role of user-preferences in computation offloading has been relatively less explored [21]. An offloading strategy that is optimized towards computation running time may perform poorly when the user expects a high BCL, while similarly, a different offloading strategy that is optimized towards energy consumption may produce high computation times when the user prefers a high QoE. Deriving an offloading strategy catering to user preferences is thus a critical challenge. Additionally, IoT devices are often heterogeneous, comprising CPU cores with varied operating frequencies, thus executing tasks locally on the IoT devices is accompanied by the additional challenge of scheduling the tasks on the appropriate core to ensure conformance to user preferences. Overall, finding the best trade-off between local execution versus MEC server execution in the presence of user preferences necessitates judicious planning and scheduling of the device-execution mapping over time as different tasks are executed by different devices. This is aggravated by the fact that some IoT devices' performance and longevity are affected by the temperature of its various components [32]. Executing a task for prolonged periods on the IoT device or on the same CPU core may often lead to device shutdowns when critical temperatures are exceeded. This often necessitates task migrations to preserve continuity of service provisioning and a steady acceptable Quality of Experience (QoE). In this paper, we derive a computation offloading strategy considering all the above scenarios. The offloading strategy serves as a wrapper to the operating system running on the IoT device, determining offloading and heterogeneous core scheduling decisions.

Our Approach: In this paper, we use formal methods for the synthesis of computation offloading strategies with probabilistic guarantees. We model the offloading policy using a Labeled Transition System (LTS) [2]. We model the IoT device and the scheduling tasks on the heterogeneous cores as another LTS. We leverage on the power of non-determinism of the LTS models to capture the various choices of offloading and scheduling. We model the resulting stochastic computation running times of executing tasks as a Discrete Time Markov Chain (DTMC) [2]. We utilize composition of the aforementioned models which yields a Markov Decision Process (MDP) capturing the non-deterministic choices of the LTSs and the stochastic nature of the DTMC. Finally, we capture the complex interactions between the MEC servers and the IoT devices as a Turn-Based Stochastic Multi-Player Game [13]. The interactions are captured using the composition of the MDP and the DTMC. Composing such models allows effective characterization of the non-deterministic choices of offloading and heterogeneous core scheduling while also characterizing the stochastic nature of task execution running times. We formalize user preferences such as high BCL or low computation times as a reward formulation and use a probabilistic model checker [13] to synthesize offloading strategies. The SMG enumerates the trade-offs such as on-device and off-device execution, heterogeneity of IoT device processor cores, IoT device battery energy consumption and battery temperature with respect to user preferences. We utilize a testbed with the Odroid XU4 board as the IoT device, with heterogeneous cores, each having different processing capabilities. We experimentally demonstrate the benefits of our adaptive SMG based preference-aware approach over conventional approaches.

2   A Motivating Example

Consider the MEC scenario depicted in Fig. 1 comprising two edge servers E_1 and E_2. The coverage areas of E_1 and E_2 are depicted by encompassing circles. An IoT device located in the coverage area of an edge site can connect to one of the associated edge servers with low latency access [15,23,24]. A user allocation policy determines which edge server an IoT device's task request is allocated to. Once the allocation is determined, the offloading policy determines whether the task is executed either locally on the IoT device or on the MEC server. We consider heterogeneous IoT devices comprising multiple cores with varied operating frequencies and power consumption. For example, in Fig. 1a, the video processing task from the IoT device, a mobile phone in this scenario, is executed locally on the mobile phone. On the other hand, in Fig. 1b, the video processing task is offloaded to the MEC server E_1 for execution. Executing the task locally versus executing the task on the MEC server is associated with varying battery energy consumption, battery temperature and resulting execution computation times. For execution on the MEC server, we consider execution computation time as the task execution running time along with the server access latency. We first perform a set of experiments to demonstrate the benefits of a preference-aware strategy. We use an ODROID XU4 board as the IoT device. We generate synthetic workloads using real-world applications from the Mibench benchmark [10,32] as explained in detail in Appendix B. We consider the scenario where the IoT device user's preference is set to low computation time and low energy consumption. We quantify this notion in Sect. 4.5. In the following, we compare the different offloading approaches.

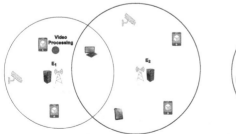   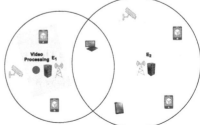

(a) Local Processing on IoT device   (b) Offload to MEC server

Fig. 1. Multi-Access Edge Computing Enabled IoT

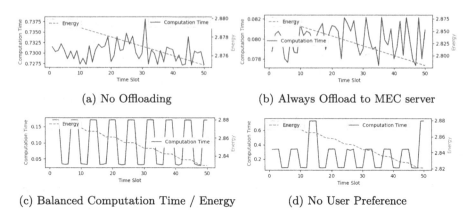

(a) No Offloading

(b) Always Offload to MEC server

(c) Balanced Computation Time / Energy

(d) No User Preference

Fig. 2. Multi-Access Edge Computing Enabled IoT

A strategy that does not offload always executes the task on the device. Such a strategy leads to lower energy consumption, however leads to higher computation times. The resulting computation times (in seconds) and remaining device energy (in Joules) are depicted in Fig. 2a. A strategy that always selects to offload the task to the MEC server incurs low computation time as observed in Fig. 2b. However, such a strategy incurs higher energy consumption as compared to the no offloading strategy. Both the strategies above execute the tasks without considering user preferences. Recently, in [21], the benefits of adapting to user preferences were demonstrated in a different context, i.e., user allocation in MEC. To overcome the limitations of the above strategies, we propose a Stochastic-Game based Strategy, discussed in detail in Sect. 4. Our SMG based strategy generates offloading strategies by considering IoT device user preferences. Figure 2c depicts the results by setting the user preferences to low computation time and low battery consumption. The benefits of incorporating user preferences into offloading decisions thus can have a critical impact on the computation time and the energy consumption. Figure 2d depicts the results when the user does not have any specific preferences, which can be handled by our approach as well.

3   Problem Formulation

We consider the following in this work:

- The MEC system comprises n MEC servers, $E = \{E_1, E_2, \ldots E_n\}$ where each server is represented by its latitude and longitude coordinates.
- Server E_j is associated with a service zone of radius r_j. IoT devices located within the zone can avail services deployed on the server with low latency.
- We consider IoT devices comprising heterogeneous cores with each core associated with different operating frequencies.

- We assume that the application for the task and its associated data are available at the MEC server in each discrete time slot.
- We assume that within an IoT core comprising multiple CPUs, the task scheduling within a specific core is carried out by the operating system.
- The MEC system follows a discrete time-slotted model [23, 24, 38].
- We consider each discrete time slot Δt of duration μ seconds.

We now describe the computation offloading strategy synthesis problem. The offloading strategy is synthesized individually for each IoT device user. Each IoT device user is associated with preferences such as low computation time or low battery energy consumption and so forth. Whenever an IoT device executes a task T, the objective of the computation offloading policy is to determine for each discrete time slot: a) whether the task should be executed on the device locally or the computation be offloaded to the MEC server and b) for local execution, determine the core to schedule the task for heterogeneous hardware enabled devices by taking into consideration the IoT device user preferences.

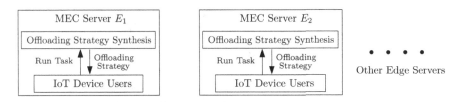

Fig. 3. Architecture of the Offloading Strategy Synthesis

We now discuss the overall architecture of our offloading strategy synthesis setup. Figure 3 depicts the overall architecture of our methodology. We use a distributed approach where each edge server is associated with an offloading strategy synthesis controller. Whenever an IoT device executes a task, the IoT device invokes the offloading controller. The IoT device provides the offloading controller with IoT user specified preferences. The offloading controller then initializes the various models as discussed in detail later in Sect. 4. The offloading model utilizes the rewards formulation to represent the IoT device preferences as discussed in detail in Sect. 4.5. The objective of strategy synthesis is specified in rPATL as discussed in Sect. 4.6. The models and the rPATL specification are then utilized as inputs to a Probabilistic Model Checker which returns the offloading strategy. The offloading is then executed in accordance with the strategy returned by the Probabilistic Model Checker. In the next section, we discuss our formal model of the computation offloading policy.

4   Formal Model of Computation Offloading

Our formal model of the offloading strategy is based on a Turn-Based SMG. We utilize a composition driven approach where we model the various components of the MEC ecosystem separately and then utilize MDP composition to obtain the SMG model. We now describe the model of each component in detail.

4.1   LTS Model of the Offloading Policy

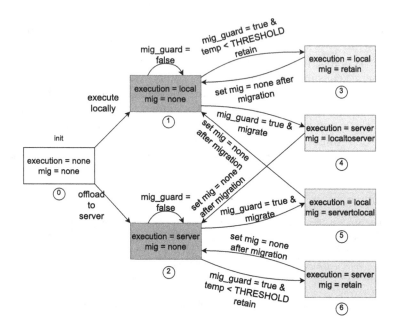

Fig. 4. LTS Model of the Offloading Policy

We model the offloading policy as an LTS. The state space of the LTS represents the various possibilities of task execution for an IoT device, i.e., local versus server execution and possible task migration across the IoT device and MEC server. Each state of the LTS is thus associated with two Atomic Propositions (APs): *execution* and *mig*, where each AP is an enumerated type denoting the various possibilities of location of task execution and task migration respectively. We model each possible execution and migration scenario using the APs. The LTS model of the offloading policy is depicted in Fig. 4. We now explain the model of the offloading policy in detail:

- *Initial State:* State ⓪ depicts the initial state of the offloading policy where both the APs are set to none representing there is no current execution and no migration being executed.
- *Choices of Execution:* We consider a discrete time slotted model where the decision to execute the task either locally or on the MEC server is executed at the beginning of each time slot. The two choices of execution are represented

with the transitions from state (0) to (1) and (2). State (1) denotes local execution of the task while (2) denotes server execution. Thus, such transitions denote the initial non-deterministic choice of local versus server execution. Note that in Fig. 4, the AP corresponding to execution is updated in states (1) and (2) accordingly.

– *Choices of Migration:* Task migration occurs when the task executing locally on the IoT device is moved to the MEC server for execution (or vice-versa). The outgoing transitions from states (1) and (2) denote the various migration possibilities. However, frequent migrations can have a derogatory effect on the running times due to the added overheads involved in executing the migration such as data transfer. Thus, in order to prevent frequent migrations, we restrict that after the allocation of the task to either the local IoT device processor or the MEC server, migration can only be executed after t_{mig} discrete timesteps. This condition is represented as a *migguard* in Fig. 4. Thus, all migration choices can only be executed when *migguard = true*. We explain the various migration possibilities in our model below:

- No migration: In this scenario, the execution state (either local execution or MEC server execution) is retained and no migration overheads are involved. Such scenarios are denoted by the transitions from (1) to (3) and (1) to (6) where *mig = retain*. Retention is only possible when the temperature of the IoT device cores is below a critical temperature THRESHOLD, which is an external parameter.

- Migrate from device to server: Migration from the device to the MEC server is denoted by the transition from (1) to (4). After successfully executing the migration, the LTS transitions to state (2) which denotes MEC server execution. The AP *mig = none* signifies migration completion.

- Migrate from MEC server to device: Migration from MEC server to the device is analogous to the above scenario with the APs adjusted accordingly. The transitions from (2) to (4) and subsequently to (1) denote the complete life-cycle of such a migration.

Note that we utilize states (3) to (6) to denote the various possibilities of migration. We account for the costs associated with the migrations by the state based reward formulation explained in detail later. In the next subsection we discuss the LTS model that we create to represent the functioning of each IoT device.

4.2   LTS Model of the IoT Device

We model each IoT device as a labeled LTS. The LTS model of the IoT device incorporates the various choices of scheduling on the IoT device locally and the resulting impact on the energy consumption, and the temperature of the IoT device. Each state of the LTS is associated with five Atomic Propositions (APs): *batteryenergy, temp, coremig, core* and *dtime*, denoting the battery State-of-Charge (SoC), the temperature of the device, the core migration status, the selected core for executing the task and discrete time slot respectively. Thus, we model each possibility of task allocation to cores and task migration across cores

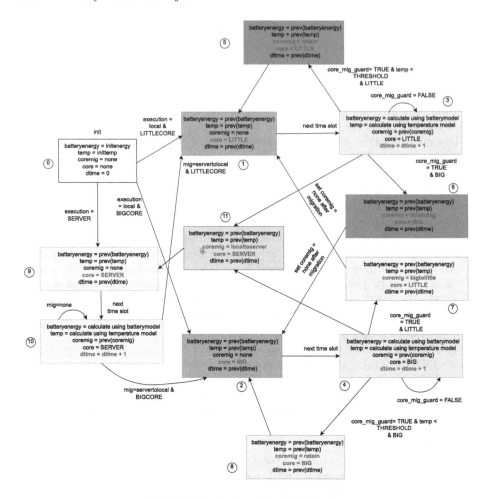

Fig. 5. LTS Model of the IoT Device

using the APs. The LTS model of the IoT device (considered as the Odroid XU4 board) is depicted in Fig. 5. We now explain the model in detail:

- *Initial State:* State ⓪ depicts the initial state of the IoT device. The AP *batteryenergy* is initialized with the remaining battery energy of the device when the task was initialized, the *temp* AP is initialized with the temperature of the device, both the *coremig* and *core* APs are initialized to none since there is no execution initially, and *dtime* is initialized with the number of discrete time slots that the model will run for.
- *Choices of Local Execution:* We consider the Odroid XU4 board as the device, which comprises two heterogeneous cores, LITTLE and BIG, each running at different clock speeds. Whenever an IoT device executes a task T and the LTS model of the offloading policy selects local execution, there are two possible

choices of scheduling the task, the LITTLE core or the BIG core. The two choices of execution are represented with the transitions from state ⓪ to ①️ and ②️. State ①️ denotes the task is executed on the LITTLE core while ②️ denotes the task is executed on the BIG core. Thus, such transitions denote the non-deterministic choice of LITTLE core versus BIG core scheduling. Note that in Fig. 5, the AP corresponding to *core* is updated in states ①️ and ②️ accordingly. For each transition between states, we utilize *prev(AP)* to depict the scenario where an AP retains the value in the previous state. When the LITTLE core is chosen as an outcome of the policy, the task may be scheduled on any of the available LITTLE cores. The same holds true for the BIG core as well. Executing the task on the LITTLE core/BIG core is associated with different energy consumption/task running times.

- *Local Execution:* On successful scheduling on either the LITTLE/BIG core, the task computation proceeds with battery energy consumption and temperature according to the core the task is being executed on. In each discrete time step, the battery energy consumption and the temperature are calculated according to the model discussed in Appendix A. The state of the LTS is then updated to reflect the battery energy consumption and the resulting temperature. Such scenarios are denoted by transitions from state ①️ to ③️ for the LITTLE core and from state ②️ to ④️ for the BIG core.
- *Choices of Migration:* Task migration across cores occurs when the task executing locally on the IoT device is moved from the LITTLE core to the BIG core (or vice-versa). The outgoing transitions from states ③️ and ④️ denote the various core migration possibilities. In order to prevent frequent migrations across cores, after the allocation of the task to either the LITTLE core or the BIG core, migration can only be executed after C_m discrete timesteps. This condition is represented as *core_mig_guard* in Fig. 5. Thus, all migrations choices can only be executed when *core_mig_guardard = true*. We explain the various core migration possibilities in our model below:
  - No migration: In this scenario, the execution state (the scheduled core) is retained and no migration overheads are involved. Such scenarios are denoted by the transitions from ③️ to ⑤️ and ④️ to ⑧️ where the *coremig* AP is set as *retain*. Retention is only possible when the temperature of the IoT device cores is below the critical temperature THRESHOLD.
  - Migrate from LITTLE to BIG core: Migration from the LITTLE core to the BIG core is denoted by the transition from ③️ to ⑥️. After successfully executing the migration, the MDP transitions to state ②️ which denotes the task is being scheduled on the BIG core. The AP *coremig* is set to *none* signifying migration completion.
  - Migrate from BIG to LITTLE core: Migration from BIG to LITTLE is analogous to the above with the APs adjusted accordingly.

Task migration from local processing on the IoT device to server execution is denoted by the transitions from ③️ to ⑪️ and ④️ to ⑪️.

– *Server Execution:* On the other hand, when the LTS model of the offloading policy selects server execution, the LTS of the IoT device transitions to state ⑨ with the core AP updated to *SERVER*. In this scenario, in each discrete time step, the battery energy consumption and the temperature are calculated according to the model similar to the earlier scenario. In this case, the energy consumption is dominated by the transmitter/receiver modules of the device. This is because in this scenario, the entire processing is carried out on the MEC server. Such scenarios are denoted by self transitions to state ⑩ until the offloading controller invokes a local migration.

Note that our model is generic and can be generalized to other IoT devices by incorporating as many non-deterministic choices of cores as supported by the device. Similarly, note that in our model we only depict *batteryenergy* and *temperature* to represent the devices's characteristics. However, our model is also generic on the characteristics front where a different characteristic such as *bandwidthconsumption* can simply be incorporated by adding a corresponding AP and the resulting updation models for *bandwidthconsumption* similar to *batteryenergy* and *temperature*. In the next subsection we discuss the DTMC model of task execution latencies.

4.3   DTMC Model of Task Execution Latencies

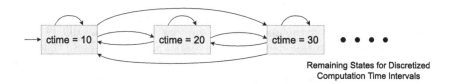

Fig. 6. DTMC Model of Task Computation Time

The LTS models of the offloading policy and the IoT device non-deterministically select the choice of task execution. As a result of selecting a particular device for execution, the resulting computation times are generated by a DTMC model representing the stochastic nature of computation times. Each state of the DTMC is represented by the *ctime* AP. In each state, *ctime* represents the computation time incurred for executing the task invoked by the IoT device during the discrete time slot. The computation time representation is discretized into equal sized sub-intervals $R_{interval}$ in the closed interval $[0, R_{max}]$. The concrete continuous valued latencies are mapped to the discretized interval computation time set as follows: computation times between 0 and $R_{interval}$ in the continuous closed-open interval $[0, R_{interval})$ are mapped to the state where the *ctime* component is $1 \times R_{interval}$, computation times in the continuous interval $[R_{interval}, 2 \times R_{interval})$ are mapped to $2 \times R_{interval}$ and so forth, where $[0, R_{interval})$ denotes the continuous interval inclusive of the lower bound 0 and exclusive of $R_{interval}$. Computation times which exceed the threshold $R^{a_k}_{max}$ are mapped to $R^{a_k}_{max}'$. Our work builds on a similar discretized interval representation used in [31, 38]. Figure 6 depicts the DTMC for task execution computation times. For task execution on the MEC server, the stochastic computation running time is inclusive of the network latency to access the MEC server. From each state of the DTMC, a transition exists to all other states. Such transitions execute in a discrete time step where the state of the DTMC is updated in accordance with the computation time encountered for task execution.

4.4 Turn-Based Stochastic Multi-player Game Model Composition

The aforementioned models systematically abstract the behaviour of the various entities involved in the MEC ecosystem. Firstly, the overall model is obtained by parallel composition of these models and is denoted as:

$$G_{MEC} = M_{offloading-policy} || M_{IoT-device} || M_{task-computation-time}$$

In order to represent the interactions between the various models, we utilize a Turn-Based SMG. In our SMG, we utilize two players, *offload&schedule* and *taskctime*. The composition of the models results in an MDP [6], since each LTS is also an MDP with only non-deterministic transitions and the DTMC is also an MDP with only stochastic transitions, where each state of the MDP is controlled by a specific player as follows:

- The player *offload&schedule* controls the choices in states corresponding to the LTS model of the offloading policy and the LTS model of the IoT device.
- The player *taskctime* controls the stochastic transitions in the states corresponding to the DTMC model of Task Execution Computation Times.

Fig. 7. Turn-Based SMG for Computation Offloading

The Game is initialized with the player *offload&schedule* and alternates in turn with player *taskctime*. Thus, when it is the turn of the player *offload&schedule*, all non-deterministic choices for local versus server execution as well as LITTLE versus BIG core are resolved. Note that the player *offload&schedule* only determines whether to execute the task on the LITTLE or the BIG core. The operating system running on the device, determines which of the CPUs within a LITTLE or BIG core to schedule the task once the non-determinism is resolved by *offload&schedule*. In the subsequent turn, the player *taskctime* generates computation time according to the defined probability distribution of the DTMC model. The Turn-Based SMG is depicted in Fig. 7.

4.5   Reward Formulation and IoT User Preferences

The Turn-Based SMG does not quantify the impact of each offloading and scheduling choice. In order to quantify each offloading and scheduling choice, we utilize the notion of rewards where each state of the SMG is associated with a quantitative reward value in accordance with the APs in the state. Such a notion of rewards is used to encode the preferences of the device user.

For each AP, we define a function $\rho_{AP} : \{AP\} \rightarrow \mathbb{R}$, a mapping from each possible value of the AP (denoted as $\{AP\}$) to a real number, indicating a quantitative measure of the values of the AP. Note that the composition yields an MDP which includes the APs from all the models [6]. The total reward of each state is thus defined as the additive reward for each AP defined by the individual rewards functions ρ_{AP}. We now discuss types of user preferences and their representation using the rewards formulation.

- No User Preferences: In such a scenario, each state of the SMG is assigned a constant positive reward denoting all states are quantitatively equivalent.
- Low Computation Time: We define a reward formulation where each *ctime* AP is assigned a constant positive value. Such values can be specified as external parameters. Each discrete computation time interval is assigned a value in accordance with the user preferences, with higher positive values assigned to computation time values with a high user preference.
- Low Energy Consumption: Represented by each state's reward value defined by the AP *batteryenergy*. Intuitively, such a formulation captures the notion of energy consumption since the AP *batteryenergy* represents the actual remaining energy of the IoT device's battery model.
- Low Computation Time and Low Energy Consumption (Balanced): Simultaneous preferences involving both computation time and energy consumption are formulated by assigning constant positive values associated with the AP *ctime* as well as assigning to each state a positive reward equal to the value of the AP *batteryenergy*. In such a scenario, the reward associated with each state is the additive value of the rewards. Additionally, since migration can involve data transfer overheads, the migration costs are incorporated by assigning rewards depending on the APs involved in Task and Core migration.

Table 1. Reward Formulation for User Preferences

State	Reward
All States	1

(a) No Preference

State	Reward
ctime = 10	100
ctime = 20	80
ctime = 30	60

(b) Low Computation Running Time

State	Reward
All States	batteryenergy

(c) Low Energy

State	Reward
ctime = 10	100
All States	batteryenergy
retain	20
mig != 0	10
tmig != 0	10

(d) Balanced

Table 1 describe four different reward encoding schemes for the different types of user preferences discussed above. Note that the reward formulation can be setup depending on the APs. The function ρ defines the reward values associated with each AP where the mapping between the AP and the defined real value is utilized to determine the reward of a state. Thus, our model is not restricted to the above particular reward formulation. Depending on the user's preferences, an appropriate reward formulation can be defined with a suitable function ρ.

4.6 Offloading Strategy Objectives

We use G_{MEC} to synthesize offloading strategies. We use temporal logic to formally specify the objective of the strategy synthesis. Specifically, to express properties for SMGs we use the logic rPATL - Probabilistic Alternating-time Temporal logic with Rewards [13]. The objective of the offloading strategy synthesis for the player *offload&schedule* is specified as:

$$\phi = \langle\langle offload\&schedule \rangle\rangle R_{\{max=?\}}[F \; dtime = TMAX]$$

In this specific scenario, the property utilizes the 'Eventually' temporal operator denoted by F. Hence, the synthesis problem for an SMG aims to find the optimal strategy π which resolves the non-deterministic choices for the *offload&schedule* player [13] for the duration when $dtime = TMAX$, where $TMAX$ denotes the number of discrete time slots for which the strategy is to be synthesized. After TMAX timeslots, the procedure is invoked again with the APs updated according to the present state. Formally the synthesis problem for G_{MEC} is defined as:

Definition 1. *Given the SMG G_{MEC}, a strategy π is a set of rules to resolve all non-deterministic choices of a player* **offload&schedule** *such that for all opponent strategies σ, where the opponent is* **taskctime**, *the resolution of the non-deterministic choices satisfies the property ϕ.*

The R_{max} operator evaluates the maximum expected value of the cumulative rewards since an SMG involves stochastic behavior. The algorithm utilized to compute the expected rewards ensures maximization of the rewards and the details are presented in [6]. Probabilistic Model Checking ensures probabilistic guarantees with respect to the specification for the expected cumulative reward [6]. We use PRISM-Games [13] to implement G_{MEC} along with the rPATL property. PRISM-Games utilizes Probabilistic Model Checking to determine the maximum numerical value of the expected reward associated with the property. Model checking systematically explores all states and transitions in the model to check whether it satisfies the given property. Thus, PRISM-Games systematically explores the search space generated by G_{MEC} considering all possible interactions between the two players and their respective choices of actions. In the next section, we describe our implementation.

5   Results and Discussion

In this section we demonstrate the effectiveness of our approach against state-of-the-art algorithms. Our experimental setup is described in Appendix B. Figure 8 provides a comparative study on the resulting temperature for the different cores simulated using the model described in Sect. A. Figure 8 also provides a comparative study of the computation time incurred versus the location of task execution (either locally on the IoT device or on the MEC server). The computation time is depicted in seconds while the energy consumption is depicted as multiplied by 10^3 for brevity. We consider preferences as considered in Sect. 4.5. For our simulation, we considered T_{env} as $32°$. Since T_{env} is a parameter in our model, any other value of T_{env} can be utilized by setting the parameter to the current environmental temperature. Figures 8a and 8b depict the computation time incurred when the preference is set to low energy consumption. In this case, our SMG based approach always schedules the device on the LITTLE core. Figures 8c and 8d depict the scheduling scenario for low computation time preferences. In this case, the SMG based strategy schedules the task always on the MEC server since the MEC server provides the lowest computation time. Figures 8e and 8f correspond to the balanced computation time and energy scenario as described in Sect. 4.5. In this scenario, our SMG based strategy schedules the tasks on the LITTLE core and the MEC server while not utilizing the BIG core. Finally, Figs. 8g and 8h depict the results for the scenario when there is no specific user preference. In such a scenario, the SMG based approach schedules the tasks on the LITTLE, BIG cores as well as MEC servers. Such variations depict the governing impact of the reward formulation on the resulting synthesized scheduling strategy. Figure 8 thus demonstrates the effectiveness of our SMG based strategy with varied user preferences.

Table 2. Model Characteristics

No. of Discrete Slots	25	40	50
Number of States	7515	25585	46515
Model Checking Time (in Seconds)			
Low Energy	0.126	1.083	1.699
Low Time	0.118	0.927	1.427
Balanced	0.134	1.236	1.934
No Preferences	0.126	1.157	1.702
Algorithm Running Time (in Seconds)			
G-ECPRA - 0.011 s, Greedy - 0.007 s			
CEFO - 0.021 s			

Table 3. Comparison With Others

	User Preference - Low Energy Consumption		User Preference - Low Computation Time	
	Energy Consumed (in Joules)	Computation Time (in Seconds)	Energy Consumed (in Joules)	Computation Time (in Seconds)
G-ECPRA [1]	756	0.224	742	0.235
SMG	761	0.212	1043	0.131
Greedy [16]	1303	0.119	1245	0.127
CEFO [35]	856	0.154	877	0.178

Table 3 depicts the comparison of our approach with G-ECPRA [1], a heuristic based algorithm that is optimized towards minimizing energy consumption and does not adapt to user preferences and CEFO [35], an offloading strategy which jointly optimizes computation time, latency and energy, however does not consider user preferences. We observe that our SMG based strategy when utilized with the reward formulation set to low energy consumption performs similar to

G-ECPRA. However, G-ECPRA when compared with the SMG strategy with the reward formulation set to low computation time, is unable to adapt to the new user preference. We compare with a greedy strategy optimized towards computation time and latency which selects the task execution location depending on the current load [16]. The greedy strategy also performs similarly as observed in Fig. 3. These approaches are geared towards a single optimization objective and cannot incorporate different user preferences as in our SMG based approach. Our approach is thus flexible catering to a variety of preference-aware scenarios. Table 2 details the number of states of the model with variation in the number of time slots with higher number of slots providing higher accuracy. We found from our experiments that selecting more than 50 discrete time slots exceeds 2 s. Additionally, the table lists the model checking times for the various types of user preferences discussed in Sect. 4.5.

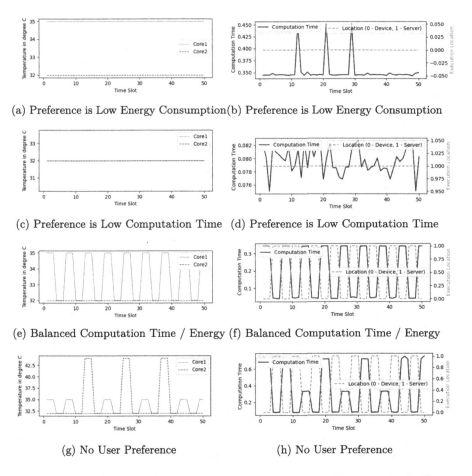

(a) Preference is Low Energy Consumption (b) Preference is Low Energy Consumption

(c) Preference is Low Computation Time (d) Preference is Low Computation Time

(e) Balanced Computation Time / Energy (f) Balanced Computation Time / Energy

(g) No User Preference   (h) No User Preference

Fig. 8. Temperature, Computation Time and Execution Location for SMG

6   Related Work

In recent times, there has been extensive research on various issues in the MEC paradigm [22,23,26,29,30,39]. Offloading in Mobile Cloud Computing [3,9,42] and Multi-Access Edge Computing [7,18,33] have both been studied extensively concerning what/when/how to offload workloads from handheld devices to the cloud or edge [41]. A number of different approaches to offloading have been proposed. [20] focuses on the energy efficiency of edge devices. Wang et al. [37] deal with minimizing each MEC server's energy consumption while satisfying QoS requirements. The authors [5] consider scenarios where migrations are allowed between edge servers while keeping the objective same as in [37]. The authors in [7] introduce a unique perspective on energy consumption with simultaneous harvesting by considering devices utilizing battery resources while harvesting other energy sources. Recently, in [36] the authors investigate the role of each user preferring an MEC server over another. However, they do not consider individual preferences such as computation time and energy. Unlike an offloading policy, service/user allocation [14] deals with determining the assignment of service requests from users to already deployed services on MEC servers. A number of allocation policies have been proposed in recent literature considering various metrics such as number of users allocated, QoE/QoS maximization, energy optimization, optimizing the number of re-allocations as users move about and so on [14,21,23,43]. Probabilistic Model Checking has been demonstrated as an effective technique for formally reasoning about the performance and reliability of systems involving stochastic behavior [12]. A number of authors have applied probabilistic model checking to provide performance guarantees in aspects of cloud and MEC systems such as auto-scaling [8], horizontal scaling [19], live migration [11] and user allocation [25,28]. However, unlike other approaches, we use probabilistic model checking to synthesize user-preference based offloading strategies in MEC enabled IoT, which makes this work novel.

7   Conclusion and Future Work

In this paper, we propose an offloading strategy for MEC enabled IoT devices derived using formal methods. We model the offloading strategy synthesis problem using a Turn-Based Stochastic Multiplayer Game as a composition of several components. We demonstrate encoding user preference objectives as the rewards formulation. Experiments on benchmarks demonstrate the effectiveness of our approach. In our work, we assume that the application and the associated data are always available at the MEC server. A possible future direction is thus to design a joint offloading and service placement strategy where the application services may not always be available at the MEC server. Another possible direction is synthesizing offloading schemes by incorporating Dynamic Voltage Frequency Scaling (DVFS) and the varied energy consumption that different DVFS levels can admit unlike our work where average power consumption is considered.

A   Battery Model

We use a battery model $(4\,\text{V}, 2{,}000\,\text{mAh})$ which is representative of IoT devices. Further, we consider an effective battery energy $(28{,}800\,\text{J})$ which has been utilized in several previous studies [32, 34, 40].

Battery Energy: For predicting the battery energy pattern, we consider the discharging cycles using the models utilized in several real-world studies. For estimating the remaining battery energy and SOC in smartphones during discharging, we use a discharging model [32]. For an accurate calculation of available charge, the model takes into account the rate capacity effect in batteries. According to the rate capacity effect of batteries, charging and discharging efficiency fall as charging and discharging currents increase [32]. Over each discrete time slot, we calculate the battery energy by monitoring the device's instantaneous power consumption and discharging current E using:

$$E(t + \Delta t) = E(t) - E_c,$$

where E_c, the energy consumption (J) over one cycle (Δt), is given by

$$E_c = \Delta t \times P_{\text{device}} + E_{\text{loss}},$$

where Δt denotes the time duration of each cycle (in second) and P_{device} is the total power consumption (Watt) of the device during Δt. In the work in [34], the authors demonstrate generation of individual power characterization of workloads on specific cores (such as the LITTLE and BIG cores) of IoT devices. We assume that such a characterization is available a priori. Thus, the value of P_{device} is calculated in accordance with the power characterization depending on where the task is executed. In our work, P_{device} refers to the average power consumption. E_{loss} is the internal loss of the battery, caused by the rate capacity effect and is calculated by

$$E_{\text{loss}} = \Delta t \times \left(i_b^2 R_{\text{total}} + i_b \cdot v_{OC} \cdot (1/\eta\,(i_b) - 1) \right),$$

where i_b is the discharging current (amp) of the battery, R_{total} is the total internal resistance (ohm) of the battery, v_{OC} is the open circuit terminal voltage (volt) of the battery, and $\eta\,(i_b)$ denotes the battery discharging efficiency, which can be approximated as $1/\left((i_b)^{k_d}\right)$, where k_d is the parameter representing the Odroid platform utilized in [32]. R_{total} and v_{OC} are calculated using

$$R_{\text{total}} = b_{21}e^{b_{22}v_{\text{soc}}+b_{23}}$$
$$v_{OC} = b_{11}e^{b_{12}v_{\text{soc}}} + b_{13}v_{\text{soc}}^4 + b_{14}v_{\text{soc}}^3 + b_{15}v_{\text{soc}}^2 + b_{16}v_{\text{soc}} + b_{17},$$

where bij are regression coefficients, and v_{SOC} is the voltage representation of the battery SOC, that is,

$$v_{SOC} = C_b/C_{b,full} \times 1\,\text{V},$$

where C_b is the remaining charge in the battery, and $C_{b,full}$ is the battery charge when it is fully charged. Note that we utilize energy consumption in our model. The State-Of-Charge is related to the energy consumption as follows:

$$\text{SOC}(t + \Delta t) = \frac{E(t + \Delta t) \times 100}{E_T}$$

The State-Of-Charge is an important factor in influencing battery life and ageing, according to [32]. Battery capacity declines with use over time owing to the loss of active components, which is known as battery ageing. As a result, in this paper, we present an offloading strategy based on the battery model used in [32]. The chosen model mimics capacity decline over time using average energy and temperature as input. In the next subsection we discuss the model of temperature utilized in our work.

Battery Temperature Model: In this paper, we employ a battery thermal model similar to the model employed in [32] that represents an IoT device. Thermal coupling between the battery and the CPU is taken into account in this model. The thermal coupling effect between the battery and the CPU plays a major role in determining the battery temperature in smartphones due to the tiny physical area. As a result, one part of a smartphone's thermal behaviour is not independent of the other. Furthermore, the CPU's thermal behaviour is strongly influenced by the application services that are running. By monitoring the power utilized by the CPU we can estimate the battery temperature in an indirect manner. The detailed model is presented below:

$$T_{bat} = T_{env} + \frac{R_{cpu-env} R_{bat-env}}{R_{bat-env} + R_{cpu-bat} + R_{cpu-env}}.P_{cpu}$$

where T_{bat} and T_{env} are battery and environment temperature, P_{cpu} and P_{bat} are CPU and battery power consumption, and R_{i-j} is the thermal resistance between i and j, where i and j can be CPU, environment, and battery. The value for these resistances are utilized from [32]. The above energy and thermal models can be replaced by any other model, as our framework is designed in a modular and replaceable manner.

B   Experimental Setup

Workload: To the best of our knowledge, there are no real-world MEC implementation workload traces that are publicly available and sufficiently representative of our problem context. Therefore, for our experiments, we generate synthetic workloads using real-world applications from the Mibench benchmark [10,32]. The Mibench benchmark suite provides benchmarks in various categories of standard applications ranging from sensor systems on simple microcontrollers to smartphones and desktops. Table 4 summarises the Mibench programmes utilised in this setup. The applications we employ in this research are in the network and security categories, both of which are pertinent to and reflective of IoT applications.

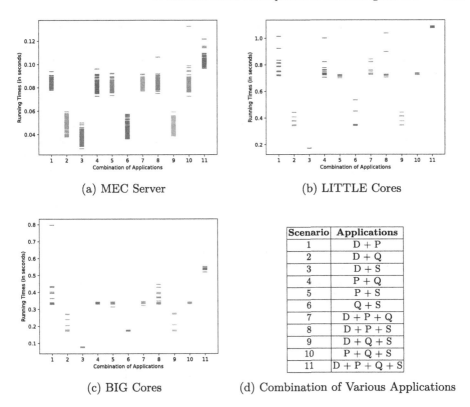

(a) MEC Server   (b) LITTLE Cores

(c) BIG Cores   (d) Combination of Various Applications

Fig. 9. Comparative Computation Running Times for Combinations of Applications on the MEC server, LITTLE core and BIG core respectively

Data Gathering: We utilize the applications described in Table 4 to characterize the running times of various tasks of the applications. We utilize the executables of the Mibench benchmark as application tasks. We utilize these running times to simulate the MEC-IoT setup. We first consider a random number of applications between 2 and 4 and execute these applications simultaneously. We consider three measurements: a) by running the applications simultaneously on the MEC server b) by running the applications simultaneously by setting the processor affinity to the "LITTLE" cores and c) by running the applications simultaneously by setting the processor affinity to the "BIG" cores. Figure 9 depicts the runtimes for the various scenarios and the corresponding application

Table 4. Mibench benchmark applications used in this work

Application	Category	Summary
Dijkstra (D)	Network	Constructs a large graph and then calculates the shortest path between every pair of nodes
Patritia (P)	Network	Creates data structure for representing routing tables in network applications
Sha (S)	Security	Secure hash algorithm
Qsort (Q)	Automotive Control	Simulate autonomous vehicle control algorithms by sorting a large array of strings

scenarios utilized with a combination of the Dijkstra, Patricia, Quick Sort and Sha applications. Each combination of applications is executed 100 times. Figures 9a, 9b and 9c depict the variations in runtimes for the different combinations of applications for 100 different executions. As can be inferred from the figures, even for the same type of application combination, for example, Dijktra and Patricia, there is a variation in the runtimes depicting the stochastic nature of application execution runtimes. Additionally, note that there is a variation in the runtimes depending on whether the application is executed on the MEC server or locally as well as on the type of core when executed locally. These variations depict the impact of the hardware specifications on the runtime of application tasks. Prior work [44] has proposed that it is possible to effectively predict the combinations of applications utilized given a specific time of day as well as locality. Our methodology is thus based on such a premise. We assume that such a characterization involving a combination of applications can be derived a priori and utilizing the setup described above their running times can be determined for a specific set of hardware. We utilize these runtimes as the values of latencies (measured in seconds) to generate the DTMC model of the Task Execution Latencies as explained in Sect. 4.3.

B.1   Simulation Setup

We simulate the MEC scenario by considering an IoT device executing tasks. Upon task execution, the models are initialized into the Probabilistic Model Checker PRISM-Games [13] along with the IoT device preference[1] The model checker returns the strategy. Each discrete time slot is then simulated by executing the strategy as returned by PRISM-Games and the data gathered from the above set of experiments is used to sample the stochastic nature of task execution times. Additionally, in the simulation, the battery energy and temperature are updated according to the models described. Our simulation thus implicitly assumes that an accurate model of the energy and thermal dynamics of the MEC ecosystem is available. However, note that our SMG based model can be utilized with any other battery and energy model as well thus providing a generic approach. We simulate the MEC scenario by considering a single IoT device since the offloading scheme is identical whenever each IoT device executes a task. When the task is executed at the MEC server, the computation time distribution is obtained by considering the running times of a combination of applications. Such a setup thus simulates several applications running on the MEC server simultaneously.

References

1. Bahreini, T., Badri, H., Grosu, D.: Energy-aware capacity provisioning and resource allocation in edge computing systems. In: Zhang, T., Wei, J., Zhang, L.-J. (eds.) EDGE 2019. LNCS, vol. 11520, pp. 31–45. Springer, Cham (2019). https://doi.org/10.1007/978-3-030-23374-7_3

---

[1] The PRISM-Games files are available at https://github.com/kauray/offloadgame.

2. Baier, C., Katoen, J.P.: Principles of Model Checking (2008)
3. Buyya, R., Yeo, C.S., Venugopal, S., Broberg, J., Brandic, I.: Cloud computing and emerging it platforms: vision, hype, and reality for delivering computing as the 5th utility. Future Gener. Comput. Syst. **25**(6), 599–616 (2009)
4. Carroll, A., Heiser, G., et al.: An analysis of power consumption in a smartphone. In: USENIX Annual Technical Conference, Boston, MA, vol. 14, p. 21 (2010)
5. Chen, L., Zhou, S., Xu, J.: Computation peer offloading for energy-constrained mobile edge computing in small-cell networks. IEEE/ACM Trans. Netw. **26**(4), 1619–1632 (2018)
6. Chen, T., Forejt, V., Kwiatkowska, M., Parker, D., Simaitis, A.: Automatic verification of competitive stochastic systems. Formal Methods Syst. Des. **43**(1), 61–92 (2013). https://doi.org/10.1007/s10703-013-0183-7
7. Chen, W., Wang, D., Li, K.: Multi-user multi-task computation offloading in green mobile edge cloud computing. IEEE Trans. Serv. Comput. **12**(5), 726–738 (2019)
8. Evangelidis, A., Parker, D., Bahsoon, R.: Performance modelling and verification of cloud-based auto-scaling policies. Future Gener. Comput. Syst. **87**, 629–638 (2018)
9. Fernando, N., Loke, S.W., Rahayu, W.: Mobile cloud computing: a survey. Future Gener. Comput. Syst. **29**(1), 84–106 (2013)
10. Guthaus, M.R., Ringenberg, J.S., Ernst, D., Austin, T.M., Mudge, T., Brown, R.B.: MiBench: a free, commercially representative embedded benchmark suite. In: Proceedings of the Fourth Annual IEEE International Workshop on Workload Characterization, WWC-4 (Cat. No. 01EX538), pp. 3–14. IEEE (2001)
11. Kikuchi, S., Matsumoto, Y.: Performance modeling of concurrent live migration operations in cloud computing systems using prism probabilistic model checker. In: IEEE CLOUD 2011, pp. 49–56 (2011)
12. Kwiatkowska, M., Norman, G., Parker, D.: PRISM: probabilistic model checking for performance and reliability analysis. ACM SIGMETRICS Perform. Eval. Rev. **36**(4), 40–45 (2009)
13. Kwiatkowska, M., Parker, D., Wiltsche, C.: PRISM-games 2.0: a tool for multi-objective strategy synthesis for stochastic games. In: Chechik, M., Raskin, J.-F. (eds.) TACAS 2016. LNCS, vol. 9636, pp. 560–566. Springer, Heidelberg (2016). https://doi.org/10.1007/978-3-662-49674-9_35
14. Lai, P., et al.: Optimal edge user allocation in edge computing with variable sized vector bin packing. In: Pahl, C., Vukovic, M., Yin, J., Yu, Q. (eds.) ICSOC 2018. LNCS, vol. 11236, pp. 230–245. Springer, Cham (2018). https://doi.org/10.1007/978-3-030-03596-9_15
15. Lai, P., et al.: Edge user allocation with dynamic quality of service. In: Yangui, S., Bouassida Rodriguez, I., Drira, K., Tari, Z. (eds.) ICSOC 2019. LNCS, vol. 11895, pp. 86–101. Springer, Cham (2019). https://doi.org/10.1007/978-3-030-33702-5_8
16. Lai, P., et al.: Cost-effective app user allocation in an edge computing environment. IEEE Trans. Cloud Comput. (2020)
17. Li, K.: Heuristic computation offloading algorithms for mobile users in fog computing. ACM Trans. Embed. Comput. Syst. (TECS) **20**(2), 1–28 (2021)
18. Mao, Y., You, C., Zhang, J., Huang, K., Letaief, K.B.: A survey on mobile edge computing: the communication perspective. IEEE Commun. Surv. Tutor. **19**(4), 2322–2358 (2017)
19. Naskos, A., et al.: Dependable horizontal scaling based on probabilistic model checking. In: 2015 15th IEEE/ACM International Symposium on Cluster, Cloud and Grid Computing, pp. 31–40. IEEE (2015)

20. Neto, J.L.D., Yu, S.Y., Macedo, D.F., Nogueira, J.M.S., Langar, R., Secci, S.: ULOOF: a user level online offloading framework for mobile edge computing. IEEE Trans. Mob. Comput. **17**(11), 2660–2674 (2018)

21. Panda, S.P., Ray, K., Banerjee, A.: Dynamic edge user allocation with user specified QoS preferences. In: Kafeza, E., Benatallah, B., Martinelli, F., Hacid, H., Bouguettaya, A., Motahari, H. (eds.) ICSOC 2020. LNCS, vol. 12571, pp. 187–197. Springer, Cham (2020). https://doi.org/10.1007/978-3-030-65310-1_15

22. Panda, S.P., Ray, K., Banerjee, A.: Service allocation/placement in multi-access edge computing with workload fluctuations. In: Hacid, H., Kao, O., Mecella, M., Moha, N., Paik, H. (eds.) ICSOC 2021. LNCS, vol. 13121, pp. 747–755. Springer, Cham (2021). https://doi.org/10.1007/978-3-030-91431-8_51

23. Peng, Q., et al.: Mobility-aware and migration-enabled online edge user allocation in mobile edge computing. In: 2019 IEEE International Conference on Web Services (ICWS), pp. 91–98. IEEE (2019)

24. Poularakis, K., Llorca, J., Tulino, A.M., Taylor, I., Tassiulas, L.: Service placement and request routing in MEC networks with storage, computation, and communication constraints. IEEE/ACM Trans. Netw. **28**(3), 1047–1060 (2020)

25. Ray, K., Banerjee, A.: Trace-driven modeling and verification of a mobility-aware service allocation and migration policy for mobile edge computing. In: IEEE ICWS (2020)

26. Ray, K., Banerjee, A.: A framework for analyzing resource allocation policies for multi-access edge computing. In: 2021 IEEE International Conference on Edge Computing (EDGE), pp. 102–110. IEEE (2021)

27. Ray, K., Banerjee, A.: Horizontal auto-scaling for multi-access edge computing using safe reinforcement learning. ACM Trans. Embed. Comput. Syst. (TECS) **20**(6), 1–33 (2021)

28. Ray, K., Banerjee, A.: Modeling and verification of service allocation policies for multi-access edge computing using probabilistic model checking. IEEE Trans. Netw. Serv. Manag. **18**(3), 3400–3414 (2021)

29. Ray, K., Banerjee, A.: Prioritized fault recovery strategies for multi-access edge computing using probabilistic model checking. IEEE Trans. Dependable Secure Comput. (2022)

30. Ray, K., Banerjee, A., Narendra, N.C.: Proactive microservice placement and migration for mobile edge computing. In: 2020 IEEE/ACM Symposium on Edge Computing (SEC), pp. 28–41. IEEE (2020)

31. Rossi, F., Nardelli, M., Cardellini, V.: Horizontal and vertical scaling of container-based applications using reinforcement learning. In: 2019 IEEE 12th International Conference on Cloud Computing (CLOUD), pp. 329–338. IEEE (2019)

32. Shamsa, E., et al.: UBAR: user- and battery-aware resource management for smartphones. ACM Trans. Embed. Comput. Syst. **20**(3), 1–25 (2021). https://doi.org/10.1145/3441644

33. Shi, W., Cao, J., Zhang, Q., Li, Y., Xu, L.: Edge computing: vision and challenges. IEEE Internet Things **3**(5), 637–646 (2016)

34. Shin, D., et al.: Online estimation of the remaining energy capacity in mobile systems considering system-wide power consumption and battery characteristics. In: 2013 18th Asia and South Pacific Design Automation Conference (ASP-DAC), pp. 59–64. IEEE (2013)

35. Shu, C., Zhao, Z., Han, Y., Min, G., Duan, H.: Multi-user offloading for edge computing networks: a dependency-aware and latency-optimal approach. IEEE Internet Things J. **7**(3), 1678–1689 (2020)

36. Tian, S., Chi, C., Long, S., Oh, S., Li, Z., Long, J.: User preference-based hierarchical offloading for collaborative cloud-edge computing. IEEE Trans. Serv. Comput. (2021)
37. Wang, F., Xu, J., Wang, X., Cui, S.: Joint offloading and computing optimization in wireless powered mobile-edge computing systems. IEEE Trans. Wirel. Commun. **17**(3), 1784–1797 (2017)
38. Wang, S., Urgaonkar, R., Zafer, M., He, T., Chan, K., Leung, K.K.: Dynamic service migration in mobile edge computing based on Markov decision process. IEEE/ACM Trans. Netw. **27**(3), 1272–1288 (2019)
39. Wu, Y., Ni, K., Zhang, C., Qian, L.P., Tsang, D.H.: Noma-assisted multi-access mobile edge computing: a joint optimization of computation offloading and time allocation. IEEE Trans. Veh. Technol. **67**(12), 12244–12258 (2018)
40. Xie, Q., Kim, J., Wang, Y., Shin, D., Chang, N., Pedram, M.: Dynamic thermal management in mobile devices considering the thermal coupling between battery and application processor. In: 2013 IEEE/ACM International Conference on Computer-Aided Design (ICCAD), pp. 242–247. IEEE (2013)
41. Xu, J., Chen, L., Zhou, P.: Joint service caching and task offloading for mobile edge computing in dense networks. In: IEEE INFOCOM 2018-IEEE Conference on Computer Communications, pp. 207–215. IEEE (2018)
42. Yao, H., Bai, C., Xiong, M., Zeng, D., Fu, Z.: Heterogeneous cloudlet deployment and user-cloudlet association toward cost effective fog computing. Concurr. Comput. Pract. Exp. **29**(16), e3975 (2017)
43. You, C., Huang, K., Chae, H., Kim, B.H.: Energy-efficient resource allocation for mobile-edge computation offloading. IEEE Trans. Wirel. Commun. **16**(3), 1397–1411 (2016)
44. Yu, D., Li, Y., Xu, F., Zhang, P., Kostakos, V.: Smartphone app usage prediction using points of interest. Proc. ACM Interact. Mob. Wearable Ubiquit. Technol. **1**(4), 1–21 (2018). Article no: 174

Analysis of an Electric Vehicle Charging System Along a Highway

Davide Cerotti[1]([✉]), Simona Mancini[1], Marco Gribaudo[2], and Andrea Bobbio[1]

[1] Università del Piemonte Orientale, Alessandria, Italy
{davide.cerotti,simona.mancini,andrea.bobbio}@uniupo.it
[2] Politecnico di Milano, Milan, Italy
marco.gribaudo@polimi.it

Abstract. To reduce carbon emission, the transportation sector evolves toward replacing internal combustion vehicles by electric vehicles (EV). However, the limited driving ranges of EVs, their long recharge duration and the need of appropriate charging infrastructures require smart strategies to optimize the charging stops during a long trip. These challenges have generated a new area of studies that were mainly directed to extend the classical Vehicle Routing Problem (VRP) to a fleet of commercial EVs. In this paper, we propose a different point of view, by considering the interaction of private EVs with the related infrastructure, focusing on a highway trip. We consider a highway where charging stations are scattered along the road, and are equipped with multiple chargers. Using Fluid Stochastic Petri Nets (FSPN), the paper compares different decision policies when to stop and recharge the battery to maximize the probability of a car to reach its destination and minimize the trip completion time.

Keywords: Electric Vehicle · Charging infrastructure · Battery charge decision policy · Fluid Stochastic Petri Nets

1 Introduction

According to the Fuel Report 2021 of the International Energy Agency [1] the transport sector is responsible for around 60% of total oil demand. Inside the transport sector, oil was the predominant energy source, providing 92% of final energy over the past decade [2]. A major way to limit carbon emission in the transport sector is to replace internal combustion engine vehicles (ICEV) by electric vehicle (EV), and many countries are introducing new regulations and incentives to push the market toward this goal.

Between the two EV technological alternatives: hybrid electric vehicle (HEV) and battery electric vehicle (BEV), we consider, in the present paper, only BEV which are exclusively powered from rechargeable batteries mounted inside the vehicle. From the carbon emission point of view, BEVs have the following benefits as compared to ICEVs [3].

Research partially supported by CNIT (Consorzio Nazionale Interuniversitario per le Telecomunicazioni).

E. Ábrahám and M. Paolieri (Eds.): QEST 2022, LNCS 13479, pp. 298–316, 2022.
https://doi.org/10.1007/978-3-031-16336-4_15

- They reduce oil consumption, and greenhouse gas emissions, improving air quality.
- They operate with minimal noise.
- Can be charged from a wide range of different primary (renewable) energy sources, reducing kilometric cost.

On the other hand, their high cost of acquisition, limited driving ranges, the need for specific charging infrastructure, and long recharge duration limit the penetration of EV in the market.

The appearance of BEVs in the private as well in the commercial sector poses new challenges for their use and for the new infrastructures they need. These challenges have generated a new area of studies denoted as green logistic [4]. The major effort in this direction was to extend the classical Vehicle Routing Problem (VRP) to a fleet of commercial EVs and is referred to as Electric Vehicle Routing Problem (EVRP) [5]. The present paper, however, assumes a different point of view, considering a flow of private vehicles traveling along a highway and we study the performability of the system formed by the BEVs and the related infrastructure.

More specifically, we consider a long stretch of a highway with a flow of cars, both BEVs and non-BEVs, driving on it. A number of charging stations is scattered along the road and each charging station has one or more chargers. The problem of optimizing the siting and sizing of the charging infrastructure has been the object of recent research [6,7], but we assume here that the stations are already located and their positions are parameters to feed the model. We tag and follow a particular BEV that enters the highway at the beginning of the stretch and drives up to the end. We study the probability that the tagged car arrives at the end of its itinerary and the distribution of the time to complete the itinerary. The battery of the BEVs discharges as a function of the time, the speed of the car and the driven kilometers, while for the charging we adopt the non-linear function discussed in [8].

We model the system by means of a Fluid Stochastic Petri Net (FSPN) [9], in which the battery is represented by a fluid place with one input and one output fluid transition representing the charging and the discharging process, respectively.

The time that the tagged car takes to drive the segment between two successive service stations is a generally distributed random variable with a know mean (determined by the average speed of the car). The BEVs arriving at a charging station queue up for charge, and we assume, in the present formulation, that their arrival and service times are exponentially distributed. When the tagged driver arrives at a charging station she must decide whether to stop and recharge or go on. This decision depends on the level of the battery, the presence and length of a queue at the station and the distance to the next station or to the destination. Different decision policies are considered and analyzed, also in view of a possible experimentation on autonomous EVs. The FSPN is solved analytically using Matlab, and a number of numerical experiments are presented to compare different decision policies.

The paper is organized as follows. In Sect. 2 we summarize the state of the art, in Section 3 the characteristics of the charging infrastructure and of the flow of cars are described. Section 4 illustrates the FSPN model, with the charging decision policies. The subsequent Sect. 5 sketches the numerical solution through a semi-discretization approach. Section 6 reports the numerical results and in Sect. 7 a discussion on the model and hints for future work close the paper.

2   State of the Art

The diffusion of EVs is limited, especially in Italy with respect to other countries (Germany, France), also as a consequence of the scarcity of charging stations along the road network, and to their uneven distribution in the national territory. Although the technological innovation continuously increases the driving range of EVs, planning the charging stops is still a critical issue due to the long charging times. The seminal paper of Erdogan & Miller-Hooks [5] has first introduced the EVRP, considering different alternative-fuel vehicles (not only powered by electricity but also by GPL, hydrogen, natural gas etc.) for which the charging stations are not widespread on the territory. The paper generated a huge interest in the scientific community, and many extensions have been proposed and studied in the following years, as documented in two recent survey papers [10, 11].

Hybrid vehicles, which can switch from the electric propulsion to a traditional fuel have been addressed in [12]. Schneider et al., [13] introduced the EVRP with Time Windows (EVRPTW) in which customers must be visited within a prefixed time-window. The concept of partial recharges to the EVRPTW was introduced in [14], while in [15] a non-linear charging function is considered. In [16], charging stations with limited capacity are addressed for the first time, and a maximum number of vehicles that can simultaneously access the station is strictly imposed. Instead, in [17], vehicles are allowed to access the station and queue up if all the charging slots are busy, so that service may start when the queue becomes empty. All the above mentioned papers deal with decision problems faced by the usage of commercial EVs in freight distribution. In the present paper, we look at the problem under a different perspective considering a private BEV driver, immersed into a flow of private vehicles, who has to cover a given trip and must decide where and when to stop to recharge her vehicle to maximize the probability of completing the trip minimizing the total trip time. We introduce, in this paper, a performability view on the interaction between the EVs and the charging infrastructure [18], since we combine the evaluation of the driver trip time with the evaluation of the probability that the trip fails and the car does not complete its itinerary. The key element in the interaction between EVs and the infrastructure is the discharging and charging process of the battery. The flow intensity of the vehicles, the traffic conditions on the road and the queue length in front of a station are non-deterministic phenomena that can be represented by a stochastic model. To combine in a single framework the continuous variation of the charging level of the battery in time and the randomness in the traffic condition, we model the system by means of a Fluid

Stochastic Petri Net (FSPN) [9,19,20] where the battery is represented by a fluid place whose fluid level is the charge, the time to travel the highway between two successive service stations is a random variable with general distribution and the queue at a station is a $M/M/\gamma$ if the station has γ parallel chargers.

FSPNs were introduced in [19] and further extended in [20]. FSPNs evolve the stochastic Petri nets framework [21] by introducing as new primitives the continuous places, which contain a fluid quantity, and the fluid arcs, which connect timed transitions to fluid places and determine the flow in and out to the fluid places. The basic FSPN formalism was enriched in [9] by introducing fluid impulses that increment the fluid level by a discrete quantity whose intensity depends both on the fluid levels and on the discrete marking of the net. FSPNs have a graphical representation that helps building the model, and then from the graphical representation we can derive the fluid stochastic equations that describe the underlying stochastic marking process. In general, the solution of these equations is a challenging task. Steady-state solution of FSPN models, with dependency on discrete places only, has been proposed in [20] using *spectral decomposition*. In the same paper, transient analysis has also been described using *upwind semidiscretization*. FSPNs have been successfully used in the literature to study systems in several technological areas, but we are not aware of applications in the field of EV routing.

3   The Infrastructure and the EV

To be concrete, we consider, the Italian motorway A1 from Bologna to Taranto (743 km), which we display in Fig. 1 with the real allocation of the service stations.

The portion of highway between two charging stations is called segment. Table 2 in Appendix shows the location of the service stations along A14 Bologna to Taranto. We assume that each service station is equipped (or will be equipped in the near future) with a charging point with a number γ of parallel chargers and that all the chargers provide the same power, so that the charging profile is the same for all the EVs in all chargers.

Along the highway runs a flow of cars, composed by BEVs and non-BEVs and by the tagged BEV that we follow from the beginning of the trip up to the end. The other BEVs in the flow may compete for the charging points, generating possible queues at the stations when the tagged car arrives for charge. The non-BEV cars are not explicitly modeled, but their presence is reflected on the average speed of the car flow including the tagged BEV. The time that the tagged BEV takes to complete one segment is a random variable whose distribution may change in each road segment. The distribution of the driving time to complete a segment is modeled by a shifted Erlang distribution whose parameters (shift, expected value and number of stages) are input data (see Sect. 4) that may depend on the traffic conditions in the segment. In this way, we allow to model fluctuations or congestion in the traffic flow in specific segments of the highway. Each station is provided with γ parallel chargers and we assume that the BEVs

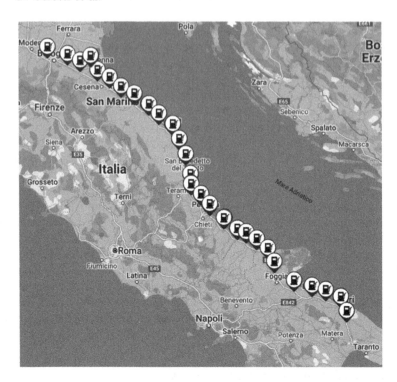

Fig. 1. The A14 from Bologna to Taranto with the actual position of the service stations. We assume that in each service station there is an EV charging point

arrive at the station according to a Poisson process of known parameter. In the present study, we assume also that all the BEVs and all the chargers have the same characteristics; furthermore we do not model the level of the residual charge of the vehicles joining the queue, hence, even if the charging profile has a nonlinear behavior (see Sect. 4.1), we approximate the charging time of the non-tagged BEVs at a station with an exponential distribution of known parameter, so that the queue in front of the station becomes a $M/M/\gamma$.

4   Fluid Stochastic Petri Net: The Scenario

Figure 2 shows the FSPN of the considered scenario. Following the customary notation for FSPNs [9, 20], the set of places $\mathcal{P}$ is partitioned in a set of discrete places $\mathcal{P}_d$ and a set of continuous places $\mathcal{P}_c$. Discrete places are drawn as single circles and may contain a discrete number of tokens, while continuous places are depicted by two concentric circles and contain a continuous quantity called fluid. The model of Fig. 2 contains a single fluid place β representing the battery whose fluid level, ranging in the interval $0 \leq \beta \leq B$, is the charge measured in km.

The set of transition $\mathcal{T}$ is partitioned into a set of timed transitions $\mathcal{T}_E$, a set of immediate transitions $\mathcal{T}_I$ and a set of fluid transitions $\mathcal{T}_F$. Timed transitions are drawn as a rectangle and are assigned a random firing time with known distribution, immediate transitions are represented by a thin bar and fire in zero time, while fluid transitions are represented by double rectangles and connect the fluid places. In Fig. 2, the timed transition $SEGMENT(\phi)$ (in gray) is assigned a general distribution, while transitions EMP_Q and $FREE$, representing the queue of BEVs at a station, have an exponentially distributed firing time. B_UP and B_DOWN are the fluid transitions modeling the charge and discharge of the battery, respectively,

The set of arcs $\mathcal{A}$ is partitioned into two subsets $\mathcal{A}_d$ and $\mathcal{A}_c$: the former is a subset of $(\mathcal{P}_d \times \mathcal{T}) \cup (\mathcal{T} \times \mathcal{P}_d)$ representing the discrete arcs and are drawn as single arrows, the latter is a subset of $(\mathcal{P}_c \times \mathcal{T}_F) \cup (\mathcal{T}_F \times \mathcal{P}_c)$ representing the fluid arcs and are drawn as double arrows. In Fig. 2, the fluid arc from B_UP to β continuously adds fluid (charge) when enabled, while the fluid arc from β to B_DOWN continuously removes fluid when enabled. Inhibitor arcs, represented with dashed lines ending by a small circle, have the usual meaning of preventing a transition to fire when the input place contains a number of tokens (or a fluid level) greater or equal to the weight. In Fig. 2, there are two inhibitor arcs: from place ϕ to transition $AVAIL\_SLOT$ which may fire only when place ϕ contains less than K tokens and from fluid place β to transition $FAIL$ which may fire when the fluid level is less than the weight $l(\phi)$. Finally, impulse arcs, which connect fluid places to discrete transitions, add or remove a finite amount of fluid during the firing event. In Fig. 2, the only impulse arc is from place β to transition $SEGMENT(\phi)$, and removes a fluid quantity $l(\phi)$ when the transition fires.

Let $\mathbf{m}_i = [\#p_i, i \in \mathcal{P}_d]$ be the discrete marking of the FSPN and let $\mathbf{x}$ be the vector of the fluid levels in the continuous places. The complete state of the fluid Petri net is given by the pair $M = (\mathbf{m}_i, \mathbf{x})$ which evolves in time, generating the stochastic marking process $\mathcal{M}(\tau) = \{(\mathbf{m}_i, \mathbf{x}), \tau \geq 0\}$. The evolution of each fluid level x depends both on a continuous component determined by the instantaneous flow rates assigned to fluid arcs and a discrete component determined by fluid impulses transferred to (or removed from) the fluid place when the impulse transition fires [9].

Place N represents the tagged car initiating segment ϕ, with $\phi = 1, ..., \mathbf{K}$, counted in place ϕ. The total number of segments to be traveled is hence denoted by $\mathbf{K}$. Transition $SEGMENT(\phi)$ represents the completion of segment ϕ and the arrival at the charging station at the end of the segment. Such transition is characterized by a state-dependent shifted-Erlang firing time, expressed by Eq. (1), where s is the number of exponential stages, t_0 the shift and ψ the rate parameter.

$$f(t) = \begin{cases} \dfrac{\psi^s(t - t_0)^{s-1}e^{-\psi(t-t_0)}}{(s-1)!} & t \geq t_0 \\ 0 & t < t_0 \end{cases} \qquad (1)$$

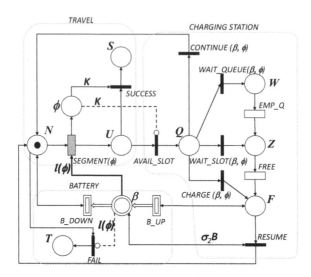

Fig. 2. Fluid Stochastic Petri Net model of the considered scenario.

Let l_ϕ be the length of segment ϕ, $v_{\max}$ the maximum speed allowed for the car, and v_{ave} the average speed on the segment. t_0 and ψ are defined as follows:

$$t_0 = \frac{l_\phi}{v_{\max}}; \quad \psi = \frac{s}{l_\phi \left(\dfrac{1}{v_{\text{ave}}} - \dfrac{1}{v_{\max}} \right)} \tag{2}$$

The distribution (1) starts at t_0, which is the traveling time when the segment is driven at the maximum speed $v_{\max}$, and the expected value is $E[t] = \frac{l_\phi}{v_{\text{ave}}}$ as required by the definition of v_{ave}. From Eqs. (2), the parameters t_0 and ψ of the distribution (1) are derived assigning $v_{\max}$ and v_{ave}, while the number of stages s is assigned independently.

The coefficient of variation c_v is lower than the corresponding non-shifted Erlang with same order s:

$$c_v = \frac{1}{\sqrt{s}} \left(1 - \frac{v_{\text{ave}}}{v_{\max}} \right) \tag{3}$$

4.1   The Battery Level

In the control screen of most EVs, the battery level is conventionally measured and displayed in kilometers. The battery is consumed during the firing time of transition $SEGMENT(\phi)$ (when a token is in place N) according to distribution in (1), and charges when a token is in place F. Battery charging and discharging are modeled by fluid transitions B_UP and B_DOWN, respectively. The discharge is a function of the length l_ϕ of segment ϕ, the average speed of the car and of the power supplied to support services such as air conditioning and music playing

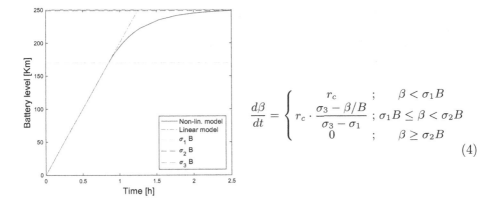

$$\frac{d\beta}{dt} = \begin{cases} r_c & ; \quad \beta < \sigma_1 B \\ r_c \cdot \dfrac{\sigma_3 - \beta/B}{\sigma_3 - \sigma_1} & ; \; \sigma_1 B \leq \beta < \sigma_2 B \\ 0 & ; \quad \beta \geq \sigma_2 B \end{cases} \tag{4}$$

Fig. 3. The non linear charging profile

(converted in equivalent km), which is a function of the time spent to drive segment ϕ. The rate of consumption of the battery increases almost linearly with the average speed [22] beyond a cruise speed of around 50 km/h.

For each segment we could assign a value for $v_{\max}$ and v_{ave} to define the distribution of the time to drive the segment. However, in the present paper we keep the same values of $v_{\max}$ and v_{ave} for all the segments, but we show how the model reacts modifying these values. The random time to drive a segment is given by t_0 (the time to drive the segment at the maximum speed), augmented by the deviation given by the Erlang component (the series of s exponential stages) (Eq. 1). During this period, the battery level β reduces at a rate $r(t, i) = -r_b$ to account for the services. At the end of the last Erlang stage, transition $SEGMENT(\phi)$ fires and a fluid impulse l_ϕ, reduces the battery level to account for the completion of segment ϕ.

When a token is in place F, the battery charges according to a non-linear model inspired by the work in [8]. The recharge model is given by Eq. (4), where r_c is the initial constant recharge rate and σ_i are the parameters needed to define the non-linear behavior. Figure 3 gives the corresponding non linear charging profile. As soon as the capacity $\sigma_2 B$ is reached, immediate transition $RESUME$ fires, putting a token in place N to prepare for driving the next segment. Whenever the battery level β becomes zero or negative, due to the reduction at rate r_b or to the effect of the negative impulses, the immediate transition $FAIL$ is triggered causing the mission to fail. This is modeled with the inhibitor arc of weight l_ϕ that connects β to FAIL.

When transition $SEGMENT(\phi)$ fires and $\phi = K$ (last segment), transition $AVAIL\_SLOT$ is disabled, immediate transition $SUCCESS$ fires and the tagged car successfully completes its itinerary. If $\phi < K$ transition $AVAIL\_SLOT$ fires and place Q becomes marked, the tagged BEV must take the decision whether to continue, or to stop and recharge. This decision is performed inside the element of the $CHARGING\ STATION$ box of Fig. 2, and depends on the current battery level β, on the position of the car ϕ, on the state of the queue and on the implemented stopping decision policy.

We have modeled the queue at the charging stations as a M/M/γ and we have computed its state probabilities. We define $p(\text{Wait})$ the probability that the chargers are busy and there is a queue in front, $p(\text{NoQueue})$ the probability that the chargers are busy, but there are no cars in the queue, and $p(\text{Free})$ the probability that there are empty slots. Such probabilities can be computed from the queue length distribution $\pi_{Q(n)}$ using the conventional formula [23]:

$$\rho = \frac{\lambda}{\gamma\mu}, \qquad \pi_{Q(n)} = \begin{cases} \dfrac{\pi_{Q(0)}}{n!} \cdot (\gamma\rho)^n & n < \gamma \\ \dfrac{\pi_{Q(0)}\gamma^\gamma \rho^n}{\gamma!} & n \geq \gamma \end{cases} \qquad (5)$$

$$\pi_{Q(0)} = \left[\frac{(\gamma\rho)^\gamma}{\gamma!} \frac{1}{1-\rho} + \sum_{k=0}^{\gamma-1} \frac{(\gamma\rho)^k}{k!} \right]^{-1} \qquad (6)$$

$$p(\text{Free}) = \sum_{n=0}^{\gamma-1} \pi_{Q(n)}, \qquad p(\text{NoQueue}) = \pi_{Q(\gamma)}, \qquad p(\text{Wait}) = 1 - \pi_F - \pi_Z \qquad (7)$$

where λ is the arrival rate of BEVs at the station and $1/\mu$ is the average time to recharge. The resulting values, together with the chosen stopping decision policy, are then used to compute the firing probabilities of the four immediate transitions *CONTINUE*, *WAIT_QUEUE*, *WAIT_SLOT* and *CHARGE*, as detailed in Sect. 4.3.

4.2   Stopping Decision Policies

In the present paper, we take into account only the waiting and service times due to the queue and we neglect the overhead in exiting the highway, pulling up to a charger, plug in, and then reentering the highway. This overhead can be easily introduced in the FSPN of Fig. 2 by adding an extra timed transition if the car decides to stop at a station.

We experiment different stopping decision policies where each policy j is characterized by three functions that return the probability of stopping depending on the current battery level β, and the distance of the next charging stations (derived from the segment ϕ): *i)* if there are free slots $u_f^{[j]}(\beta, \phi)$ (token in place F), *ii)* if the slots are busy, but there are no cars queuing, $u_z^{[j]}(\beta, \phi)$ (token in place Z) and *iii)* if there are already other cars queuing for a charging facility to become free $u_w^{[j]}(\beta, \phi)$ (token in place W).

We have tested five different stopping policies:

1. **Stop only when absolutely needed**: stop only if the remaining battery level is not enough to reach the next charging station. This policy is implemented with the following functions:

$$u_f^{[1]}(\beta, \phi) = \mathbf{1}(\beta < l_{\phi+1})$$
$$u_z^{[1]}(\beta, \phi) = \mathbf{1}(\beta < l_{\phi+1})$$
$$u_w^{[1]}(\beta, \phi) = \mathbf{1}(\beta < l_{\phi+1})$$

where $\mathbf{1}(\bullet)$ is the indicator function that returns 1 if proposition $\bullet$ is true, 0 otherwise.

2. **Blind probabilistic stopping**: stop at any station, independently of the queue, with a given probability p:

$$u_f^{[2]}(\beta, \phi) = p$$
$$u_z^{[2]}(\beta, \phi) = p$$
$$u_w^{[2]}(\beta, \phi) = p$$

3. **Informed probabilistic stopping**: stop at any station, with a given high probability p_e if there is at least an empty slot, or with a lower probability p_w if there is to wait:

$$u_f^{[3]}(\beta, \phi) = p_e$$
$$u_z^{[3]}(\beta, \phi) = p_w$$
$$u_w^{[3]}(\beta, \phi) = p_w$$

4. **Avoid waiting**: We define a safety threshold of η km. Whenever the battery level β is such that $(\beta < \eta)$, stop if there is at least a free charger, otherwise retry at the next station. However, if the remaining battery level is less than the length of the next segment, stop anyway.

$$u_f^{[4]}(\beta, \phi) = \mathbf{1}(\beta < \eta)$$
$$u_z^{[4]}(\beta, \phi) = \mathbf{1}(\beta < l_{\phi+1})$$
$$u_w^{[4]}(\beta, \phi) = \mathbf{1}(\beta < l_{\phi+1})$$

5. **Skip long queues**: This policy relaxes the previous one, by allowing the car to stop with probability p, even if all chargers are busy, but there are no other cars in the queue.

$$u_f^{[5]}(\beta, \phi) = \mathbf{1}(\beta < \eta)$$
$$u_z^{[5]}(\beta, \phi) = p \cdot \mathbf{1}(\beta < \eta)$$
$$u_w^{[5]}(\beta, \phi) = \mathbf{1}(\beta < l_{\phi+1})$$

Of the proposed policies, the first three are just for comparison purpose, since they are not realistic; the ones on which this work really focuses are the last two.

4.3   Computing Decision Policy Probability

As transition $AVAIL\_SLOT$ fires depositing a token in place Q four immediate competing transitions are enabled determining the next move of the tagged car. The weights of the four immediate transitions are a function of the state of the queue and the stopping policy j

$$w_{CONTINUE} = \tag{8}$$
$$p(\text{Free})(1 - u_f^{[j]}) + p(\text{NoQueue})(1 - u_z^{[j]}) + p(\text{Wait})(1 - u_w^{[j]})$$

$$w_{WAIT\_QUEUE} = p(\text{Wait})u_w^{[j]} \tag{9}$$

$$w_{WAIT\_SLOT} = p(\text{NoQueue})u_z^{[j]} \tag{10}$$

$$w_{CHARGE} = p(\text{Free})u_f^{[j]} \tag{11}$$

In Eqs. (8) to (11) the dependencies on β and ϕ have been omitted to simplify the presentation. If there is a queue of cars already waiting to be served the waiting time is modelled by transition $EMP\_Q(\gamma)$, which, according to queuing theory [24], is exponentially distributed with rate:

$$q_{\text{EMP\_Q}(\gamma)} = \gamma\mu - \lambda \tag{12}$$

If all slots are full, but there are no other cars in the queue, the waiting time represented by transition $FREE(\gamma)$, is exponentially distributed with rate:

$$q_{\text{FREE}(\gamma)} = \gamma\mu \tag{13}$$

When a token arrives in place F, the charging of the battery begins. If, instead, the decision is to continue, the immediate transition $CONTINUE$ fires, moving the tagged BEV to the next segment.

5   Solution Equation

Since there is only one fluid place with fluid level β, the stochastic marking process of the FSPN of Fig. 2 can be written as (see Sect. 4) $\mathcal{M}(t) = \{(\mathbf{m}_i, \beta), t \geq 0\}$. The non exponential distribution of transition $SEGMENT(\phi)$ is modeled with a phase-type approach, and the discrete marking $\mathbf{m}_i$ includes both the s stages of the shifted Erlang distribution and the state of the queue.

Then, let us define $\pi_i(t, \beta)$ as the probability density of finding the system in state $\mathbf{m}_i$ with fluid level β at time t, q_{ij} as the transition rates from state $\mathbf{m}_i$ to state $\mathbf{m}_j$ and d_{ij} as the fluid impulse transferred to the fluid place at the transition firing from state $\mathbf{m}_i$ to state $\mathbf{m}_j$. In our case, q_{ij} corresponds either to the rates of the Erlang stages given in Eq. (2), to the rates given in Eqs. (12) or (13), or to a Dirac's delta on the fluid component to represent the battery level dependent jumps caused by either end of charging or failure. The fluid impulse terms are $d_{ij} = -l_\phi$ for the transitions at the end of the Erlang stages, and $d_{ij} = 0$ otherwise.

The transient behavior of the model illustrated in Sect. 4 follows the system of partial differential equations:

$$\frac{\partial \pi_i(t, \beta)}{\partial t} - \frac{\partial \left(r(t, i, \beta) \cdot \pi_i(t, \beta)\right)}{\partial \beta}$$

$$= \sum_{\mathbf{m}_j \in \mathcal{M}_d, \mathbf{m}_j \neq \mathbf{m}_i} q_{ji} \pi_i(t, \beta + d_{ij}), \quad \forall \mathbf{m}_i \in \mathcal{M}_d. \tag{14}$$

We apply a semi-discretization of (14) in the coordinate direction β using a first-order upwind method [20]. Since the fluid place of the battery is bounded at a maximum level B, its fluid level can be discretized at a finite number of equidistant points $\beta_i = i\Delta\beta$ with $0 \leq i \leq \lfloor \frac{B}{\Delta\beta} \rfloor$, where $\Delta\beta$ is the size of the discretization interval of the battery level.

From the semi-discretization we obtain the linear system of ordinary differential equations:

$$\frac{d\tilde{\boldsymbol{\pi}}(t)}{dt} = \tilde{\boldsymbol{\pi}}(t)(\tilde{\boldsymbol{Q}} + \tilde{\boldsymbol{W}}), \tag{15}$$

where $\tilde{\boldsymbol{\pi}}(t)$ is the vector of the probabilities that the system is in a discrete marking at different points of the discretized fluid range, $\tilde{\boldsymbol{Q}}$ is the state transition matrix representing the discrete part of the net and $\tilde{\boldsymbol{W}}$ is the discretization of the space derivative multiplied by the flow rates; more details can be found in [20].

Such equations can be integrated and solved by any standard method. In particular, we resorted to the ode23() function of Matlab to implement the proposed scheme with adaptive step size integration with $\Delta X = 5$ km or $\Delta X = 2$ km. Solution required between 20 s to 10 min on a standard MacBook Pro 2016 laptop.

Table 1. Model parameters

Param.	Description	Value
γ	Num. charging slots	2
s	Erlang stages	4
$v_{\max}$	Maximum speed	130 km/h
v_{ave}	Average speed	100 km/h
λ	Electric cars arrival rate	1 car/h
B	Battery capacity (km)	250 km
r_b	Basic energy consumption (km/h)	10 km/h
r_c	Charging rate (km/h)	200 km/h
$\sigma_1, \sigma_2, \sigma_3$	Parameters of the charging model	$0.68, 0.995, 1.0$
η	Threshold for the battery level	80 km

6 Numerical Results

We fix the number of slots to $\gamma = 2$ for every charging station. To study how the model captures the evolution of the system, we start focusing on policy 4), assigning a threshold $\eta = 80$ km and setting the other parameters to the values in Table 1. Figure 4 shows the main state probabilities: charging, waiting, success and failure. On Fig. 4 we have also reported the average battery percentage $\bar{\beta}(t)$ at time t, and the average number of segments $\bar{\phi}(t)$ already driven at time t, where $\phi(i)$ is the highway segment corresponding to state $\mathbf{m}_i$. The two quantities $\bar{\beta}(t)$ and $\bar{\phi}(t)$ are expressed in Eq. (16) and are conditioned on the fact that the tagged BEV reaches its destination, and on the probability that it is still traveling at time t. Let us call $\pi_{dest}(t)$ the probability that the car has successfully reached its destination at time t, and $\pi_{fail}(t)$ the probability that the car has run out of battery while traveling on a segment. Then we have:

$$\bar{\beta}(t) = \frac{1}{B} \left[\sum_i \int_{\beta=0}^{B} \beta\, \pi_i(t, \beta)\, d\beta \right] (1 - \pi_{dest}(t) - \pi_{fail}(t))^{-1}$$

$$\bar{\phi}(t) = \frac{1}{K} \left[\sum_i \int_{\beta=0}^{B} \phi(i)\pi_i(t, \beta)\, d\beta \right] (1 - \pi_{dest}(t))^{-1} \qquad (16)$$

From the plot, it is possible to visualize when the car stops, and how long it has to wait for charging the battery. It is interesting to note that the probability of charging the battery becomes negligible as the probability of reaching the destination becomes higher. The "staircase" trend of "Ave. Battery" and "Ave.

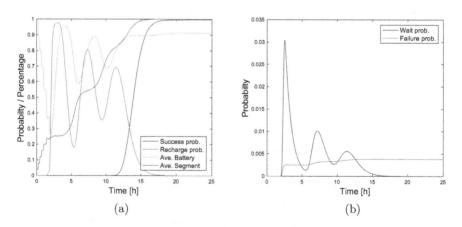

(a)   (b)

Fig. 4. States evolution: a) success probability, charging probability, average battery percentage and average number of segments driven at time; b) waiting probability and failure probability

segment" curves, observable at time t < 2h, is due to the embedded process behavior. In particular to the impulsive battery level reduction l_ϕ that occurs at discrete times when the vehicle reaches the next service station, modeled by the firing of transition $SEGMENT(\phi)$ in Fig 2.

We then start comparing the various policies, showing the probability of unsuccessful arrival at destination in Fig. 5a), and the trip duration distribution conditioned on a successful arrival in Fig. 5b). Policy 2), which stops the car at every charging point, has basically no probability of failing, but it experiences an extremely long trip duration. Conversely, policy 1) that stops the car only when absolutely needed has the worst reliability. Its response time is not one of the best either, since it stops independently on the queue at the station where the battery level becomes too small. Policy 3), as expected, places itself between the two. The more advanced policies 4) and 5) have a much shorter traveling time, but policy 5), which anticipates the recharge, shows a lower failure probability than policy 4).

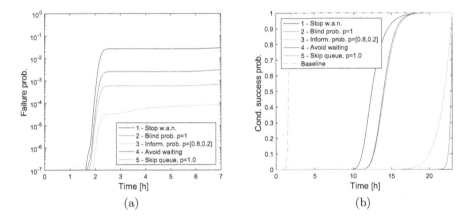

Fig. 5. a) Failure probability and b) conditioned success probability, as function of time for the 4 considered policies

Figure 6a) shows the average battery level as a function of time. for the different decision policies. Policies 2) and 3) induce frequent stops and recharges and the battery power is not well exploited since the charge is almost always at the maximum level. Policy 1) depletes the battery almost completely before charging. For Policies 4) and 5) (whose curves are almost indistinguishable) the battery depletion and then the stops to charge are well visible; the two policies perform an early charge, more or less in the same station.

Figure 6b) for the only Policy 4) shows the effect on the battery cycles of the speed of the car. The increased average speed has two competing effects. The battery depletion rate increases almost linearly with the speed [22], but since driving

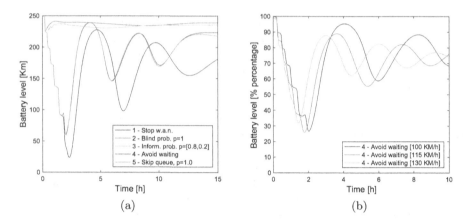

Fig. 6. a) Average battery level versus time for the considered policies and b) Average battery level versus average speed for Policy 4)

a segment takes a shorter time, also the battery consumption due to the services (AC, sound, etc.) is lower. The combined effect in Fig. 6b) shows that at increased speed the recharge must be done more frequently. The curves in Fig. 6b) have been obtained by setting $v_{\max} = 140$ km/h and $v_{ave} = 100, 115, 130$ km/h.

As explained in Sect. 4.2, Policies 4) and 5) require a safety threshold η to decide their stopping interval. Figure 7 shows the average trip time and the failure probability for policies 4) and 5) under different thresholds $\eta \in [20 \ldots 120]$ km. While the failure probability increases as the margin decreases, as expected, the average trip time has a non-linear and non-monotonic behavior, even if it tends to increase with η. This tendency is motivated from the fact that by stopping earlier there is a higher chance of requiring an additional stop. Non-monotonic behavior is instead caused by the non uniform position of service stations along the road. In some cases, increasing the margin can have a positive effect: arriving at a station with a larger remaining battery capacity decreases the stopping time. If this does not increase the required number of stops, it has the effect of reducing the traveling time. By combining both measures with appropriate goal-dependent weights, the proposed model can thus be used to find the value of η which gives the best trade-off between traveling speed and failure probability depending on the position of the charging points along the trip.

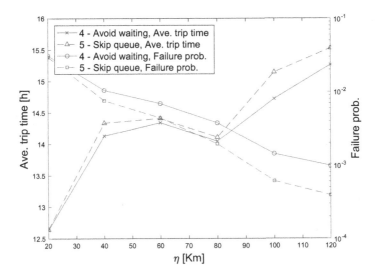

Fig. 7. Average traveling time and failure probability for policies 4) and 5) under different margins $\eta \in [20 \dots 120]$.

7   Conclusions

This paper has explored a new line of research by analyzing a system composed by electric vehicles, immersed in a traffic flow, on a real highway, and their interaction with the related charging infrastructure. The balance between the probability of successfully completing the itinerary versus the time required to complete the trip, that is influenced by the number of stops and the state of the queue at the charging stations, has been examined under different charging decision policies. The described system has been modeled and analyzed using stochastic fluid models, and in particular, Fluid Stochastic Petri Nets. Although results are still preliminary, the paper shows the appropriateness of the considered technique for studying the performance and the reliability of the proposed system.

The model is predisposed to be extended in different directions. More realistic traffic condition, in which the speed of the car flow can be modulated according to the traffic intensity, and the battery depletion rate is sensitive to the speed of the tagged BEV. The incorporation of chargers with different characteristics that modify the nonlinear charging profile of the battery. The inclusion of the probability of the charging station to be up or down, and the adaptation of the stopping decision policies. Further the decision policy can benefit by the availability of system level controllers (i.e., an app) that informs the driver of unavailable or free slots along the way.

The model investigated in this paper can be proposed as a building block for an optimization strategy aimed at finding the best policy parameters for a given road. Further, the analysis of charging decision policies can provide suitable algorithms to be implemented on board, mainly in view of the possible new generation of autonomous EVs.

Appendix

In Table 2 the location of the service stations along A14 Bologna to Taranto is shown. In particular: the name of the service stations (column 2), the length of the segments (column 3) together with the progressive distances from the start (column 4) and to the end (column 5).

Table 2. Location of service stations along A14 Bologna to Taranto

	Station	Distances [km]		
ϕ	Name	segment	from start	to end
0	Bologna	0	0	743.4
1	La Pioppa	2.3	2.3	741.1
2	Sillaro	35.1	37.4	706
3	Santerno	22.1	59.5	683.9
4	Bevano	30	89.5	653.9
5	Rubicone	21.8	111.3	632.1
6	Montefeltro	22.3	133.6	609.8
7	Foglia	25.3	158.9	584.5
8	Metauro	27.3	186.2	557.2
9	Esino	22.5	208.7	534.7
10	Conero	30.3	239	504.4
11	Chienti	24.9	263.9	479.5
12	Piceno	26.9	290.8	452.6
13	Tortoreto	32.9	323.7	419.7
14	Vomano	16.6	340.3	403.1
15	Torre Cerrano	22.8	363.1	380.3
16	Alento	30.8	393.9	349.5
17	Sangro	34.9	428.8	314.6
18	Trigno	29.8	458.6	284.8
19	Torre Fantine	15	493.5	269.8
20	S.Trifone	19.9	517.51	249.9
21	Gargano	48.73	542.23	201.17
22	Le Saline	44.94	587.17	156.23
23	Canne Battaglia	33.19	620.36	123.04
24	Dolmen Di Bisceglie	24.05	644.41	98.99
25	Murge	27.03	671.44	71.96
26	Le Fonti	26.18	697.5	45.78
27	Taranto	36.78	743.4	0

References

1. International Energy Agency. Global energy review 2021 (2021). https://www.iea.org/reports/global-energy-review-2021. Accessed No 2021
2. International Energy Agency. Global EV outlook 2020 (2020). https://www.iea.org/reports/global-ev-outlook-2020. Accessed Nov 2021
3. García-Villalobos, J., Zamora, I., San Martín, J.I., Asensio, F.J., Aperribay, V.: Plug-in electric vehicles in electric distribution networks: a review of smart charging approaches. Renew. Sustain. Energy Rev. **38**, 717–731 (2014)
4. Sbihi, A., Eglese, R.W.: Combinatorial optimization and green logistics. 4OR, **5**, 99–116 (2014)
5. Erdoğan, S., Miller-Hooks, E.: A green vehicle routing problem. Transp. Res. Part E Logist. Transp. Rev. **48**(1), 100–114 (2012)
6. Khaksari, A., Tsaousoglou, G., Makris, P., Steriotis, K., Efthymiopoulos, N.S., Varvarigos, E.: Sizing of electric vehicle charging stations with smart charging capabilities and quality of service requirements. Sustain. Cities Soc. **70**, 102872 (2021)
7. Dupont, A., Hayel, Y., Jiménez, T., Beaude, O., Wan, C.: Coupled queueing and charging game model with energy capacity optimization. In: Ballarini, P., Castel, H., Dimitriou, I., Iacono, M., Phung-Duc, T., Walraevens, J. (eds.) EPEW/ASMTA -2021. LNCS, vol. 13104, pp. 325–344. Springer, Cham (2021). https://doi.org/10.1007/978-3-030-91825-5_20
8. Al-Karakchi, A., Putrus, G., Das, R.: Smart EV charging profiles to extend battery life. In: 2017 52nd International Universities Power Engineering Conference (UPEC), pp. 1–4 (2017)
9. Gribaudo, M., Sereno, M., Horváth, A., Bobbio, A.: Fluid stochastic Petri nets augmented with flush-out arcs: modelling and analysis. Discrete Event Dy. Syst. **11**(1/2), 97–117 (2001). https://doi.org/10.1023/A:1008339216603
10. Erdelić, T., Carić, T.: A survey on the electric vehicle routing problem: variants and solution approaches. J. Adv. Transp. **2019**, 1–48 (2019). Article ID 5075671
11. Kucukoglu, I., Dewil, R., Cattrysse, D.: The electric vehicle routing problem and its variations: a literature review. Comput. Ind. Eng. **161**, 107650 (2021)
12. Mancini, S.: The hybrid vehicle routing problem. Transp. Res. Part C. Emerg. Technol. **78**, 1–12 (2017)
13. Schneider, M., Stenger, A., Goeke, D.: The electric vehicle-routing problem with time windows and recharging stations. Transp. Sci. **48**(4), 500–520 (2014)
14. Bruglieri, M., Pezzella, F., Pisacane, O., Suraci, S.: A variable neighborhood search branching for the electric vehicle routing problem with time windows. Electron. Notes Discrete Math. **47**, 221–228 (2015)
15. Montoya, A., Guéret, C., Mendoza, J.E., Villegas, J.G.: The electric vehicle routing problem with nonlinear charging function. Transp. Res. Part B Methodol. **103**, 87–110 (2017)
16. Bruglieri, M., Mancini, S., Pezzella, F., Pisacane, O.: A path-based solution approach for the green vehicle routing problem. Comput. Oper. Res. **103**, 109–122 (2019)
17. Keskin, M., Laporte, G., Çatay, B.: Electric vehicle routing problem with time-dependent waiting times at recharging stations. Computers & Operations Research **107**, 77–94 (2019)
18. Meyer, J.F.: On evaluating the performability of degradable systems. IEEE Trans. Comput. **C-29**, 720–731 (1980)

19. Trivedi, K., Kulkarni, V.: FSPNs: fluid stochastic petri nets. In: Proceedings of the 14th International Conference on Application and Theory of Petri Nets, pp. 24–31, Chicago (1993)
20. Horton, G., Kulkarni, V.G., Nicol, D.M., Trivedi, K.S.: Fluid stochastic petri nets: theory, applications, and solution techniques. Eur. J. Oper. Res. **105**(1), 184–201 (1998)
21. Bobbio, A.: System modelling with petri nets. In: Colombo, A.G., Saiz de Bustamante, A., (ed.) System Reliability Assessment, pp. 103–143. Kluwer Academic (1990)
22. Grée, F., Laznikova, V., Kim, B., Garcia, G., Kigezi, T., Gao, B.: Cloud-based big data platform for vehicle-to-grid (V2G). World Electr. Veh. J. **11**(2), 30 (2020)
23. Trivedi, K., Bobbio, A.: Reliability and Availability Engineering: Modeling, Analysis, and Application. Cambridge University Press, Cambridge (2017)
24. Sztrik, J.: Basic queueing theory: foundations of system performance modeling. University of Debrecen, Faculty of Informatics, vol. 193 (2016)

Verifier's Dilemma in Ethereum Blockchain: A Quantitative Analysis

Daria Smuseva[1], Ivan Malakhov[1], Andrea Marin[1], Aad van Moorsel[2], and Sabina Rossi[1](✉)

[1] Università Ca' Foscari Venezia, Venice, Italy
{daria.smuseva,ivan.malakhov,marin,sabina.rossi}@unive.it
[2] Newcastle University, Newcastle upon Tyne, UK
aad.vanmoorsel@ncl.ac.uk

Abstract. A blockchain is an immutable ledger driven by a distributed consensus protocol. In public blockchains such as Bitcoin and Ethereum consensus is established through a computational effort called Proof-of-Work (PoW). Special users called *miners* contribute to the PoW computational effort in exchange for a fee and also verify the data stored in blocks mined by the other miners. Here is where the Verifier's Dilemma emerges. To maximise their profits, miners are incentivized to invest their resources in PoW, because they do not receive any incentives for the verification phase. In this paper, we study the Verifier's Dilemma using a quantitative model based on PEPA. The analysis demonstrates the circumstances under which non-verifying miners gain fees higher than that of verifying miners. Moreover, we consider a mitigation approach consisting of the injection of invalid blocks to disturb the mining process of non-verifying miners. The model allows us to derive the optimal rate at which invalid blocks must be injected, so that skipping the verifying phase becomes economically disadvantageous while the throughput of the blockchain is only minimally reduced. The impact on miners' rewards and overall performance is also assessed.

Keywords: Blockchain · Stochastic Process Algebra · Verifier's Dilemma

1 Introduction

A blockchain is a decentralized distributed network with an immutable and time ordered ledger whose records are stored in blocks. Blockchain users submit their transactions to the system, where other special users called *miners* verify the transactions and include them in blockchain blocks.

Every blockchain network follows a certain consensus *protocol*. Particularly, Ethereum [5] (the second largest blockchain project after Bitcoin) is driven by the Proof-of-Work (PoW) consensus protocol. PoW consists of a race among miners to solve a certain computationally intensive problem. However, once a solution is announced, other miners can verify its correctness quickly. Winning

© Springer Nature Switzerland AG 2022
E. Ábrahám and M. Paolieri (Eds.): QEST 2022, LNCS 13479, pp. 317–336, 2022.
https://doi.org/10.1007/978-3-031-16336-4_16

the race is important because only the miner who announces the new block is rewarded for its efforts. One of the main strengths of PoW is the fact that any modification to transactions in the ledger has a high computational cost that miners must face. In particular, since blocks are linked, modifications of consolidated blocks requires to re-mine all the subsequent blocks, and this has such a high computational cost that is commonly considered infeasible.

Miners do not only work on the solution of the PoW, but they also have to verify the transactions that they want to include in their candidate block and those contained in the blocks announced by others. Ethereum transactions can be of two types: financial and smart contract. The latter ones are essentially transactions encompassing a piece of computing code in them. This naturally implies that verification of such transactions is connected to the execution of the corresponding code pieces which, indeed, requires extra computational effort with respect to the financial transaction verification. It should be noted that the language used to write smart contracts (usually Solidity) is Turing complete and hence, the execution may be computationally quite demanding. While the verification process delays miners in the race of mining the next block, there is no miners' incentive mechanism behind this process nor a way to check if miners performed this task. As a result, miners face the Verifier's Dilemma: *use resources for verification to support the blockchain honestly or avoid the verification process to have more time to mine new blocks in favour of maximizing their revenues.*

Clearly, in a healthy blockchain, the large majority of miners should be fair, i.e., they verify other miners' blocks. For this reason, methods to incentivize fair miners or disincentivize unfair ones are being investigated in the literature. Among the most promising approaches to the mitigation of the Verifier's Dilemma problem is the intentional production of invalid blocks [1]. The solution refers to the implementation of a special node that generates invalid blocks, i.e., blocks that contain at least one invalid transaction. Such a block will be rejected by fair miners but accepted by others that will immediately start to compute the PoW of the subsequent block. While this unethical behaviour gives an advantage to the unfair miners upon a valid block announcement, it is harmful when an invalid block is announced. In fact, unfair miners will try to consolidate a new block after an invalid one and, even if they succeed, they will never get any reward because all fair miners will never recognize their blockchain branch. If the miner does not verify an invalid block, they realise that the network of peers has not accepted it only when a new valid block is announced, and hence the waste of time could be much larger than the advantage of skipping the verification.

Injecting invalid blocks has some drawbacks. First, it diverts some computational power of fair miners to the verification of meaningless blocks. Second, the invalid block must still exhibit a valid PoW that will not be used to secure the immutability of the ledger. For these reasons, the rate of injection of invalid blocks is a crucial problem for the system designers. *This should be the minimum rate that makes miners' unfairness non- or dis-advantageous.*

In this paper, we use the Performance Evaluation Process Algebra (PEPA) [10], for modelling the salient aspects of Ethereum network allowing us to study the Verifier's Dilemma and the above described mitigation approach. The choice of PEPA is motivated by its simplicity and compositionality properties. Moreover, it is quite interesting to observe that the solution of the PoW is known to be a memoryless process, i.e., the delays between two subsequent blocks are independent and exponentially distributed, as it is exponentially distributed the time required by a single miner to solve the PoW (see, e.g., [4]).

The main contributions of the paper are:

- First, we introduce a base PEPA model to understand, investigate and validate our approach to the Verifier's Dilemma problem. In particular, the model quantitative indices are confirmed by the simulations performed in [1] based on real data from the Ethereum blockchain. The model, thanks to an aggregation of states [2,13,14] is analytically tractable.
- The second contribution is the introduction of a PEPA model that takes into account the invalid block injection mechanism. Using this model, we are able to investigate this important countermeasure from the perspective of the single miner and that of the entire network. The model allows us to determine the waste in terms of rewards both of fair miners (that spend more time in verifying) and of unfair miners (that spend time in mining possibly invalid blocks because subsequent to an injected one). A counter-intuitive observation following this model analysis is that the optimal rate of injection of invalid blocks grows with the percentage of fair miners. We will explain the reason from a system perspective in the following sections.
- The analysis of the model allows us to define a numerical algorithm for the computation of the optimal injection rate which is provably correct under mild conditions. The advantage of this approach is that it can be used to dynamically control the optimal injection rate without requiring the solution of the model for each change in its parameters.

The paper is structured as follows. In Sect. 2 we provide a literature review. Section 3 gives the necessary background. Section 4 proposes a model for the Verifier's dilemma. Section 5 proposes a model for the invalid block injection approach. In Sect. 6, we study the optimal configuration of the system using the invalid block injection. Finally, Sect. 7 concludes the paper.

2   Related Work

The Verifier's Dilemma remains the subject of discussion in many Ethereum-related works. The authors of [12] firstly spotted this problem. They have shown from a game theoretical perspective that rational miners would omit the verification process to maximise their rewards. Based on this idea, other authors [9] analyzed the case in which malicious miners design dummy smart contracts which are computationally expensive to verify in order to slow down the mining process of all other miners and hence increase the possibilities of winning the PoW race. In addition, [1,9,12] showed the profitability of skipping the verification process in scenarios with computationally intensive smart contracts.

Various authors [3,6,8,17] introduced the parallel computation of the smart contracts transactions in order to mitigate the drawback of long verification time. Particularly, in [6] the on-chain protocol that allows miners to delay verification of transactions in a block by up to the certain number of blocks ahead was introduced. The authors showed that thanks to this implementation the Ethereum-like blockchains become less vulnerable to the problems caused by the growth of the verification time.

Alternatively, in [1,7,11,16] a number of off-chain solutions have been proposed for efficient computation of complex smart contracts. Thus, they suggested that the verification of complex contracts is not anymore performed by all the nodes, but only a small subset of them. Those nodes earn reward only if they perform the verification correctly. Furthermore, in [12], the authors proposed to divide complex transactions into a set of transactions with smaller size so that it becomes possible to include them in special blocks.

A closely related work is [1]. The authors introduced a simulation framework that allows them to investigate the network behaviour when a part of the miners do not verify the blocks. They showed that the single non-verifying miner can be rewarded approximately 23% higher than the network's fair share as total gain of all other fair nodes is reduced respectively. They also proposed the injection of invalid blocks as a solution to the Verifier's Dilemma problem. The authors propose to use a special node for intentionally generating invalid blocks as a way to punish non-verifying miners. Furthermore, they assume this node to perform full verification of all blocks generated by other miners. With respect to our work, we have used the simulation estimates of [1] to validate our PEPA model and proved its importance in determining the optimal rate of injection of invalid blocks. In fact, it would be very computationally expensive to run stochastic simulations searching the optimal value of this rate while, thanks to our model, the problem is reduced to the application of the bisection algorithm to find the root of a certain polynomial.

3 Background

3.1 Verifier's Dilemma

Similarly to most blockchains, in Ethereum consensus is fully distributed and miners contribute to the life of the blockchain in two ways: (i) they verify the transactions stored in the blocks and (ii) they compete to solve the PoW problem.

Smart contracts are stored in Ethereum as contract accounts, which contain the code and some storage space to support their execution [5]. Users of the blockchain control the so called externally owned accounts. External entities and contract accounts cooperate with each other through two types of transactions: financial and contract-based. The former is responsible for moving the Ethereum cryptocurrency, namely Ether, among the accounts. Contract transactions control all interactions with contract accounts: either to publish a new smart contract or to execute an existing one.

While the verification of financial transactions is very quick since its heaviest activity is to verify that no double spending has been done, smart contracts pose more problems. Smart contracts are written with a Turing complete language and the verification of a transaction containing the output of their execution requires all miners to re-execute the same program on the same input [12].

Miners are rewarded for their work according to a probabilistic process. In fact, only the miner that solves first the PoW problem receives the reward. When all miners behave fairly, the probability of winning the race for PoW is equal to the proportion of computational power invested by a miner. This feature leads miners to face the Verifier's Dilemma. Some miners tend to skip the verification process to save computational resources for profitable mining instead of supporting the blockchain fairly.

One of the solution is to inject invalid blocks in the network to punish unfair miners. The idea is that if an unfair miner does not verify the announced blocks, it would mine its candidate block after an invalid one and hence does not gain a fee in case of success. So the unfair miners must face a trade-off between the advantage of skipping the verification phase and the risk to mine a new block after an invalid one.

At the state of the art, there has not been a rigorous analysis of this mitigation solution using quantitative analysis techniques. Our goal is to investigate the impact of the Verifier's Dilemma and to detect the minimum rate of invalid blocks injection that makes the unfair policy less lucrative than the fair one.

3.2 PEPA

We use the Performance Evaluation Process Algebra (PEPA) [10] to investigate the Verifier's Dilemma in Ethereum and its impact on miners' behaviour. In this section, we give a brief overview of PEPA. The advantages of using PEPA lie primarily in the fact that PEPA is compositional, every PEPA specification has an underlying stochastic process and, under given assumptions, this stochastic process is a continuous time Markov process. Moreover PEPA is supported by a tool that can be applied to Eclipse, namely PEPA Eclipse Plug-in.

A PEPA system specification consists of a collection of active agents or components cooperating with each other through activities to achieve the system behaviour. The syntax for PEPA terms is defined by the following grammar:

$$P ::= P \underset{L}{\bowtie} P \mid P/L \mid S \qquad S ::= (\alpha, r).S \mid S + S \mid A$$

where S denotes a *sequential component*, while P denotes a *model component* which can be obtained as the cooperation of sequential terms. The meaning of the operators is the following: $(\alpha, r).S$ performs the activity (α, r) with action type α and rate r and subsequently behaves as P. $P+Q$ specifies a system which may behave either as P or as Q. The component P/L behaves as P except that any activity of type within the set L are relabelled with the *unknown type* τ. The meaning of a constant A is given by a defining equation such as $A \overset{def}{=} P$

which gives the constant A the behaviour of the component P. The cooperation combinator $\bowtie_L$ represents an interaction between two components, which is determined by the set of action types L, namely *cooperation set*. Activities with action types in the cooperation set L, called *shared* activities, require the synchronisation of the components. It is assumed that each component proceeds independently with the activities whose types do not occur in the cooperation set L. The duration of a shared activity is reflected by the rate of the slower participant. If in a component an activity has the *unknown rate* $\top$, then the rate of the shared activity will be that of the other component.

4   Analysis of the Verifier's Dilemma

In this section, we present a base PEPA model for the Ethereum network with fair and unfair miners. The validation of the model is done with the estimates obtained by stochastic simulation in [1].

4.1   Model Description

Let N be the number of miners and λ be the total computational power of the network measured in expected number of PoW problems that can be solved per unit of time. We assume that all miners have the same computational power $\gamma = \lambda/N$. If this is not the case, it will be easy to aggregate some miners to obtain multiples of γ, while the change on the verification process brings to negligible effects on the overall model behaviour.

Among the N miners, a fraction $p \in [0, 1]$ is fair and verifies the new incoming blocks, while $1 - p$ is unfair and skip this verification.

Since the number of miners is in practice very high, the detailed representation of the state of each of them would incur into a state space explosion problem. Thus, we resort to a representation based on the aggregation of all fair (unfair) miners with the exception of one. Leaving one fair and one unfair miner outside the aggregation is important because it allows us to study their detailed behaviours and consider situations such as that in which all miners are fair with the exception of one, that will result to be crucial to determine our main practical result in Sect. 6.

Table 1 shows the PEPA description of our Base Model (BM). The fair (i.e., correctly operating) environment is defined as E_F and represents the behaviour of all verifying miners that perform the verification step V_{E_F}. E_U corresponds to the unfair environment where miners avoid the verification part. Notice that, $p\lambda$ and $(1 - p)\lambda$ are the fractions of the hash power of verifying miners and non-verifying ones, respectively. The *Network* component is obtained as the cooperation of all components over a set of actions L.

Let us inspect the behaviours of single fair and unfair miners; those of the environments will follow trivially (see Table 7 in Appendix A). Component M_F represents a single fair (verifying) miner. The miner mines a new block with action type m_F and rate γ. Then, it returns to its initial state and it starts to

Table 1. PEPA specification of BM where $L = \{m_F, m_U, m_{E_F}, m_{E_U}\}$.

M_F	$\overset{def}{=}$	$(m_F, \gamma).M_F + (m_{E_F}, \top).V_F + (m_{E_U}, \top).V_F + (m_U, \top).V_F$
V_F	$\overset{def}{=}$	$(\tau, \beta).M_F + (m_{E_F}, \top).V_F + (m_{E_U}, \top).V_F + (m_U, \top).V_F$
M_U	$\overset{def}{=}$	$(m_U, \gamma).M_U + (m_{E_F}, \top).M_U + (m_F, \top).M_U + (m_{E_U}, \top).M_U$
E_F	$\overset{def}{=}$	$(m_{E_F}, p\lambda).V_{E_F} + (m_{E_U}, \top).V_{E_F} + (m_F, \top).V_{E_F} + (m_U, \top).V_{E_F}$
V_{E_F}	$\overset{def}{=}$	$(\tau, \beta).E_F + (m_{E_U}, \top).V_{E_F} + (m_F, \top).V_{E_F} + (m_U, \top).V_{E_F}$
E_U	$\overset{def}{=}$	$(m_{E_U}, (1-p)\lambda).E_U + (m_{E_F}, \top).E_U + (m_F, \top).E_U + (m_U, \top).E_U$
$Network$	$\overset{def}{=}$	$(M_F \underset{L}{\bowtie} M_U) \underset{L}{\bowtie} (E_F \underset{L}{\bowtie} E_U)$

mine another block. Once verifying miner M_F gets a block from the network, it starts the verification process V_F. Meanwhile, the non-verifying, or unfair, miner M_U skips this step and adds the incoming block directly to its copy of the ledger. The verification of a block occurs with rate β.

It is worth of notice that, while the race policy of PoW requires us to sum the rates of all miners belonging to an environment, the verification phase is not governed by a race policy and hence it is performed with rate β as it happens for the single miner. Intuitively, the amount of work required for the verification is assumed to be exponentially distributed but, since all miners have the same computational power and must perform the same operations, we can say that the aggregated speed remains β. Another interesting aspect is that since the time for solving PoW is independent and exponentially distributed by design, the aggregation of the rates is exact.

The underlying derivation graph of the model is depicted in Fig. 1. Each fair miner that is in the verification phase can still get new blocks from the rest of the network. It means that, in real systems, miners must maintain a queue of incoming blocks. However, the required time for verification is usually much lower than the block interval time. This observation allows us to simplify V_F and V_{E_F} by abstracting out the queue of blocks to verify. Experiments confirmed that this simplification does not significantly affect the type of analyses that we perform.

To illustrate the described model, consider a network with $N = 100$ miners, each controlling an equal amount of hash power. We assume that the network consists of large blocks, such that each block is full and contains only contract-based transactions since financial transactions take less time to verify and do not contribute to the Verifier's Dilemma. Block verification time is, on average, $3.18\,s$ as an extreme case scenario [1]. Block interval time is, on average, $12\,s$. Thus, $\beta = 1/3.18 = 0.314$ and $\lambda = 1/12 = 0.083$. Since all miners have equal hash power, $\gamma = \lambda/N = 8.3 \times 10^{-4}$. Assume, furthermore, that there are 10 of the 100 miners who do not verify. As a result, the fraction of fair miners is $p = 0.9$. These parameters are summarized in Table 2.

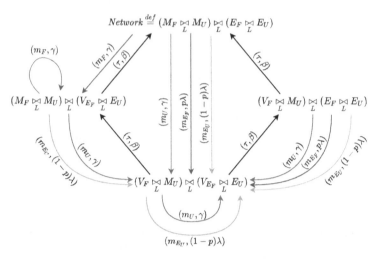

Fig. 1. Derivation graph of BM.

We assume that the reward that miners gain is proportional to the *effective throughput*, which describes what part of invested computational power is used for mining. Since unfair miners do not spend time for verification, the effective throughput for a single miner is equal to its invested power $\gamma = 8.3 \times 10^{-4}$. The effective throughput of the single fair miner is 6.6×10^{-4} since it has to verify all blocks. Thus, the received reward of the single unfair miner is $\approx 25.75\%$ more than the fair miner's one. These results confirm those in [1] for the same scenario.

4.2 Model Assessment

The Continuous Time Markov Chain (CTMC) underlying BM is depicted in Fig. 2, while its states are described in Table 3.

The symbolic expressions of the steady-state probabilities are reported in Appendix B. If we use the parameters of Table 2, we obtain the following values:

$$\pi_1 \approx 0.69518, \quad \pi_2 \approx 0.09568, \quad \pi_3 \approx 0.11691, \quad \pi_4 \approx 0.09223.$$

We can then compute the throughput of action types m_F, m_U, m_{E_F} and m_{E_U}, namely, the effective throughput and, hence, the profits of fair and unfair miners from the individual and aggregated perspective.

Figure 3a shows the total throughput of all fair (green line) and all unfair (orange line) miners as a function of the fraction of verifying miners. Indeed, the higher number of fair miners implies a reduction of the total throughput of unfair miners and consequently leads to a rise of the throughput for those who follow the protocol. Notice that the equilibrium point, where the two lines intersect each other, corresponds to 55% of verifying miners in the network. Thus, in case of equal proportion of these two classes of miners the non-verifying ones gain an

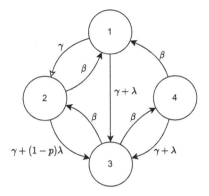

Fig. 2. Markov chain of BM.

Table 2. BM rate parameter assignment.

Parameter	Value
λ	0.083 blocks/s
N	100
$\gamma = \lambda/N$	8.3×10^{-4} blocks/s
β	0.314 s^{-1}
p	0.9

Table 3. BM Markov chain description.

State	Short description
1	All miners are mining
2	The fair environment is in verification step (V_{E_F}) and the rest of the network is mining
3	All fair components are verifying (V_{E_F} and V_F) and the rest of the network is mining
4	The single fair miner is verifying (V_F) and the rest of the network is mining

advantage since their total throughput, and so the total reward, is approximately 27% higher than that of fair miners.

Figure 3b shows a comparison between the effective throughput for a single fair and unfair miner as a function of the fraction of verifying miners in the network (blue and red lines, respectively). The throughput of the single unfair miner remains equal to γ since it omits the verification step using the available resources only for mining. Meanwhile, the throughput of the single fair miner is reduced by $\approx$19% with respect to the network with all verifying miners. This is due to the fact that the prevalence of non-verifying miners as well as their constantly higher total throughput lead to a higher production rate of blocks that every fair miner has to verify.

In conclusion, the unfair behaviour of miners has two consequences on fair miners: first, it causes unfairness in the network, i.e., the reward obtained by a miner is not proportional to its computational power. Second, and probably even worse, when a large amount of the network is unfair, the throughput of blocks is higher and fair miners will spend more time in the verification phase hence reducing their absolute throughput. Finally, while the throughput of an unfair miner is insensitive to the throughput of blocks in the network, its relative advantage over all other miners is highest when it is the only unfair miner (indeed in a network of all unfair miners, it would have no advantage at all).

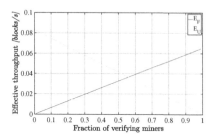 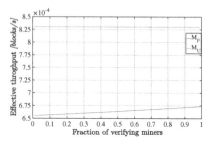

(a) Effective throughput of the fair environment E_F and the unfair environment E_U

(b) Effective throughput of a single fair miner M_F and a single unfair miner M_U

Fig. 3. Effective throughput as function of the fraction of verifying miners p.

5 Analysis of the Invalid Block Injection Policy

In the literature, several solutions have been proposed to mitigate the Verifier's Dilemma. In this section, we discuss one mitigation solution, namely the intentional injection of invalid blocks, which was proposed to mitigate the problem by Alharby et al. in [1].

In general, miners benefit from skipping the verification step because almost all blocks are valid anyway. Intentional production of invalid blocks aims to punish unfair miners, since they could end up working on new blocks on top of the invalid ones, thus not gaining a fee for the new blocks. Invalid blocks are supposed to be generated by a special node, named *creator*.

When an invalid block is injected it introduces a 'soft fork' of the chain, with the unfair miners working on an invalid branch and fair ones staying on a correct branch. We assume that, thanks to a blacklist of invalid block hashes, fair miners always recognize the correct branch. On the other hand, unfair miners realise that they are working on the wrong branch as soon as a fair miner announces a new block. Basically, since an unfair miner knows that they could be working on a wrong branch of the chain, it is rational for them to work on the last block announced (recall that PoW is a memoryless operation and hence this does not cause a waste of work).

5.1 Model Description

To study and optimize the invalid block injection approach, we extend the BM with more components. We assume that there are no malicious miners in the network, so all blocks are valid with the exception of those intentionally injected. If this is not the case, the rate of invalid block creation from malicious miners should be added to that of the creator. The PEPA model of the system with invalid block injection is called *Invalid Block Injection Model (IBIM)* and is reported in Table 4.

Table 4. PEPA specification of IBIM where $L = \{m_F, m_U, m_{E_F}, m_{E_U}, m_{fake}\}$.

$$
\begin{aligned}
C &\stackrel{def}{=} (m_{fake}, \epsilon).C + (m_F, \top).C + (m_U, \top).C + (m_{E_F}, \top).C + (m_{E_U}, \top).C \\
M_F &\stackrel{def}{=} (m_F, \gamma).M_F + (m_{E_F}, \top).V_F + (m_{E_U}, \top).V_F + (m_U, \top).V_F + (m_{fake}, \top).V_F \\
V_F &\stackrel{def}{=} (\tau, \beta).M_F + (m_{E_F}, \top).V_F + (m_{E_U}, \top).V_F + (m_U, \top).V_F + (m_{fake}, \top).V_F \\
M_U &\stackrel{def}{=} (m_U, \gamma).M_U + (m_{E_F}, \top).M_U + (m_F, \top).M_U + (m_{E_U}, \top).M_U + (m_{fake}, \top).M_{U_{INV}} \\
M_{U_{INV}} &\stackrel{def}{=} (\tau, \gamma).M_{U_{INV}} + (m_{E_F}, \top).M_U + (m_{E_U}, \top).M_{U_{INV}} + (m_F, \top).M_U + (m_{fake}, \top).M_{U_{INV}} \\
E_F &\stackrel{def}{=} (m_{E_F}, p\lambda).V_{E_F} + (m_{E_U}, \top).V_{E_F} + (m_F, \top).V_{E_F} + (m_U, \top).V_{E_F} + (m_{fake}, \top).V_{E_F} \\
V_{E_F} &\stackrel{def}{=} (\tau, \beta).E_F + (m_{E_U}, \top).V_{E_F} + (m_F, \top).V_{E_F} + (m_U, \top).V_{E_F} + (m_{fake}, \top).V_{E_F} \\
E_U &\stackrel{def}{=} (m_{E_U}, (1-p)\lambda).E_U + (m_F, \top).E_U + (m_{E_F}, \top).E_U + (m_U, \top).E_U + (m_{fake}, \top).E_{U_{INV}} \\
E_{U_{INV}} &\stackrel{def}{=} (\tau, (1-p)\lambda).E_{U_{INV}} + (m_F, \top).E_U + (m_{E_F}, \top).E_U + (m_U, \top).E_{U_{INV}} + (m_{fake}, \top).E_{U_{INV}} \\
Network' &\stackrel{def}{=} (M_F \underset{L}{\bowtie} M_U) \underset{L}{\bowtie} (E_F \underset{L}{\bowtie} E_U) \underset{L}{\bowtie} C
\end{aligned}
$$

The new component C models the behaviour of the invalid block creator. It produces invalid blocks with action type m_{fake} and rate ϵ. The creator recognises and rejects all invalid blocks and always works on the valid blockchain copy.

Furthermore, to model the behaviour of non-verifying miners M_U and E_U, we introduce the components $M_{U_{INV}}$ and $E_{U_{INV}}$ representing the same components working on an invalid blockchain copy. Once miner M_U, which is working on the valid chain, receives a fake block from the creator, its state changes to $M_{U_{INV}}$. The same explanation can be applied to the states of the unfair environment E_U. A description of the new components is presented in Table 8 of Appendix A. The underlying derivation graph of the model is depicted in Fig. 4.

5.2 Model Assessment

The CTMC underlying the model is shown in Fig. 5, while the description of its states is shown in Table 5.

The plots in Fig. 6 describe the relation between the effective throughput and the rate of invalid block injection ϵ in a network of 100 miners. Figures 6a and 6b, 6c and 6d, 6e and 6f study the network with a percentage of fair miners or $p = 90\%$, $p = 50\%$ and $p = 10\%$, respectively. Figures 6a, 6c and 6f show the impact of invalid block creation on both the environments E_F and E_U in terms of effective throughput. Figures 6b, 6d and 6f are crucial for our reasoning and show the impact of invalid block injection rate ϵ on a single fair and unfair miner. The importance of these plots rely on the fact that rational miners decide if a fair or unfair behaviour is convenient based on the maximisation of the expected profits. Hence, we desire to choose ϵ in such a way that the effective throughput of a fair miner is higher than that of an unfair one. We know from the base model that for $\epsilon = 0$, rational miners choose not to verify. However, as ϵ grows, we observe a point in which it is more convenient to be fair. What stops us to choose a very large ϵ? This is explained in the plots of left-hand side column of Fig. 6. Indeed, when ϵ grows the throughput of both fair and unfair miners decrease (for different reasons). This leads us to conclude that *the optimal value*

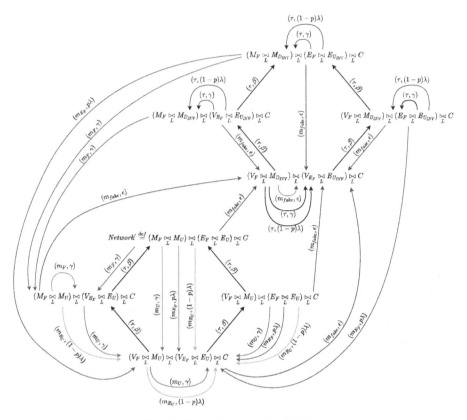

Fig. 4. Derivation graph of IBIM.

for ϵ is that at which the effective throughput of a single fair miner is equal to that of a single unfair one.

Another important observation is that the optimal value of ϵ is higher when there are many fair miners. This seems to be quite counter-intuitive, but the explanation is connected to the way that forks due to the injection of invalid blocks are resolved. In fact, the presence of many verifying miners helps the unfair ones because they reveal, by announcing their blocks, that the previous one was invalid. Thus, the unfair miner knows earlier that it is wasting its work. As a consequence, to make the unfair behaviour inconvenient the creator has to choose a value of ϵ higher than what would happen if the network had more unfair miners. This observation is proved for small valued of γ in Sect. 6.

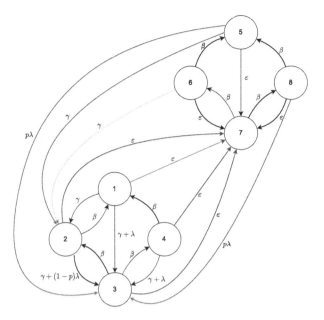

Fig. 5. Markov chain of IBIM.

Table 5. IBIM Markov chain description.

State	Short description
1	The creator and all miners are mining
2	The fair environment is in verification step (V_{E_F}) and the rest of the network (including the creator) is mining
3	All fair components are verifying (V_{E_F} and V_F) and the rest of the network is mining
4	The single fair miner is verifying (V_F) and the rest of the network is mining
5	The creator and all miners are mining
6	The fair environment is in verification step (V_{E_F}) and the rest of the network is mining
7	All fair components are verifying (V_{E_F} and V_F) and the rest of the network is mining
8	The single fair miner is verifying (V_F) and the rest of the network is mining

6   Optimal Invalid Block Injection Rate

In this section, we show that the analysis of the Markov chain depicted in Fig. 5 allows us to derive the symbolic expressions of the steady-state probabilities (see Appendix C). Hence, we can also derive the expression of the throughput for a single fair (T_F) and unfair miner (T_U).

$$T_F = (\pi_1 + \pi_2 + \pi_5 + \pi_6)\gamma, \quad T_U = (\pi_1 + \pi_2 + \pi_3 + \pi_4)\gamma.$$

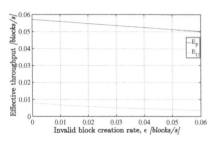

(a) Effective throughput of 90 fair miners in E_F and 10 unfair miners in E_U

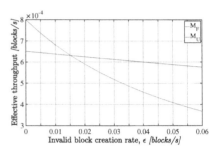

(b) Effective throughput of a single fair miner M_F and a single unfair miner M_U in the network of 90 fair miners and 10 unfair miners.

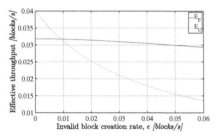

(c) Effective throughput of 50 fair miners in E_F and 50 unfair miners in E_U

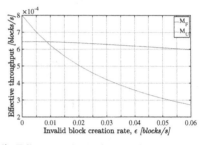

(d) Effective throughput of a single fair miner M_F and a single unfair miner M_U in the network of 50 fair miners and 50 unfair miners.

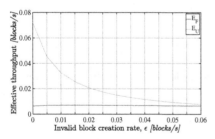

(e) Effective throughput of 10 fair miners in E_F and 90 unfair miners in E_U

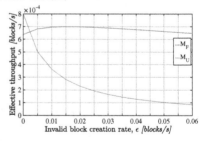

(f) Effective throughput of a single fair miner M_F and a single unfair miner M_U in the network of 10 fair miners and 90 unfair miners.

Fig. 6. Effective throughput as a function of the rate of invalid block creation.

The optimal injection rate is the rate at which $T_F = T_U$, that is $\pi_5 + \pi_6 = \pi_3 + \pi_4$. To simplify our analysis, we assume that $\gamma \to 0$, i.e., the single miner computational power is much lower than the entire network. This assumption is very realistic especially in public blockchains. Therefore, we need to find the real zeros of a rational function. These correspond to the roots of the polynomial

Table 6. Coefficients of $P(\epsilon)$.

Coefficient	Expression	Sign
c_6	$-\beta$	−
c_5	$-7\beta^2 - 3\beta\lambda$	−
c_4	$-19\beta^3 - (17-p)\beta^2\lambda - (3-p^2+p)\beta\lambda^2$	−
c_3	$-25\beta^4 - (35-4p)\beta^3\lambda - (13-2p^2-p)\beta^2\lambda^2 - (1-2p^2+2p)\beta\lambda^3$	−
c_2	$-16\beta^5 - (31-5p)\beta^4\lambda - (18-10p-p^2)\beta^3\lambda^2 - (3+2p^3-4p^2-p)\beta^2\lambda^3 - (1-p)p\beta\lambda^4$	−
c_1	$-4\beta^6 + 2\beta^5\lambda + (12p-8)\beta^4\lambda^2 - (2+3p^3-3p^2-9p)\beta^3\lambda^3$	−
c_0	$4p\beta^5\lambda^2 + 6p\beta^4\lambda^3 + (-4p^2+4p+2)\beta^3\lambda^4 + 2(1-p)p^2\beta^2\lambda^5$	+

$$P_p(\epsilon) = \sum_{i=0}^{6} c_i\epsilon^i,$$

where the coefficients are shown in Table 6.

Since the polynomial has degree higher than 4, we cannot find a general closed form expression for its roots. However, we can use a bisection based algorithm whose applicability is ensured by the following Theorem:

Theorem 1. $P_p(\epsilon)$ *has exactly one positive real root provided that* $\lambda < \beta$.

Proof. By Descartes' rule of signs [15], the number of sign changes between consecutive (nonzero) coefficients is 1 and then the number of positive roots is exactly 1. This can be easily observed by inspecting the coefficients of $P_p(\epsilon)$ in Table 6 with some care for the coefficient of the linear term that is negative provided that $\lambda < \beta$. □

The condition in Theorem 1 is straightforwardly satisfied by the actual models since it requires that the expected verification time must be lower than the expected block generation time.

At this point, we can easily derive the optimal injection rate thanks to the bisection method with arbitrary accuracy. Indeed, it is easy to observe that $P_p(0) > 0$ while $P_p(\infty) < 0$. Therefore, once an interval $[0, b]$ is found such that the $P_p(b) < 0$, the standard bisection algorithm can be applied.

In general, it is impossible to know the correct value of p, i.e., the fraction of fair miners. However, the following result formally supports the experimental observations of Sect. 5.

Theorem 2. *If* $\lambda < \beta$, *the optimal invalid block injection rate is monotonically increasing with the fraction of fair miners* p *and hence it reaches its maximum for* $p = 1$.

Proof. The proof relies on the fact that, if $\lambda < \beta$, it holds that, for $\epsilon \geq 0$:

$$p > q \implies P_p(\epsilon) > P_q(\epsilon).$$

This can be proved by computing the polynomial $Q_{p,q}(\epsilon) = P_p(\epsilon) - P_q(\epsilon)$ and by observing that the coefficients of $Q_{p,q}(\epsilon)$ are all non-negative and hence $Q_{p,q}(\epsilon) > 0$ for all $\epsilon \geq 0$. Since $P_q(0) > 0$ and, by Theorem 1, $P_p(0)$ must have a positive root, this implies that the root of $P_p(\epsilon)$ is strictly greater than that of $P_q(\epsilon)$. □

In conclusion, it is safe for the network to compute the optimal injection rate assuming $p = 1$ and maintain the same rate unless a change on rates λ or β are measured. In this way, p will not change even in presence of rational miners since verifying the blocks is more convenient than cheating.

7 Conclusion

In this paper, we have investigated Ethereum Verifier's Dilemma by means of quantitative Markovian models expressed in PEPA. The models allowed us to study the problem by showing that the profits of an unfair (i.e., non-verifying) miner can be significantly higher than those of fair ones and that these depend on the amount of unfair miners. Indeed, the optimal scenario for an unfair miner is that it is the only one that skips the verification of announced blocks. Then, we have proposed a model for the countermeasure that requires the injection of invalid blocks to encourage unfair miners to abandon their unethical behaviour by making it economically disadvantageous. The second model has two purposes: (i) it allows us to determine the optimal injection rate, i.e., the minimum rate at which the average reward of an unfair miner is lower than that of a fair one, and (ii) it can be used to estimate the impact of this countermeasure on the overall network throughput and miners' rewards. More specifically, regarding (i), we have shown that the optimal rate can be computed numerically as the unique positive root of a certain polynomial thanks to a simple bisection method. Moreover, we observed that, since an unfair miner has the maximum advantage when all others are fair, the optimal injection rate reaches its maximum when there is only one unfair miner. Thus, in practice, this is the scenario that must be considered to configure the invalid block injection.

Acknowledgements. This work has been partially supported by the Project PRIN 2020 "Nirvana - Noninterference and Reversibility Analysis in Private Blockchains" - N. 20202FCJMH and by the Project GNCS 2022 "Proprietà qualitative e quantitative di sistemi reversibili".

Appendix

A Tables of Notations

The following tables provide a description of the notations used in the PEPA specification of the BM and IBIM models. In particular they describe the notations used to model the behaviours of single fair and unfair miners and those used for the environments.

Table 7. BM component description.

Symbol	Description
M_F	Single fair miner which mines a block with action type m_F and rate γ. When it receives a block, then it starts the verification V_F.
V_F	Verification step of a single fair miner.
M_U	Single unfair miner which mines a block with action type m_U and rate γ. When it receives a block, then it returns to the initial state M_U.
E_F	Correctly operating environment representing the fraction of fair miners mining blocks with action type m_{E_F} and rate $p\lambda$.
V_{E_F}	Verification step for the fair environment.
E_U	Unfair environment representing the fraction of non-verifying miners mining blocks with an action type m_{E_U} and rate $(1-p)\lambda$.

Table 8. IBIM component description.

Symbol	Description
C	Invalid blocks creator who creates invalid blocks with action type m_{fake} and rate ϵ.
$M_{U_{INV}}$	Single unfair miner with a fake block in its blockchain copy. It mines blocks with action type τ and rate γ. When it receives a block from M_F or E_F, then it obtains the valid blockchain copy and returns to the initial state M_U.
$E_{U_{INV}}$	Unfair environment with an invalid blockchain copy. It almost always remains on the same state and returns to E_U only when the ledger is synchronised with M_F and E_F.

B   Steady State Probabilities for BM

We report the symbolic expressions of the steady-state probabilities for the BM model. In particular, the steady-state probabilities for the CTMC depicted in Fig. 2 are as follows:

$$\pi_1 = \frac{(\beta^2(2(\beta + \gamma + \lambda) - \lambda p))}{K}$$

$$\pi_2 = \frac{(\beta((\gamma + \lambda)(2\gamma + \lambda) + \beta(3\gamma + \lambda)))}{K}$$

$$\pi_3 = \frac{(\beta(\beta(\gamma + \lambda) + (2\gamma + \lambda)(\gamma + \lambda - \lambda p)))}{K}$$

$$\pi_4 = \frac{((\beta + \gamma + \lambda)(\beta(\gamma + \lambda) + (2\gamma + \lambda)(\gamma + \lambda - \lambda p)))}{K}$$

where $\sum_{i=1}^{4} \pi_i = 1$ and K is the normalising constant whose expression is

$$K = (2\beta^3 + \beta^2(7\gamma - \lambda(-5 + p)) + \beta(7\gamma^2 + \gamma\lambda(11 - 4p)$$
$$- 2\lambda^2(-2 + p)) + (\gamma + \lambda)(2\gamma + \lambda)(\gamma + \lambda - \lambda p)).$$

C   Steady-State Probabilities for IBIM

We report the symbolic expressions of the steady-state probabilities for the IBIM model. In particular, the steady-state probabilities for the CTMC depicted in Fig. 5 are as follows:

$$\pi_1 = \frac{\beta^2 \left(\lambda^3 r^2 \left(-2\beta^2(r - 2) - \beta\varepsilon(r - 2)\right) + \lambda^2 r \left(3\beta^2\varepsilon(r + 2) + 2\beta^3(r + 2)\right.}{K}$$

$$\frac{+\beta\varepsilon^2(r + 2)\right) + \lambda r(\beta + \varepsilon)\left(6\beta^2\varepsilon + 4\beta^3 + 2\beta\varepsilon^2\right))}{K}$$

$$\pi_2 = \frac{\beta \left(\beta^2 \lambda r(\varepsilon + \lambda) \left(4\varepsilon^2 + 3\varepsilon\lambda(r + 1) + 2\lambda^2 r\right) + 2\beta^4 \lambda r(\varepsilon + \lambda)+}{K}$$

$$\frac{+\beta^3 \lambda r(\varepsilon + \lambda)(5\varepsilon + 2\lambda(r + 1)) + \beta\varepsilon\lambda r(\varepsilon + \lambda)^2(\varepsilon + \lambda r))}{K}$$

$$\pi_3 = \frac{\beta(\beta + \varepsilon + \lambda) \left(2\beta^3 \lambda r(\varepsilon + \lambda) + \beta^2 \lambda r(\varepsilon + \lambda)(5\varepsilon + 2\lambda)\right.}{K}$$

$$\frac{+\beta\lambda r(\varepsilon + \lambda) \left(4\varepsilon^2 + 3\varepsilon\lambda - 2\lambda^2(r - 1)r\right) + \varepsilon\lambda r(\varepsilon + \lambda)(\varepsilon + \lambda - \lambda r)(\varepsilon + \lambda r))}{K}$$

$$\pi_4 = \frac{\beta^2 \left(2\beta^3 \lambda r(\varepsilon + \lambda) + \beta^2 \lambda r(\varepsilon + \lambda)(5\varepsilon + 2\lambda)\right.}{K}$$

$$\frac{+\beta\lambda r(\varepsilon + \lambda) \left(4\varepsilon^2 + 3\varepsilon\lambda - 2\lambda^2(r - 1)r\right) + \varepsilon\lambda r(\varepsilon + \lambda)(\varepsilon + \lambda - \lambda r)(\varepsilon + \lambda r))}{K}$$

$$\pi_5 = \frac{\beta^2\varepsilon(2\beta + 2\varepsilon + \lambda r)}{(\beta + \varepsilon)(2\beta + \varepsilon)(\varepsilon + \lambda r)(\beta + \varepsilon + \lambda r)}$$

$$\pi_6 = \frac{\beta\varepsilon}{(\beta + \varepsilon)(2\beta + \varepsilon)}$$

$$\pi_7 = \frac{\varepsilon}{2\beta + \varepsilon}$$

$$\pi_8 = \frac{\beta\varepsilon}{(2\beta + \varepsilon)(\beta + \varepsilon + \lambda r)}$$

where $\sum_{i=1}^{8} \pi_i = 1$ and K is

$$K = (\beta + \varepsilon)(2\beta + \varepsilon)(\varepsilon + \lambda r)(\beta + \varepsilon + \lambda r)\left(2\beta^3 + \beta^2(5\varepsilon + 5\lambda - \lambda r)\right.$$
$$\left. + 2\beta(\varepsilon + \lambda)(2\varepsilon + 2\lambda - \lambda r) + (\varepsilon + \lambda)^2(\varepsilon + \lambda - \lambda r)\right).$$

References

1. Alharby, M., Lunardi, R.C., Aldweesh, A., van Moorsel, A.: Data-driven model-based analysis of the ethereum verifier's Dilemma. In: 2020 50th Annual IEEE/IFIP International Conference on Dependable Systems and Networks (DSN), pp. 209–220. IEEE (2020)
2. Alzetta, G., Marin, A., Piazza, C., Rossi, S.: Lumping-based equivalences in Markovian automata: algorithms and applications to product-form analyses. Inf. Comput. **260**, 99–125 (2018)
3. Anjana, P.S., Kumari, S., Peri, S., Rathor, S., Somani, A.: An efficient framework for optimistic concurrent execution of smart contracts. In: 2019 27th Euromicro International Conference on Parallel, Distributed and Network-Based Processing (PDP), pp. 83–92. IEEE (2019)
4. Balsamo, S., Marin, A., Mitrani, I., Rebagliati, N.: Prediction of the consolidation delay in blockchain-based applications. In: ICPE 2021: ACM/SPEC International Conference on Performance Engineering, Virtual Event, France, 19–21 April 2021, pp. 81–92 (2021)
5. Buterin, V., et al.: Ethereum: A next-generation smart contract and decentralized application platform (2014)
6. Das, S., Awathare, N., Ren, L., Ribeiro, V.J., Bellur, U.: Tuxedo: maximizing smart contract computation in PoW blockchains. Proc. ACM Meas. Anal. Comput. Syst. **5**(3), 1–30 (2021)
7. Das, S., Ribeiro, V.J., Anand, A.: YODA: Enabling computationally intensive contracts on blockchains with byzantine and selfish nodes. arXiv preprint arXiv:1811.03265 (2018)
8. Dickerson, T., Gazzillo, P., Herlihy, M., Koskinen, E.: Adding concurrency to smart contracts. Distrib. Comput. **33**(3), 209–225 (2020). https://doi.org/10.1007/s00446-019-00357-z
9. Fiz Pontiveros, B.B., Ferreira Torres, C., State, R.: Sluggish mining: profiting from the verifier's dilemma. In: Bracciali, A., Clark, J., Pintore, F., Rønne, P.B., Sala, M. (eds.) FC 2019. LNCS, vol. 11599, pp. 67–81. Springer, Cham (2020). https://doi.org/10.1007/978-3-030-43725-1_6
10. Hillston, J.: A Compositional Approach to Performance Modelling. Cambridge University Press, Cambridge (1996)
11. Kalodner, H., Goldfeder, S., Chen, X., Weinberg, S.M., Felten, E.W.: Arbitrum: scalable, private smart contracts. In: 27th USENIX Security Symposium (USENIX Security 18), pp. 1353–1370 (2018)
12. Luu, L., Teutsch, J., Kulkarni, R., Saxena, P.: Demystifying incentives in the consensus computer. In: Proceedings of the 22nd ACM SIGSAC Conference on Computer and Communications Security, pp. 706–719 (2015)
13. Marin, A., Rossi, S.: On the relations between lumpability and reversibility. In: MASCOTS, pp. 427–432. IEEE Computer Society (2014)

14. Marin, A., Rossi, S.: On the relations between Markov chain lumpability and reversibility. Acta Informatica **54**(5), 447–485 (2016). https://doi.org/10.1007/s00236-016-0266-1

15. Olver, F.W.J., Lozier, D.W., Boisvert, R.F., Clark, C.W.: The NIST Handbook of Mathematical Functions. Cambridge University Press, Cambridge (2010)

16. Teutsch, J., Reitwießner, C.: A scalable verification solution for blockchains. arXiv preprint arXiv:1908.04756 (2019)

17. Yu, L., Tsai, W.T., Li, G., Yao, Y., Hu, C., Deng, E.: Smart-contract execution with concurrent block building. In: 2017 IEEE Symposium on Service-Oriented System Engineering (SOSE), pp. 160–167. IEEE (2017)

Comparing Statistical and Analytical Routing Approaches for Delay-Tolerant Networks

Pedro R. D'Argenio[1,2], Juan A. Fraire[1,3],
Arnd Hartmanns[4(✉)], and Fernando Raverta[1,2]

[1] CONICET, Córdoba, Argentina
[2] Universidad Nacional de Córdoba, Córdoba, Argentina
[3] Inria, Lyon, France
[4] University of Twente, Enschede, The Netherlands
a.hartmanns@utwente.nl

Abstract. In delay-tolerant networks (DTNs) with uncertain contact plans, the communication episodes and their reliabilities are known a priori. To maximize the end-to-end delivery probability, a bounded network-wide number of message copies are allowed. The resulting multi-copy routing optimization problem is naturally modelled as a Markov decision process with distributed information. The two state-of-the-art solution approaches are statistical model checking with scheduler sampling, and the analytical RUCoP algorithm based on probabilistic model checking. In this paper, we provide an in-depth comparison of the two approaches. We use an extensive benchmark set comprising random networks, scalable binomial topologies, and realistic ring-road low Earth orbit satellite networks. We evaluate the obtained message delivery probabilities as well as the computational effort. Our results show that both approaches are suitable tools for obtaining reliable routes in DTN, and expose a trade-off between scalability and solution quality.

1 Introduction

Delay-tolerant networks (DTNs) are time-evolving networks lacking continuous and instantaneous end-to-end connectivity [11,18]. The DTN domain comprises deep-space [9] and near-Earth communication [10], airborne networks [27], vehicular ad-hoc networks [5], mobile social networks [32], Internet of things scenarios [6], and underwater networks [40], among many others. A *bundle layer* overcomes the delay and disruption in DTNs by means of (i) a persistent storage on each DTN node and by (ii) assuming no immediate response from neighboring nodes [41]. As a result, *bundles* of data (a data unit in the Bundle Protocol [47])—and status information about the rest of the network—flow in a

Authors are ordered alphabetically. This work was supported by Agencia I+D+i grants PICT-2017-1335 and PICT-2017-3894 (RAFTSys), DFG grant 389792660 as part of TRR 248, the European Union's Horizon 2020 research and innovation programme under the Marie Skłodowska-Curie grant agreement 101008233 (MISSION), NWO VENI grant 639.021.754, and SeCyT-UNC grant 33620180100354CB (ARES).

© Springer Nature Switzerland AG 2022
E. Ábrahám and M. Paolieri (Eds.): QEST 2022, LNCS 13479, pp. 337–355, 2022.
https://doi.org/10.1007/978-3-031-16336-4_17

store-carry-and-forward fashion as transmission opportunities become available. Connectivity in DTNs is represented by means of *contacts*: an episode of time when a node is able to transfer data to another node.

Where contacts can be accurately predicted, the DTN is *scheduled* [22]; in *probabilistic* DTNs, the contact patterns can be dynamically inferred; no assumptions on future contacts can be made in *opportunistic* DTNs [11]. Recent work extended this classification to also consider *uncertain* DTNs, in which forthcoming connectivity can be described by probabilistic schedules available *a priori* [17,23,38,39,44,45]. Instead of a guaranteed contact plan, uncertain contact plans include information on the reliability (i.e. failure probability) of planned links. In other words, the materialization of contacts can differ from the original plan with a probability that can be computed/estimated in advance. Uncertain DTNs describe a plethora of practical scenarios: unreliable space networks [23], public vehicle networks with uncertain mobility patterns [35], interference-sensitive communication links in cognitive radio [46], or networks based on third-party carriers with limited but well-known availability [33].

This work summarises and compares existing routing solutions for uncertain DTNs. The state-of-the-art techniques are *lightweight scheduler sampling* (LSS) [17] and *routing under uncertain contact plans* (RUCoP) [44]. Both leverage Markov decision processes (MDPs), allow a bounded network-wide number of message copies to maximize the delivery probability, and properly assume that uncertain DTN nodes can only act on limited local knowledge. However, they are different in nature: LSS exploits simulation and statistical model checking techniques [1] whereas RUCoP is based on an analytical solution that exhaustively explores the MDP akin to probabilistic model checking [3,4]. Both are off-line approaches, as a central node is assumed to pre-compute the routing in advance and then distribute the required information to the DTN nodes.

We provide an extensive benchmarking framework to evaluate LSS and RUCoP comprising random networks (random contact assignment), binomial networks (multi-level tree contact topologies with controllable complexity), and realistic ring-road low-Earth orbit satellite networks. In these scenarios, we compare the resulting message delivery probability and computational effort in terms of time and memory consumption. Our results highlight the performance-cost trade-off between these two state-of-the-art routing techniques for uncertain DTN. We also report on our enhancements to encoding DTNs for use with LSS that significantly improve the cost/performance ratio of the approach.

Section 2 of this paper revises the background of DTNs, MDP, and modelling the routing problem. We dive into the details of the LSS and RUCoP techniques in Sect. 3, including a summary of our improvements to LSS for DTNs. We present, apply and analyze the benchmark framework in Sect. 4.

2 Background

This section describes the concept and context of DTNs and explains how to encode DTN routing with global and local information as MDP.

Scheduled vs. Uncertain DTNs. The term "DTN" was introduced in the context of interplanetary communication to designate time-evolving networks lacking continuous and instantaneous end-to-end connectivity [9]. The concepts and mechanisms devised to deal with the delays and disruptions of interplanetary communications can readily be applied to other domains characterized by long signal propagation time, frequent node occlusion, high node mobility, and/or reduced communication range and resources [24] such as airborne, vehicular, social, IoT, underwater and space networks [5,6,10,21,27,32,40]. DTN protocols like the Bundle Protocol [13,47] address the delays and disruptions by implementing the principles of *store-carry-and-forward* and *minimal end-to-end messaging exchange* for control or feedback [11]. The time-evolving and partitioned nature of DTNs favors representing connectivity by *contacts*: episodes of time where a node can transfer data to another node. Contacts can be classified [11] as *opportunistic* (no assumptions can be made on future contacts), *probabilistic* (contact patterns can be inferred from history, e.g. in social networks), and *scheduled* (contacts can be accurately predicted and documented in a *contact plan*).

A contact plan comprises the set of forthcoming contacts, and is a central element in scheduled DTN routing. The routing process is typically divided into *planning* (future episodes of communication are estimated to form the contact plan), *routing* (the plan is used to compute routes, either in a centralized (offline) or decentralized (on-line) fashion [20]), and *forwarding* (effectively enqueuing the data for the correct next-hop node). Contact graph routing (CGR) [2] is the de-facto standard decentralized routing algorithm when a contact plan is available. It is the sole routing approach that has been flight-validated in deep-space [48] and near-Earth networked missions [34]. CGR optimizes delivery *time* by leveraging adaptations of Dijkstra's algorithm to the time dynamics of DTNs.

The limitation of the contact plan structure and associated routing algorithms like CGR is that they assume that connectivity episodes are guaranteed. Instead, an *uncertain contact plan* comprises contacts whose materialization can differ from the original plan with a given probability available a priori [45]. Reasons include well-known failure modes of the DTN nodes, or an incomplete/inaccurate knowledge of the system status by the time the schedule was computed. Uncertain contact plans gave raise to a new type of DTNs coined *uncertain DTNs* [17,38,39,44,45] that exploit time-dependent probabilistic information of the forthcoming communication opportunities. Instead of a single copy sent via the fastest path like CGR, uncertain DTNs can use the uncertainty information in the contact plan to optimally route *multiple copies* of the data to increase its successful delivery probability (SDP).

Markov decision processes (MDPs) provide a mathematical framework capturing the interaction between non-deterministic and probabilistic choices [19,42], making them appropriate for modelling decision making under probabilistically quantified uncertainty. In its simplest form, an MDP $\mathcal{M}$ is a tuple $(S, Act, \mathbf{P}, s_0)$ where S is a finite set of states with initial state $s_0 \in S$, Act is a finite set of actions, and $\mathbf{P} : S \times Act \times S \to [0, 1]$ is a transition probability function such that $\sum_{s' \in S} \mathbf{P}(s, \alpha, s') \in \{0, 1\}$ for all $s \in S$ and $\alpha \in Act$. If $\sum_{s' \in S} \mathbf{P}(s, \alpha, s') = 1$,

α is *enabled* in s, and $\mathbf{P}(s, \alpha, s')$ gives the probability that the next state is s' conditioned on the system being in state s and action α being chosen.

A *reachability problem* is characterized as follows: given a set of *goal states* $B \subseteq S$, maximize the probability that a state in B is reached from the initial state s_0. That is, we want to calculate $Pr_{s_0}^{\max}(reach(B))$. In our application, B is the set of states in which a bundle has been successfully delivered. Moreover, we are also interested in determining the decisions—namely, the *policy* or *scheduler*—that lead to such a maximizing value. A *scheduler* is a function $\pi : S \to Act$ that defines the decision that resolves a possible non-determinism. This problem can be solved e.g. by using value iteration on the Bellman equations [4].

Encoding Uncertain Contact Plans. Consider the example contact plan with nodes A, B, C, and D in Fig. 1. It spans a window of five time slots, t_0 to t_4. We also assume an ending time t_5. The possible contacts in each slot are depicted by an arrow labelled with the contact failure probability. In time slot t_1, for instance, node C is in reach of node B with transmission failure probability of 0.1 (and success probability of 0.9).

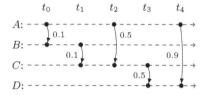

Fig. 1. Uncertain contact plan.

Suppose we want to transmit a bundle from A to D. To increase the probability of success, we allow two copies throughout the network. A state of the MDP consists of the number of copies that each node holds at a given time slot. Initially, at the beginning of t_0, node A has the two copies while the others have none, represented by state $[A^2 B^0 C^0 D^0 \mid t_0]$ in Fig. 2. At this point, node A has three options: (i) sending only one copy to node B, represented by action "$A \xrightarrow{1} B$" leaving from state $[A^2 B^0 C^0 D^0 \mid t_0]$, (ii) sending two copies to B (action "$A \xrightarrow{2} B$"), or (iii) keeping the two copies (action "A stores"). In the first case, the successful transmission leads to state $[A^1 B^1 C^0 D^0 \mid t_1]$ where A has kept one copy and the other has reached B. Since success probability is 0.9, we have

$$\mathbf{P}([A^2 B^0 C^0 D^0 \mid t_0], A \xrightarrow{1} B, [A^1 B^1 C^0 D^0 \mid t_1]) = 0.9.$$

Failing to transmit moves us to the next time slot without altering the number of copies in each node. Therefore

$$\mathbf{P}([A^2 B^0 C^0 D^0 \mid t_0], A \xrightarrow{1} B, [A^2 B^0 C^0 D^0 \mid t_1]) = 0.1.$$

Action $A \xrightarrow{1} B$ is the black transition out of $[A^2 B^0 C^0 D^0 \mid t_0]$ in Fig. 2 where the solid line represents the successful transmission while the dotted arrow represents the failing event. The situation is

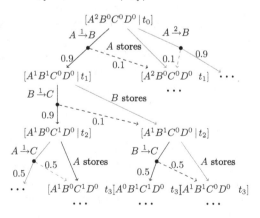

Fig. 2. MDP for Fig. 1. (Color figure online)

analogous for action $A \xrightarrow{2} B$ (red transition on the right), while for storing the two bundles there is no possibility of failure, so we have

$$\mathbf{P}([A^2 B^0 C^0 D^0 \,|\, t_0], A \text{ stores}, [A^2 B^0 C^0 D^0 \,|\, t_1]) = 1.$$

The construction is similar for the rest of the MDP. Figure 2 depicts it partially; we indicate with "..." where the MDP needs to continue.

We assume that the sending node can determine whether a transmission was successful or not; in case of success, it deletes the transmitted number of copies, while in case of failure, it keeps them. This ensures that the entire network contains the intended number of copies at any time, and is possible and typical in LEO constellations. We refer to this assumption as *acknowledged communication* (a.k.a. custody transfer in the Bundle Protocol [24]). The alternative is *fully unreliable* communication where transmitted copies are lost upon failure, which is natural in deep-space communication.

Global and Local Information. For the MDP described above, the maximizing scheduler for goal set $B = \{ [A^a B^b C^c D^d \,|\, t_5] \mid d \geq 1 \}$ describes the optimal routing decisions. This scheduler, however, is based on a *global* view of the system: decisions are taken based on the current state of the whole network. This implies that distributed nodes need to know where all copies are in the network at any moment, including remote and potentially disconnected nodes. This is impossible to achieve in practice in highly partitioned DTNs. Nodes must therefore decide based on partial local knowledge. To illustrate, consider time slot t_2 in the example of Fig. 1. Here, node A has two possible decisions: storing or forwarding to C. Consider precisely the situation in which A has one copy and the second copy is already on its way. A's optimal decision depends on whether the other copy is on B or C at time t_2, reflecting the optimal decisions on Fig. 2: A stores if C already has the other copy and A forwards to C if B has the copy. However, it is most likely that A is not able to know whether the second copy is in B or C, in which case A's decision should be the same regardless if it is in state $[A^1 B^1 C^0 D^0 \,|\, t_2]$ or $[A^1 B^0 C^1 D^0 \,|\, t_2]$. This type of problem, in which decisions in an MDP associated to a distributed system may only be based on *local* knowledge, is known as *distributed scheduling* [12,25,26].

3   Routing in Uncertain DTNs

The optimal *global* scheduler can be computed using any probabilistic model checker such as PRISM [36], STORM [16], or MCSTA of the MODEST TOOLSET [30]: we compactly describe the MDP and the goal set in the tool's higher-level input language; then the tool generates and stores in memory the MDP's entire state space, solves the reachability problem by solving the linear program induced by the Bellman equations or by using an iterative algorithm such as a sound variant of value iteration [28,31,43], and writes the induced scheduler to file. Probabilistic model checkers, however, are generic tools that solve arbitrarily structured MDP without optimizations for the DTN routing application. For complex networks, they will quickly encounter the state space explosion problem and run out of memory (see [44]). Furthermore, none of them provides a solution for the local-information problem. We now summarize the two existing MDP-based approaches for optimal DTN routing under uncertain contact plans, RUCoP and LSS. Both can also produce schedulers based on *local* information only, and approach the routing process in an off-line fashion: the routing decisions are pre-computed in a centralized node.

3.1   RUCoP

RUCoP [45] (<u>r</u>outing under <u>u</u>ncertain <u>c</u>ontact <u>p</u>lans) provides an analytical solution to find the routing decisions optimising SDP for an uncertain contact plan.

The first observation exploited by RUCoP is that, due to the inclusion of the current time slot value in the states, the MDP for an uncertain contact plan is acyclic. RUCoP thus only constructs the "optimal" part of the MDP by following the Bellman equations backwards. In our example from the previous section, it starts at any state in t_5 in which D contains at least one copy. It then walks backwards in the contact plan, selecting only the maximizing transitions according to the Bellman equations. In its general form, RUCoP (i) considers the possibility that multiple nodes can transmit to each other in one time slot, which may produce a cycle in the MDP. However, since cyclic transmission would only lower the SDP, RUCoP can break all such cycles and keep the MDP acyclic. It also in general (ii) takes a target node and builds the optimal part of the MDP for any possible transmitting source rather than restricting to a single source node as in our example. The full RUCoP algorithm is in 2-EXPTIME: its runtime is exponential in the number of nodes and doubly exponential in the number of copies. This makes RUCoP highly expensive in time and memory. However, for memory optimization, RUCoP not only constructs the optimal part of the MDP backwards in an on-the-fly manner, but also writes all information that is not going to be necessary for further calculations to disk. In particular, only the states at the current time slot are necessary for calculating the states at the preceding time slot and the respective connecting optimal transitions.

RUCoP delivers optimal routing decisions for acknowledged communication in general. However, it is based on a global view of the system. To find local-information schedulers, we need to use its L-RUCoP (<u>l</u>ocal RUCoP) variant. It works as follows: Suppose that, to increase reliability, n copies of the bundle are used. L-RUCoP builds a table $T(N, c, t_i)$ that assigns to each node N holding c copies ($1 \leq c \leq n$) at time t_i the best decision based on local knowledge. This decision is taken from running RUCoP on c copies (instead of n), which basically amounts to supposing that N holds c copies and no copy is on the other nodes. Thus, for our example, the decision for states $[A^1 B^1 C^0 D^0 \,|\, t_2]$ and $[A^1 B^0 C^1 D^0 \,|\, t_2]$ will be both taken from $T(A, 1, t_2)$ which in turn is obtained from the decision in state $[A^1 B^0 C^0 D^0 \,|\, t_2]$ derived from running RUCoP with one single copy. On top of this basic idea, L-RUCoP also exploits extra knowledge that may be available in certain occasions. For instance, at time t_1 in our example, A knows if B holds a copy depending on whether the transmission at time t_0 was successful or not. In this case, L-RUCoP looks ahead using the appropriate RUCoP instance on the state with the *available* knowledge where, just like before, all information about the *other* (unknown) copies is assumed to be 0. In the example, at time t_1, the entry $T(A, 1, t_1)$ will be filled with the information retrieved from RUCoP for two copies on state $[A^1 B^1 C^0 D^0 | t_1]$ since A knows B has received the copy. The interested reader may find the details of L-RUCoP as well as the full specification of RUCoP in [45].

3.2 LSS

Given a discrete-time Markov chain (DTMC), i.e. an MDP where every state has at most one enabled action, Monte Carlo simulation or *statistical model checking* (SMC [1]) can be used to estimate the probabilities for reachability problems: We (pseudo-)randomly sample n paths—*simulation runs*—through the DTMC, identify each success (that reaches a goal state) with 1 and every failure with 0, and return the average as an estimate of the reachability probability. The result is correct up to a statistical error and confidence depending on n. Compared to probabilistic model checking, SMC needs only constant memory, assuming that we can effectively simulate the MDP from a high-level description so that we do not need to store its entire state space. As a simulation-based approach, SMC is easy to parallelize and distribute on multi-core systems and compute clusters.

Lightweight scheduler sampling [37] (LSS) extends SMC to MDP: Given an MDP M, it (i) randomly selects a set Σ of m schedulers, each identified by a fixed-size integer (e.g. of 32 bits), (ii) employs some heuristic (that involves simulating the DTMCs $M|_\sigma$ resulting from combining M with a scheduler $\sigma \in \Sigma$) to select the $\sigma_{max} \in \Sigma$ that appears to induce the highest probability, and finally (iii) performs a standard SMC analysis on $M|_{\sigma_{max}}$ to provide an estimate $\hat{p}_{\sigma_{max}}$ for $Pr_{s_0}^{max}(reach(B))$. However, note that—unless we are lucky and Σ happens to include an optimal scheduler and the heuristic identifies it as such—$\hat{p}_{\sigma_{max}}$ is an underapproximation of $Pr_{s_0}^{max}(reach(B))$ only, and subject to the statistical error of the SMC analysis. The effectiveness of LSS depends on the probability mass of the set of near-optimal schedulers among the set of all schedulers that we sample Σ from: It works well if a randomly selected scheduler is somewhat likely to be near-optimal, but usually fails in cases where many decisions need to be made in exactly one right way in order to get a successful path at all. We use the *smart sampling* [15] approach to select σ_{max} in step (ii): We start by performing 1 simulation run for each of the m schedulers, then discard the $\lceil \frac{m}{2} \rceil$ worst of them; in the next round, we perform 2 runs for each of the approx. $\frac{m}{2}$ remaining schedulers, and again discard the worst half. We continue until only one scheduler remains, which is σ_{max}. In this way, the number of simulation runs, and thus the runtime, needed for LSS grows only logarithmically in m.

The key to LSS is the constant-memory representation of schedulers as (32-bit) integers. It enables LSS' constant memory usage in the size of the MDP, which sets it apart from simulation-based machine learning techniques such as reinforcement learning, which need to store learned information (e.g. Q-tables) for each visited state. Let $i \in \mathbb{Z}_{32}$ identify scheduler σ_i. Then, upon encountering a state s with $k > 1$ enabled actions while simulating $M|_{\sigma_i}$, LSS selects the $(\mathcal{H}(i.s) \bmod k)$-th action, where $i.s$ is the concatenation of the binary representations of s and i, and $\mathcal{H}$ is a (usually simple non-cryptographic) hash function that maps its inputs to a fixed-size integer so that, ideally, the resulting values are uniformly distributed over the output space. This selection procedure is deterministic, so we can reproduce the decision for state s at any time knowing i. For nontrivial $\mathcal{H}$, it is also highly unpredictable: changing i, e.g. by modifying a certain bit, may result in a different decision for many states.

Local Information. As described above, LSS produces global-information schedulers. However, it can be adapted to sample from local-information schedulers only [17]: When having to make a decision on node N, instead of feeding $i.s$ into $\mathcal{H}$, we use $N.i.s|_N$ instead, where $s|_N$ contains only the locally available information: the number of locally-stored copies and the current time slot. To avoid conflicts where two nodes need to make a decision at the same time, certain restrictions apply to the high-level modelling of the MDP as a system of multiple independently executing nodes; we refer the interested reader to [17] for details. We refer to LSS with local-information schedulers as L-LSS.

A scheduler found to be good via L-LSS can in principle be implemented, e.g. on the satellites themselves, by simply replicating the L-LSS decision procedure: each node knows its identifier N, the number of copies it stores, and can translate the current time into a time slot in the contact plan. The only data that needs to be transmitted to the node is the integer identifying the scheduler.

Our Improvements to LSS for DTN. For our comparison in Sect. 4, we use the implementation of DTN routing with LSS and L-LSS of [17]. It consists of two parts: a CP2MODEST Python script that converts a contact plan into a high-level description of the MDP as described in Sect. 2 in the MODEST modelling language [29], and an implementation of LSS and L-LSS in the MODES simulator/statistical model checker [7] of the MODEST TOOLSET. We use the latter as-is, but have added preprocessing based on decisions already implemented in RUCoP to the former in order to produce more succinct MDP models as follows:

1. *Useful contacts only.* A contact may be useless for transmitting a message from the source to the target node because it leads to a dead-end, i.e. a situation where a message copy is transmitted to node X in time slot t but there is no sequence of contacts reaching the target from X after t. Similarly, there may not be any sequence of contacts from the source to X before t: then X is guaranteed not to have any copies in t. We analyse the contact graph for such situations and drop all useless contacts. This reduces the amount of decisions in the MDP, and thus the number of schedulers to sample from, without excluding any scheduler with positive message delivery probability. Consequently, (near-)optimal schedulers are more likely to be sampled.
2. *Forcing to send.* With the same motivation, when we are in node X's last useful contact, it would be useless to keep any copies. Thus, for such contacts, the only option that we generate now is to send all available copies.
3. *Forcing to receive.* Like a node deciding to store all copies at a contact, i.e. choosing not to send, the previous translation allowed the receiving node to

ignore the incoming transmission (which would consequently look like a failure to the sender). While this allowed some interesting collaborations between nodes to share non-local information [17, Sect. 5.3], we are not interested in such special behaviours, and consequently omit the option to ignore an incoming message. This again reduces the scheduler sampling space.

4. *Skipping empty slots.* The previous translation generated a "clock tick" action to advance time from t to $t+1$ in all nodes for every time slot, even if that slot had no contacts. To improve simulation runtime, we now omit these actions for empty slots and directly skip ahead to the next slot with a contact.

All combined, these improvements eliminate many useless schedulers from the sample space, making (L-)LSS noticeably more likely to find good ones; they also simplify the model, improving the runtime and memory consumption of MODES. We will showcase the difference on one of our benchmarks in Sect. 4.

4   Evaluation

In order to evaluate the performance-cost trade-off of LSS and RUCoP in uncertain DTNs, we created a benchmark set consisting of three use cases to compute SDP metrics and the associated computational cost.

4.1   Benchmark Set

Random networks use a uniform distribution of contacts among a configurable number of network nodes and contact plan duration. We use 10 random topologies with 8 nodes, each covering a duration of 100 s. Time is discretized into episodes of 10 s. In each episode, the connectivity between nodes (i.e. the presence of contacts) is decided based on a contact density parameter of 0.2, similar to [39]. We assume an all-to-all traffic pattern, run each of the routing algorithms 100 times on each of the 10 networks, and report the averages.

Binomial Networks. To gain insights into how increasing the topological complexity affects the routing algorithms, we devised a family of contact plans with a binomial topology. They are easy to scale up in a controlled manner that preserves the characteristics of the topology. The topology is a binomial tree. The higher the number of levels in the tree, the more complex the routing problem is to solve. Specifically, a binomial topology with L levels implies: (i)

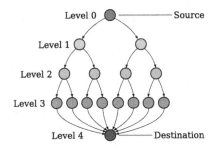

Fig. 3. Binomial tree.

$1+2^{L-2}$ nodes have contacts with two neighbors; (ii) $\sum_{i=1}^{i<L-2} 2^i$ nodes have contacts with three neighbors; and (iii) 1 final destination node has 2^{L-2} contacts. The resulting tree is illustrated in Fig. 3. Contacts between consecutive levels

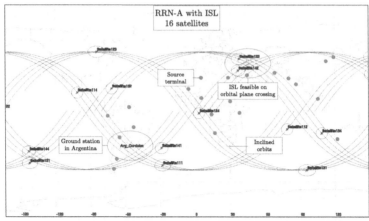

RRN with ISL	- 16 LEO satellites - 1 ground station (dst) - 22 ground terminals (src)				
Name	T. Anomaly [deg]	Altitude [km]	Arg. Perigee [deg]	Inclination [deg]	RAAN [deg]
Satellite111,12,13,14	0, 90, 180, 270	500	0	50	0
Satellite121,22,23,24	23, 113, 203, 293	500	0	50	90
Satellite131,32,33,34	45, 135, 225, 315	500	0	50	180
Satellite141,42,43,44	68, 158, 248, 338	500	0	50	270
Name	Latitude [deg]	Longitude [deg]	Altitude [km]		
Arg_Cordoba	-31.5242	-64.4636	0.724		

Fig. 4. RRN satellite constellation topology, parameters and orbital tracks [23].

are also consecutive in the time dimension, that is, the order of the contacts corresponds to enumerating the arrows in Fig. 3 left-to-right, top-to-bottom. A node on the i-th level will have a total of 2^{L-2-i} paths to the destination. Therefore, the larger the level count, the more nodes are in the network and the more paths per node have to be evaluated. For example, a binomial topology of 6 levels results in 32 nodes with up to 32 simple paths. When considering the forwarding of 3 copies, a total of 91000 possible actions need to be considered.

Ring Road Networks. Finally, we use a realistic satellite topology exported from high-precision orbital propagators. Specifically, we consider a low-Earth orbit Walker constellation of 16 satellites as proposed and described in [23]. Satellites act as data mules by receiving data from 22 isolated ground terminals, storing the data, and delivering it to a ground station placed in Argentina. We use an all-to-one traffic pattern. The satellites are equipped with inter-satellite links (ISLs), so contacts are possible in orbit. The dynamics of the topology and the specific orbital and ground parameters are depicted in Fig. 4. Routes can involve multiple hops between satellites and ground terminals. The scenario spans 24 h and is sliced into 1440 time slots, each of 60 s. Within a time slot, we consider a contact feasible if communication is possible for more than 30 s.

4.2   Analysis

Our evaluation results present compelling evidence of the trade-off between the LSS and RUCoP approaches, both in their global (LSS and RUCoP) and local

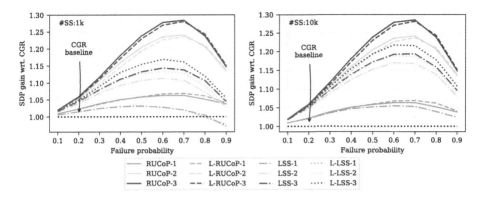

Fig. 5. SDP gain over CGR in random networks.

versions (L-LSS and L-RUCoP). We evaluate them in terms of the SDP of the
computed scheduler, and the computational resources used: processing time and
memory consumption. Plain single-copy CGR is used as a baseline. We write
"(L-)RUCoP-c" and "(L-)LSS-c" for the respective method when allowing c
copies. We used an Intel Core i5-5300U (2 cores, 4 threads, 2.3–2.9 GHz) system
with 12 GB of memory running 64-bit Ubuntu 18.04.5 for all experiments.

Random Networks. The SDPs we obtained for random networks are illustrated
in Fig. 5. To facilitate the interpretation of the outcomes, we plot the curves
with respect to the SDP delivered by CGR. Indeed, CGR is the baseline of
comparison as it assumes a perfect contact plan that does not drift from reality. As the contact plan becomes more uncertain, the RUCoP- and LSS-based
schemes provide increasingly better SDPs. This holds up to the point where the
failure probability is such that the partitioning of the topology dominates (i.e.
$p_f \approx 0.8$), a situation in which delivery of data becomes much more difficult.
Still, in these cases, RUCoP and LSS perform noticeably better than CGR.

We ran LSS and L-LSS in two configurations, one sampling $m = 1000$ and
one sampling $m = 10000$ schedulers. We indicate m as "#SS", the number of
sampled schedulers, in our figures. From Fig. 5, we observe that increasing m
from 1000 to 10000 does not improve the SDP drastically in these random networks. In particular, averaged along all failure probabilities, sampling $m = 10000$
schedulers improves SDP by ≈1.8%, with ≈5.8% being the maximum gain registered at $p_f = 0.7$. We explain this limited improvement with the simplicity of
the random topologies, which are easily explored with few schedulers.

When compared to L-RUCoP, L-LSS is, on average, 3% and 1% worse in
terms of SDP, for 1000 and 10000 schedulers, respectively. The larger difference
is observed at $p_f \approx 0.7\%$ and 3 copies, where L-RUCoP outperforms L-LSS by
10%. We observe that the lower the number of copies, the smaller the difference
between L-RUCoP and L-LSS, with the single-copy case almost identical in
SDP. Interestingly, the single-copy case provides limited or no gain with respect

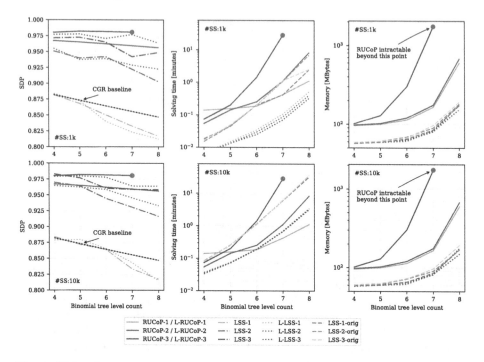

Fig. 6. SDP, solving time, and memory for binomial networks with varying complexity.

to the CGR baseline in these simple topologies. A similar effect was reported for Opportunistic CGR in [8].

Regarding the processing and memory footprint for random networks, all the techniques we study always complete in less than 20 s, using less than 20 MB of memory. Also, we observed that the runtime and memory values were rather stable and independent of the failure probability. In the following, we thus leverage the more complex binomial and ring-road topologies for a more detailed time and memory consumption assessment.

Binomial Networks Analysis. The results obtained for binomial networks are plotted in Fig. 6. All links in the topology were set to a failure probability of 0.1 in this case. Instead, we vary the tree level count from 4 to 8 (i.e. 8 to 128 nodes, and 13 to 449 paths), to evaluate the performance of RUCoP and LSS with increasing topological complexity, and thus, increasing routing decision making difficulty. Results are expressed, from left to right in the figure, in terms of SDP, solving time, and required memory.

In the binomial topologies, the CGR baseline is always equal to RUCoP with one copy (RUCoP-1) since the path with the earliest delivery time is also the one with highest SDP. On the other hand, the global view of RUCoP can be directly implemented with a limited local view. This is because each node can only reach two exclusive neighbors, which means that the local information is already enough to take a globally-optimal decision (i.e. the amount of copies to send to one of the two next hop nodes). As a result, L-RUCoP and RUCoP plots in Fig. 6 are presented in a single curve (solid line).

On the one hand, the SDP plots show that LSS is rather close to RUCoP when leveraging 10000 schedulers, especially for low level counts (with less than 0.01% difference). In the worst-case scenario with 8 levels, L-LSS is only 3% below L-RUCoP for the single and dual copy scenarios. However, due to memory exhaustion, RUCoP (and thus L-RUCoP) fails to deliver a valid routing schedule for 8 levels and 3 copies (its limit highlighted by the red circle in Fig. 6). We verify that for this case, more than 15 million actions need to be considered in the MDP. Another observation from these plots is that the delivery probability when using dual copies increases from ≈ 0.88 to ≈ 0.97 (i.e. by 10%) for 4 levels and from ≈ 0.85 to ≈ 0.96 (i.e. by 13%) for 8 levels. However, due to the binomial nature of the topology, having a third copy provides limited or no advantage.

Regarding the time and memory requirements in the binomial topologies, RUCoP proves to be by far the most demanding approach. In the worst case solved for 3 copies (7 levels), RUCoP needs 28 min of computation time, compared to less than 10 s for LSS with 1000 schedulers, or 1 min with 10000 schedulers. This is a notable difference considering the similar performance in terms of SDP. Solving time and memory plots of the original LSS as in [17], i.e. without the improvements described in Sect. 3.2, are also plotted in Fig. 6, in gray dashed lines. These improvements reduce LSS runtime by up to $\approx 600\%$ (from 117 down to 17 s). A reduction of $\approx 6\%$ in memory is also achieved. Indeed, in memory utilization, RUCoP quickly escalates up to more than 1 GB to keep track of the MDP decision tree, while lightweight schedulers never require more than 100 MB, even for the most complex binomial topologies.

In summary, for binomial topologies, LSS and L-LSS with 10k schedulers closely follow RUCoP and L-RUCoP in delivery probability and solving effort for simple trees. As the topology's complexity rises (notably for more than 7 levels), RUCoP exhausts the available memory. Even in these challenging cases, LSS is able to deliver a valid solution with minimal runtime and memory footprint.

Ring Road Networks Analysis. We have evaluated all downlink source-destination pairs in the realistic RRN network. Figure 7 present some representative cases for the different behaviors we observed. In this figure, node 38 as the destination stands for the mission control center on ground, while node 1 and 7 are remote nodes sending data via the ring-road satellites[1]. For these nodes, we present the computation of the routing schedule for varying contact plan sizes, spanning durations from 1 to 3 h (plots from top to bottom). The #SS parameter is again varied to 1000 and 10000 schedulers, to gain sensitivity on the improvement of the sampling technique (plots from left to right).

The SDP plots in Fig. 7 show that the longer the contact plan, the more noticeable the difference between the analytic and statistical approaches (i.e. curves separate progressively). In particular, there is barely any difference for any failure probabilities for the shorter contact plan with 1 h of scheduling horizon.

---

[1] Nodes 1 and 7 correspond to nodes 8 and 15 in the contact plan used in [23].

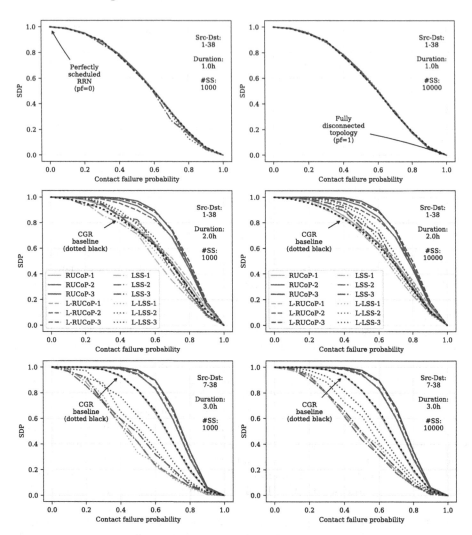

Fig. 7. SDP for RRN for different source-target nodes and plan durations.

However, we observe that L-RUCoP is notably superior to L-LSS for the 2 h and 3 h plans, especially for failure probabilities between 0.4 and 0.8. Specifically, we observe that the gap between RUCoP and LSS can be as large as ≈60%, for failure probabilities of ≈0.6, and contact plans of 3 h. Interestingly, the gap is reduced to ≈30% if we raise the number of schedulers to 10000 in LSS, indicating that this case is right on the boundary of what can effectively be solved via LSS. Nevertheless, both LSS and L-LSS perform worse than the CGR baseline even when leveraging multiple copies in schedules larger than 2 h. This is compelling evidence that the uninformed sampling strategy of LSS may not be fully adequate for realistic RRN topologies, even though it performed pretty well in generic

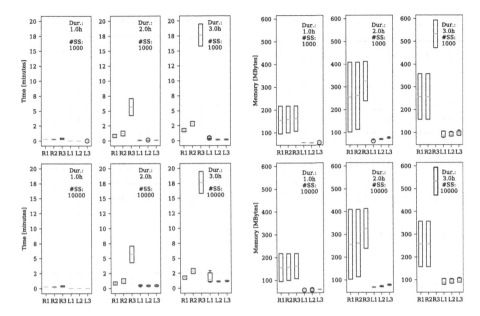

Fig. 8. Solving time (left) and memory (right) for RRN for different source-target nodes, contact plan durations, and numbers of schedulers sampled (R: RUCoP, L: LSS).

binomial and random topologies, and may need to be adapted to a variant yet more specifically tailored to the DTN routing application.

Also, we observe that LSS and L-LSS are typically close, but L-LSS frequently presents better SDP than the global LSS. This was also observed in Fig. 5, but in a much more subtle manner. We explain this phenomenon with the fact that L-LSS has a reduced space of schedulers to be sampled from, which increases the chances of finding a better routing policy.

Figure 8 presents the computational resources required to obtain the discussed SDP results for ring-road networks. This figure is computed based on the computational effort of solving several downlinking node pairs (instead of the two example pairs discussed in Fig. 7). The results confirms once again that RUCoP is able to deliver network performance at the expense of significantly higher memory and runtime. In particular, the runtimes for the analytical approach can reach up to ≈20 min (for the 3-h contact plan, with 3 copies), while LSS typically delivers a result in less than 1 min. We thus postulate that the 3 h contact plan is as challenging for RUCoP as the 7-level binomial topology, i.e. that larger contact plans are likely intractable for RUCoP. Memory-wise, we observe similar ratios. While RUCoP needs as much as 600 MB of memory for the worst-case scenario, LSS consistently uses about 100 MB. Again, this is due to the simulation nature of LSS, where no decision trees need to be stored as in RUCoP. Interestingly, LSS also showed a limited computational cost sensitivity to increasing L-LSS from 1000 to 10000. This is likely due to the possibility of using multiple CPU threads concurrently to perform the exploration in

LSS. Indeed, LSS can exploit parallelization intensively: each scheduler can be evaluated independently in separate threads. However, in RUCoP, the calculations for each time slot strongly depend on the successor time slot, which limits parallelization.

In summary, the evaluation over realistic ring-road networks showed that there is still room for improvement on scheduler sampling techniques to cope with more heterogeneous or application-specific topologies. In our particular satellite constellation, L-RUCoP provided delivery probabilities up to 60% higher than LSS, at higher computational costs. The reported runtimes and memory usages anyway appear reasonable for this kind of satellite application. In particular, since satellites revisit ground stations at most every ≈90 min [22], solving times of 20 min, as measured for RUCoP, are by all means acceptable.

5   Conclusions

This paper provides the first extensive comparison of the state-of-the-art analytical and statistical routing approaches for uncertain DTNs. While both RUCoP and LSS leverage MDP models, the former performs an exhaustive and optimal exploration of the solution space whereas the latter exploits SMC with sampling for optimization. We improved the DTN models for LSS for efficiency. We thoroughly compared the two approaches in a new benchmarking framework comprising random, binomial, and realistic satellite network topologies.

The outcomes provided quantitative evidence of the performance of the global- and local-information flavors of RUCoP and LSS. On the one hand, both schemes provide routes that deliver up to 1.8 times the data volume achievable by the baseline CGR approach. However, we touched the tractability limits of RUCoP in binomial networks of 8 levels. While RUCoP failed to deliver, LSS was able to solve the problem with just 5% of the memory footprint. We attribute part of this success to the improvements made to LSS for DTNs in this paper. Last but not least, the analysis on realistic satellite networks showed that despite the good performance of LSS, its applicability to case-specific topologies could enjoy further refinement. Such work is indeed needed seeing that RUCoP already stressed the computational resources for 3-h contact plans.

Even though LSS and RUCoP stand on the frontier of the state-of-the-art of routing in uncertain DTNs, a few challenges remain to be tackled. On the one hand, both approaches assume non-congested links: routing in uncertain *and* congested DTNs is an open research topic. Also the integration of uncertain and Opportunistic CGR [8] is appealing future work. Finally, the evaluation of the routing schedules obtained from the presented use cases in realistic DTN protocol simulations is currently being investigated by the authors.

Data Availability. A dataset with the models and tools needed to replicate our experimental evaluation is archived and available at *DOI* 10.4121/20334687 [14].

References

1. Agha, G., Palmskog, K.: A survey of statistical model checking. ACM Trans. Model. Comput. Simul. **28**(1), 6:1-6:39 (2018). https://doi.org/10.1145/3158668
2. Araniti, G., et al.: Contact graph routing in DTN space networks: overview, enhancements and performance. IEEE Comms. Mag. **53**(3), 38–46 (2015). https://doi.org/10.1109/MCOM.2015.7060480
3. Baier, C., de Alfaro, L., Forejt, V., Kwiatkowska, M.: Model checking probabilistic systems. In: Handbook of Model Checking, pp. 963–999. Springer, Cham (2018). https://doi.org/10.1007/978-3-319-10575-8_28
4. Baier, C., Katoen, J.: Principles of Model Checking. MIT Press, Cambridge (2008)
5. Benamar, N., Singh, K.D., Benamar, M., Ouadghiri, D.E., Bonnin, J.M.: Routing protocols in vehicular delay tolerant networks: a comprehensive survey. Comput. Commun. **48**, 141–158 (2014). https://doi.org/10.1016/j.comcom.2014.03.024
6. Benhamida, F.Z., Bouabdellah, A., Challal, Y.: Using delay tolerant network for the Internet of Things: Opportunities and challenges. In: 2017 8th International Conference on Information and Communication Systems (ICICS), pp. 252–257, April 2017. https://doi.org/10.1109/IACS.2017.7921980
7. Budde, C.E., D'Argenio, P.R., Hartmanns, A., Sedwards, S.: An efficient statistical model checker for nondeterminism and rare events. Int. J. Softw. Tools Technol. Transf. **22**(6), 759–780 (2020). https://doi.org/10.1007/s10009-020-00563-2
8. Burleigh, S., Caini, C., Messina, J., Rodolfi, M.: Toward a unified routing framework for DTN. In: 2016 IEEE International Conference on Wireless for Space and Extreme Environments (WiSEE), pp. 82–86, Sept 2016
9. Burleigh, S., et al.: Delay-tolerant networking: an approach to interplanetary internet. Comm. Mag. **41**(6), 128–136 (2003). https://doi.org/10.1109/MCOM.2003.1204759
10. Caini, C., Cruickshank, H., Farrell, S., Marchese, M.: Delay- and disruption-tolerant networking (DTN): an alternative solution for future satellite networking applications. Proc. IEEE **99**(11), 1980–1997 (2011). https://doi.org/10.1109/JPROC.2011.2158378
11. Cerf, V., et al.: Delay-tolerant networking architecture. RFC 4838, RFC Editor, April 2007. http://www.rfc-editor.org/rfc/rfc4838.txt
12. Cheung, L., Lynch, N.A., Segala, R., Vaandrager, F.W.: Switched PIOA: parallel composition via distributed scheduling. Theor. Comput. Sci. **365**(1–2), 83–108 (2006). https://doi.org/10.1016/j.tcs.2006.07.033
13. Consultative Committee for Space Data Systems (CCSDS): CCSDS bundle protocol specification (blue book, recommended standard CCSDS 734.2-B-1), September 2015. https://public.ccsds.org/Pubs/734x2b1.pdf
14. D'Argenio, P.R., Fraire, J.A., Hartmanns, A., Raverta, F.: Comparing statistical and analytical routing approaches for delay-tolerant networks (artifact). 4TU.ResearchData (2022). https://doi.org/10.4121/20334687
15. D'Argenio, P., Legay, A., Sedwards, S., Traonouez, L.-M.: Smart sampling for lightweight verification of Markov decision processes. Int. J. Softw. Tools Technol. Transf. **17**(4), 469–484 (2015). https://doi.org/10.1007/s10009-015-0383-0
16. Dehnert, C., Junges, S., Katoen, JP., Volk, M.: A STORM is coming: a modern probabilistic model checker. In: Majumdar, R., Kunčak, V. (eds) CAV 2017. LNCS, vol. 10427, pp. 592–600. Springer, Cham (2017). https://doi.org/10.1007/978-3-319-63390-9_31

17. D'Argenio, P.R., Fraire, J.A., Hartmanns, A.: Sampling distributed schedulers for resilient space communication. In: Lee, R., Jha, S., Mavridou, A., Giannakopoulou, D. (eds.) NFM 2020. LNCS, vol. 12229, pp. 291–310. Springer, Cham (2020). https://doi.org/10.1007/978-3-030-55754-6_17

18. Fall, K.: A delay-tolerant network architecture for challenged internets. In: Proceedings of the 2003 Conference on Applications, Technologies, Architectures, and Protocols for Computer Communications, SIGCOMM 2003, pp. 27–34. ACM, New York (2003). https://doi.org/10.1145/863955.863960

19. Filar, J., Vrieze, K.: Competitive Markov Decision Processes. Springer-Verlag, Heidelberg (1996). https://doi.org/10.1007/978-1-4612-4054-9

20. Fraire, J., Gasparini, E.: Centralized and decentralized routing solutions for present and future space information networks. IEEE communication Magazine, SI on Space Information Networks: Technological Challenges, Design Issues and Solutions (2021, in Press)

21. Fraire, J.A., Feldmann, M., Burleigh, S.C.: Benefits and challenges of cross-linked ring road satellite networks: a case study. In: 2017 IEEE International Conference on Communications (ICC), pp. 1–7 (2017)

22. Fraire, J.A., Finochietto, J.M.: Design challenges in contact plans for disruption-tolerant satellite networks. IEEE Commun. Mag. **53**(5), 163–169 (2015). https://doi.org/10.1109/MCOM.2015.7105656

23. Fraire, J.A., et al.: Assessing contact graph routing performance and reliability in distributed satellite constellations. Hindawi J. Comput. Netw. Commun. (2017). https://doi.org/10.1155/2017/2830542

24. Fraire, J.A., De Jonckère, O., Burleigh, S.C.: Routing in the space internet: a contact graph routing tutorial. J. Netw. Comput. Appl. **174**, 102884 (2021). https://doi.org/10.1016/j.jnca.2020.102884

25. Giro, S.: On the Automatic Verification of Distributed Probabilistic Automata with Partial Information. Ph.D. thesis, Universidad Nacional de Córdoba, Argentina (2010)

26. Giro, S., D'Argenio, P.R., Fioriti, L.M.F.: Distributed probabilistic input/output automata: expressiveness, (un)decidability and algorithms. Theor. Comput. Sci. **538**, 84–102 (2014). https://doi.org/10.1016/j.tcs.2013.07.017

27. Gupta, L., Jain, R., Vaszkun, G.: Survey of important issues in UAV communication networks. IEEE Commun. Surv. Tutor. **18**(2), 1123–1152 (2015)

28. Haddad, S., Monmege, B.: Interval iteration algorithm for MDPs and IMDPs. Theor. Comput. Sci. **735**, 111–131 (2018). https://doi.org/10.1016/j.tcs.2016.12.003

29. Hahn, E.M., Hartmanns, A., Hermanns, H., Katoen, J.P.: A compositional modelling and analysis framework for stochastic hybrid systems. Formal Methods Syst. Des. **43**(2), 191–232 (2013). https://doi.org/10.1007/s10703-012-0167-z

30. Hartmanns, A., Hermanns, H.: The Modest Toolset: an integrated environment for quantitative modelling and verification. In: Ábrahám, E., Havelund, K. (eds.) TACAS 2014. LNCS, vol. 8413, pp. 593–598. Springer, Heidelberg (2014). https://doi.org/10.1007/978-3-642-54862-8_51

31. Hartmanns, A., Kaminski, B.L.: Optimistic value iteration. In: Lahiri, S.K., Wang, C. (eds.) CAV 2020, Part II. LNCS, vol. 12225, pp. 488–511. Springer, Cham (2020). https://doi.org/10.1007/978-3-030-53291-8_26

32. Hom, J., Good, L., Yang, S.: A survey of social-based routing protocols in delay tolerant networks. In: 2017 International Conference on Computing, Networking and Communications (ICNC), pp. 788–792, January 2017. https://doi.org/10.1109/ICCNC.2017.7876231

33. Hwang, C., Tillman, F.A., Lee, M.: System-reliability evaluation techniques for complex/large systems: a review. IEEE Trans. Reliab. **30**(5), 416–423 (1981)
34. Jenkins, A., Kuzminsky, S., Gifford, K.K., Pitts, R.L., Nichols, K.: DTN: flight test results from the international space station. In: 2010 IEEE Aerospace Conference, pp. 1–8, March 2010
35. Kalaputapu, R., Demetsky, M.J.: Modeling schedule deviations of buses using automatic vehicle-location data and artificial neural networks. In: Transportation Research Record, pp. 44–52 (1995)
36. Kwiatkowska, M., Norman, G., Parker, D.: PRISM 4.0: verification of probabilistic real-time systems. In: Gopalakrishnan, G., Qadeer, S. (eds.) CAV 2011. LNCS, vol. 6806, pp. 585–591. Springer, Heidelberg (2011). https://doi.org/10.1007/978-3-642-22110-1_47
37. Legay, A., Sedwards, S., Traonouez, L.-M.: Scalable verification of Markov decision processes. In: Canal, C., Idani, A. (eds.) SEFM 2014. LNCS, vol. 8938, pp. 350–362. Springer, Cham (2015). https://doi.org/10.1007/978-3-319-15201-1_23
38. Madoery, P., Raverta, F., Fraire, J., Finochietto, J.: On the performance analysis of disruption tolerant satellite networks under uncertainties. In: Proceedings of the 2017 XVII RPIC Workshop, September 2017
39. Madoery, P.G., Raverta, F.D., Fraire, J.A., Finochietto, J.M.: Routing in space delay tolerant networks under uncertain contact plans. In: 2018 IEEE International Conference on Communications (ICC), pp. 1–6, May 2018. https://doi.org/10.1109/ICC.2018.8422917
40. Partan, J., Kurose, J., Levine, B.N.: A survey of practical issues in underwater networks. SIGMOBILE Mob. Comput. Commun. Rev. **11**(4), 23–33 (2007). https://doi.org/10.1145/1347364.1347372
41. Pöttner, W.B., Morgenroth, J., Schildt, S., Wolf, L.: Performance comparison of DTN bundle protocol implementations. In: Proceedings of the 6th ACM Workshop on Challenged Networks, pp. 61–64. ACM (2011)
42. Puterman, M.L.: Markov Decision Processes: Discrete Stochastic Dynamic Programming, 1st edn. Wiley, New York (1994)
43. Quatmann, T., Katoen, J.-P.: Sound value iteration. In: Chockler, H., Weissenbacher, G. (eds.) CAV 2018, Part I. LNCS, vol. 10981, pp. 643–661. Springer, Cham (2018). https://doi.org/10.1007/978-3-319-96145-3_37
44. Raverta, F.D., Demasi, R., Madoery, P.G., Fraire, J.A., Finochietto, J.M., D'Argenio, P.R.: A Markov decision process for routing in space DTNs with uncertain contact plans. In: 2018 6th IEEE International Conference on Wireless for Space and Extreme Environments (WiSEE), pp. 189–194, December2018. https://doi.org/10.1109/WiSEE.2018.8637330
45. Raverta, F.D., Fraire, J.A., Madoery, P.G., Demasi, R.A., Finochietto, J.M., D'Argenio, P.R.: Routing in delay-tolerant networks under uncertain contact plans. Ad Hoc Netw. **123**, 102663 (2021). https://doi.org/10.1016/j.adhoc.2021.102663
46. Sahai, A., Tandra, R., Mishra, S.M., Hoven, N.: Fundamental design tradeoffs in cognitive radio systems. In: Proceedings of the First International Workshop on Technology and Policy for Accessing Spectrum, p. 2. ACM (2006)
47. Scott, K., Burleigh, S.: Bundle protocol specification. RFC 5050, RFC Editor, November 2007. http://www.rfc-editor.org/rfc/rfc5050.txt
48. Wyatt, J., Burleigh, S., Jones, R., Torgerson, L., Wissler, S.: Disruption tolerant networking flight validation experiment on NASA's EPOXI mission. In: First International Conference on Advances in Satellite and Space Communications, 2009. SPACOMM 2009, pp. 187–196, July 2009. https://doi.org/10.1109/SPACOMM.2009.39

Automata Theory and Applications

Mirrors and Memory in Quantum Automata

Carla Piazza and Riccardo Romanello$^{(\boxtimes)}$

Department of Mathematics, Computer Science and Physics, University of Udine,
Udine, Italy
{carla.piazza,riccardo.romanello}@uniud.it

Abstract. In this paper we start from the simplest form of Quantum
Finite Automata (QFAs), namely Measure-Once QFAs with cut-point.
First we elaborate on a variant of their semantics that can be obtained
through a shift from the Schrödinger to the Heisenberg picture of Quantum Mechanics. In the Schrödinger picture states evolve in time while
observables remain constant, while in the Heisenberg one states are constant and observables evolve. Interestingly, in the case of a QFA such
shift reverts time-evolution. However, the equivalence of the two pictures over the class of QFAs holds thanks to the closure of the class with
respect to language mirroring. Since the expressive power of such class
of automata remains limited to infinite languages, we then consider their
extension with bounded (multi-letter QFAs) and unbounded memory.
Unfortunately, while bounded memory enhances the expressive power,
the unbounded memory approach does not behave as one would expect.

Keywords: Quantum Automata · Heisenberg Picture · Language
Mirroring · Memory in Quantum Automata

1 Introduction

Deterministic and Nondeterministic Finite State Automata (DFA/NFA) are the
building blocks of classical computation. They are the models at the basis of
Verification Techniques such as Temporal Logic Model Checking [15].

A shift to their probabilistic and stochastic counter-parts is necessary whenever the evolution of the computation depends on probabilities and rates. In this
context models such as Probabilistic/Stochastic Automata, Discrete/Continuous
Time Markov Chains, and Probabilistic/Stochastic Process Algebra have been
described (e.g., [20,21,24]). Their formal analysis involves performance metrics,
behavioural equivalences, and extensions of temporal logics.

A currently emerging field in the context of Quantitative Computation and
Performances Evaluation is Quantum Computation, where again extensions of

This work is partially supported by PRIN MUR project Noninterference and Reversibility Analysis in Private Blockchains (NiRvAna) - 20202FCJM and by GNCS INdAM
project LESLIE.

E. Ábrahám and M. Paolieri (Eds.): QEST 2022, LNCS 13479, pp. 359–380, 2022.
https://doi.org/10.1007/978-3-031-16336-4_18

automata, Markov chains, and temporal logics constitute a starting point for understanding properties of the computations (e.g., [3,17,19]).

Even though Quantum Automata have been studied since the end of the nineties, still today there is not a unique widely accepted definition of Quantum Finite Automata (QFAs). Moore and Crutchfield [27] introduced the idea of General Quantum Automata and characterized the properties of Quantum Regular Languages. The model they introduced was named *Measure-Once Quantum Finite Automata* (MO-QFAs) because the result can be observed (measured) only when the read of the input string has terminated. In the same years, Kondacs and Watrous in [23] introduced a different model of QFAs in which measurements can be used at each step of the computation. For this reason, these are called *Measure-Many Quantum Finite Automata* (MM-QFAs). Moreover, similarly to what happens on probabilistic automata [32], a key role in the expressive power of such models is played by the acceptance condition. The two most adopted conditions are called *cut-point* and *bounded error*.

The expressive power of both MO-QFA and MM-QFA has been deeply investigated in [2,8,14]. The expressive power of MO-QFAs does not include all languages accepted by DFAs. As a consequence, different extensions have been considered. In [1] a model called *Latvian* QFAs was considered. Bertoni *et al.* [9] introduced MO-QFAs *with control language* which are able to recognize regular languages with bounded error. The same behavior can be found in a formalism in which a MO-QFAs are used together with a classical set of states [31]. Another model that can at least recognize regular languages was presented in [29] were the concept of *Ancilla* qubits is used.

In this paper we are interested in the most simple of these models, e.g., MO-QFAs with cut-point acceptance condition. In the case of Quantum Circuits the principle of *deferred measurements* states that measurements can always be moved from an intermediate stage to the final step. This is not true in the case of Quantum Automata, since MO-QFAs and MM-QFAs are not equivalent. So, a Measure-Once condition is more in the spirit of a basic model. As for the acceptance condition, bounded error ensures the possibility of arbitrarily improving the precision. Consequently, it has been largely studied in the literature. However, it is not the "equivalent" of what happens in experimental disciplines such as biology and medicine, where cut-offs have to be arbitrarily chosen and no separation is guaranteed between positive and negative answers.

First, in this paper we analyse whether it is possible to increase the expressive power of MO-QFAs with cut-point without enriching their syntax, but simply moving to an alternative semantics. Such semantics from the point of view of physics is as natural as the one which has been considered in the literature so far. We are talking of a shift from the Schrödinger picture of Quantum Mechanics to the Heisenberg one. We will not obtain a positive answer in terms of increase of the expressive power, but our investigation provides a closure property of MO-QFAs with respect to mirror images which is new. In other terms, the mirror closure proves that not only each internal step of a MO-QFA is reversible, but its computation as a whole is. Such result was not granted because of the asymmetric use in MO-QFAs of final states and measurement.

As a second step, we are interested in considering another semantics for MO-QFAs. This is only inspired by the Heisenberg picture and at first sight it seems to provide an unbounded quantity of memory to the automata. In particular, at each point of the computation all the prefix that has been read so far is involved in the choice of the evolution. However, as it usually happens in the quantum realm, our intuition is cheated and such unbounded quantity of memory is less expressive than expected. Again, the path which leads us to such "negative" result is interesting by itself. We quantify the minimum amount of memory necessary for accepting finite languages and provide a pumping lemma for a class of QFAs which have been studied in the literature with bounded error, but not with cut-point [6,30].

The paper is organized as follows. In Sect. 2 we give a brief presentation of the notation and the basic concepts that are useful throughout the paper. In Sect. 3 we introduce MO-QFAs and we briefly survey the state of the art about their expressive power and realizations. These results will be useful in Sect. 4 where we define *Heisenberg Quantum Finite Automata* (HQFAs) and compare them with MO-QFAs. In Sect. 5 we study a class of *Heisenberg inspired* automata which we call *Unbounded Memory Quantum Automata* (UMQFAs) and we compare them with a *bounded memory* counter-part. The proofs of the main results of this paper can be found in the Appendix.

2   Preliminaries

2.1   Strings and Languages

An alphabet Σ is a set of symbols. We always refer to finite alphabets. A string $\mathbf{x} = x_1 x_2 \ldots x_m$ of length m over Σ is a finite sequence of symbols $x_i \in \Sigma$. The empty string ϵ is the only string of length 0. With Σ^i we indicate the set of all strings of length i over Σ, while $\Sigma^{\leq i} = \cup_{j=0}^{i} \Sigma^j$ is the set of all strings of length at most i. $\Sigma^* = \cup_{i \in \mathbb{N}} \Sigma^i$ is the set of all finite length strings we can build on Σ.

Given a string $\mathbf{x} = x_1 x_2 \ldots x_m$ we denote by $\overleftarrow{\mathbf{x}}$ its mirror image, i.e., the string $\overleftarrow{\mathbf{x}} = x_m x_{m-1} \ldots x_1$. Given an index $1 \leq j \leq n$ we denote by $\mathbf{x}_j$ the prefix of $\mathbf{x}$ from x_1 to x_{j-1}, i.e., $\mathbf{x}_j = x_1 x_2 \ldots x_{j-1}$. If $j = 1$, then $\mathbf{x}_j$ is the empty string. Moreover, for $h \in \mathbb{N}$ we denote by $\mathbf{x}_j^h$ the sub-string of $\mathbf{x}$ ranging from x_{j-h} to x_{j-1} if $j - h > 0$, and the prefix $\mathbf{x}_j$ otherwise. In other terms, $\mathbf{x}_j^h$ is the sub-string of $\mathbf{x}$ ranging from x_k to x_{j-1}, where k is the maximum between 1 and $j - h$. Notice that $\mathbf{x}_j^h$ has length either h or $j - 1$.

A language L is a set of strings over an alphabet Σ, i.e., $L \subseteq \Sigma^*$. Given a language L, we denote by $\overleftarrow{L}$ the mirror image of L, i.e., $\overleftarrow{L} = \{ \overleftarrow{\mathbf{x}} \mid \mathbf{x} \in L \}$.

2.2   Quantum Computing

The most used model of Quantum Computation relies on the formalism of state vectors, unitary operators, and projectors. At high level we can say that state vectors evolve during the computation through unitary operators, then projectors remove part of the uncertainty on the internal state of the system.

The state of the system is represented by a unitary vector over the Hilbert space $\mathbb{C}^d$ with $d = 2^k$ for some $k \in \mathbb{N}$. The concept of bit of classical computation is replaced by that of qubit. While a bit can have value 0 or 1 a qubit is a unitary vector of $\mathbb{C}^2$. When the two components of the qubit are the complex numbers $\alpha = x + iy$ and $\beta = z + iw$, the squared norms $|\alpha|^2 = x^2 + y^2$ and $|\beta|^2 = z^2 + w^2$ represent the probabilities of measuring the qubit thus reading 0 and 1, respectively. In the more general case of k qubits the unitary vectors range in $\mathbb{C}^d$ with $d = 2^k$. Adopting the standard Dirac notation we denote a column vector $v \in \mathbb{C}^d$ by $|v\rangle$, and its conjugate transpose $v^\dagger$ by $\langle v|$. A *quantum state* is a unitary vector:

$$|\psi\rangle = \sum_{h=1}^{d} c_h |v_h\rangle$$

for some basis $\{|v_h\rangle\}$. In this case we also say that $|\psi\rangle$ is a *superposition* with coefficients $\{c_h\}$ over the basis $\{|v_h\rangle\}$. When not specified, we refer to the *canonical basis* denoted by $\{|0\rangle, |1\rangle, \ldots, |n-1\rangle\}$, where for each $q \in [0, d-1]$ the vector $|q\rangle$ is the unitary vector having 1 as $q+1$-th component and all its other components are 0. Moreover, usually $|q\rangle$ is written using the binary representation of q of length m. The canonical basis is an ortonormal basis for $\mathbb{C}^d$. Further details can be found in [28].

Unitary operators are a particular class of reversible linear operators. They preserve both the angles between vectors and their lengths. In other terms, unitary operators are transformation from one orthonormal basis to another. Hence, they are represented by unitary matrices. Let U be a square matrix over $\mathbb{C}$. U is said to be *unitary* iff $UU^\dagger = U^\dagger U = I$. We describe the application of a unitary matrix U to a state $|\psi\rangle$ by writing:

$$|\psi'\rangle = U |\psi\rangle$$

meaning that the state $|\psi\rangle$ becomes $|\psi'\rangle$ after applying the operator U.

In order to extract informations from a quantum state $|\phi\rangle$ a *measurement*, also called *observation*, must be performed. Projectors are the most common measurements/observables. Let $|u\rangle$ be a vector. The *projector* operator P_u along the direction of the unitary vector $|u\rangle$ is the linear operator defined as:

$$P_u = |u\rangle\langle u|$$

where $|u\rangle\langle u|$, being the product between a column vector and a row one both of size d, returns a matrix of size $d \times d$. Given a set of directions $F = \{|u_1\rangle, \ldots |u_f\rangle\}$ specified by unitary vectors the projector operator associated to F is defined as:

$$P_F = \sum_{u \in F} |u\rangle\langle u|$$

3   Measure-Once Quantum Finite Automata

Quantum Finite Automata (QFA) are the quantum counterpart of Finite Automata. Two models of Quantum Automata were independently introduced in the literature: Measure-Once QFAs (MO-QFAs) [27] and Measure-Many QFAs (MM-QFAs) [23]. The difference between the two definitions is about the number of observations that are made. While a MM-QFA is measured after reading each letter from the input, in a MO-QFA only one measurement is made after the whole input has been read.

We focus on MO-QFAs. Therefore, for sake of readability, we refer to MO-QFAs with just QFAs.

Let $\mathbb{C}^d$ be a finite dimension Hilbert space and $Q = \{|0\rangle, |1\rangle, \ldots |d-1\rangle\}$ be its canonical basis. Usually in quantum computation it holds that $d = 2^k$ for some $k \in \mathbb{N}$, where k is the number of involved qubits. However, we refer here to a generic dimension d. It is not difficult to embed all the definitions and results we present into a space of dimension $2^{k'} > d$, thus using k' qubits, whenever it is necessary in the implementations.

Definition 1 (QFA). *A QFA is a 5-tuple $M = (Q, \Sigma, \mathscr{U}, |\psi\rangle, F)$ where:*

- *Q–the set of states– is the finite canonical basis of $\mathbb{C}^d$ for some $d \in \mathbb{N}$;*
- *Σ is a finite alphabet;*
- *$\mathscr{U} = \{U_\sigma\}_{\sigma \in \Sigma}$ is a finite set of unitaries of dimension $\mathbb{C}^d \times \mathbb{C}^d$;*
- *$|\psi\rangle \in \mathbb{C}^d$ is a unitary vector representing the initial superposition of M;*
- *$F \subseteq Q$ is the set of final states.*

In the literature the standard semantics attributed to QFA is based on the *Schrödinger picture* of quantum mechanics in which states evolve in time. We will come back to this in Sect. 4, when we will compare this interpretation with other possible ones. However, in the remaining of this section we will use the letter S of Schrödinger to refer to a generic QFA.

A generic configuration for a QFA S is a unitary vector of $\mathbb{C}^d$, i.e., it is a vector of the form:

$$|\varphi\rangle = \sum_{|q\rangle \in Q} \alpha_q |q\rangle$$

Let $|\varphi\rangle$ be the current configuration of S and $\sigma \in \Sigma$ be the current input symbol. $|\varphi\rangle$ evolves as follows:

$$|\varphi'\rangle = U_\sigma |\varphi\rangle$$

The computation starts from $|\psi\rangle$ and evolves reading the symbols of the string $\mathbf{x}$. At the end of the computation, i.e., when all the symbols of $\mathbf{x}$ have been read, a measurement is performed on the obtained state of S using the matrix $P_F = \sum_{|q\rangle \in F} |q\rangle\langle q|$. The probability of S accepting a string $\mathbf{x}$ is:

$$p_S(\mathbf{x}) = \|P_F U_\mathbf{x} |\psi\rangle\|^2 = \langle\psi| U_\mathbf{x}^\dagger P_F^\dagger P_F U_\mathbf{x} |\psi\rangle = \sum_{|q\rangle \in F} |\langle q| U_\mathbf{x} |\psi\rangle|^2$$

where $U_{\mathbf{x}}$–the evolution matrix accumulated along the read of $\mathbf{x}$– is defined as:

$$U_{\mathbf{x}} = U_{x_n} U_{x_{n-1}} \cdots U_{x_1}$$

We consider two different acceptance conditions. The first one is called with *cut-point* and it recalls the acceptance condition of probabilistic automata [32].

Definition 2 (Cut-point QFA). *A language $L \subseteq \Sigma^*$ is* accepted by a QFA S with cut-point λ *if and only if $L = \{\mathbf{x} \in \Sigma^* \mid p_S(\mathbf{x}) > \lambda\}$.*

A language $L \subseteq \Sigma^*$ is said to be accepted by a QFA with cut-point *if and only if there exist a QFA S and $\lambda \geq 0$ such that $L \subseteq \Sigma^*$ is accepted by S with cut-point λ.*

The second one is called with *certainty*. In this case we mimic the acceptance of a deterministic automata (DFA).

Definition 3 (Certainty QFA). *A language $L \subseteq \Sigma^*$ is said to be accepted by a QFA S with* certainty *if the following holds:*

$$\mathbf{x} \in L \text{ iff } p_S(\mathbf{x}) = 1 \text{ and } \mathbf{x} \notin L \text{ iff } p_S(\mathbf{x}) = 0$$

It is straightforward to see that an acceptance with certainty implies an acceptance with cut-point $1 - \epsilon$, $\forall \epsilon \in (0, 1]$. The converse is trivially false.

The class of languages accepted by QFAs with *cut-point* was introduced and characterized in [14]. Such class is called *Unrestricted Measure-Once*, UMO. One of the main contribution to the characterization of such class is the connection with the languages accepted by Probabilistic Automata:

Theorem 1 ([14]). *Let L be a language accepted by a QFA S with cut-point λ. There exists a Probabilistic Finite Automaton that accepts L with cut-point λ', for some λ'.*

The class UMO was further investigated in [8,27], with the introduction of the following *pumping lemma*.

Theorem 2 ([27]). *Let $L \subseteq \Sigma^*$ be the language accepted by a QFA S with cut-point λ. $\forall \mathbf{x} = \mathbf{uv} \in L$ and $\forall \mathbf{y} \in \Sigma^*$, there exists $k \in \mathbb{N}^+$ such that $\mathbf{uy}^k\mathbf{v} \in L$.*

A straightforward consequence of the above theorem is that finite languages cannot be accepted by QFA.

Corollary 1. *QFAs can accept only languages that are either empty or infinite.*

Notice that the theorem holds for any possible split of the string $\mathbf{x}$ into two strings $\mathbf{u}$ and $\mathbf{v}$. So, either $\mathbf{u}$ or $\mathbf{v}$ could be empty. In particular, taking $\mathbf{v}$ empty we get that languages whose elements have a fixed suffix cannot be recognized.

Corollary 2. *Let $\Sigma = \{a, b\}$ and $L = \{\mathbf{x} \mid \mathbf{x} \text{ ends with } a\}$. L cannot be accepted by any QFA S with cut-point.*

Proof. Suppose such a S exists. Let $\mathbf{x} \in L$ and $\mathbf{y} = b$, by Theorem 2 it holds that $\exists k \in \mathbb{N}^+$ such that $\mathbf{x}\mathbf{y}^k \in L$. This contradicts the definition of L. □

The above corollary also gives an example of a regular language that cannot be accepted by QFAs.

Despite being unable of accepting finite languages, QFAs can accept languages that are not regular. Let $\mathbf{x} \in \Sigma^*, \sigma \in \Sigma$. We denote by $|\mathbf{x}|_\sigma$ the number of occurrences of σ in $\mathbf{x}$. It was proven in [14] that there exists a QFA that accepts the language $L = \{\mathbf{x} \in \{a, b\}^* : |\mathbf{x}|_a \neq |\mathbf{x}|_b\}$ with cut-point 0.

The equivalent of UMO in the case of MM-QFAs is denoted by UMM (*Unrestricted Measure-Many*) and it was introduced in [14]. It was then characterized and eventually further investigated in the literature (see, e.g., [2]). Results on Quantum Automata descriptional complexity can be found in [12]. Recently in [18] the expressive power of Quantum Automata over the unary alphabet under different acceptance condition has been investigated. A physical realization of Quantum Automata has been presented in [26]. Undecidability results have been proved in [5]. In [11] it has been proved that languages accepted by MO-QFAs with bounded error are not definable in Linear Time Temporal Logics, while it is definable in the case of Measure-Many. A recent review can be found in [10].

Even more recently, Quantum Automata minimization has been studied in [22], while succinctness has been described and implemented in [25]. Physical realizations of Quantum Computing algorithms always require to consider the noise introduced by non-perfect gates. In [13] the aim is to implement QFAs on noisy devices.

4   Heisenberg Quantum Finite Automata

The most widely adopted formulation of the Copenaghen interpretation of quantum mechanics is the *Schrödinger representation*. It is based on the idea that there is a *state vector* in an Hilbert space that completely describes the configuration of the system. This state vector evolves through time according to the *Schrödinger equation*. In particular, at each time instant a unitary operator is applied to the state vector. So, in the Schrödinger picture the state vector is time-dependent, while the unitaries and the observables remain unchanged.

There exists another representation known as *Heisenberg picture* in which the state vector is time-independent and always remains fixed to its value at time 0. Therefore, the time-dependency is *shifted* on the observables.

A third representation, named *Dirac picture*, also known as *Interaction picture*, "distributes" time dependencies over both states and operators.

Even though a mathematical equivalence between Schrödinger and Heisenberg representations has been proved by Von Neumann in [33], divergencies were pointed by Dirac in [16].

In terms of Quantum Finite Automata all the models described in the literature so far rely on the Schrödinger picture, where the initial state evolves through time using unitaries, while the observables never change[1].

In this section we shift to the Heisenberg picture and we formalize a new semantics for QFAs, named *Heisenberg Quantum Finite Automata* (HQFAs). The idea is that while the string **x** is read the state is unchanged, but there is an effect on the projector. At the end of the read such modified projector is applied to the initial state to obtain the final result. The way in which the observable gets modified is in a sense arbitrarily chosen. In our definition we try to keep such choice as close as possible to that of QFAs. In particular, in quantum mechanics when one shifts from the Schrödinger picture to the Heisenberg one a transformation of the states of the form $U |\varphi\rangle$ is mapped into a transformation of the observables/projectors of the form $U^\dagger P U$, where the meaning is that $U^\dagger$ has been applied to P. As a consequence HQFAs have exactly the same definition of QFAs, while the difference is in the acceptance condition, i.e., in the semantics.

Let P be the current observable of a HQFA H and $\sigma \in \Sigma$ be the current input symbol. P evolves as follows:

$$P' = U_\sigma^\dagger P U_\sigma$$

The computation starts from the observable P_F and evolves reading the symbols of **x**. At the end of the read a measurement is performed using the resulting projector and the probability of accepting **x** is:

$$\rho_H(\mathbf{x}) = \| U_{\overleftarrow{\mathbf{x}}}^\dagger P_F U_{\overleftarrow{\mathbf{x}}} |\psi\rangle \|^2 = \langle \psi | U_{\overleftarrow{\mathbf{x}}}^\dagger P_F^\dagger P_F U_{\overleftarrow{\mathbf{x}}} |\psi\rangle$$

where consistently with the definition given in Sect. 3 the evolution matrix $U_{\overleftarrow{\mathbf{x}}}$ is defined as:

$$U_{\overleftarrow{\mathbf{x}}} = U_{x_1} U_{x_2} \cdots U_{x_n}$$

The acceptance condition with cut-point for an HQFA now inolves ρ_H.

Definition 4 (Cut-point HQFA). *A language $L \subseteq \Sigma^*$ is accepted by an HQFA H with cut-point λ if and only if $L = \{\mathbf{x} \in \Sigma^* \mid \rho_H(\mathbf{x}) > \lambda\}$.*

A language $L \subseteq \Sigma^*$ is said to be accepted by an HQFA with cut-point *if and only if there exist an HQFA H and $\lambda \geq 0$ such that $L \subseteq \Sigma^*$ is accepted by H with* cut-point λ.

Example 1. Let $Q = \{|0\rangle, |1\rangle\}$ be the canonical basis of $\mathbb{C}^2$. Let $\Sigma = \{a, b\}$. Consider the two unitary matrices $U_a = X$ (the negation gate) and $U_b = H$ (the Hadamard gate), i.e.:

$$U_a = \begin{pmatrix} 0 & 1 \\ 1 & 0 \end{pmatrix} \qquad U_b = \frac{1}{\sqrt{2}} \begin{pmatrix} 1 & 1 \\ 1 & -1 \end{pmatrix}$$

Let $|\psi\rangle = \frac{1}{\sqrt{2}}(|0\rangle + |1\rangle) = |+\rangle$ and $F = \{|0\rangle\}$.

---

[1] Some work has been done for Quantum Cellular Automata, where the equivalence between Schrödinger model and Heisenberg model has been proved (e.g., [4]).

If we consider M as a QFA, i.e., we endow M with the Schrödinger semantics, we get that the probability for the sting ab is:

$$p_M(ab) = \| \, |0\rangle \langle 0| \, U_b U_a \, |+\rangle \, \|^2 = \| \, |0\rangle \langle 0| \, U_b \, |+\rangle \, \|^2 = \| \, |0\rangle \langle 0| \, |0\rangle \, \|^2 = \| \, |0\rangle \, \|^2 = 1$$

This means that no matter which is λ, the string ab is accepted.

If we consider the string abb we have to apply again U_b before projecting. Hence, we obtain $p_M(abb) = \| \, |0\rangle \langle 0| \, U_b \, |0\rangle \, \|^2 = \| \, |0\rangle \langle 0| \, |+\rangle \, \|^2 = 1/2$.

On the other hand, if we look at M as a HQFA, i.e., we apply to M the Heisenberg semantics, the probability for the string ab is:

$$\rho_M(ab) = \| U_b^\dagger U_a^\dagger \, |0\rangle \langle 0| \, U_a U_b \, |+\rangle \, \|^2 = \| U_b^\dagger \, |1\rangle \langle 1| \, U_b \, |+\rangle \, \|^2 = \| \, |-\rangle \langle -|+\rangle \, \|^2 = 0$$

where $|-\rangle = \frac{1}{\sqrt{2}}(|0\rangle - |1\rangle)$. This means that no matter which is λ, the string ab is not accepted. Instead, if we consider the string ba we obtain:

$$\rho_M(ba) = \| U_a^\dagger U_b^\dagger \, |0\rangle \langle 0| \, U_b U_a \, |+\rangle \, \|^2 = \| U_a^\dagger \, |+\rangle \langle +| \, U_a \, |+\rangle \, \|^2 = \| \, |+\rangle \langle +|+\rangle \, \|^2 = 1$$

As a matter of fact, in this simple example one can notice that for all $\mathbf{x} \in \Sigma^*$ the behaviour of S on $\mathbf{x}$ is equivalent to the behaviour of H on its mirror image $\overleftarrow{\mathbf{x}}$, i.e., $p_S(\mathbf{x}) = \rho_H(\overleftarrow{\mathbf{x}})$. In the following we prove this result in the general case, for any automaton. $\qquad\square$

Theorem 3. *Let M be a QFA over an alphabet Σ. For each $\mathbf{x} \in \Sigma^*$ it holds that*

$$p_M(\mathbf{x}) = \rho_M(\overleftarrow{\mathbf{x}})$$

Proof. Let $\mathbf{y} = \overleftarrow{\mathbf{x}}$. We have that $\overleftarrow{\mathbf{y}} = \mathbf{x}$. So, $\rho_M(\mathbf{y}) = \langle \psi | \, U_{\overleftarrow{\mathbf{y}}}^\dagger P_F^\dagger P_F U_{\overleftarrow{\mathbf{y}}} \, |\psi\rangle = \langle \psi | \, U_{\mathbf{x}}^\dagger P_F^\dagger P_F U_{\mathbf{x}} \, |\psi\rangle = p_M(\mathbf{x})$. $\qquad\square$

Intuitively, when we shift to the Heisenberg picture, the effect of the first character of $\mathbf{x}$ is close to the observable instead of being close to the initial state. So, the word is read in the usual way from left to right by the automaton and the effects of the read are accumulated on the observable. However, when we look to such effects on the state, it is like if the word is read from right to left. In a sense it seems that the flow of time is reverted in the Heisenberg picture.

One could argue that we could have avoided the *mirror effect* by using in the Heisenberg definition the inverse unitary operators. Since the inverse of a unitary operator is its transposed conjugate, this would have meant to define the evolution of an observable P after reading a symbol σ as $P' = U_\sigma P U_\sigma^\dagger$. In the following example we show that such choice does not help in avoiding the mirroring.

Example 2. Let us consider again the automaton M defined in Example 1. The two matrices $U_a = X$ and $U_b = H$ coincide with their transposed conjugate, i.e., $U_a^\dagger = U_a$ and $U_b^\dagger = U_b$. So, the automaton M' defined using the transposed conjugate coincides with M. Hence, $p_M(\mathbf{x}) = \rho_{M'}(\overleftarrow{\mathbf{x}})$, for any string $\mathbf{x}$. $\qquad\square$

As a consequence of Theorem 3, the languages accepted by Heisenberg semantics are exactly the mirror images of those accepted by Schrödinger one.

Corollary 3. *Let $L \subseteq \Sigma^*$. L is accepted by a QFA with cut-point λ if and only if $\overleftarrow{L}$ is accepted by an HQFA with cut-point λ.*

So, now the question is whether the two formalisms have the same expressive power. As a consequence of the above corollary this is equivalent to check whether QFAs are closed under mirror images. By Example 1 we already know that it is not true that each language recognized by a QFA is closed under mirror images. However, it can be the case that whenever a language L is recognized by a QFA S, the language $\overleftarrow{L}$ is recongnized by a QFA S'.

Invoking Von Neumann's proof of equivalence of Schrödinger and Heisenberg pictures is not satisfactory by many point of views. First, Von Neumann's result has been proved in a general setting, while here we are confined in a restricted model, where there is a single initial state, while the final states are many. Moreover, only one projective measurement can be used and only at the end of the read. Second, we are not dealing with a single quantum system, but with an infinite set of systems, one for each string **x**. The input **x** does not affect the initial state, but the sequence of unitary transformations. In a sense it affects the hamiltonian of the system. Third, it would be interesting to have either a constructive proof of equivalence or a counter-example in this specific setting.

The following result shows that QFAs are closed under mirror images. We provide a constructive proof. Given a QFA for a language L, we build a QFA for the language $\overleftarrow{L}$. Intuitively, the asymmetry between a single initial state and a set of final ones is solved through an opportune increase in the state space size.

Theorem 4 (Mirror Closure of QFAs). *Let $L \subseteq \Sigma^*$. L is accepted by a QFA with cut-point if and only if $\overleftarrow{L}$ is accepted by a QFA with cut-point.*

So, we can conclude that HQFAs do not increase QFAS expressive power.

Corollary 4 (Equivalence between QFAs and HQFAs). *L is accepted by a QFA with cut-point if and only if L is accepted by an HQFA with cut-point.*

Proof. Let L be accepted by a QFA with cut-point. By Theorem 4 $\overleftarrow{L}$ is accepted by a QFA with cut-point. As a consequence of Corollary 3 L is accepted by an HQFA with cut-point.

On the other hand, let L be accepted by an HQFA with cut-point. By Corollary 3 $\overleftarrow{L}$ is accepted by a QFA with cut-point. By Theorem 4 L is accepted by a QFA with cut-point. □

5   Heisenberg Inspired Automata: (Un)bounded Memory

The Heisenberg semantics introduced in the previous section has the same expressive power of the Schrödinger one introduced in the literature. However, we can take inspiration from Heisenberg proposal and analyse what happens if each time a character is read all the unitary matrices are transformed, i.e., instead of changing at each step the observables we modify the unitaries associated to the single characters. We do such changes by exploiting the characters that have already been read.

In particular, given an automaton $M = (Q, \Sigma, \{U_\sigma\}_{\sigma \in \Sigma}, |\psi\rangle, F)$, after reading the prefix $\mathbf{x}_j$ of the string $\mathbf{x} = x_1 x_2 \dots x_n$ the unitary matrix associated to a character σ has evolved into:

$$\mathbb{W}_\sigma^{\mathbf{x}_j} = U_{\mathbf{x}_j} U_\sigma$$

where $U_\epsilon = Id$ is the identity transformation. So, if the current configuration after reading $\mathbf{x}_j$ is $|\varphi\rangle$ and we read x_j, then the state evolves as follows:

$$|\varphi'\rangle = \mathbb{W}_{x_j}^{\mathbf{x}_j} |\varphi\rangle = U_{\mathbf{x}_j} U_{x_j} |\varphi\rangle$$

The computation starts from $|\psi\rangle$ and evolves reading the symbols of the string $\mathbf{x}$. The state reached at the end of the read is:

$$\mathbb{W}_{\mathbf{x}} |\psi\rangle$$

where $\mathbb{W}_{\mathbf{x}}$–the evolution matrix accumulated along the read of $\mathbf{x}$– is defined as:

$$\mathbb{W}_{\mathbf{x}} = \mathbb{W}_{x_n}^{\mathbf{x}_n} \mathbb{W}_{x_{n-1}}^{\mathbf{x}_{n-1}} \cdots \mathbb{W}_{x_2}^{\mathbf{x}_2} \mathbb{W}_{x_1}^{\mathbf{x}_1}$$

As in the case of QFAs the projector P_F is finally applied to obtain the probability of accepting a string $\mathbf{x}$, denoted by $\omega_M(\mathbf{x})$:

$$\omega_M(\mathbf{x}) = \| P_F \mathbb{W}_{\mathbf{x}} |\psi\rangle \|^2 = \langle \psi | \mathbb{W}_{\mathbf{x}}^\dagger P_F^\dagger P_F \mathbb{W}_{\mathbf{x}} |\psi\rangle$$

Example 3. Let us consider again the automaton of Example 1. If we consider the string abb the evolution matrix that is applied to the initial state is:

$$[(U_b U_a) U_b][(U_a) U_b][(Id) U_a]$$

where we use the parenthesis to emphasized the single steps. In particular, the squared parenthesis enclose the read of a single character, while the rounded ones enclose the transformations due to the read of the prefix accumulated so far. Instantiating U_a, U_b and $|\psi\rangle$ as in Example 1 we obtain that the state reached at the end of the read is $- |1\rangle$. So, since $F = \{|0\rangle\}$, we get:

$$\omega_M(abb) = (- \langle 1|) P_F^\dagger P_F (- |1\rangle) = 0$$

Example 4. Let us now consider a simpler example in which $\Sigma = \{a\}$, $U_a = X$, the initial state is $|\psi\rangle = |0\rangle$, and $F = \{|0\rangle\}$. It is immediate to see that when a string of the form a^k is read the evolution matrix has the form:

$$X^k X^{k-1} \cdots X^2 X = X^{\frac{(k+1)k}{2}}$$

This means that a string of length k is accepted by the automaton if and only if $(k + 1)k$ is a multiple of 4. $\square$

As in the case of HQFAs, for these automata, that we call UMQFAs (*Unbounded Memory Quantum Finite Automata*), the syntactic definition is the same as for QFAs, while the accepting condition is different.

Definition 5 (Cut-point UMQFA). *A language $L \subseteq \Sigma^*$ is accepted by a UMQFA M with cut-point λ if and only if $L = \{\mathbf{x} \in \Sigma^* \mid \omega_M(\mathbf{x}) > \lambda\}$.*

A language $L \subseteq \Sigma^*$ is said to be accepted *by a UMQFA with cut-point if and only if there exist a UMQFA M and $\lambda \geq 0$ such that $L \subseteq \Sigma^*$ is accepted by M with* cut-point λ.

Notice that we arbitrarily decided to rely on a single set of matrices. One could have considered a more general definition. The only important point is that when a character is read the unitary matrix that is applied depends also on all the characters that have been read before. However, such dependency have to be defined in a finitary way, i.e., relying on a finite initial set of matrices.

So the question now becomes: is this semantics increasing the expressive power of QFAs? In order to analyse such question we first take a step back and study what happens when, instead of using all the characters that have been read so far, we only use a bounded amount of them. On the one hand, such step back makes the situation more similar to what happen in the case of classical automata, which have a finite amount of memory. On the other hand, this naturally allows to give a more general definition, where a larger set of unitaries is used.

5.1   Bounded Memory

The most natural way to instantiate the above semantics in order to take care only of a bounded quantity of characters is to fix $h \geq 0$ and to refer to $\mathbf{x}_j^h$ instead of $\mathbf{x}_j$ (see Sect. 2.1). The sub-string $\mathbf{x}_j^h$ takes into account at most h symbols that precede x_j in the string $\mathbf{x}$. Since there exists a finite number of strings of length at most h, the matrices $W_\sigma^{\mathbf{y}}$, with $\mathbf{y}$ of length at most h, can be directly specified in the definition of the automaton. Such automata have been already defined in the literature [6,30], using an equivalent notation, and called *Multi-letter Quantum Finite Automata (MQFA)*. However, as we will discuss a different acceptance condition was used. Let $h \in \mathbb{N}$.

Definition 6 (h-MQFA). *An h-MQFA is a 5-tuple $M = (Q, \Sigma, \mathcal{W}, |\psi\rangle, F)$ where:*

- *Q–the set of states– is the finite canonical basis of $\mathbb{C}^d$ for some $d \in \mathbb{N}$;*
- *Σ is a finite alphabet;*
- *$\mathcal{W} = \{W_\sigma^{\mathbf{y}}\}_{\sigma \in \Sigma, \mathbf{y} \in \Sigma^{\leq h}}$ is a finite set of unitaries of dimension $\mathbb{C}^d \times \mathbb{C}^d$;*
- *$|\psi\rangle \in \mathbb{C}^d$ is a unitary vector representing the initial superposition of M;*
- *$F \subseteq Q$ is the set of final states.*

Let $\mathbf{x} = x_1 x_2 \ldots x_n$ be an input string for an h-MQFA $M = (Q, \Sigma, \mathcal{W}, |\psi\rangle, F)$. The computation starts in the state $|\psi\rangle$. Let us assume that after reading the first $j-1$ symbols of $\mathbf{x}$ a state $|\varphi\rangle$ is reached. When x_j is read the states evolves according to the following law:

$$|\varphi'\rangle = W_{x_j}^{\mathbf{x}_j^h} |\varphi\rangle$$

The computation starts from $|\psi\rangle$ and evolves reading the symbols of the string $\mathbf{x}$. At the end of the computation, a measurement is performed on the state of M through the projector P_F. The probability of M accepting $\mathbf{x}$ is:

$$\mu_M(\mathbf{x}) = \| P_F W_{\mathbf{x}} |\psi\rangle \|^2$$

where $W_{\mathbf{x}}$ is defined as:

$$W_{\mathbf{x}} = W_{x_n}^{\mathbf{x}_n^h} W_{x_{n-1}}^{\mathbf{x}_{n-1}^h} \cdots W_{x_2}^{\mathbf{x}_2^h} W_{x_1}^{\mathbf{x}_1^h}$$

The acceptance condition with cut-point for an h-MQFA is based on μ_M.

Definition 7 (Cut-point h-MQFA). *A language $L \subseteq \Sigma^*$ is accepted by an h-MQFA M with cut-point λ if and only if $L = \{\mathbf{x} \in \Sigma^* \mid \mu_M(\mathbf{x}) > \lambda\}$.*

A language $L \subseteq \Sigma^*$ is said to be accepted by an h-MQFA with cut-point *if and only if there exist an h-MQFA M and $\lambda \geq 0$ such that $L \subseteq \Sigma^*$ is accepted by M with cut-point λ.*

Intuitively, h-MQFAs have bounded memory h in the sense that at each point of the computation the preceding h characters are used for choosing the evolution. Notice that QFAs coincide with 0-MQFAs. Moreover, each h'-MQFA can be embedded into a h-MQFA with $h > h'$ by simply defining $\mathscr{W}$ in such a way that if $\mathbf{x}$ and $\mathbf{y}$ are two strings of length at most h that coincide on the suffix of length h', then $W_\sigma^{\mathbf{x}} = W_\sigma^{\mathbf{y}}$, for each $\sigma \in \Sigma$.

Example 5. Let $\Sigma = \{a, b\}$ and $L = \{a, b\}^* b$, i.e., the language of strings that end with b. The pumping lemma for QFAs ensures that this language cannot be accepted by a QFA with cut-point. Consider instead the 1-MQFA $M = (Q, \Sigma, \mathscr{W}, |\psi\rangle, F)$ where $Q = \{|0\rangle, |1\rangle\}$, $\Sigma = \{a, b\}$, $|\psi\rangle = |0\rangle$, $F = \{|1\rangle\}$. The set $\mathscr{W}$ is defined as follows:

$$W_b^\epsilon = W_b^a = W_a^b = X \qquad W_b^b = Id$$

and all the other matrices are the identity. The above matrices exactly simulate the behaviour of the following deterministic automaton, interpreting $|i\rangle$ as q_i:

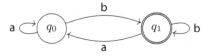

Fig. 1. Deterministic automaton accepting $\{a, b\}^* b$.

So, M accepts L with certainty, hence also any cut-point $\lambda \geq 0$ is fine.

In [6, 30] properties of this class of automata have been studied in the case of isolated cut-point acceptance condition, also called bounded error. It was shown that the expressive power of h-MQFAs is strictly dependent on the parameter h. The set of languages recognized by h-MQFAs with bounded error coincides with those recognized by h-Group Finite Automata and are a subset of regular languages. As a consequence in [30] it has been proved that the set of languages accepted by a h'-MQFAs with bounded error is strictly included in the set of languages accepted by h-MQFAs with bounded error, for $h' < h$. This is consistent with our intuition that more memory increases the computation power. Moreover, in [30], it was proved that if the minimal DFA accepting a language L contains a particular forbidden structure, then L cannot be accepted by h-MQFAs with bounded error, for any $h \geq 0$. This is a structural characterization of languages that cannot be accepted by h-MQFAs with bounded error.

In this section, as in the rest of this paper, we focus on cut-point acceptance condition which is less demanding than bounded error and has not been studied in the literature for h-MQFAs. We start presenting a *pumping lemma* which provides a structural characterization of the languages that are accepted from h-MQFAs with cut-point. Then we investigate on the expressive power of h-MQFAs with respect to h. Differently from QFAs, h-MQFAs can also recognize finite languages and still they constitute a proper hierarchy.

In the proof of the pumping lemma we exploit the following lemma which is also at the basis of the pumping lemma for QFAs. The norm $\|A\|$ of a matrix A is defined as:

$$\|A\| = \sup_{\langle u|u \rangle = 1} \{\|A\,|u\rangle\,\|\}$$

Lemma 1 ([8]). *Let $V \in \mathbb{C}^d \times \mathbb{C}^d$ be a unitary matrix let $Id \in \mathbb{C}^d \times \mathbb{C}^d$ be the identity matrix of dimension d. For any $\varepsilon > 0$ there exists $k \in \mathbb{N}^+$ such that:*

$$\|Id - V^k\| \leq \varepsilon$$

The pumping lemma for h-MQFAs states that if we consider a sufficiently long suffix of a string which is inside the accepted language, then we can pump such suffix for an opportune number of times and fall again inside the language.

Theorem 5 (Pumping Lemma for h-MQFAs). *Let $L \subseteq \Sigma^*$ be the language accepted by an h-MQFA. Then, $\forall \mathbf{x} = \mathbf{uv} \in L$ with $|\mathbf{v}| \geq h$ there exists $k \in \mathbb{N}^+$ such that $\mathbf{xv}^k \in L$.*

Notice that differently from Theorem 2, the above pumping lemma does not prevent finite languages to be accepted by h-MQFAs. As a matter of fact, if all the strings accepted by an h-MQFAs are shorter than h, then it is not possible to find a suffix that can be pumped.

Theorem 6 (Singleton/Finite Languages). *Let $L = \{\mathbf{w}\}$ with $\mathbf{w} \in \Sigma^{h-1}$ and $h - 1 > 0$. Then there exists an h-MQFA that accepts L with certainty.*

Let L be a finite language whose elements have length less than h. There exists an h-MQFA that accepts L with cut-point.

We can exploit our pumping lemma to prove that the amount of memory we provided to the h-MQFA in the above theorem is the minimum.

Lemma 2. *Let $L = \{\mathbf{w}\}$ with $\mathbf{w} \in \Sigma^{h-1}, h-1 > 0$. Then, there is no h'-MQFA, with $h' < h$ that accepts L with cut-point.*

Proof. Assume by contradiction that there exists an h'-MQFA that accepts L with $h' < h$. Since $|\mathbf{w}| = h - 1 \geq h'$, the string $\mathbf{w}$ can be written as $\mathbf{uv}$ with $\mathbf{v} \in \Sigma^{h'}$. By Theorem 5, it holds that there exists a $k \in \mathbb{N}^+$ such that $\mathbf{wv}^k \in L$. So, we have a contradiction.   □

Exploiting Theorems 5 and 6, together with Lemma 2 we have that the set of languages accepted by h'-MQFAs is a proper subset of the set of languages accepted by h-MQFAs, with $h' < h$. The inclusion immediately follows from the definition of h-MQFAs and our results show that the inclusion is proper by exhibiting as witnesses all the singleton languages of strings of length h'.

Corollary 5. *The set of languages accepted by h'-MQFAs with cut-point is a proper subset of those accepted by h-MQFAs with cut-point, when $h' < h$.*

The hierarchy result proved in [30] concerns sets of languages which are all included in the set of regular languages, while our hierarchy includes already at level 0 non-regular languages.

5.2 Unbounded Memory

QFAs and also h-MQFAs fail to recognize many classical regular languages, since unitary transformations introduce a notion of memory which is quite different from the classical one.

On the one hand, it is easy to define a classical automaton for a finite language by using the finite set of states of the automaton to store the finite quantity of memory that is necessary. This cannot be achieved in QFAs and in h-MQFAs, when h is not large enough, as a consequence of the following property of unitary matrices that has been stated in Lemma 1:

$$\forall \varepsilon > 0 \; \exists k \in \mathbb{N}^+ \; \|Id - V^k\| \leq \varepsilon$$

This is the key ingredient of the pumping lemmas for QFAs and h-MQFAs.

On the other hand, it is possible to define a QFA that accepts the non-regular language of strings having a different number of a and b characters. Classical automata do not have enough memory for this language, since it is necessary to count an unbounded number of characters.

We started this section introducing the Heisenberg inspired automata called UMQFAs hoping to increase the expressive power of QFA and h-MQFAs still relying on a finite set of unitaries and a single measurement at the end of the read. It is time to draw some conclusions about this. The automaton described in Example 4 pointed out that in UMQFAs we are not able to replicate the use of a unitary matrix V for any possible $k \in \mathbb{N}^+$, i.e., we cannot exploit Lemma 1. For instance in the example the matrix X can only occur with an exponent of the form $(k+1)k/2$, i.e., all the possible values assumed by a polynomial $p(k)$ when k ranges in $\mathbb{N}^+$. The proof of Lemma 1 in [8] is based on Cauchy sequences and cannot be easily generalized. However, there is another proof of the same result in [14] that ultimately relies on the following algebraic property:

for each $\alpha \in \mathbb{R} \setminus \mathbb{Q}$ the set of fractional parts of the multiples of α, i.e., $\{k\alpha - \lfloor k\alpha \rfloor \mid k \in \mathbb{N}\}$, is dense in $[0,1]$.

This results generalizes to polynomials having irrational coefficients and to multiple dimensions (e.g., [7]). As a consequence we have a language that can be accepted by h-MQFAs, but not by UMQFAs.

Theorem 7. *Let $\Sigma = \{a\}$ and $L = \{\epsilon, a\}$. There is a 2-MQFA that accepts L with cut-point and there is not a UMQFAs that accepts L with cut-point.*

There are technical ingredients in the proof of the above result that are somehow interesting. We had to carefully choose the language L in order to obtain homogeneous polynomials. Otherwise the eigenvalues related to rotations that are rational multiples of π would have given troubles. Moreover, the interplay between some eigenvalues could be favorable for constructing UMQFAs that *approximate* h-MQFAs, since the distribution of the wrong strings accepted by the UMQFA is not uniform.

Beside these technical considerations, the result shows that the unbounded memory we tried to introduce does not generalize the bounded one, and it does not seem easy to find a natural generalization with a finitary description.

6   Conclusions

Quantum Computing is becoming a more and more investigated subject thanks to phenomena like quantum speed-up. Using the properties of quantum mechanics it is possible to design algorithms that polynomially solve problems that require exponential time with classical computation [28]. However, when one looks at basic models of computation such as automata the rules of quantum mechanics, imposing unitary evolutions along the computation, constitute more an obstacle to the expressive power, than an advantage. Informally, we can say that the unitaries cause a loss of memory in the automata. As a matter of fact, a simple language including only one string cannot be accepted by MO-QFAs.

In our work we tried to better understand the role of unitaries and measurements in MO-QFAs. We proved that for any MO-QFA there is a "reversed" MO-QFA that accepts the mirror language. Then we analysed the effect of playing with the unitaries. We forced a sort of stuttering behaviour hoping to gain expressive power. We obtained a first negative result which however gives some suggestions for further investigations. For example, there may be other definitions for the Unbounded Memory case that lead to larger expressive power.

A. Appendix: Proofs of Main Theorems

A.1 Proof of Theorem 4

Let $M = (Q, \Sigma, \{U_\sigma\}_{\sigma \in \Sigma}, |\psi\rangle, F)$ be a QFA accepting L with cut-point λ. We recall that Q is the canonical basis of $\mathbb{C}^d$, for some $d \in \mathbb{N}$. Without loss of generality, let $F = \{q_0, q_1, \ldots q_{m-1}\}$. Let $\mathbf{x} = x_1 x_2 \ldots x_n$ be an input string. By definition, the acceptance probability of M for $\mathbf{x}$ is:

$$p_M(\mathbf{x}) = \sum_{i=0}^{m-1} |\langle q_i| U_{\mathbf{x}} |\psi\rangle|^2$$

We now define $\overleftarrow{M} = (\overleftarrow{Q}, \Sigma, \{V_\sigma\}_{\sigma \in \Sigma}, |\overleftarrow{\psi}\rangle, \overleftarrow{F})$, where $\overleftarrow{Q}$ is the canonical basis of $\mathbb{C}^{dm}$ and:

$$V_\sigma = \sum_{i=0}^{m-1} |i\rangle \langle i| \otimes U_\sigma^\dagger, \quad |\overleftarrow{\psi}\rangle = \frac{1}{\sqrt{m}} \sum_{i=0}^{m-1} |i\rangle \otimes |q_i\rangle, \quad \overleftarrow{F} = \{|i\rangle \otimes |\psi\rangle \mid i \in [0, m-1]\}$$

We have that $V_{\mathbf{x}} = V_{x_n} V_{x_{n-1}} \cdots V_{x_1}$ and $U_{\overleftarrow{\mathbf{x}}} = U_{x_1} U_{x_2} \ldots U_{x_n}$. By definition of QFA we get:

$$
\begin{aligned}
p_{\overleftarrow{M}}(\mathbf{x}) &= \|P_{\overleftarrow{F}} V_{\mathbf{x}} |\overleftarrow{\psi}\rangle\|^2 \\
&= \|P_{\overleftarrow{F}} \frac{1}{\sqrt{m}} \sum_{i=0}^{m-1} |i\rangle \otimes U_{x_n}^\dagger U_{x_{n-1}}^\dagger \cdots U_{x_1}^\dagger |q_i\rangle\|^2 \\
&= \|\frac{1}{\sqrt{m}} \sum_{i=0}^{m-1} |i\rangle \otimes \left(|\psi\rangle \langle\psi| U_{x_n}^\dagger U_{x_{n-1}}^\dagger \cdots U_{x_1}^\dagger |q_i\rangle \right)\|^2 \\
&= \|\frac{1}{\sqrt{m}} \sum_{i=0}^{m-1} \left(\langle\psi| U_{x_n}^\dagger U_{x_{n-1}}^\dagger \cdots U_{x_1}^\dagger |q_i\rangle \right) |i\rangle \otimes |\psi\rangle\|^2 \\
&= \frac{1}{m} \sum_{i=0}^{m-1} |\langle\psi| U_{x_n}^\dagger U_{x_{n-1}}^\dagger \cdots U_{x_1}^\dagger |q_i\rangle|^2 \\
&= \frac{1}{m} \sum_{i=0}^{m-1} |(\langle q_i| U_{\overleftarrow{\mathbf{x}}} |\psi\rangle)^*|^2 = \frac{1}{m} \sum_{i=0}^{m-1} |\langle q_i| U_{\overleftarrow{\mathbf{x}}} |\psi\rangle|^2 = \frac{1}{m} p_M(\overleftarrow{\mathbf{x}})
\end{aligned}
$$

Let $\overleftarrow{\lambda} = \frac{\lambda}{m}$. We have that:

$$\overleftarrow{\mathbf{x}} \in L \text{ iff } p_M(\overleftarrow{\mathbf{x}}) > \lambda \text{ iff } p_{\overleftarrow{M}}(\mathbf{x}) > \overleftarrow{\lambda} \text{ iff } \mathbf{x} \in \overleftarrow{L}$$

$\square$

A.2 Proof of Theorem 5

For sake of readability we prove the result for $|\mathbf{v}| = h$. For $|\mathbf{v}| > h$ the idea is the same, just the notation would be much heavier.

Let $L \subseteq \Sigma^*$ be a language and let $M = (Q, \Sigma, \mathscr{W}, |\psi\rangle, F)$ be an h-MQFA that accepts L with *cut-point* λ.

Let $\mathbf{x} = \mathbf{uv} = u_1 \dots u_a v_1 \dots v_h$ be a string of L. First we write the matrix $W_{\mathbf{xv}^j}$, for a generic $j \in \mathbb{N}$, in order to make explicit its relationship with $W_{\mathbf{x}}$. In particular, by applying the definition of h-QMFA we have that $W_{\mathbf{xv}^j} = V^j W_{\mathbf{x}}$, where V is defined as:

$$V = W_{v_h}^{v_h v_1 \cdots v_{h-1}} W_{v_{h-1}}^{v_{h-1} v_h v_1 \cdots v_{h-2}} \dots W_{v_1}^{v_1 v_2 \cdots v_h}$$

Having represented a vector $|v\rangle$ in the canonical basis and being $|q\rangle$ be an element of the canonical basis, let $(|v\rangle)_q$ be the q-th component of $|v\rangle$. For any $j \in \mathbb{N}$ it holds:

$$|\mu_M(\mathbf{x}) - \mu_M(\mathbf{xv}^j)| = \left| \sum_{q \in F} \left(|(W_{\mathbf{x}} |\psi\rangle)_q|^2 - |(W_{\mathbf{xv}^j} |\psi\rangle)_q|^2 \right) \right|$$
$$\leq 2 \sum_{q \in F} \left| |(W_{\mathbf{x}} |\psi\rangle)_q| - |(W_{\mathbf{xv}^j} |\psi\rangle)_q| \right| \leq 2 \sum_{q \in F} |(W_{\mathbf{x}} |\psi\rangle)_q - (W_{\mathbf{xv}^j} |\psi\rangle)_q|$$
$$= 2 \sum_{q \in F} \left| (W_{\mathbf{x}} |\psi\rangle)_q - (V^j W_{\mathbf{x}} |\psi\rangle)_q \right| = 2 \sum_{q \in F} \left| \langle q| (Id - V^j) W_{\mathbf{x}} |\psi\rangle \right|$$
$$\leq 2 \sum_{q \in F} \|Id - V^j\| = 2|F| \|Id - V^j\|$$

Since $\mathbf{x} \in L$, we have $\mu_M(x) - \lambda = \Delta > 0$. By Lemma 1 there exists $k \in \mathbb{N}^+$ such that:

$$\|Id - V^k\| \leq \frac{\Delta}{4|F|}$$

which yields to $|\mu_M(\mathbf{x}) - \mu_M(\mathbf{xv}^k)| \leq \frac{\Delta}{2}$. Therefore, $\mathbf{xv}^k \in L$, since:

$$\mu_M(\mathbf{xv}^k) - \lambda \geq \mu_M(\mathbf{x}) - \frac{\Delta}{2} - \lambda \geq \frac{\Delta}{2} \geq 0$$

□

A.3 Proof of Theorem 6

Let $M = (Q, \Sigma, \mathscr{W}, |\psi\rangle, F)$ be a h-MQFA, where $Q = \{|0\rangle, |1\rangle\}$, $|\psi\rangle = |0\rangle$, $F = \{|1\rangle\}$. The states $|0\rangle$ and $|1\rangle$ are such that $|1\rangle = X|0\rangle$ and $|0\rangle = X|1\rangle$. Since $\mathbf{w}$ has length $h - 1 > 0$, $\mathbf{w} = \mathbf{u}\alpha$ with $\mathbf{u} \in \Sigma^{h-2}, \alpha \in \Sigma$. We define:

$$W_\alpha^{\mathbf{u}} = X$$
$$W_\sigma^{\mathbf{w}} = X \quad \forall \sigma \in \Sigma$$

while all the other matrices inside $\mathscr{W}$ are the identity matrix. We must now prove that the language accepted by M is exactly L.

If $\mathbf{w}$ is the input for M, then the computation evolves as follows:

$$W_{\mathbf{w}} = W_m^{\mathbf{w}_m^h} W_{w_{m-1}}^{\mathbf{w}_{m-1}^h} \dots W_{w_1}^{\mathbf{w}_1^h}$$

Since all the matrices we set to be different from the identity concern strings with length that is at least $h - 1$, it holds that $W_{w_j}^{\mathbf{w}_j^{h+1}} = I, \forall j \in \{1, 2, \ldots m - 1\}$ Therefore,

$$W_{\mathbf{w}} = W_{w_m}^{\mathbf{w}_m^{h+1}} = W_\alpha^{\mathbf{u}} = X$$

Since the initial state is $|0\rangle$, then $\| P W_{\mathbf{w}} |0\rangle \|^2 = \| P |1\rangle \|^2 = 1$.

Otherwise, suppose $\mathbf{x} \neq \mathbf{w}$ is the input for M. If the string $\mathbf{x}$ does not contain the sub-string $\mathbf{w}$, then clearly $W_{\mathbf{x}} = Id$, and $\mathbf{x}$ is refused. If $\mathbf{x}$ has $\mathbf{w}$ as proper prefix, then $\mathbf{x}$ is of the form $\mathbf{ws}$, with $\mathbf{s} = \sigma_1 \ldots \sigma_j$, $j \geq 1$. In this case, we have that $W_{\mathbf{x}}$ is as follows:

$$W_{\mathbf{x}} = W_{\sigma_j}^{\mathbf{x}_{h-j-1}^h} W_{\sigma_{j-1}}^{\mathbf{x}_{h-j-2}^h} \cdots W_{\sigma_1}^{\mathbf{x}_{h-1}^h} W_{\mathbf{w}}$$
$$= W_{\sigma_j}^{\mathbf{x}_{h-j-1}^h} W_{\sigma_{j-1}}^{\mathbf{x}_{h-j-2}^h} \cdots W_{\sigma_1}^{\mathbf{w}} W_{\mathbf{w}} = Id \cdots X X = Id$$

since all the matrices of the form $W_\sigma^{\mathbf{y}}$, with $\mathbf{y}$ of length h are the identity matrix. So, $\mathbf{x}$ is refused. The last case we need to consider is when $\mathbf{w}$ occurs as a proper sub-string of $\mathbf{x}$, but it is not a proper prefix of $\mathbf{x}$. This means that the input $\mathbf{x}$ is of the form $\mathbf{x} = \mathbf{vws}$, with $\mathbf{v} \neq \epsilon$ and $\mathbf{vw}$ which does not have $\mathbf{w}$ as prefix. In this case, the key point is that since $|\mathbf{w}| = h - 1$, but $\mathbf{w}$ is now preceded by at least one character the matrices $W_\alpha^{\mathbf{u}}$ and $W_\sigma^{\mathbf{w}}$ do not occur in $W_{\mathbf{x}}$. So, $W_{\mathbf{x}} = Id$ and $\mathbf{x}$ is refused. Notice that the automaton we defined accepts with certainty.

Let $L = \{\mathbf{x}_1, \ldots, \mathbf{x}_\ell\}$ be a finite language whose elements have length less than h. For each element $\mathbf{x}_j$ there exists an h_j-MQFA that accepts only $\mathbf{x}_j$ with certainty. As already observed any h'-MQFA can be embedded into an h-MQFA that accepts the same language with the same cut-point, if $h \geq h'$. Let h be greater than the length of the longest string in L. We have that for each element $\mathbf{x}_j$ of L there exists an h-MQFA M_j that accepts only $\mathbf{x}_j$ with certainty. The tensor product M of the M_j's automata, whose construction is similar to that used in the proof of Theorem 4 accepts the language L. The tensor product automaton does not accept with certainty but with cut-point λ, with λ any number in the interval $(0, 1/\ell)$. □

A.4 Proof of Corollary 5

Let $h, h' \in \mathbb{N}^+$ with $h' < h$. From Lemma 2 we know that there exist languages accepted by h−MQFAs, but not by h'−MQFAs.

We must prove that all the languages accepted by h'−MQFAs are also accepted by h−MQFAs.

Let $M = (Q, \Sigma, \mathcal{W}, |\psi\rangle, F)$ be a h'−MQFA accepting a language L. We can build an h−MQFA $M' = (Q, \Sigma, \mathcal{W}', |\psi\rangle, F)$ accepting the same language setting $\mathcal{W}' = \mathcal{W}$ (eventually completing with identity matrices). □

A.5 Proof of Theorem 7

By Theorem 6 there is a 2-MQFA that accepts $L = \{\epsilon, a\}$ with cut-point.

Let us assume by contradiction that there exists $M = (Q, \Sigma, \mathscr{U}, |\psi\rangle, F)$ UMQFA that accepts $L = \{\epsilon, a\}$ with cut-point λ. Since the string ϵ is in L it has to be:

$$\omega_M(\epsilon) = \|P_F |\psi\rangle\|^2 = \|P_F U_a |\psi\rangle\|^2 = \lambda + \Delta > \lambda$$

Any other string a^k, with $k > 1$ over the alphabet Σ would instead give:

$$\omega_M(a^k) = \|P_F \mathbb{W}_{a^k} |\psi\rangle\|^2 = \|P_F U_a^{\frac{k(k+1)}{2}} |\psi\rangle\|^2$$

Let us consider a generic unitary matrix V and study the sequence:

$$\{V^{\frac{k(k+1)}{2}}\}_{k>1}$$

As observed in [14], V can be diagonalized and $V^h = RD^hR^{-1}$, where R is unitary and D is the diagonal matrix of the eigenvalues of V. Let $e^{i\pi v_j}$ be the j-th eigenvalue of V.

If all the r_js are rational, then let $n = 4\Pi_j q_j$, where the q_js are the denominators of the r_js. We have that $D^{\frac{n(n+1)}{2}} = Id$, and hence $V^{\frac{n(n+1)}{2}} = Id$.

If m of the r_j are irrational, and ℓ of them are rational, we can safely assume that the first m are the irrational ones. Let again n be defined as above considering only the rational coefficients. If we consider the sub-sequence:

$$\{V^{\frac{nk(nk+1)}{2}}\}_{k>1}$$

we have that all the rational eigenvalues have always values 1 in the subsequence. On the other hand, the remaining eigenvalues take values of the form $e^{i\pi r_j \frac{nk(nk+1)}{2}}$ in the sub-sequence. Let $p : \mathbb{N} \to \mathbb{R}^m$ be defined as $p(k) = (r_1 4nk(4nk+1), \ldots, r_m 4nk(4nk+1))$. These are quadratic polynomials in the variable k with irrational coefficients. The fractional parts of each of these polynomials are dense and uniformly distributed over $[0,1]$ (e.g., [7]). This means that each of these polynomials is infinitely many times arbitrarily close to a multiple of 4. This implies that each of the values $e^{i\pi r_j \frac{nk(nk+1)}{2}}$ is infinitely many times arbitrarily close to 1. As far as the whole polynomial function p is concerned it is uniformly distributed over $[0,1]^m$ if the irrational r_js are independent. When some of the of the irrational r_js are linear combinations of the others the uniform distribution is no more ensured, but the density in $(0, 0, \ldots, 0)$ is preserved, since by making the fractional parts of the independent ones arbitrary small we can ensure that also the fractional parts of their linear combination are small enough.

Hence, for any unitary matrix V, and for each ε there exists $k > 1$ such that:

$$\|Id - V^{\frac{k(k-1)}{2}}\| \leq \varepsilon$$

As a consequence working as in the proof of Theorem 5 on the string ϵ which is in L and using $U_a^{\frac{k(k-1)}{2}}$ we obtain that there exist $k > 1$ such that a^k is accepted by M. This is a contradiction. $\qquad\square$

References

1. Ambainis, A., Beaudry, M., Golovkins, M., Kikusts, A., Mercer, M., Thérien, D.: Algebraic results on quantum automata. Theory Comput. Syst. **39**(1), 165–188 (2006)

2. Ambainis, A., Freivalds, R.: 1-way quantum finite automata: strengths, weaknesses and generalizations. In: Proceedings 39th Annual Symposium on Foundations of Computer Science (Cat. No. 98CB36280), pp. 332–341. IEEE (1998)

3. Anticoli, L., Piazza, C., Taglialegne, L., Zuliani, P.: Towards quantum programs verification: from Quipper circuits to QPMC. In: Devitt, S., Lanese, I. (eds.) RC 2016. LNCS, vol. 9720, pp. 213–219. Springer, Cham (2016). https://doi.org/10.1007/978-3-319-40578-0_16

4. Arrighi, P.: An overview of quantum cellular automata. Nat. Comput. **18**(4), 885–899 (2019)

5. Bell, P.C., Hirvensalo, M.: Acceptance ambiguity for quantum automata. In: 44th International Symposium on Mathematical Foundations of Computer Science (MFCS 2019). Schloss Dagstuhl-Leibniz-Zentrum fuer Informatik (2019)

6. Belovs, A., Rosmanis, A., Smotrovs, J.: Multi-letter reversible and quantum finite automata. In: Harju, T., Karhumäki, J., Lepistö, A. (eds.) DLT 2007. LNCS, vol. 4588, pp. 60–71. Springer, Heidelberg (2007). https://doi.org/10.1007/978-3-540-73208-2_9

7. Bergelson, V., Leibman, A.: Distribution of values of bounded generalized polynomials. Acta Mathem. **198**(2), 155–230 (2007)

8. Bertoni, A., Carpentieri, M.: Analogies and differences between quantum and stochastic automata. Theort. Comput. Sci. **262**(1–2), 69–81 (2001)

9. Bertoni, A., Mereghetti, C., Palano, B.: Quantum Computing: 1-way quantum automata. In: Ésik, Z., Fülöp, Z. (eds.) DLT 2003. LNCS, vol. 2710, pp. 1–20. Springer, Heidelberg (2003). https://doi.org/10.1007/3-540-45007-6_1

10. Bhatia, A.S., Kumar, A.: Quantum finite automata: survey, status and research directions. arXiv preprint arXiv:1901.07992 (2019)

11. Bhatia, A.S., Kumar, A.: On relation between linear temporal logic and quantum finite automata. J. Logic Lang. Inf. **29**(2), 109–120 (2020)

12. Bianchi, M.P., Mereghetti, C., Palano, B.: Quantum finite automata: advances on Bertoni's ideas. Theoret. Comput. Sci. **664**, 39–53 (2017)

13. Birkan, U., Salehi, Ö., Olejar, V., Nurlu, C., Yakaryılmaz, A.: Implementing quantum finite automata algorithms on noisy devices. In: Paszynski, M., Kranzlmüller, D., Krzhizhanovskaya, V.V., Dongarra, J.J., Sloot, P.M.A. (eds.) ICCS 2021. LNCS, vol. 12747, pp. 3–16. Springer, Cham (2021). https://doi.org/10.1007/978-3-030-77980-1_1

14. Brodsky, A., Pippenger, N.: Characterizations of 1-way quantum finite automata. SIAM J. Comput. **31**(5), 1456–1478 (2002)

15. Clarke, E.M., Henzinger, T.A., Veith, H., Bloem, R., et al.: Handbook of Model Checking, vol. 10. Springer, Cham (2018). https://doi.org/10.1007/978-3-319-10575-8

16. Dirac, P.A.M.: Lectures on quantum field theory. Am. J. Phy. **37**, 233–233 (1969)

17. Feng, Y., Yu, N., Ying, M.: Model checking quantum Markova chains. J. Comput. Syst. Sci. **79**(7), 1181–1198 (2013)

18. Gainutdinova, A., Yakaryılmaz, A.: Unary probabilistic and quantum automata on promise problems. Quant. Inf. Process. **17**(2), 1–17 (2018)

19. Gay, S.J., Nagarajan, R., Papanikolaou, N.: QMC: a model checker for quantum systems. In: Gupta, A., Malik, S. (eds.) CAV 2008. LNCS, vol. 5123, pp. 543–547. Springer, Heidelberg (2008). https://doi.org/10.1007/978-3-540-70545-1_51

20. Hermanns, H., Herzog, U., Katoen, J.P.: Process algebra for performance evaluation. Theoret. Comput. Sci. **274**(1–2), 43–87 (2002)

21. Katoen, J.P.: The probabilistic model checking landscape. In: Proceedings of the 31st Annual ACM/IEEE Symposium on Logic in Computer Science, pp. 31–45 (2016)

22. Khadiev, K., et al.: Two-way and one-way quantum and classical automata with advice for online minimization problems. Theoret Comput. Sci. **920**, 76–94 (2022)

23. Kondacs, A., Watrous, J.: On the power of quantum finite state automata. In: Proceedings 38th Annual Symposium on Foundations of Computer Science, pp. 66–75 (1997)

24. Kwiatkowska, M., Norman, G., Parker, D.: Probabilistic model checking: advances and applications. In: Drechsler, R. (ed.) Formal System Verification, pp. 73–121. Springer, Cham (2018). https://doi.org/10.1007/978-3-319-57685-5_3

25. Mereghetti, C., Palano, B.: Guest column: Quantum finite automata: From theory to practice. ACM SIGACT News **52**(3), 38–59 (2021)

26. Mereghetti, C., Palano, B., Cialdi, S., Vento, V., Paris, M.G., Olivares, S.: Photonic realization of a quantum finite automaton. Physical Review Research **2**(1), 013089 (2020)

27. Moore, C., Crutchfield, J.P.: Quantum automata and quantum grammars. Theoret. Comput. Sci. **237**(1–2), 275–306 (2000)

28. Nielsen, M.A., Chuang, I.: Quantum Computation and Quantum Information. Springer, Heidelberg (2002). https://doi.org/10.1007/3-540-33133-6

29. Paschen, K.: Quantum finite automata using ancilla qubits. Technical report, Universität Karlsruhe (TH) (2000). https://doi.org/10.5445/IR/1452000

30. Qiu, D., Yu, S.: Hierarchy and equivalence of multi-letter quantum finite automata. Theoret. Comput. Sci. **410**(30–32), 3006–3017 (2009)

31. Qiu, D., Mateus, P., Sernadas, A.: One-way quantum finite automata together with classical states. arXiv preprint arXiv:0909.1428 pp. 3006–3017 (2009)

32. Rabin, M.O.: Probabilistic automata. Inf. Control **6**(3), 230–245 (1963)

33. Von Neumann, J.: Mathematical foundations of quantum mechanics. In: Mathematical Foundations of Quantum Mechanics. Princeton University Press, Princeton (2018)

Monte Carlo Tree Search for Priced Timed Automata

Peter Gjøl Jensen, Andrej Kiviriga[⊠], Kim Guldstrand Larsen, Ulrik Nyman,
Adriana Mijačika, and Jeppe Høiriis Mortensen

Aalborg University, Selma Lagerløfs Vej 300, 9220 Aalborg, Denmark
{pgj,kiviriga,kgl,ulrik}@cs.aau.dk

Abstract. Priced timed automata (PTA) were introduced in the early 2000s to allow for generic modelling of resource-consumption problems for systems with real-time constraints. Optimal schedules for allocation of resources may here be recast as optimal reachability problems. In the setting of PTA this problem has been shown decidable and efficient symbolic reachability algorithms have been developed. Moreover, PTA has been successfully applied in a variety of applications. Still, we believe that using techniques from the planning community may provide further improvements. Thus, in this paper we consider exploiting Monte Carlo Tree Search (MCTS), adapting it to problems formulated as PTA reachability problems. We evaluate our approach on a large benchmark set of PTAs modelling either Task graph or Job-shop scheduling problems. We discuss and implement different complete and incomplete exploration policies and study their performance on the benchmark. In addition, we experiment with both well-established and our novel MTCS-based optimizations of PTA and study their impact. We compare our method to the existing symbolic optimal reachability engines for PTAs and demonstrate that our method (1) finds near-optimal plans, and (2) can construct plans for problems infeasible to solve with existing symbolic planners for PTA.

Keywords: Priced Timed Automata (PTA) · Model-checking · Monte Carlo Tree Search (MCTS) · Planning · Upper confidence bounds for trees (UCT)

1 Introduction

The world is full of planning and scheduling problems that have impact on the real world. Finding optimal solutions for such problems can be of great importance for profit maximization or resource minimization, affecting financial success and sustainable development. In general such problems do not just have one solution, but many solutions – with varying cost. These scheduling problems are one sub-field within operations research, and lots of effort has been put into finding both optimal and near optimal solutions to them.

One technique that has been successfully applied to planning is that of model checking, e.g. BDD based model checking [19]. For optimal planning problems involving timing constraints, the notion of priced timed automata was introduced in the early

ⓒ Springer Nature Switzerland AG 2022
E. Ábrahám and M. Paolieri (Eds.): QEST 2022, LNCS 13479, pp. 381–398, 2022.
https://doi.org/10.1007/978-3-031-16336-4_19

2000s, with initial decidability results [4,7] based on so-called *corner-point regions* and later with efficient symbolic forward reachability algorithms using so-called *priced zones* made available in the tool UPPAAL CORA. Here a generic and highly expressive modeling formalism is provided, extending the classical notion of timed automata [3] with a cost-variable (to be optimized), but also providing support for discrete variables over structured (user-defined) types, as well as user-specified procedures [8]. In fact, the notion of PTA allows for an extension of Planning Domain Definition Language (PDDL) 2.1 at level 3 towards duration-dependent and continuous effects to be encoded as demonstrated by [18]. Most recently so-called *extrapolation* techniques have been introduced for more efficient analysis of PTA, implemented in the tool TiaMo [13].

Applications of PTA and UPPAAL CORA are several and from a variety of areas [14], e.g. power optimization of dataflow applications [2], battery scheduling [27], planning of nano-satellites [23,30], grape harvest logistic [33], programmable logic controllers [35], smart grids [21], service oriented systems [17], and optimal multicore mapping of spreadsheets [11] to mention a few.

Despite the success of PTA and UPPAAL CORA, we still believe that the performance may be improved by exploiting advances made by the planning community. Thus, we consider in this paper various ways of exploiting Monte Carlo Tree Search (MCTS) to further improve performance of PTA optimization. MCTS is a powerful technique that has seen application in many domains requiring (near-) optimal planning, including problem instances where the size of the search-space makes symbolic and complete methods infeasible. In particular, MCTS [16] has already been applied directly to Job-shop [5] scheduling problems. We benchmark our implementations of MCTS based analysis of PTA on Job-shop and Task graph problems and compare against the two tools UPPAAL CORA [9] and TiaMo [13].

The rest of the paper is organized as follows: First we formally define Priced Timed Automata, then we introduce a general formalization of Monte Carlo Tree Search along with specific PTA policies. Finally we discuss additional enhancements and present our experimental evaluation.

2   Priced Timed Automata

The priced timed automaton [6] is an extension of timed automaton [3] with prices on both locations and transitions. Delaying in locations entails a price growth based on fixed price (cost) rate, while taking transitions is associated with a fixed price. We now present the formal definition of priced time automaton and its semantics based on [9].

Let $\mathbb{C}$ be a set of clocks. The set of constraints over clocks $\mathbb{C}$, $B(\mathbb{C})$, are defined as the set of conjunctions of atomic constraints of the form $x \bowtie n$, where $x \in \mathbb{C}$, $\bowtie \in \{<, \leq, =, >, \geq\}$ and $n \in \mathbb{N}_{\geq 0}$. Such constraints – guards and invariants – allow to restrict the behavior w.r.t. the values of clocks. The power set of $\mathbb{C}$ is denoted as $2^{(\mathbb{C})}$.

Definition 1 (Priced Timed Automaton). *A Priced Timed Automaton (PTA) over clocks $\mathbb{C}$ and actions Act is represented as a tuple $A = (L, l_0, E, I, P)$ where:*

- *L is a finite set of locations,*
- *$l_0 \in L$ is the initial location,*
- *$E \subseteq L \times B(\mathbb{C}) \times Act \times 2^{(\mathbb{C})} \times L$ is a set of edges where an edge connects two locations and contains a guard, an action, and a set of clocks to be reset,*

- $I: L \rightarrow, \mathcal{B}(\mathbb{C})$ is a set of location invariants, and
- $P: (L \cup E) \rightarrow \mathbb{N}$ assigns cost rates and cost increments to locations and edges, respectively.

In the case of $(l, g, a, r, l') \in E$, we write $l \xrightarrow{g,a,r} l'$. A clock valuation v over $\mathbb{C}$ is a mapping $v: \mathbb{C} \rightarrow \mathbb{R}_{\geq 0}$ and $\mathbb{R}^{\mathbb{C}}$ denotes a set of all clock valuations. The semantics of a PTA is defined in terms of a priced transition system:

Definition 2 (Priced Transition System). *A Priced Transition System (PTS) over actions Act is a tuple $T = (S, s_0, \Sigma, \rightarrow)$ where:*

- *S is a set of states*
- *s_0 is an initial state,*
- *$\Sigma = Act \cup \mathbb{R}_{\geq 0}$ is the set of labels, and*
- *$\rightarrow \subseteq (S \times \Sigma \times \mathbb{R}_{\geq 0} \times S)$ is a set of labelled and priced transitions. We write $s \xrightarrow{a}_p s'$ whenever $(s, a, p, s') \in \rightarrow$.*

Now a PTA $A = (L, l_0, E, I, P)$ defines a PTS $T_A = (S, s_0, \Sigma, \rightarrow)$, where the set of states S are pairs (l, v), with $l \in L$ is a location and v is a clock valuation s.t. the invariant $I(l)$ of l is satisfied by v, denoted $v \models I(l)$.

There are two possible types of transitions between states: *action transitions* and *delay transitions*. Action transitions are the result of following an enabled edge in the PTA A. As a result, the destination location is activated and the clocks in the reset set are set to zero, and the price of the transition is given by the cost of the edge. Formally:

$$(l, v) \xrightarrow{a}_p (l', v') \text{ iff } \exists (l, g, a, r, l') \in E, \text{ such that}$$
$$v \models g \wedge v' = v[r] \wedge v' \models I(l) \wedge p = P((l, g, a, r, l'))$$

where $v[r]$ is the valuation given by $v[r](x) = 0$ if $x \in r$ and $v[r](x) = v(x)$ otherwise.

Delay transitions allow the time to pass resulting in an increase of the value of all clocks, but with no change of the location. The cost of a delay transition is the product of the duration of the delay and the cost rate of the active location. Formally:

$$(l, v) \xrightarrow{d}_p (l, v') \text{ iff } v' = v + d \wedge v \models I(l) \wedge v' \models I(l) \wedge p = d \cdot P(l)$$

where $v + d$ is the valuation given by $(v + d)(x) = v(x) + d$ for all x. Finally, the initial state is $s_0 = (l_0, v_0)$, where l_0 is the initial location, and $v_0(x) = 0$ for all clocks x. For networks of priced timed automata we use vectors of locations and the cost rate of a vector is the sum of the cost rates of individual locations.

An example of a PTA is shown in Fig. 1 with clocks x and y and five locations – ℓ_0 (initial), ℓ_1, ℓ_2, ℓ_3, and ℓ_g (goal), with cost rates $P(\ell_0) = +5$, $P(\ell_2) = +10$ and $P(\ell_3) = +1$, and the cost of the edge from ℓ_2 (ℓ_3) to ℓ_g is +1 (+7). Note that the invariant $y = 0$ in ℓ_1 enforces that the location must be left immediately. Below we show two example traces for the automaton:

$$\pi_1 = (\ell_0, x = 0, y = 0) \rightarrow_0 (\ell_1, x = 0, y = 0) \rightarrow_0 (\ell_3, x = 0, y = 0)$$
$$\xrightarrow{2}_2 (\ell_3, x = 2, y = 2) \rightarrow_7 (\ell_g, x = 2, y = 2)$$
$$\pi_2 = (\ell_0, x = 0, y = 0) \xrightarrow{1.5}_{7.5} (\ell_0, x = 1.5, y = 1.5) \rightarrow_0 (\ell_1, x = 1.5, y = 0)$$
$$\rightarrow_0 (\ell_2, x = 1.5, y = 0) \xrightarrow{0.5}_5 (\ell_2, x = 2, y = 0.5) \rightarrow_1 (\ell_g, x = 2, y = 0.5)$$

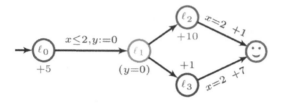

Fig. 1. Priced Timed Automata example

We see that π_1 reaches ℓ_g with a total cost of $2 + 7 = 9$, whereas the reachability cost of π_2 is $7.5 + 5 + 1 = 13.5$. In fact, among the infinitely many traces that reach ℓ_g, π_1 has the minimum cost. The question of cost-optimal reachability was shown decidable by [7] and later proven to be PSPACE-complete [12]. Here, extending the result for reachability of TAs in [15], it is observed that a PTS semantics with natural-valued delays is complete for PTAs with non-strict guards. Moreover, if k is the maximum constant to which clocks are compared to in guards and invariants, it suffices to consider delays no greater than $k + 1$. In short, in Definition 2 it suffices to consider finite-state PTS with $\Sigma = Act \cup \mathbb{N}_{\leq k+1}{}^1$ – as in the PTA of Fig. 1, where $k = 2$.

These observations are crucial for our developments of non-symbolic MCTS-based methods for optimal reachability of PTA as we shall see.

3   Monte Carlo Tree Search

Monte-Carlo Tree Search (MCTS) is a family of algorithms that has been intensely studied in the last decades due to its high success in a range of domains, in particular - game playing. MCTS works on a search tree that grows in asymmetric fashion and in accordance to the results of random samples (or heuristics) that are used to estimate the reward (potential) of the action taken. The tree is iteratively expanded starting from the root node according to four steps:

- Selection: Descend down the tree by selecting the best child according to the chosen policy and until a first unexplored node is met. The selection process typically tries to balance between exploration (visiting promising nodes) and exploitation (visiting nodes with least visits).
- Expansion: Generate a successor of the given state according to the chosen policy.
- Simulation: Estimate the reward of the expanded node by performing simulations, aka roll-outs until the terminal node is reached. Typically, the performance of the algorithm can be drastically improved by a smart simulation strategy.
- Backpropagation: The estimated reward is "backed up" through the tree to update reward estimates.

The first two steps (selection and expansion) are often referred to as *tree policy*, whereas the simulation (roll-out) step is called *default policy*. The algorithm does not

---

^{1} $\mathbb{N}_{\leq k+1}$ are all natural numbers less than or equal to $k + 1$.

have a predefined termination condition and is typically running until either a computational budget (time, memory, etc.) is reached or some different, domain-specific condition is met.

Some of the characteristics that have made MCTS popular in other domains are particularly relevant in the setting of PTA. Tree policy allows to favor more promising regions of the model which over time leads to *asymmetric* tree growth. This helps alleviate the state-space explosion – the most prominent obstacle in model-checking. Moreover, MCTS being *aheuristic* – easily applicable without the need for domain-specific knowledge – it can be applied to any problem domain as long as it can be modelled as PTA.

We now introduce the formal definition of MCTS and then give the pseudocode of the algorithm – both adapted for the setting of PTA with non-strict guards. Recall that for PTA A with non-strict guards and with maximum constant k (to which clocks are compared) it suffices to consider the *finite* set of labels $\Sigma = Act \cup \mathbb{N}_{\leq k+1}$ to get a finite and complete PTS F_A. We let Σ^* denote the language of finite (natural-valued and bounded) timed strings over Σ and let $\epsilon \in \Sigma^*$ denote the empty string.

By convention we let $|\epsilon| = 0$ and otherwise define $|a_0 \ldots a_n| = n$ to be the length of a word. We denote by $w_i \in \Sigma$ the i'th index of the word $w \in \Sigma^*$.

A timed word $w \in \Sigma^*$ of a PTS $T = (S, s_0, \Sigma, \rightarrow)$ is valid iff for $n = |w|$ we have:

$$s_0 \xrightarrow{w_0} s_1 \xrightarrow{w_1} \ldots \xrightarrow{w_n} s_{n+1}$$

We let the function $O : \Sigma^* \rightharpoonup S$ denote the outcome of such a valid trace w be $O(w) = s_{n+1}$. By convention we let $O(\epsilon) = s_0$.

Definition 3 (Search Tree). *We define $\Upsilon_T = (N, n_0, \Rightarrow)$ to be the search-tree for a natural- and bounded-valued PTS $T = (S, s_0, \Sigma, \rightarrow)$ as follows:*

- *$N = \Sigma^*$ is set of nodes,*
- *$n_0 = \epsilon$ is the root node, and*
- *$\Rightarrow \subseteq N \times \Sigma \times N$ is the transition relation such that $(n, b, n') \in \Rightarrow$ if and only if $nb = n'$ with $b \in \Sigma$ and $(O(n), b, O(n')) \in \rightarrow$.*

We delimit our attention to the most popular MCTS algorithm – the upper confidence bound for trees (UCT) [29]. UCT uses upper confidence bound (UCB1) formula as the tree policy, which addresses the exploration-exploitation dilemma of selecting the most promising paths by treating it as a multiarmed bandit problem. UCB1 makes a good candidate since it is guaranteed to be within a constant factor of the best bound for regret.

Let us define the global functions of the MCTS algorithm. Let $V : N \rightarrow \mathbb{N}$ assign the number of node visits, $Q : N \rightarrow \mathbb{R}$ assign the accumulative reward of the node, $P : N \rightarrow N$ maps to the parent of a node s.t. $P(n) = n'$ where $n' = n\alpha$ and $(n, \alpha, n') \in \Rightarrow$, and $Y_X : N \rightarrow \mathcal{P}(N)$ defines all children of the node that are valid according to the policy transition relation $\underset{X}{\Rightarrow}$, s.t. $Y_X(n) = \{n' \mid n \underset{X}{\Rightarrow} n'\}$. The definitions for each policy and respective transition relations are given in the following sections. Children are also partitioned into unexplored (Y^U) and explored (Y^E) ones s.t. $Y_X(n) = Y_X^U(n) \cup Y_X^E(n)$ and $Y_X^U(n) \cap Y_X^E(n) = \emptyset$.

Algorithm 1. The UCT Algorithm. This is a PTA-adapted redefinition of the Algorithm from [16].

1: **function** UCTSEARCH(An initial state s_0, a set of goal-states $\mathcal{G}$, an empty set of solved nodes $\mathcal{S}$, an empty set of dead nodes $\mathcal{D}$, and a C_p constant)

2:     $n_0 \leftarrow s_0$

3:     **while** budget remaining **do**

4:         $n \leftarrow$ TREEPOLICY$(n_0, \mathcal{G}, C_p, \mathcal{S}, \mathcal{D})$

5:         $\Delta \leftarrow$ DEFAULTPOLICY$(n, \mathcal{G})$

6:         BACKUP(n, Δ)

7:         **if** $O(n) \in \mathcal{G}$ **then**

8:             MARKSOLVED$(n, \mathcal{S})$

9:         **if** $O(n) \notin \mathcal{G}$ **and** $Y(n) = \emptyset$ **then**

10:             PRUNE$(n, \mathcal{D})$

11:     **return** BESTCHILD$(n_0, 0, \emptyset, \mathcal{D})$

12: **function** TREEPOLICY$(n, \mathcal{G}, C_p)$

13:     **while** $O(n) \notin \mathcal{G}$ **do**

14:         **if** $Y_X^U(n) \neq \emptyset$ **then**

15:             **return** EXPAND(n)

16:         **else**

17:             $n \leftarrow$ BESTCHILD$(n, C_p, \mathcal{S}, \mathcal{D})$

18:     **return** n

19: **function** EXPAND(n)

20:     sample $n' \in Y_X^U(n)$

21:     $V(n') = Q(n') = 0$

22:     $Y_X^E(n') = \emptyset$

23:     add n' to $Y_X^E(n)$

24:     **return** n'

25: **function** BESTCHILD$(n, C_p, \mathcal{S}, \mathcal{D})$

26:     **return** $\underset{n' \in Y_X^E(n) \setminus (\mathcal{S} \cup \mathcal{D})}{\mathrm{argmax}} \; Q_B \frac{V(n')}{Q(n')} + C\sqrt{\frac{\ln V(n)}{V(n')}}$

27: **function** DEFAULTPOLICY$(n, \mathcal{G})$

28:     **while** $n \notin \mathcal{G}$ **and** within roll-out budget **and**

29:     $Y_X(n) \neq \emptyset$ **do**

30:         sample $n' \in Y_X(n)$ uniformly

31:         $n \leftarrow n'$

32:     **return** reward for n

33: **function** BACKUP$(n, reward)$

34:     **while** $n \neq \epsilon$ **do**

35:         $V(n) \leftarrow V(n) + 1$

36:         $Q(n) \leftarrow Q(n) + reward$

37:         $n \leftarrow P(n)$

38: **function** MARKSOLVED$(n, \mathcal{S})$

39:     **while** $n \in \mathcal{G}$ **or** $n' \in \mathcal{S}$ for all $n' \in Y_X(n)$ **do**

40:         $\mathcal{S} \leftarrow \mathcal{S} \cup \{n\}$

41:         $n \leftarrow P(n)$

42: **function** PRUNE$(n, \mathcal{D})$

43:     **if** $n \neq \epsilon$ **and** $Y_X(n) = \emptyset$ **then**

44:         PRUNE$(P(n))$

45:         $\mathcal{D} \leftarrow \mathcal{D} \cup \{n\}$

Algorithm 1 gives a pseudocode for our PTA-adapted version of the UCT algorithm. The selection strategy used is a standard UCT formula (line 26). The expected reward of a node, determined by the exploitation factor $Q_B \frac{V(n')}{Q(n')}$, is inversely proportional to the average cost found so far which is normalized according to the currently best solution Q_B. The normalization ensures the reward value to be in range between 0 and 1 and thus supports domain (cost range) independence and eliminates the need for any prior knowledge about the reward distribution, which is also apriori unknown for PTAs. The significance of the exploration term is controlled by the value of C constant.

Once a solution is found, we mark the given node terminal to avoid re-exploration (lines 7 and 38–41). As long as the underlying search-tree is complete (determined by the variant of $\underset{X}{\Rightarrow}$), the algorithm is guaranteed to (eventually) provide an optimal solution given that one exists.

4   General PTA Challenges

Infinite Transition Sequences: MCTS algorithms have in large parts been developed for game playing, probabilistic planning or other, typically finite, state-space problems. However, in the setting of PTA, infinite transition sequences are possible, e.g. due to loops in the model. First and foremost it means that traditional roll-outs, directed at reward estimation, might never come to a halt. To overcome this problem we introduce a maximum budget for a roll-out (line 28). An example of the budget is an upper bound on maximum allowed steps that can be done in the default policy before the simulation is terminated.

Reward Evaluation: In turn, capped roll-out length can pose a problem by introducing the need to evaluate non-terminal states. Fortunately, PTA contains all the necessary information needed to evaluate the current cost of any state, including non-terminals. We evaluate and back-propagate the reward regardless of whether the rollout has reached a terminal state.

'Dead' States: Apart from infinite transition sequences, it is possible to encounter states with no possible successors in PTA. In most MCTS algorithm domains such no successor states are also terminal states; however, it is not necessarily the case for PTA. This is an issue for UCT as it is not equipped to deal with such *dead* states. In UCT, a dead state can be encountered either during expansion or simulations step. For the latter we simply terminate the roll-out upon reaching a dead state (line 29). In case of the former, if UCT expands into a dead state, it must have highest so far expected reward. Simulating from a dead state will not generate any new information, resulting in that state being the best-so-far. To avoid computational overhead, we prune dead states and their parent states from the search tree (lines 9 and 42–45) until no dead states remain in the current branch of the tree.

5   Policies

In MCTS, the structure of the search tree is decided by the unfolding mechanism of the tree policy. The same unfolding strategy is also used during the simulation process of the default policy. In this section we discuss different unfolding strategies that we refer to as *policies*. The specific choice of policy can have a dramatic effect on the performance of MCTS (as we shall demonstrate in the experiments). In particular, for PTA, the search-tree transition function $\Rightarrow$ for the PTA in Fig. 1 would for the state $(\ell_0, x = 0, y = 0)$ contain both the delay-action of 2 time units and the delay-action of 1 time unit (which would be repeatable), leading to the exact same configuration with the same total cost, namely $(\ell_0, x = 2, y = 2)$ at cost 10.

We thus explore both incomplete and complete policies, all restrictions over the full search-tree transition function $\Rightarrow$, with the latter category quarantining the existence of at least one optimal trace. Here, an incomplete policy does not retain the entire search-tree and does not guarantee preservation of an optimal solution. As the first policy, we introduce the Unit Delay Policy.

Definition 4 (Unit Delay Policy). *The transition function* $\underset{UDP}{\Rightarrow}$ *is given directly by*
$$\underset{UDP}{\Rightarrow} = (N \times (Act \cup \{1\}) \times N) \cap \Rightarrow.$$

While the UDP policy streamlines the application of delays, we observe a decreasing probability to pick larger delays. A child node (in tree and default policies) is chosen randomly between all available actions from that state and a delay of a single time unit; consequently, the probability for sequential choice of d unit-delay transitions at state s, i.e. delaying d time units, can be captured as follows:

$$Pr(s, d) = \left(\frac{1}{|Act_s| + 1} \right)^d$$

where $s \in S$, $d \in \mathbb{N}$ and assuming that all actions Act_s are available from state s at all times. If a state has actions that are only valid after a certain amount of time, then those actions are considerably less likely to be explored. We anticipate that such a skewed construction of the tree severely affects the ability of MCTS to find optimal solutions.

To alleviate this, we introduce a *Delay Sampling* policy (DSP) that allows to choose delays according to a more favorable probability distribution by enforcing a particular structure where delay and action transitions are always alternated. We also use this node layer alternation in the policies following the DSP policy giving a clear cut between transitioning by delay or action. Let $X : S \to \mathcal{P}(\mathbb{N})$ be a function that given a state returns a set of natural-valued delays w.r.t. to location-based constants, which includes the smallest possible delay, the largest possible delay, and a certain amount of delays from in between the bounds. We include only a subset of possible delays, which is limited to contain at most 100 values and at most 30% of the number of possible values (excluding bounds). The set of possible delays is selected in an attempt to reduce potentially huge branching factor due to delay-actions as to direct the search towards more cost-promising paths. Notice that X may change with each subsequent execution of the algorithm, but will not change during. Formally, DSP is defined as follows.

Definition 5 (Delay Sampling Policy). *The DSP policy* $\underset{DSP}{\Rightarrow}$ *is defined s.t. if* $(n, \alpha, n') \in \Rightarrow$ *then* $(n, \alpha, n') \in \underset{DSP}{\Rightarrow}$ *iff:*

– $n' = na$, $a \in Act$, $n = n''d$, $d \in \mathbb{N}$, *or*
– $n' = nd$, $d \in X(O(n))$, $a \in Act$ *and either* $n = \epsilon$ *or* $n = n''a$.

The policy solves the issue of uneven probability distribution for larger delays. However, it is incomplete in the function X not guaranteeing preservation of key delay values. In addition, we note that the policy still considers a fair degree of delay values (up to 100), quickly leading to a significant degree of branching in the search-tree.

As an alternative, we introduce a policy with the behavior inspired by *Non-lazy schedules* of [1]. The idea behind non-laziness is to avoid unnecessary simultaneous idling of both jobs and corresponding resources. If the resource is available, the job should claim the resource unless some other job can also use it. In the latter case, the first job can be delayed to 'pass' the resource to the second job. We do not give the formal definition of Non-lazy schedules here to maintain readability and refer the interested reader to the mentioned paper for more details.

We introduce our Non-Lazy policy with delays restricted to being either zero, to mimic no delay, or a non-lazy delay, representing the smallest non-zero delay leading to some action becoming enabled, similarly to non-lazy schedules. In comparison to DSP this drastically reduces the breadth of the search tree to at most 2 children and in part alleviates the state-space explosion problem. Let $NLD : S \rightarrow \mathcal{P}(\mathbb{N})$ give a set of zero and non-lazy delay, and $A' = \{\alpha \in \Sigma \mid s \not\xrightarrow{\alpha}\}$ be a set of actions that are not immediately enabled from a given state.

$$NLD(s) = \{0 \mid \exists \alpha \in \Sigma \text{ s.t. } s \xrightarrow{\alpha} s'\} \cup$$

$$\{d' \mid d' = \underset{d \in \mathbb{N}_{>0}}{\arg\min}\{\exists \alpha \in A' \text{ s.t. } s \xrightarrow{d} s'' \xrightarrow{\alpha} s'\}\}$$

We now give a formal definition of the policy.

Definition 6 (Non Lazy Policy). *The NLP policy* $\underset{NLP}{\Rightarrow}$ *is defined s.t. if* $(n, \alpha, n') \in \Rightarrow$ *then* $(n, \alpha, n') \in \underset{NLP}{\Rightarrow}$ *iff:*

– $n' = na$, $a \in Act$, $n = n''d$, $d \in \mathbb{N}$, *or*
– $n' = nd$, $d \in NLD(O(n))$, $a \in Act$ *and either* $n = \epsilon$ *or* $n = n''a$.

In [1] it is shown that non-lazy schedulers preserve optimal solutions for Job-shop scheduling problems; however, this is not the case for all problems expressible as PTA – implying that the method is incomplete for general PTAs.

Lastly we introduce a policy inspired by *Randomized Reachability Analysis* heuristics from [28]. The idea is to consider action transitions and select delays based on availability range of the chosen action transition. This supports an equal probability distribution to traverse each individual action transition irrespective of its availability range in terms of delays and overall provides a 'fair' exploration. The authors of this heuristics demonstrated its efficiency in finding rare events. We here adapt the idea for finding cost-optimal plans under the heuristic that taking only the smallest possible delay for each transition will often lead to a lower cost.

We now give a formal definition of the Enabled Transition policy. Let $LB : S \times \Sigma \to \mathbb{N}$ give the lower bound of the transition's availability range over the actions of a given PTS. Simply put, LB gives the smallest delay after which a certain action can be taken. Formally:

$$LB(s, \alpha) = \begin{cases} 0 \text{ if } \nexists \, d \in \mathbb{N} \text{ s.t. } s \xrightarrow{d} s' \xrightarrow{\alpha} s'' \\ \underset{d \in \mathbb{N}}{\arg\min} \, s \xrightarrow{d} s_1 \xrightarrow{\alpha} s_2 \text{ otherwise} \end{cases}$$

Definition 7 (Enabled Transition Policy). *The ETP policy $\underset{ETP}{\Rightarrow}$ is defined s.t. if $(n, \alpha, n') \in \Rightarrow$ then $(n, \alpha, n') \in \underset{ETP}{\Rightarrow}$ iff:*

- *$n' = na$, $a \in Act$, $d \in \mathbb{N}$, $n = n''d$, $d = LB(O(n''), a)$, or*
- *$n' = nd$, $a \in Act$, $d \in \{LB(O(n), a') \mid a' \in Act\}$ and either $n = n''a$ or $n = \epsilon$.*

Similarly to NLP, ETP is also an incomplete policy but with more relaxed conditions allowing it to consider all eventually enabled (either now or after delay) actions from a given state.

6   Enhancements

To improve on the performance of the MCTS algorithm, we propose the following modifications over the standard MCTS algorithm presented in Algorithm 1.

Building Rollouts. The standard UCT algorithms uses rollouts to estimate the reward of a node, but strictly in a way s.t. the tree is not expanded, as to preserve memory. We propose to add a rollout to the tree under two conditions: if 1. a roll-out reaches the terminal state, and 2. it does so with the so-far-best cost. We denote such configuration as BR.

Tree Pruning with Steps. It can be beneficial to perform a step (advance the root) once 'enough' information has been gathered to ensure near-optimal action choice in the root of the search-tree. Two domain-independent techniques – *Absolute pruning* and *Relative pruning* – have been introduced in [25]. They have shown that the Absolute pruning in fact preserves the optimality of the search tree, but concluded that rather few nodes are actually being pruned due to pruning conditions being too strict. We will thus only study the Relative pruning technique.

We briefly recall the condition for Relative pruning (RP), which is dependent on the tunable parameter μ.

Condition 1 (Relative pruning condition). *Node n_i can be pruned if $\exists j$ such that $V(n_j) > V(n_i) + \mu$, where $i \in \{1, \cdots, k\}, j \in \{1, \cdots, k\}, i \neq j$ and for all i we have $(n, \alpha, n_i) \in \Rightarrow$ with $\alpha \in \Sigma$.*

We also propose a simpler method of pruning based on a constant *stepping value*, i.e. a number of samples required in the current root-node before advancing the root of the tree. We denote this pruning technique Stepping pruning (SP).

7 Experiments

We perform experiments on three benchmarks:

1. Job-shop scheduling[2] problems,
2. Task graph scheduling[3] problems of [34] translated to PTA by [20], and
3. satellite mission scheduling problems [10,31].

We select 120 Task graph models (of thousands) and use all 162 Job-shop models, and all of the satellite models. The largest Job-shop model contains 100 jobs using 20 machines and the largest Task graph consists of 300 tasks (83 chains) executed on 16 machines. To account for randomness of the MCTS and random-search methods, we report the average of 10 executions. For symbolic methods (which are deterministic) we only conduct one execution. All experiments are limited to 10 min and the best found solution is reported (if any). The experiments are conducted on AMD Opteron 6376 processors with frequency-scaling disabled running Debian with a Linux 5.8 kernel and limited to 8 GB of memory (except for experiments with TiaMo which is given sufficient memory).

Solving Using PDDL (Planning Domain Definition Language) Planners. As a consequence of our restriction to natural-valued delays, it is possible to compile the PTA models into (classical, deterministic) planning problems and apply well-studied classic planning algorithms. To study this, we convert the Job-shop PTA models to PDDL 2.2 with action costs from PDDL 3.1 and use the Fast Downward[4] planner to find cost-efficient plans. We apply some classical algorithms, e.g. greedy best-first search with the FF heuristic for sub-optimal plans [24] and A^* with LM-Cut for optimal plans [22]. However, the so-called grounding phase never terminates within the time and memory limit, even for the smallest Job-shop model consisting of 6 jobs and 6 machines. Scaling down the models further (by gradually removing jobs) reveals that the complexity of the model with 3 jobs already surpasses the capabilities of the planner to find a solution in allotted time. It is well-known that if the parameter-space of the actions in PDDL encoding grows large, which is the case for our models, the state-space suffers from an exponential explosion. We thus refrain from comparing to classical planners in the remainder of this section and leave comparison to more complex planners (e.g. temporal planning algorithms) to future work.

Presentation of Results. In our graphs we present the relative performance of a method against *Best Known Solutions* (BKS) which is known for the Job-shop and Task graph problems. A 0% deviation indicates that the BKS was found and a 10% deviation denotes a solution that is 110% of the BKS. We refer to the BKS as the reference value. For all but the last experiment we present the results over both benchmarks in one single plot. Figures 2, 3, 4, 5, 6, 7, 8 and 9 are plotted as "Cactus" or "Survival" plots. The y-axis shows the quality of the solution as "% worse than the BKS" (Fig. 2, 3, 4, 5, 6 and 7). Each method is sorted individually, resulting in monotonically increasing

---

[2] https://github.com/tamy0612/JSPLIB.

[3] https://github.com/marmux/spreadsheets.

[4] https://www.fast-downward.org/HomePage.

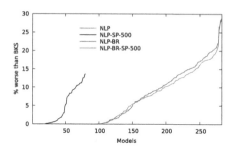

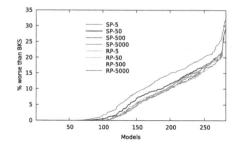

Fig. 2. The effect of BR and SP on the NLP policy. (Color figure online)

Fig. 3. Comparison of stepping values for NLP using BR and with $C_p = \sqrt{2}$.

lines. Therefore, data-points from different methods for a given x-value can be produced by different models, showcasing the general trend of each individual method over the benchmark.

We conduct the following sets of experiments:

- **Building Rollouts** where we construct the search-tree if a terminal node is found during rollout,
- **Impact of Stepping** where we experiment with pruning techniques,
- C_p **Sensitivity** where we vary the exploration constant,
- **Policy Study** where we compare the proposed policies, and
- **Comparison w. Existing Methods** where we compare our best performing method with existing solvers for PTA, and
- a study of the methods on a set of more general PTA models stemming from the domain of satellite mission planning.

Building Rollouts. We initially study the impact of the BR enhancement as any configuration without this enhancement is unable to yield results for a significant portion of the benchmarks. As a representative configuration we here present the results with the NLP policy both with and without the SP pruning and the exploration constant C fixed to $\sqrt{2}$. Other configurations demonstrate a similar tendency. We observe in Fig. 2 that only versions with the BR optimization manage to find a solution to all the instances. In particular, we see that the version without both SP and BR produces no results at all (red line). We witness the effect of BR from the plot and see that the best performing configurations are deviating no more than 30% from the reference. In addition, for roughly 50% of the models, this deviation is less than 5%.

Impact of Stepping. In Fig. 3 we compare different stepping sizes for SP and different upper-bounds number of visits (μ) for RP. We here restrict the reported results to the BR variant of the NLP policy. We observe that SP is highly sensitive to the stepping size and see that the smallest step sizes result in worse performance due to a too rapid progression of the root-node while too high values fail to reduce the search-space to a feasible size. We observe a similar tendency with RP wrt. the sensitivity of the μ-value, albeit to a lesser degree. Importantly we observe that SP (using a stepsize of 500) and

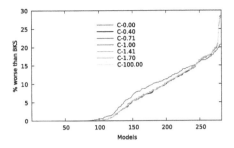

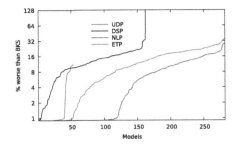

Fig. 4. Comparison of different C_p values effect on NLP with BR and SP-500 options.

Fig. 5. Comparison of UDP, DSP, NLP, ETP policies with $C_p = \sqrt{2}$ and the best enhancements used: BR and SP-500.

RP (with $\mu = 5$) perform similarly well – and we delimit ourselves to reporting only on variants using SP in subsequent experiments.

While using $C_p = \sqrt{2}$ is often considered a good value to strike a balance between exploration and exploitation, we here study the sensitivity to changes in the C_p-value, in particular as our setting is a single-player setting. Specifically we can in Fig. 4 observe the difference in performance when $C_p \in \{0, 0.4, \frac{1}{\sqrt{2}}, 1, \sqrt{2}, 1.70, 100\}$ where the value 100 is chosen arbitrarily as "a sufficiently large value" to force the algorithm to focus purely on exploration. From Fig. 4 we observe that apart from $C_p = 0$, the choice of C_p has little to no impact on the performance – likely due to the fact that our setting is a single player setting. Regarding $C_p = 0$, we conjecture that the effect observed stems from an intensive search around the initially found solution. For instances with a positive effect we believe that a (near-)optimal solution is found within the vicinity of *any* solution, where a negative effect indicate a larger difference between local minima in the search-space. While a small set of models clearly favor $C_p = 0$, we use $C_p = \sqrt{2}$ for the remainder of the experiments as it provides overall good performance and is the value recommended by literature.

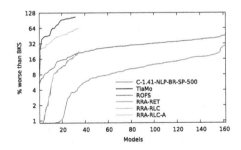

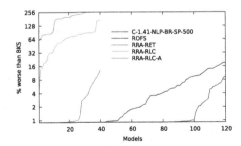

Fig. 6. Job-shop overview.

Fig. 7. Task graph overview.

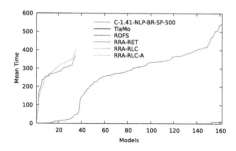 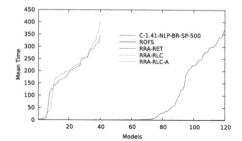

Fig. 8. Job-shop runtime overview.   **Fig. 9.** Task graph runtime overview.

Policy Study. The summary on the performance of different policies is shown in Fig. 5. Here we fix the configuration to use the BR and SP enhancements with a step-size of 500. We observe that UDP has the worst performance with less than 20% of problem instances solved within the given time-frame - and significantly worse quality solutions. We believe this to be due to the low probability of selecting larger delays and the state-space explosion of having to consider all possible delays. While DSP is an improvement over UDP, it suffers from a similar problem in that the branching factor can explode leading to a performance degradation. Both NLP and ETP were able to solve all problem instances with near-optimal solutions of at most 28.88% and 35.42% away from the reference value, respectively, however with a clear advantage to NLP.

Comparison w. Existing Methods. Lastly we perform a comparison of our best config-uration with other existing state-of-the-art solvers for PTA, namely UPPAAL CORA and TiaMo. In addition, we have also adapted the *Randomized Reachability Analysis* (RRA) methods of [28] to search for optimal schedules rather than rare events. We experiment with several of the techniques proposed for RRA (RET, RLC and RLC-A) to search for optimal solutions. We refer the interested reader to the mentioned paper for more details.

In the case of CORA we use both the complete and optimal search-method as well as the incomplete *Random Optimal First Search* (ROFS) approach, which allows for a very lightweight search in a depth-first manner while choosing the most optimal action at each step but providing no guarantee wrt. optimality of the returned solution. It is important to note that both CORA (except for the ROFS version) and TiaMo are com-plete and able to find an optimal solution if given enough time and memory - and that both methods are relying on a symbolic representation of the search-space.

Figure 6 gives an overview of all the methods for Job-shop scheduling benchmark compared against BKS from [26]. Note that CORA has not managed to solve any instance for either of the benchmarks, primarily limited by the fact that it is a piece of 32bit software only capable of utilizing 4 GB of memory. Unfortunately CORA does not provide anytime solutions in its current distribution. Both TiaMo and RRA methods solve less than 20% of the instances, with TiaMo delivering sub-par solutions as it never completes the search within the time-limit, and thus provides only any-time solutions as they are found.

Table 1. Results for different PTA models of satellite problems. MCTS policies executed with $C_p = \sqrt{2}$, BR and SP-500 enhancements enabled. (oom = out of memory)

		DSP	NLP	ETP	ROFS	Cora
gomx3-1day	Mean cost	**186,007** (±0.00%)	188,408 (±1.95%)	**186,007** (±0.00%)	198,292 (±0.00%)	**186,007**
	Time	40.2	49.8	61.5	**0.05**	5.12
gomx3-2day	Mean cost	442,190 (±0.04%)	442,218 (±0.01%)	**442,080** (±0.06%)	478,002 (±0.17%)	oom
	Time	223.3	268.0	230.7	**0.05**	-
5sat	Mean cost	5,072,861 (±4.12%)	5,961,014 (±1.72%)	**3,548,824** (±0.77%)	3,739,730 (±1.82%)	oom
	Time	267.5	366.7	295.8	**0.25**	-
10sat	Mean cost	**5,632,414** (±0.57%)	6,130,961 (±0.68%)	nf	5,687,131 (±2.47%)	oom
	Time	232.4	232.6	600.0	**0.56**	-
MaxData626	Mean cost	nf	nf	nf	**7,458,522** (±2.17%)	oom
	Time	600.0	600.0	600.0	**0.62**	-

The ROFS algorithm of CORA outperforms both the random search and TiaMo in terms of solved instances, while having a drawback with the quality of the produced plans when compared to our MCTS implementation. In terms of time (Fig. 8 and 9), the ROFS algorithm is the fastest overall, completing its search within single-digit seconds. We note that the overall quality of the schedules found by ROFS is within a surprisingly reasonable distance from the optimal, indicating that a greedy search strategy is well suited for the given benchmark. We observe the best performance of the proposed MCTS configuration using NLP, SP, BR and $C_p = \sqrt{2}$ and see a deviation of up to 28.88% of the BKS - with a median of deviations of no more than 10.3%. However, investigating the computation time, we can see that the best found solution is in the median produced at 289 s and peaking at 546 s.

The overview for the Task graph scheduling benchmark compared against BKS from [32] is shown in Fig. 7. Due to its limited support of the PTA syntax, the TiaMo tool was not applicable. For over 80% of the benchmark (100/120 models) the solutions found by NLP are (near-)optimal with the quality of solutions of at most 1% away from BKS. For the rest of the benchmark the performance of NLP slightly worsens reaching at most 9.11% deviation from BKS. In general, the trends for different methods are very similar: RRA methods solve around 33% of models only, while ROFS finds solutions near instantly, but their quality degrades with increased model complexity.

Satellite Models. Additionally, we experiment with two satellite cases - GomX-3 and Ulloriaq - designed, delivered, and operated by Danish satellite manufacturer GomSpace. The PTA models for these satellites have been developed in [10] and [31] studies, respectively, and analyzed with UPPAAL CORA (including ROFS). We show the results in Table 1, but exclude UDP as it produces no results within the time limit. For all models (but one) MCTS provides the best mean cost across all the methods; however, ROFS finds solutions up to 4 orders of magnitude faster and with a modest reduction of quality (up to 10% from the best MCTS method). We believe this is due to a generally small variance in the quality of solutions in the solution-space and the fact that ROFS performs only a single traversal of the model, immediately reporting the result upon reaching the terminal state. For "MaxData626" model MCTS methods timeout without a solution. Further experiments with an increased time-limit of 5 h do not yield additional results indicating issues with the incompleteness of the methods rather than missing computation-time. The relative efficiency of the ROFS method demonstrates a potential for extending the MCTS method in the direction of a symbolic search, allowing for an efficient and complete MCTS tree-search method, and overcoming the current limitations of the discretized equivalents studied in this paper.

8   Conclusion

We have adapted the Monte Carlo Tree Search (MCTS) algorithm for the setting of problems described as Priced Timed Automata (PTA) – a formalism that can capture the behavior of a wide range of optimization problems such as resource-consumption or -allocation problems. PTA is a very versatile modeling formalism, facilitating more direct modeling of a problem domain. We introduced a number of complete and non-complete policies that act as unfolding mechanism and decide the structure of the tree. Some domain-independent enhancements to improve the performance and coverage of the algorithm are suggested.

We have evaluated the performance of our MCTS algorithm adapted to PTA on three benchmarks of Job-shop, Task graph and satellite mission scheduling problems and compared it against other state-of-the-art methods and tools. For the first two benchmarks, the results indicate that MCTS is able to find near-optimal solutions for all investigated problem instances. In general, we observed an up to 28.88% and 9.11% deviation (on average) from the best known solution in a set of Job-shop and Task graph scheduling problems, respectively. For satellite models, MCTS methods have found the best cost across all tested methods except for one model where only ROFS was able to produce results, hinting at issues with the incompleteness of MCTS methods.

All this suggests that MCTS is a promising alternative that copes well with the state-space explosion problem where other existing, exhaustive and complete methods perform poorly or fail. We note that the Random Optimal First Search strategy of the tool UPPAAL CORA performs well, even when compared to MCTS. The study of more symbolic approaches to MCTS for PTA is left as future work.

Data Availability Statement. A reproducibility artifact, which contains binaries, models and scripts to reproduce results can be found at https://doi.org/10.6084/m9.figshare.19772926.

References

1. Abdeddaïm, Y., Maler, O.: Job-shop scheduling using timed automata? In: Berry, G., Comon, H., Finkel, A. (eds.) CAV 2001. LNCS, vol. 2102, pp. 478–492. Springer, Heidelberg (2001). https://doi.org/10.1007/3-540-44585-4_46

2. Ahmad, W., Hölzenspies, P.K.F., Stoelinga, M., van de Pol, J.: Green computing: power optimisation of VFI-based real-time multiprocessor dataflow applications. In: DSD 2015, pp. 271–275. IEEE Computer Society (2015). https://doi.org/10.1109/DSD.2015.59

3. Alur, R., Dill, D.: The theory of timed automata. In: de Bakker, J.W., Huizing, C., de Roever, W.P., Rozenberg, G. (eds.) REX 1991. LNCS, vol. 600, pp. 45–73. Springer, Heidelberg (1992). https://doi.org/10.1007/BFb0031987

4. Alur, R., La Torre, S., Pappas, G.J.: Optimal paths in weighted timed automata. In: Di Benedetto, M.D., Sangiovanni-Vincentelli, A. (eds.) HSCC 2001. LNCS, vol. 2034, pp. 49–62. Springer, Heidelberg (2001). https://doi.org/10.1007/3-540-45351-2_8

5. Banharnsakun, A., Sirinaovakul, B., Achalakul, T.: Job shop scheduling with the best-so-far ABC. Eng. Appl. Artif. Intell. 25(3), 583–593 (2012). https://doi.org/10.1016/j.engappai.2011.08.003

6. Behrmann, G., Fehnker, A., Hune, T., Larsen, K., Pettersson, P., Romijn, J.: Efficient guiding towards cost-optimality in UPPAAL. In: Margaria, T., Yi, W. (eds.) TACAS 2001. LNCS, vol. 2031, pp. 174–188. Springer, Heidelberg (2001). https://doi.org/10.1007/3-540-45319-9_13

7. Behrmann, G., et al.: Minimum-cost reachability for priced time automata. In: Di Benedetto, M.D., Sangiovanni-Vincentelli, A. (eds.) HSCC 2001. LNCS, vol. 2034, pp. 147–161. Springer, Heidelberg (2001). https://doi.org/10.1007/3-540-45351-2_15

8. Behrmann, G., Larsen, K.G., Rasmussen, J.I.: Optimal scheduling using priced timed automata. SIGMETRICS Perform. Eval. Rev. 32(4), 34–40 (2005). https://doi.org/10.1145/1059816.1059823

9. Behrmann, G., Larsen, K.G., Rasmussen, J.I.: Priced timed automata: algorithms and applications. In: de Boer, F.S., Bonsangue, M.M., Graf, S., de Roever, W.-P. (eds.) FMCO 2004. LNCS, vol. 3657, pp. 162–182. Springer, Heidelberg (2005). https://doi.org/10.1007/11561163_8

10. Bisgaard, M., Gerhardt, D., Hermanns, H., Krčál, J., Nies, G., Stenger, M.: Battery-aware scheduling in low orbit: the GomX-3 case. Formal Aspects Comput. 31(2), 261–285 (2019). https://doi.org/10.1007/s00165-018-0458-2

11. Bøgholm, T., Larsen, K.G., Muñiz, M., Thomsen, B., Thomsen, L.L.: Analyzing spreadsheets for parallel execution via model checking. In: Margaria, T., Graf, S., Larsen, K.G. (eds.) Models, Mindsets, Meta: The What, the How, and the Why Not? LNCS, vol. 11200, pp. 27–35. Springer, Cham (2019). https://doi.org/10.1007/978-3-030-22348-9_3

12. Bouyer, P., Brihaye, T., Bruyère, V., Raskin, J.: On the optimal reachability problem of weighted timed automata. Formal Methods Syst. Des. 31(2), 135–175 (2007). https://doi.org/10.1007/s10703-007-0035-4

13. Bouyer, P., Colange, M., Markey, N.: Symbolic optimal reachability in weighted timed automata. In: Chaudhuri, S., Farzan, A. (eds.) CAV 2016. LNCS, vol. 9779, pp. 513–530. Springer, Cham (2016). https://doi.org/10.1007/978-3-319-41528-4_28

14. Bouyer, P., Fahrenberg, U., Larsen, K.G., Markey, N.: Quantitative analysis of real-time systems using priced timed automata. Commun. ACM 54(9), 78–87 (2011). https://doi.org/10.1145/1995376.1995396

15. Bozga, M., Maler, O., Tripakis, S.: Efficient verification of timed automata using dense and discrete time semantics. In: Pierre, L., Kropf, T. (eds.) CHARME 1999. LNCS, vol. 1703, pp. 125–141. Springer, Heidelberg (1999). https://doi.org/10.1007/3-540-48153-2_11

16. Browne, C.B., et al.: A survey of Monte Carlo tree search methods. IEEE Trans. Comput. Intell. AI Games 4(1), 1–43 (2012). https://doi.org/10.1109/TCIAIG.2012.2186810

17. Čaušević, A., Seceleanu, C., Pettersson, P.: Checking correctness of services modeled as priced timed automata. In: Margaria, T., Steffen, B. (eds.) ISoLA 2012. LNCS, vol. 7610, pp. 308–322. Springer, Heidelberg (2012). https://doi.org/10.1007/978-3-642-34032-1_29

18. Dirks, H.: Finding optimal plans for domains with restricted continuous effects with UPPAAL CORA. In: ICAPS 2005. American Association for Artificial Intelligence (2005)
19. Edelkamp, S.: Heuristic search planning with BDDs. In: PuK 2000 (2000). http://www.puk-workshop.de/puk2000/papers/edelkamp.pdf
20. Ejsing, A., Jensen, M., Muñiz, M., Nørhave, J., Rechter, L.: Near optimal task graph scheduling with priced timed automata and priced timed Markov decision processes (2020)
21. Geuze, N.: Energy management in smart grids using timed automata. Master's thesis, University of Twente (2019)
22. Helmert, M., Domshlak, C.: Landmarks, critical paths and abstractions: what's the difference anyway? In: Nineteenth International Conference on Automated Planning and Scheduling (2009)
23. Hermanns, H., Krcál, J., Nies, G.: How is your satellite doing? Battery kinetics with recharging and uncertainty. Leibniz Trans. Embed. Syst. **4**(1), 04:1–04:28 (2017). https://doi.org/10.4230/LITES-v004-i001-a004
24. Hoffmann, J., Nebel, B.: The FF planning system: fast plan generation through heuristic search. J. Artif. Intell. Res. **14**, 253–302 (2001)
25. Huang, J., Liu, Z., Lu, B., Xiao, F.: Pruning in UCT algorithm. In: 2010 International Conference on Technologies and Applications of Artificial Intelligence, pp. 177–181 (2010). https://doi.org/10.1109/TAAI.2010.38
26. Jain, A., Meeran, S.: Deterministic job-shop scheduling: past, present and future. Eur. J. Oper. Res. **113**(2), 390–434 (1999). https://doi.org/10.1016/S0377-2217(98)00113-1
27. Jongerden, M.R., Haverkort, B.R., Bohnenkamp, H.C., Katoen, J.: Maximizing system lifetime by battery scheduling. In: IEEE/IFIP International Conference on DSN 2009, pp. 63–72. IEEE Computer Society (2009). https://doi.org/10.1109/DSN.2009.5270351
28. Kiviriga, A., Larsen, K.G., Nyman, U.: Randomized reachability analysis in UPPAAL: fast error detection in timed systems. In: Lluch Lafuente, A., Mavridou, A. (eds.) FMICS 2021. LNCS, vol. 12863, pp. 149–166. Springer, Cham (2021). https://doi.org/10.1007/978-3-030-85248-1_9
29. Kocsis, L., Szepesvári, C.: Bandit based Monte-Carlo planning. In: Fürnkranz, J., Scheffer, T., Spiliopoulou, M. (eds.) ECML 2006. LNCS (LNAI), vol. 4212, pp. 282–293. Springer, Heidelberg (2006). https://doi.org/10.1007/11871842_29
30. Korvell, A., Degn, K.: Designing a tool-chain for generating battery-aware contact plans using UPPAAL. Master thesis, Aalborg University (2019)
31. Kørvell, A., Degn, K.: Designing a tool-chain for generating battery-aware contact plans using UPPAAL (2019)
32. Kasahara Laboratory: Standard task graph set. https://www.kasahara.cs.waseda.ac.jp/schedule/index.html
33. Saddem-yagoubi, R., Naud, O., Godary-dejean, K., Crestani, D.: Model-checking precision agriculture logistics: the case of the differential harvest. Discrete Event Dyn. Syst. **30**(4), 579–604 (2020). https://doi.org/10.1007/s10626-020-00313-1
34. Tobita, T., Kasahara, H.: A standard task graph set for fair evaluation of multiprocessor scheduling algorithms. J. Sched. **5**(5), 379–394 (2002). https://doi.org/10.1002/jos.116
35. Vulgarakis, A., Čaušević, A.: Applying REMES behavioral modeling to PLC systems. In: 2009 XXII International Symposium on Information, Communication and Automation Technologies, pp. 1–8. IEEE (2009)

Author Index

Printed in the United States
by Baker & Taylor Publisher Services